2/04

99.95

Absolutism
and the
Scientific Revolution,
1600–1720

Recent Titles in
The Great Cultural Eras of the Western World

Absolutism and the Scientific Revolution, 1600–1720

A Biographical Dictionary

Edited by
CHRISTOPHER BAKER

The Great Cultural Eras of the Western World
Ronald H. Fritze, Series Adviser

GREENWOOD PRESS
Westport, Connecticut • London

Library of Congress Cataloging-in-Publication Data

Absolutism and the scientific revolution, 1600–1720: a biographical dictionary / edited by
Christopher Baker.
      p.   cm.—(The great cultural eras of the Western world, ISSN 1534–9527)
    Includes bibliographical references and index.
    ISBN 0–313–30827–6 (alk. paper)
    1. Europe—History—17th century—Biography—Dictionaries.   2. Europe—History—18th
century—Biography—Dictionaries.   3. Europe—Intellectual life—17th century—Biography—
Dictionaries.   4. Europe—Intellectual life—18th century—Biography—Dictionaries.
5. Enlightenment—Europe—18th century—Biography—Dictionaries. 6. Authoritarianism—
Europe—History—17th century. 7. Authoritarianism—Europe—History—18th century.
I. Baker, Christopher. II. Series.
D205.A27   2002
940.2'52—dc21   2002017109

British Library Cataloguing in Publication Data is available.

Library of Congress Catalog Card Number: 2002017109
ISBN: 0–313–30827–6
ISSN: 1534–9527

First published in 2002

Greenwood Press, 88 Post Road West, Westport, CT 06881
An imprint of Greenwood Publishing Group, Inc.
www.greenwood.com

Printed in the United States of America

The paper used in this book complies with the
Permanent Paper Standard issued by the National
Information Standards Organization (Z39.48–1984).

10 9 8 7 6 5 4 3 2 1

For Barbara and Matt

Contents

Introduction

Despite the truism that every age is an age of transition, the hindsight of history reveals certain eras to have been periods of radically decisive intellectual change. Such was the seventeenth century in Western Europe. By 1600, the momentum of the Renaissance, whether defined as the birth of a new spirit of creativity or as the slow death of medievalism, was peaking, and the current of religious reformation was in full tide, soon to make itself felt as well in the North American colonies. These and other aspects of the sixteenth century have led to a recent tendency to replace the term "Renaissance" with "Early Modern," a label that, while raising problematic issues of its own, nevertheless cautions against artificially rigid divisions in the flux of historical development. The boundaries of historical periods are necessarily permeable, and those of this volume are no different; yet it is fair to say that by 1720 the adjective "modern" no longer requires that we modify it with "early."

One could perhaps point to 1642, the year of Galileo's death and Newton's birth, as the last chronological juncture between the "old world" and the "new," or possibly 1674, the year of poet John Milton's death, the last English literary voice of the Renaissance, as the point of transition from "early modern" to "modern." However, paraphrasing Virginia Woolf's only half facetious comment upon seeing the 1901 London exhibition of Postimpressionist art, events in 1648 make it tempting to assert that in that year "human nature changed." It saw two occurrences that demonstrated a decisive alteration in the assumptions of Christian humanism and the Protestant reformation, which had marked the sixteenth century. The Treaty of Westphalia ended the Thirty Years' War; though it may have been a struggle that, in C.V. Wedgewood's phrase, "need not have happened and settled nothing worth settling," it was the last European conflict in which religious differences were believed important enough to go to battle over. Issues of politics, class, and national interest were key causes as well, but the conflict drew heavily upon the tensions between Calvinists and

Roman Catholics in almost every European nation as it ground tediously through its Bohemian, Danish, Swedish, and Swedish-French phases. England was un-affected by the widespread carnage, but its own mid-century civil war grappled with similar issues within its own shores. After the Peace of Westphalia was signed, Catholics and Calvinists agreed to disagree, and denominational differ-ences would no longer emerge as a major cause of such continental struggle, a mark of the growing secularization of Europe.

If 1648 signaled the end (or the beginning of the end) of organized religion's power to motivate armies, it also marked the expansion of an organization ded-icated to what would become the secular faith of modernism, a skeptical em-piricism. Contemporary English mathematician John Wallis noted that in about 1648 a group of men whom Robert Boyle called "our invisible college or the philosophical college" who had been meeting in London extended their meetings to Oxford. Boyle's "college" had been in existence since 1645, but it was clear that three years later the nucleus of what would become the Royal Society of Scientific Inquiry had taken on a life of its own. Informal yet organized scientific research in England had begun, with Robert Hooke being designated the group's "Curator of Experiments." Groundbreaking inventions and discoveries appeared in Europe with increasing frequency, starting at the very beginning of the cen-tury with Gilbert's discovery of the earth's magnetic field (1600), Kepler's first two laws of motion (1609), Oughtred's slide rule (about 1630), and Harvey's revelations of the circulation of the blood (1638). A scientific organization in Rome counted Galileo as a member in 1603, the Royal Society was chartered in 1662, the French Academy of Science in 1666, and one in Berlin in 1701. The language for this new model of nature, mathematics, developed as vigor-ously, with the appearance of Napier's logarithms, Newton's and Leibniz's de-velopment of integral and differential calculus, and Descartes' pathbreaking work in analytical geometry.

Where religion had proceeded deductively from theological axioms delivered through revelation, science would arrive at its conclusions inductively, beginning with hypotheses that were then tested by means of quantifiable data gained through a process of methodical experimentation. Significant roots of this major change in how knowledge was acquired, and in the very definition of knowledge itself, lay in the work of, among others, the previous century's heroes of as-tronomy, Copernicus and Brahe, but it remained for Francis Bacon to publicly advocate a reorganization of all knowledge around an inductive method. Bacon and later Isaac Newton thus advanced a revolutionary change in how the phys-ical world was perceived and investigated, a development now termed a "par-adigm shift." As Alexander Pope would later write in his "Epitaph Intended for Isaac Newton," "Nature and Nature's laws lay hid in night: / God said, Let Newton be! And all was light." Induction was to become the dominant logical method of science, although Descartes had pointed out as well the virtues of deduction, positing specific cases from general principles rather than observed data. The era thus fed a growing confidence in an inquiring rationality that laid

the basis for the eighteenth century's fame as the Age of Reason, or the Enlightenment. The broad current of philosophical rationalism of which the scientific method was one aspect caused a change in religious sensibility as well.
God was now seen to be revealed more persuasively in the ordered and observable operation of natural processes than in the biblical testimonies of prophets
and patriarchs. Deism, the name given this rationalized faith, led to the worship
of a "divine clockmaker" as the prevailing form of religious faith among emerging rationalist intellectuals. Distrust of mystical or "enthusiastic" religious feeling grew, until by the late eighteenth century Thomas Jefferson would feel no
hesitation in snipping from his New Testament all of the miracles of Jesus,
thereby producing a much more reasonable savior.

The seventeenth century also saw a steady increase in the desire of monarchs
to strengthen their central authority. The tendency for rulers to acquire more
direct control over ever-broadening areas of policy and procedure, with fewer
prerogatives permitted for nobility, parliaments, or church, became known as
absolutism. This process too, like the scientific revolution, had begun in the
previous century as Henry VIII and Elizabeth I in England, Philip II in Spain,
and Francis I in France forcefully influenced policies that weakened traditional
competitors to the crown. To varying degrees, absolutism affected kingdoms
from England to Russia. Two examples may outline its impact.

England's absolutist period occurred early in the seventeenth century, with
James I, who ruled from 1603 to 1625, announcing his God-given authority, or
"divine right," to govern. With the Reformation's success in challenging the
Pope's authority as sole earthly head of the church, monarchs tended to impose
their individual religious preferences as state policy according to the dictum
"cuius regio, eius religio": "whose the region, his the religion." James, a vigorous Anglican, provoked strong Puritan opposition that intensified against his
more inflexible son Charles I (1625–1649), who ruled without Parliament for
eleven years, and against the Archbishop of Canterbury, William Laud, who
strenuously enforced Anglican practices. Armed opposition to Charles broke out
in the rebellion of the Scots Presbyterians in 1639, and civil war between
Charles's royalist forces and Oliver Cromwell's Puritan army erupted in 1642.
Seven years later, the Puritans beheaded Charles, and Cromwell set up his alternative to the monarchy, the Protectorate. But the Puritan political experiment
lasted only until 1660, when the monarchy was restored under Charles II. He
avoided absolutist measures during his reign (1660–1685), unlike his successor
and brother James II, whose attempt to restore Catholicism lasted only three
years (1685–1688). His opponents persuaded his Dutch Protestant son-in-law,
William of Orange, to ascend the throne; the reign of William and Queen Mary
(1688–1702) was the outcome of this so-called "Glorious Revolution." Queen
Anne (1702–1714) and George I (1714–1727) worked closely with Parliament
through their ministers, forging a more participatory method of governance, and
the English experience of absolutism faded into history.

In France, the impact of absolutism was equally decisive, if in a different way. In 1610, King Louis XIII ascended the throne at the age of ten; his wily chief minister, Cardinal Richelieu, extended the monarch's authority into the provinces, weakened the power of the nobility, and took over the stronghold of the Protestant Huguenots at La Rochelle in 1628, permitting them freedom of worship but denying them any political influence. He also renewed French opposition to the Catholic Hapsburgs, joining forces with the Swedish and German Protestants in the Thirty Years' War. One year after Richelieu's death in 1642, the five-year-old Louis XIV became king, his minister Cardinal Mazarin continuing Richelieu's policies. Despite a series of popular revolts (the Fronde) against royal policies from 1648 until 1653, Louis's power grew stronger, and after Mazarin's death in 1661, he ruled with unparalleled authority until his death in 1715. Imbued with his absolute control, he confidently declared *"L'état, c'est moi"* ("I am the state"). The theoretical basis for his rule was provided by Jacques Bossuet, whose *Politics Drawn from Holy Scripture* (ca. 1670) asserted the biblical justification for complete obedience to the king's divinely sanctioned authority. (Arguing from a more political than religious standpoint, Thomas Hobbes's *Leviathan* [1651] had argued for absolutism on the grounds that individuals are so selfishly contentious by nature that only an absolute monarch can enforce the laws needed for social stability.) Eager to also exert administrative control in church affairs, Louis closed the Calvinist-leaning Jansenist monastery at Port Royal and rescinded the 1598 Edict of Nantes, stripping the Huguenots of the freedom of religion it had granted. Succeeding monarchs had to confront the staggering debts his various wars had incurred. The inexorably growing opposition of both nobility and commoners to absolutism and its resulting national insolvency eventually contributed to the French Revolution.

The dynamic changes affecting science, religion, and politics were matched by those in the arts. Change itself or the portrayal of contending forces and shifting tensions were features of the period's most inclusive (though not necessarily most precise) aesthetic style—the Baroque. Taken either from *barroco*, the Portuguese word for an oddly shaped pearl, or (less likely) from *baroco*, an Italian Renaissance term for overly intricate scholastic philosophical arguments, the term has come to describe a style that depicts the breaking of boundaries, the movement of forceful energy, and the blending of contrasting values. Originally used pejoratively in 1733 to mock the music of Rameau, "baroque" later evolved into a much broader concept to describe the dominant aesthetic context (especially in art and architecture) for the seventeenth century; for the nineteenth-century art historian Heinrich Wölflinn, it lasted from roughly 1570 until the start of the "Sturm und Drang" period in Germany in about 1765. Wölflinn proposed five characteristics of the Baroque: It favored the "painterly" rather than the linear image; it emphasized expansiveness rather than limits; it favored open, unlimited forms rather than enclosed spaces; it strove for an integral unity among its parts; and it presented only a relative, rather than an absolute, distinction among its parts. Thus, Bernini's canopy over the altar in St. Peter's Basilica in the Vatican rises on four massive columns that, rather than being straight, are energetically spiraled, while his statue *St. Teresa in Ecstasy* portrays the saint in a moment of mystical transport, enveloped in flowing robes and apparently levitating in midair, not unlike his bust of Louis XIV at Versailles. His vast colonnaded piazza in front of St. Peter's controls a tremendous area within its countless columns. (Bernini's work also points to the complex relationship between the Baroque and the Counter Reformation.) Churriguera in Spain, Jones and Wren in England, Mansart in France, to name only a few, designed churches, country homes, and public buildings that tended toward vast designs that lorded over their surroundings, displaying imposing facades that demanded a single visual vantage point for their full effect. Baroque painting revealed similar principles operating within two dimensions. In El Greco's unusual, often disturbing, color combinations; in Rembrandt's chiaroscuro, in which figures emerge from shadowed backgrounds while seeming to remain part of them; or in the broad canvases of Rubens crowded with voluminous nudes, we note a consistent pulse of energy, whether extroverted or introspective, contained within the restraining force of an overall thematic intent. Yet in a reaction to these baroque extremes (and in this very opposition thus somewhat baroque itself) is the Calvinist restraint upon popular art, which existed in Holland, resulting in realistic portrayals of daily bourgeois life, eschewing explicitly classical, mythological, or erotic themes. The "genre" paintings of Steen and De Hooch, the jolly villagers of Hals, and the sometimes allegorical landscapes of Van Ruisdael and Cuyp reveal a fine attention to detail often combined with a moral theme stressing the vanity of earthly life. The premier painter of domestic, private scenes was Vermeer, who combines sensitivity of human expression with a heightened perception of light and color.

The Baroque in music was likewise an age of varying innovations, with the vast architectural forms of a Bernini finding their counterpart in compositions such as Handel's oratorio *The Messiah* or Purcell's opera *Dido and Aeneas*. As Rembrandt was in painting, so Bach, born at the end of the seventeenth century, was the most commanding and versatile figure of the Baroque, creating a profusion of works from the magisterial Mass in B Minor to the intricately meditative Goldberg Variations for keyboard. One measure of the era's musical genius lies in the fact that many of the genres now enjoyed were either created or perfected in the seventeenth century: opera, cantata, oratorio, suite, sonata, sinfonia, concerto, various types of chamber music, and the art of improvisation. This was the golden age of the harpsichord, and fine instrument makers such as the Amati and Stradivari families produced stringed instruments of higher quality that could sustain the new compositional forms. Composers often benefited from the patronage of royalty, whose courts were centers not only of political power but of prolific creativity. Gabrieli's music for brass choirs conveys a grandeur that is both worldly and sublime.

It has proved to be somewhat more problematic to apply the categories of the Baroque to seventeenth-century literature, but European poetry of the period, whether sacred or secular, often displays an increased fondness for the colloquial voice, the startling metaphor, or the self-conscious turn of phrase that distinguishes it from sixteenth-century works. Whether these qualities stamp this literature as merely the decadent remnant of the fading Renaissance or, contrastingly, as the energetic experimentation of a more artistically adventurous era is a subject for debate. In England, the poetry of Donne, Herbert, Crashaw, Marvell, and others earned the name *metaphysical* for its blending of intricately reasoned content with vividly realized emotion. In Italy, the term *marinism* implied that the nervously ornate style of Marino was so unique that it could only be described in terms of itself, while in Spain, *Gongorism*, after the poetry of Góngora, names his innovative, Latinate manipulation of diction and syntax. His style, displaying features also known as *culturanismo*, is further distinguished from *conceptismo*, which stressed the ingenious presentation of complex ideas, as in the works of Quevedo, Gracián, or Calderón. The Baroque qualities in the work of these poets would later earn the disfavor of eighteenth-century neoclassical critics, whose attitudes in fact were already being anticipated in the French theater of Racine and Corneille and in the critical essays of Boileau. Apart from the early poems of Malherbe and some works by Scarron and Corneille, France was less influenced by the Baroque mood than was Italy or Spain; the more classical tendency of its literature was aided by the founding of the Académie Française on 10 July 1637, which sought to ensure the purity and clarity of the French language; for example, when it examained Malherbe's *Stances sur le roi allant en Limousin*, it declared only one stanza of the poem to be perfect.

An obvious influence upon literature was also the climate of protracted religious debate between Reformation and Counter-Reformation factions. It is not surprising that much of the period's prose was shaped by the doctrinal disputes of the time, as Anglicans, Puritans, Arminians, Jansenists, Quietists, Gallicans, Jesuits, Neoplatonists, and a host of dissenting sects exchanged their views in a blizzard of tracts, books, letters, and sermons. In England especially this was an era of impressive prose writers who blended theology and style in memorable ways, whether in fiction (Bunyan), treatise (Hooker), sermon (Andrewes, Donne), or essay (Browne). Yet the taste of the public was drawn as much to the secular as to the sacred, as French, English, and Spanish drama reveals. Spaniards enjoyed not only the one-act religious plays (*autos sacramentales*) of Calderón, but also the romantic dramas of Lope de Vega, over 500 of whose works remain today, and the decadent mood of such Jacobean tragedians as Webster and Tourneur was balanced toward the century's end by the social wit in the plays of Congreve or Molière.

The 400 entries in this volume, prepared by 73 contributors, are designed for both ready reference and as initial points for further study on each figure. To facilitate additional inquiry, each entry is followed by a bibliographical supplement; a general bibliography for the volume, arranged by subject, provides for broader consideration of topics within the historical period. The timeline comparatively places significant events and achievements within their chronological contexts. Cross-referenced entries (indicated by an *asterisk) will guide the reader to figures of related importance, as will the index. Appendixes group the entries by both subject and country. Individuals were chosen according to their contribution to the social, artistic, and intellectual milieu of seventeenth-century Europe. Political or military figures are included only when they play major roles in the broader cultural life of the period; hence, Oliver Cromwell is present as the most prominent exemplar of the English experiment in Puritan government, whereas his able general, George Monck, is not. The increasing role of women in Europe, suggested by the female artists and writers included here and in the bibliography of studies on women of the period, suggests their prominence in its cultural life. However, despite the increasing importance of women as producers of culture (as, for example, in the rapid increase in the number of English actresses), the tremendous expansion of science and technology during this age remained, as Londa Schiebinger has discussed in her study, an almost exclusively male phenomenon.

Completion of this volume would not have been possible without the cooperation, suggestions, and patience of its contributors, especially those who agreed to complete additional entries beyond their initial assignments. I also received valuable assistance from the Faculty Support staff of the Computer Information Services Office at Armstrong Atlantic State University, especially William R. Bates, Pamela Culberson, Sandy M. Hart, Nancy Luke, Kim Saade-Simshauser, Kathy Schaefer, and Gregory Surrette, Sr. I am also grateful to the Office of the Vice President for Academic Affairs for funding two faculty research grants that significantly aided the completion of this work and to Robert Parham, Head of the Department of Languages, Literature, and Philosophy, for his interest and help. Special thanks are due as well to series adviser Ronald H. Fritze and to Cynthia Harris and Frank Saunders, my editors at Greenwood, for their continued support and encouragement. My greatest debt is recorded in the dedication.

Chronology

Dates are approximate for some entries

1600   Thomas Dekker, *The Shoemaker's Holiday*
Pedro Caldéron de la Barca born
William Shakespeare, *Hamlet*
William Gilbert, *De Magnete*
Thomas Morley, *First Book of Ayres*
Michelangelo Merisi da Carravagio, *Doubting Thomas*

1601   William Shakespeare, *Troilus and Cressida*
Pierre de Fermat born
Thomas Morley, *Triumphs of Oriana*
Carlo Gesualdo, *Madrigals*
Michelangelo Merisi da Carravagio, *Conversion of St. Paul*

1602   John Marston, *The History of Antonio and Mellida*
Ben Jonson, *Poetaster*
Thomas Campion, *Observations in the Art of English Poesie*
William Lilly born
Otto von Guericke born
Bodleian library established at Oxford University
Lope Félix de Vega, *La hermosura de angélica*
William Shakespeare, *All's Well That Ends Well*

1603   William Gilbert dies
Claudio Monteverdi, *Fourth Book of Madrigals*

Roger Williams born

Ben Jonson, *Sejanus* produced

Samuel Daniel, *A Defence of Rhyme*

Francisco Gomez de Quevedo y Villegas, *La vida de buscón*

Sir Walter Ralegh imprisoned for treason

1604   Johann Rudolf Glauber born

Michelangelo Merisi da Carravagio, *The Deposition*

Lope Félix de Vega, *Comedias*

John Marston, *The Malcontent*

William Shakespeare, *Measure for Measure*

Jacques Auguste de Thou, *Historiae sui temporis*

George Chapman, *Bussy D'Ambois*

1605   Tomás Luis de Victoria, *Officium de Functorum*

John Dowland, *Lachrymae*

Claudio Monteverdi, *Fifth Book of Madrigals*

Miguel de Cervantes, *Don Quixote, Part I*

George Chapman, *All Fooles*

Samuel Daniel, *Philotas*

Michael Drayton, *Poems*

Thomas Randolph born

William Shakespeare, *King Lear, Macbeth*

1606   Rembrandt Harmenszoon van Rijn born

William Shakespeare, *Antony and Cleopatra*

Pierre Corneille born

William Davenant born

Ben Jonson, *Volpone*

John Marston, *Parasitaster*

Madeleine de Scudéry born

1607   John Norden, *The Surveyors' Dialogue*

Claudio Monteverdi, *Orfeo*

William Byrd, *Gradualia*

Cyril Tourneur, *The Revenger's Tragedy*

William Shakespeare, *Coriolanus, Timon of Athens*

John Marston, *What You Will*

Thomas Heywood, *A Woman Killed with Kindness* printed

Paul Gerhardt born

Honoré d'Urfé, *Astree*

1608   John Smith, *A True Relation of Virginia*

        Giovanni Borelli born

        El Greco, *Golgotha*

        Francis de Sales, *Introduction to the Devout Life*

        Joseph Hall, *Characters of Vertues and Vices*

        Thomas Middleton, *A Mad World My Masters*

        John Milton born

        William Shakespeare, *Pericles*

1609   Johannes Kepler, *Astronomia Nova*

        Thomas Ravenscroft, *Pammelia*

        El Greco, *Brother Paravicino*

        Francis Beaumont and John Fletcher, *The Knight of the Burning Pestle*

        Thomas Dekker, *The Guls Hornebooke*

        Paul Fleming born

        Ben Jonson, *Epicoene, or the Silent Woman*

        William Shakespeare, *Cymbeline*

        Francis Bacon, *De Sapientia Veterum*

        Hugo Grotius, *Mare Liberum*

        Garcilaso de la Vega, *History of Peru* (completed in 1617)

1610   Ben Jonson, *The Alchemist*

        Paul Scarron born

        William Shakespeare, *The Winter's Tale*

        Peter Paul Rubens, *Raising of the Cross*

        Michael Praetorius, *Musae Sioniae*

        John Webster, *The White Devil*

        George Chapman, *The Revenge of Bussy D'Ambois*

1611   John Speed, *Theatre of the Empire of Great Britaine; Historie of Great Britaine*

        Henry Hudson dies

        Thomas Ravenscroft, *Melismata*

        Tomás Luis de Victoria dies

        Peter Paul Rubens, *Descent from the Cross*

        Authorized Version of the English Bible (King James Version)

        William Byrd, John Bull, Orlando Gibbons, *Parthenia*

1612   Samuel Butler born

        Michael Drayton, *Poly-Olbion, Part I*

        Giovanni Gabrieli dies

        Orlando Gibbons, *First Set of Madrigals and Motets*

El Greco, *Baptism of Christ*

Jakob Boehme, *Aurora*

John Webster, *The White Devil*

1613   Samuel Purchas, *Purchas His Pilgrimage*

William Shakespeare, *Henry VIII*

William Browne, *Britannia's Pastorals*

Miguel de Cervantes, *Exemplary Tales*

Richard Crashaw born

François de La Rochefoucauld born

Lope Félix de Vega, *Fuenteovejuna*

Francisco Suárez, *Defensio fidei catholicae* . . .

Guido Reni, *Aurora* frescoes

Thomas Bodley dies

1614   Ben Jonson, *Bartholomew Fair*

Sir Thomas Overbury, *Characters*

Henry More born

Sir Walter Ralegh, *The History of the World*

El Greco dies

Girolamo Frescobaldi, *Toccate di Cembalo*

John Napier, *Mirifici Logarithmorum Canonis Descriptio*

Pocahontas marries John Rolfe

1615   Miguel de Cervantes, *Don Quixote, Part II*

George Chapman, translation of *Odyssey*

John Denham born

William Camden, *Annales Rerum Anglicarum*, vol. 1

Gian Lorenzo Bernini, *Amalthea*

Peter Paul Rubens, *The Battle of the Amazons*

Thomas Middleton, *The Witch*

1616   Francis Beaumont dies

Miguel de Cervantes dies

Andreas Gryphius born

Ben Jonson, first folio of his works

William Shakespeare dies

Johann Valentin Andreae, *Chymische Hochzeit Christiani Rosenkreutz anno 1459*

Francis de Sales, *Treatise on the Love of God*

Frans Hals, *Banquet of the Officers of the Saint George Civic Guard Company of Haarlem*

	Inigo Jones designs Queen's House
	Peter Paul Rubens, *The Lion Hunt*
	William Baffin discovers Baffin Bay
	John Smith, *A Description of New England*
1617	Théophile de Viau, *Pyramus et Thisbe*
	Ben Jonson becomes poet laureate
	Christian Hoffmann von Hoffmannswaldau born
	Francisco Suárez dies
	Bartolome Esteban Murillo born
	Guido Reni, *The Deeds of Hercules*
	Anthony Van Dyck, *A Study of Four Negro Heads*
	Peter Lely born
	John Napier dies
	Pocahontas dies
	Domenico Domenichino, *Diana's Hunt*
	Sir Walter Ralegh travels to Guiana in search of El Dorado
1618	Francis Bacon made Lord Chancellor
	Abraham Cowley born
	John Fletcher, *The Humourous Lieutenant*
	Richard Lovelace born
	Gian Lorenzo Bernini, *Aeneas, Anchises, and Ascanius*
	Johannes Kepler, *Harmony of the Worlds*
	Sir Walter Ralegh executed
1619	Francis Beaumont and John Fletcher, *The Maid's Tragedy*
	Samuel Daniel dies
	Savinien de Cyrano de Bergerac born
	Johann Valentin Andreae, *Christianopolis*
	Jakob Boehme, *On the Principles of Christianity*
	Hugo Grotius, *De veritate religionis Christianae*
	Nicholas Hilliard dies
	Inigo Jones designs banqueting house, Whitehall
	Diego Rodríguez de Silva Velázquez, *Adoration of the Kings; The Water Carrier of Seville*
	Jan Sweelinck, *Cantiones*
	William Harvey discovers circulation of the blood
1620	Pilgrims land in Massachusetts
	Thomas Campion dies
	Francis Bacon, *Instauratio Magna*

Gian Lorenzo Bernini, *Neptune and Triton*

Aelbert Cuyp born

Peter Paul Rubens, *Fall of the Damned*

Anthony van Dyck, *St. Sebastian*

Claudio Monteverdi, *Seventh Book of Madrigals*

Michael Praetorius, *Syntagma musicum*

Edmund Gunter, *Canon triangulorum, or the Table of Artificial Sines and Tangents*

1621 Jean de la Fontaine born

John Fletcher, *The Wild Goose Chase*

Cardinal Robert Bellarmine dies

Robert Burton, *The Anatomy of Melancholy*

Anthony Van Dyck, *Rest on the Flight into Egypt*

Michael Praetorius dies

Jan Sweelinck dies

Johannes Kepler, *Epitome astronomiae Copernicanae*

1622 Michael Drayton, *Poly-Olbion, Part II*

Gian Lorenzo Bernini, *Pluto and Proserpina*

Jean-Baptiste Molière born

Henry Vaughan born

Francis Bacon, *History of the Reign of Henry VIII*

Jakob Boehme, *De signatura rerum*

St. Francis de Sales dies

Guido Reni, *Job*

Peter Paul Rubens, *The Medici Cycle*

William Baffin dies

1623 Blaise Pascal born

Gian Lorenzo Bernini, *David*

William Shakespeare, publication of first folio of his works

Inigo Jones builds Queen's Chapel, Westminster

François Mansart builds St. Marie de la Visitation, Paris

Guido Reni, *Baptism of Christ*

Anthony Van Dyck, *Cardinal Bentivoglio*

Diego Rodríguez de Silva Velázquez becomes royal painter for King Philip IV

William Byrd dies

1624 Thomas Middleton, *A Game at Chess*

Martin Opitz, *Buch von der Deutschen Poeterey*

Jakob Boehme dies

Edward Herbert, Baron of Chirbury, *De veritate*

John Donne, *Devotions Upon Emergent Occasions*

Arnold Geulincx born

George Fox born

Frans Hals, *The Laughing Cavalier*

Nicolas Poussin, *Rape of the Sabine Women*

John Smith, *The General Historie of Virginia, New England and the Summer Isles*

Thomas Sydenham born

Johannes Baptista van Helmont first uses "gas" as scientific term

1625   Glauber's salt isolated from spring water by Johann Rudolf Glauber

Gian Lorenzo Bernini, *Apollo and Daphne*

Giovanni Domenico Cassini born

Orlando Gibbons dies

Heinrich Schütz, *Cantiones sacrae*

Jan Breughel (the elder) dies

Inigo Jones designs Covent Garden Church, London

Nicolas Poussin, *Parnassus*

Hugo Grotius, *De Jure Belli ac Pacis*

Ben Jonson, *The Staple of News*

Giambattista Marino dies

Joost van den Vondel, *Palamedes*

Martin Opitz made poet laureate of Vienna

Hans Jacob Christoffel von Grimmelshausen born

John Webster dies

1626   John Aubrey born

Francis Bacon dies

Joseph Hall, *Contemplations*

François Mansart designs Château de Balleroy

Peter Paul Rubens, altarpiece, *Assumption of the Virgin*

Jan Steen born

Peter Minuit buys Manhattan

1627   Luis de Góngora y Argote dies

Francisco Gómez de Quevedo y Villegas, *Los Sueños*

Thomas Middleton dies

Robert Boyle born

Frans Hals, *The Merry Drinker*

Rembrandt van Rijn, *The Money Changer*

Peter Paul Rubens, *Mystic Marriage of St. Catherine*

Heinrich Schütz, *Dafne*

1628   John Bunyan born

Juan Ruiz de Alarcon y Mendoza, *La Verdad Sospechosa*

Charles Perrault born

Sir William Temple born

John Amos Comenius, *Informatorium der Mutterschul*

René Descartes, *Régles pour la direction de l'esprit*

William Laud becomes bishop of London

Frans Hals, *Gypsy Woman*

Nicolas Poussin, *Martyrdom of St. Erasmus*

Jacob Issackszoon van Ruisdael born

Diego Rodríguez de Silva Velázquez, *Christ on the Cross*

John Bull dies

John Ray born

1629   Pedro Calderón de la Barca, *La dama duende*

Pierre Corneille, *Mélite*

John Ford, *The Lover's Melancholy*

Lancelot Andrewes, *XCVI Sermons*

Gian Lorenzo Bernini begins work on St. Peter's Basilica, Rome

Pieter de Hooch born

Anthony Van Dyke, *Rinaldo and Armida*

Francisco de Zurbaran, *St. Bonaventura*

Heinrich Schütz, *Sinfoniae sacrae*

Albert Gerard uses brackets in mathematical notation

Christiaan Huygens born

1630   John Winthrop arrives in Massachusetts

Tirso de Molina, *El burlador de Sevilla*

Thomas Middleton, *A Chaste Mayde in Cheapside*

Jusepe de Ribera, *Archimedes*

Peter Paul Rubens, *Blessings of Peace*

Girolamo Frescobaldi, *Arie musciale*

Johannes Kepler dies

1631   John Donne dies

Michael Drayton dies

John Dryden born

Thomas Heywood, *The Fair Maid of the West*

Ben Jonson, *The Devil Is an Asse*

Diego Rodríguez de Silva Velázquez, *Infanta Maria, Queen of Hungary*

William Oughtred denotes multiplication by the symbol "x"

Théophraste Renaudot establishes Paris *Gazette*

1632   Giovanni Battista Basile dies

Thomas Dekker dies

Philip Massinger, *The City Madam*

John Locke born

John Selden, *Mare Clausum*

Baruch de Spinoza born

Luca Giordano born

Nicolaes Maes born

Johannes Vermeer born

Christopher Wren born

Jean-Baptiste Lully born

Antony van Leeuwenhoek born

1633   Abraham Cowley, *Poetical Blossoms*

John Ford, *'Tis Pity She's a Whore*

George Herbert dies

Philip Massinger, *A New Way to Pay Old Debts*

Samuel Pepys born

John Cotton arrives in Boston

Anthony Van Dyke, *Charles I*

Jacopo Peri dies

1634   George Chapman dies

Marie-Madeleine de Lafayette born

John Ford, *Perkin Warbeck*

John Marston dies

John Milton, *Comus*

Anne Hutchinson arrives in Massachusetts

Rembrandt van Rijn, *Artemisia*

Francisco de Zurbaran, *The Defense of Cadiz*

Jean Nicolet arrives in Wisconsin

1635   Pedro Calderón de la Barca composes *La vida es sueño*

George Etherege born

Philippe Quinault born

	Thomas Randolph dies
	François Mansart builds Château Blois
	Robert Hooke born
	Françoise d'Aubigne, Marquise de Maintenon born
1636	Pierre Corneille composes *Le Cid*
	Roger Williams exiled from Massachusetts
	Joseph Glanvill born
	Harvard College founded
	Marin Mersenne, *Harmonie Universelle*
	Anthony Van Dyke, *Charles I on Horseback*
1637	René Descartes, *Discours de la méthode*
	Jan Swammerdam born
	Dietrich Buxtehude born
	Jusepe de Ribera, *Pietà*
	John Milton, *Lycidas*
	Ben Jonson dies
1638	Anne Hutchinson exiled from Massachusetts
	Claudio Monteverdi, *Eighth Book of Madrigals*
	Pieter Breughel the Younger dies
	Peter Paul Rubens, *The Three Graces*
	Nicolas Malebranche born
	Cornelius Otto Jansen dies
1639	Jeremiah Horrocks observes his predicted path of Venus
	William Gascoigne invents micrometer
	Francisco de Zurbaran, *St. Francis in Meditation*
	Académie Français creates French dictionary
	Increase Mather born
	Juan Ruiz de Alarcón y Mendoza dies
	Thomas Carew dies
	Martin Opitz dies
	Jean-Baptiste Racine born
1640	John Bull dies
	Rembrandt van Rijn, *Self-portrait*
	Peter Paul Rubens dies
	William Wycherly born
	Philip Massinger dies
	John Ford dies

Paul Fleming dies

Aphra Behn born

1641   French settlers arrive in Michigan

Anthony Van Dyck dies

Claude Lorraine, *Embarkation of St. Ursula*

René Descartes, *Meditationes de prima philosophia*

John Evelyn begins his diary

1642   Christian Weise born

English Puritans close theaters

John Amos Comenius, *A Reformation of Schooles*

Guido Reni dies

Rembrandt van Rijn, *Night Watch*

Galileo Galilei dies

Isaac Newton born

Abel Janszoon Tasman reaches Tasmania and New Zealand

1643   Evangelista Torricelli invents barometer

Girolamo Frescobaldi dies

Claudio Monteverdi dies

Diego Rodríguez de Silva Velázquez, *Venus and Cupid*

John Milton, *The Doctrine and Discipline of Divorce*

1644   Willam Penn born

Pierre Corneille, *Rodogune*

René Descartes, *Principia philosophicae*

John Milton, *Areopagitica*

Gian Lorenzo Bernini, sculpture *Ecstasy of St. Teresa*

Antonio Stradivari born

1645   Paul Scarron, *Jodolet* produced

Hugo Grotius dies

Rembrandt van Rijn, *The Rabbi*

1646   Henry Vaughan, *Poems*

James Shirley, *Poems*

Gottfried Wilhelm von Leibniz born

Godfrey Kneller born

Athanasius Kircher builds projection lantern

1647   Henry More, *Philosophical Poems*

Pierre Bayle born

Peter Lely, *The Children of Charles I*

	Bonaventura Cavalieri dies
	Evangelista Torricelli dies
1648	Marin Mersenne dies
	Johann Rudolf Glauber isolates hydrochloric acid
	John Blow born
	Jusepe de Ribera, *The Holy Family with St. Catherine*
	Society of Friends (Quakers) established by George Fox
	Tirso de Molina dies
	Paul Scarron, *Le Roman comique*
1649	William Drummond of Hawthornden dies
	René Descartes, *Les Passions de l'Âme*
	Diego Velázquez, *Pope Innocent X*
1650	Christoph Scheiner dies
	Pierre Corneille, *Andromède*
	Richard Baxter, *The Saints' Everlasting Rest*
	René Descartes dies
	Nicolas Poussin, *Self-Portrait*
1651	William Davenant, *Gondibert*
	Pedro Calderón de la Barca ordained priest
	Thomas Hobbes, *Leviathan*
	Jeremy Taylor, *Holy Dying*
	Rembrandt van Rijn, *Girl with a Broom*
	Nell Gwyn born
1652	Thomas Otway born
	Nahum Tate born
	Inigo Jones dies
	Otto von Guericke invents air pump
	William Byrd the elder born
1653	Jacob Issackszoon van Ruisdael, *Castle of Bentheim*
	Arcangelo Corelli born
	Johann Pachelbel born
	Théophraste Renaudot dies
1654	Joost van den Vondel, *Lucifer*
	Rembrandt van Rijn, *Portrait of Jan Six*
	Jacques Bernoulli born
	Blaise Pascal and Pierre Fermat announce probability theory
1655	Daniel Heinsius dies

Nicolas Poussin, *The Four Seasons*

John Newton discovers binomial theorem

1664   Heinrich Schütz, *Christmas Oratorio*

Samuel Butler, *Hudibras, Part II*

Francisco de Zurbaran dies

Johannes Vermeer, *The Lacemaker*

Andreas Gryphius dies

Jean-Baptiste Molière, *Tartuffe*

Matthew Prior born

John Vanbrugh born

1665   Pierre de Fermat dies

Robert Hooke, *Micrographia, or some Physiological Descriptions of Minute Bodies*

Nicolas Poussin dies

John Bunyan, *The Holy City*

François de la Rochefoucauld, *Réflexions ou sentences et maximes morales*

1666   Jean-Baptiste Molière, *Le Misanthrope*

John Bunyan, *Grace Abounding to the Chief of Sinners*

Frans Hals dies

François Mansart dies

Antonio Stradivari begins violin-making

1667   John Arbuthnot born

Abraham Cowley dies

John Milton, *Paradise Lost*

Jonathan Swift born

George Wither dies

1668   Aphra Behn, *Oroonoko*

William Davenant dies

François Couperin born

Johann Rudolf Glauber dies

Isaac Newton builds reflecting telescope

1669   Jan Swammerdam, *History of the Insects*

Rembrandt van Rijn dies

Matthew Locke, *The Treasury of Musick*

John Denham dies

Samuel Pepys ends his *Diary*

1670   William Congreve born

Jean-Baptiste Racine, *Bérénice*

	John Amos Comenius dies
	Johannes Vermeer, *The Pearl Necklace*
1671	Colley Cibber born
	John Milton, *Paradise Regained*; *Samson Agonistes*
	Anthony Ashley Cooper, Third Earl of Shaftesbury born
	Rob Roy MacGregor born
1672	Joseph Addison born
	Heinrich Schütz dies
	Richard Steele born
1673	Jacques Marquette and Louis Joliet discover source of Mississippi
	John Dryden, *Marriage À-la-Mode*
	Jean-Baptiste Molière dies
1674	William Byrd the Younger born
	Bartoleme Esteban Murillo, *St. Francis*
	Isaac Watts born
	Thomas Traherne dies
	John Milton dies
	Nicolas Boileau-Despréaux, *L'Art poètique*
1675	William Wycherly, *The Country Wife*
	Baruch de Spinoza completes *Ethics*
	Johannes Vermeer dies
	Antonio Vivaldi born
	Isaac Newton, *Opticks*
1676	George Etherege, *Sir Fopling Flutter, or the Man of Mode*
	Hans Jacob Christoffel von Grimmelshausen dies
1677	Jean-Baptiste Racine, *Phèdre*
	Angelus Silesius dies
	Baruch de Spinoza dies
	Wenceslaus Hollar dies
1678	George Farquhar born
	Andrew Marvell dies
	John Bunyan, *Pilgrim's Progress*
1679	Joost van den Vondel dies
	Christian Hoffmann von Hoffmannswaldau dies
	Thomas Hobbes dies
	Jan Steen dies
1680	Samuel Butler dies

François de La Rochefoucauld dies

Gian Lorenzo Bernini dies

Peter Lely dies

Jan Swammerdam dies

1681   Georg Philipp Telemann born

Pedro Calderón de la Barca dies

John Dryden, *Absalom and Achitophel*

1682   Claude Lorrain dies

Bartolome Esteban Murillo dies

Jacob Issackszoon van Ruisdael dies

John Dryden, *Absalom and Achitophel, Part 2*

Thomas Otway, *Venice Preserv'd* produced

1683   William Penn, *A General Description of Pennsylvania*

Pieter de Hooch dies

Jean Philippe Rameau born

Roger Williams dies

1684   John Bunyan, *Pilgrim's Progress, Part II*

Pierre Corneille dies

Increase Mather, *Illustrious Providences*

Jean-Antoine Watteau born

1685   John Gay born

Thomas Otway dies

J. S. Bach born

Georg Friedrich Handel born

Domenico Scarlatti born

1686   Otto von Guericke dies

Gabriel Daniel Fahrenheit born

1687   Nell Gwyn dies

Jean-Baptiste Lully dies

Isaac Newton, *Philosophiae naturalis principia mathematica*

1688   John Bunyan dies

Phillipe Quinault dies

Ralph Cudworth dies

1689   Aphra Behn dies

Henry Purcell, *Dido and Aeneas*

Jean-Baptiste Racine, *Esther*

Thomas Sydenham dies

1690	John Locke, *An Essay Concerning Human Understanding*; *Two Treatises on Government*
	Charles Le Brun dies
1691	George Etherege dies
	Robert Boyle dies
	Aelbert Cuyp dies
1692	Thomas Shadwell dies
	Henry Purcell, *The Fairy Queen*
1693	Marie-Madeleine de Lafayette dies
	John Locke, *Thoughts Concerning Education*
	Nicolaes Maes dies
	Alessandro Scarlatti, *Teodora*
1694	William Congreve, *The Double Dealer*
	Christopher Wren designs Greenwich Hospital
1695	Jean de La Fontaine dies
	Henry Vaughan dies
	John Locke, *The Reasonableness of Christianity*
	Henry Purcell dies
1696	Madame de Sévigné dies
	Thomas Southerne, *Oronooko*
	John Ray describes peppermint
1697	John Aubrey dies
	John Vanbrugh, *The Relapse*
1698	Society for Promoting Christian Knowledge (SPCK) established
1699	François de Salignac de la Mothe Fénelon, *Télémaque*
	Jean-Baptiste Racine dies
	Gilbert Burnet, *Exposition of the Thirty-nine Articles*
1700	William Congreve, *The Way of the World*
	Joseph Sauveur analyzes musical tones
1701	Charles Sedley dies
	Benjamin Whichcote, *Several Discourses*
	Antoine de la Mothe Cadillac establishes Detroit
1702	George Farquhar, *The Twin Rivals*
1703	Samuel Pepys dies
	Cotton Mather, *Magnalia Christi Americana*
	Charles Perrault dies
1704	John Locke dies
	Isaac Newton, *Optics*

1705	Luca Giordano dies
	John Ray dies
	Jacques Bernoulli dies
1706	Johann Pachelbel dies
	John Evelyn dies
	George Farquhar, *The Recruiting Officer*
1707	George Farquhar dies
	Jean Mabillon dies
	Dietrich Buxtehude dies
1708	Christian Weise dies
	John Blow dies
1709	Joseph Addison and Richard Steele's *The Tatler* appears
1710	Thomas Betterton dies
	Cotton Mather, *Essays to Do Good*
	Gottfried Wilhelm Leibniz, *Théodicé*
1711	Joseph Addison and Richard Steele's *The Spectator* appears
	Anthony Ashley Cooper, *Characteristicks of Men, Manners, Opinions, Times*
1712	Arcangelo Corelli, *Twelve Concerti Grossi*
1713	Arcangelo Corelli dies
	François Fénelon, *Traité de l'existence et des attributs de Dieu*
1714	Nicholas Rowe's edition of Shakespeare published
	Gottfried Wilhelm Leibniz, *Monadologie*
1715	Nahum Tate dies
	William Wycherly dies
	Matthew Prior, *Solomon, or the Vanity of the World*
1716	Gottfried Wilhelm von Leibniz dies
1717	Smallpox innoculation introduced into England
1718	Nicholas Rowe dies
	William Penn dies
1719	Joseph Addison dies
	John Flamsteed dies
1720	John Gay, *Collected Poems*
	Nicholas Hawksmoor designs St. George's, London

Absolutism
and the
Scientific Revolution
1600–1720

A

ADDISON, JOSEPH (1672–1719). The quality of Addison's essays was such that the great eighteenth-century critic Samuel Johnson declared, "Whoever wishes to attain an English style must give his days and nights to the volumes of Addison." The son of Lancelot Addison, later Dean of Lichfield, Joseph Addison was admitted first to Queen's College, Oxford, and then to Magdalen, taking an M.A. in 1693. An accomplished classical scholar at Oxford, he composed Latin poems that appeared in *Musae Anglicanae* (1691–1699) and one entitled *Pax Gulielmi* (1697) for *William III upon the Treaty of Ryswick, which ended the War of the Grand Alliance and declared William legitimate king of England; the poem earned him a pension. In 1699, he departed on a grand tour of Europe, visiting Italy, Switzerland, Germany, and Holland, returning in 1703 and publishing his poem *A Letter from Italy*, addressed to Lord Halifax, his Whig patron. In 1704, his poem *The Campaign*, in honor of the Duke of Marlborough's Blenheim victory, earned him a government sinecure, while he also held several diplomatic posts and served as a member of Parliament, becoming one of the most influential Whig spokesmen in the nation.

In 1710, he started a short-lived periodical called *The Whig Examiner*, but it was his association with Richard Steele in *The Tatler* from 1709 to 1711 that began his rise to fame. In March 1711, Addison and Steele began publication of *The Spectator* with Addison writing over half its issues; unlike *The Tatler*, it appeared daily, running until 6 December 1712. Addison renewed it in June 1714 by himself for another eighty issues, ending on 20 December 1714. The essays focused on a group of fictional characters who commented, with their compatriot "Mr. Spectator," on issues of the day. Owing something to the earlier "character essays" of *John Earle and others, these were more highly individualized sketches with realistically detailed personae. "Captain Sentry" spoke from a military perspective, "Sir Andrew Freeport" was the voice of the merchant class, and "Sir Roger de Coverley" was the staid country squire, among others. The essays aimed "to enliven morality with wit and to temper wit with morality"; coming after the more licentious excesses of *Charles II's reign, they anticipated the moral guidance offered in later Victorian literature. They tended to steer a middle course between Whig and Tory political extremes, instead focusing on broad social criticism and satire in engagingly detailed glimpses of daily English life. The essays still live for modern readers owing to Addison's masterfully crafted "middle style" of prose, which flows as easily as spoken speech, yet retains a cultivated precision without stylistic excess.

Addison later wrote a dramatic tragedy *Cato*, which premiered in 1713, was dedicated to the Duchess of Marlborough, and ran for thirty-five performances. Though translated into German and French, it was thought to be a Whig propaganda piece and was condemned by audiences for the very quality for which Voltaire praised it, a strongly classical dramatic design. Addison in later life was the subject of acrimonious satire by Alexander Pope, who had been a contributor to *The Spectator*. He held several governmental posts under George I, and in 1716 he married the dowager Countess of Warwick, an apparently unhappy match. Addison died on 17 June 1719, a victim of dropsy (edema).

Bibliography: E. Bloom and L. Bloom, *Joseph Addison's Sociable Animal*, 1971; P. Smithers, *Joseph Addison*, 1954.

Christopher Baker

ALARCÓN Y MENDOZA, JUAN RUIZ DE (1581–1639).

Born in Mexico, Alarcón is one of the greatest playwrights of the Spanish Golden Age, together with *Lope, *Tirso and *Calderón. His characters have a psychological depth that is exceptional in Spanish theater of the period and anticipates later theater.

Alarcón was born of Spanish parents in Mexico (New Spain). His father was from noble origins, and his mother was from the powerful Mendoza family. Alarcón went to Spain to study law at the University of Salamanca (1600). Later, he returned to Mexico to continue his studies at the University of Mexico. He lived in Spain from 1614 until his death. He held several offices to which his noble origins entitled him. He lived in Madrid during the acme of Spanish *comedia*, a popular genre that Lope had transformed to an enormous success; every afternoon, one of Madrid's several open air theaters performed a new play. Alarcón was very active in this world, but his plays did not have the popular success of Lope's. He was mocked in the literary circles for this, as well for his physical deformity—he was a bow-legged hunchback.

He wrote twenty plays that were published in two volumes (1628 and 1634). These plays encompassed all forms of the *comedia* genre (religious, historical, cloak and dagger, etc.). In spite of not being popular with the audiences, his plots are more carefully crafted than Lope's. Alarcón's are thesis plays that chastise moral vices through the creation of character types who incarnate these vices. These carefully created characters present psychological depth, a feature missing in the *comedias* of the period. His theater influenced important European writers such as Molière and Goldoni. His most famous play, *La Verdad Sospechosa* (Suspicious Truth), inspired *Corneille's *Le Menteur*. In this play, Alarcón narrates the misadventures of a young man whose own lies to conquer a young lady end up entangling him. At the end, everything concludes happily, as was customary in the genre, and the young man learns an important lesson about the importance of being truthful.

Bibliography: E. Claydon, *Juan Ruiz de Alarcón, Baroque Dramatist*, 1970; W. Poesse, *Juan Ruiz de Alarcón*, 1972.

Enrique Fernández

ALEXANDER VII, POPE (1599–1667). Pope from 7 April 1655 to 22 May 1667, Fabio Chigi excelled at his studies as a youth, eventually earning a doctorate in theology from the University of Siena in 1626. He rose through the ranks from Bishop of Nardo to Inquisitor at Malta to nuncio at Cologne. He became Vatican secretary of state in 1651, a cardinal one year later, and pope four years after that. Alexander staunchly opposed the doctrines of *Cornelius Jansen, condemning the five tenets of Jansenism as they were expressed in Jansen's *Augustinus*. He thus supported Innocent X's previous censure of this movement and answered the defense of *Antoine Arnauld who had argued that, though the propositions were heretical, they were not contained in that tract. *Louis XIV supported the pope's stricture against Jansen, though king and pope quarreled over a conflict between the king's embassy and the pope's Corsican guards, as well as over Louis's failure to aid Alexander's anti-Turkish policies. Despite ill health, Alexander also renovated the papal archiving system and commissioned *Bernini to complete the colonnade at St. Peter's; Bernini also designed his tomb.

Bibliography: L. Pastor, *The History of the Popes from the Close of the Middle Ages*, vol. 31, 1938–1961.

Christopher Baker

AMATI, NICOLO (1596–1684). Nicolo Amati was the grandson of Andrea Amati (before 1511—before 1580), founder of a family of stringed instrument makers who lived and worked in the north Italian city of Cremona from the sixteenth through the late seventeenth centuries. Andrea Amati is generally credited with the foundation of the Cremonese school of violin making and also with developing the forms of the violin, viola, and cello, which later became the standards for modern instruments. His "classic" violin shape incorporated a relatively small and flat body in comparison to that of the earlier viol, as well as delicately rounded sound holes and scrolls. Andrea also pioneered the use of a characteristic golden-brown or amber-colored varnish, which eventually came into common use. There are only sixteen instruments made by Andrea Amati still in existence; they are known for their elegant proportional design and beautiful finishing details, as well as for their relative lightness and warm quality of tone.

Nicolo was the son of Girolamo Amati (1561–1630), who, with his brother

Antonio (1560–1649), continued the family tradition of instrument construction. Antonio and Girolamo worked together in the family shop through 1630. Their instruments are not considered to be as fine as those of their father. Nicolo is regarded as the greatest instrument maker among the Amati. Nicolo's instruments, which include violins, violas, cellos, and bass viols, are known for their fine tone quality and responsiveness and were only surpassed in their beauty of design and tone by the instruments of his pupil *Antonio Stradivari.

Bibliography: C. Bonetti, *A Genealogy of the Amati Family of Violin Makers, 1500–1740*, 1989; D. Boyden, *The History of Violin Playing from its Origins to 1765*, 1965.

Maria Archetto

AMYRAUT, MOÏSE (1596–1664). Amyraut (Moses Amyraldus) was the most prominent theologian of the moderate school of Calvinist divinity associated most closely with the Protestant academy at Saumur and the Reformed congregation of Charenton near Paris. He began his study of theology at Saumur's Protestant Académie in 1618, and there one of his professors was the Scots divine John Cameron, who had an enormous and decisive impact on Amyraut's thinking. In 1626, Amyraut became Saumur's minister and soon began lecturing at the Académie.

During 1631, Amyraut's first major publication appeared, an apologetical work that responded to the attacks of Catholic scepticism on Protestant theology by affirming humankind's limited but real rationality. He became a central actor in a controversy that was to greatly trouble Calvinism when in 1634 he was so disturbed by the Reformed debates over predestination that he published a treatise in which he followed Cameron in advocating hypothetical universalism, the position that God's offer of saving grace to all of humankind was antecedent to his decision to predestine most souls to eternal damnation. This stance emphasized God's mercy and the inscrutability of some of his decrees to human reason, but was distasteful to many Protestant scholastics, who found it illogical or Arminian. However, an attempt to convict Amyraut of heresy at the French Reformed national synod of 1637 failed, thanks to Amyraut's powerful supporters and fears of schism within what remained of French Protestantism. The dispute continued intermittently for more than two decades; Amyraut insisted on the pastoral weaknesses of his opponents' attitude and the importance of following Jean Calvin in basing theology on scriptural exegesis, not Aristotelian logic. Consistent with his pastoral concerns, Amyraut published a treatise in French of several thousand pages devoted to ethics and many sermons, the latter demonstrating his mastery of pulpit oratory and command of a Ciceronian rhetoric, which in many respects foreshadowed the style of Classical French Catholic preaching. Friendly with Catholics and the French court, Amyraut was a staunch royalist.

Bibliography: B. Armstrong, *Calvinism and the Amyraut Heresy*, 1969; P. Bayler, ed., *Selected Sermons of the French Baroque*, 1983; F. Laplanche, *L'Ecriture, le sacré, et l'histoire*, 1986.

Matthew Koch

ANDREAE, JOHANN VALENTIN (1586–1654). Johann Valentin Andreae is generally considered to be the founder of Rosicrucianism. He was one of the most original and interesting literary writers of the German Baroque. Andreae was born on 17 August 1586 in Herrenburg near Tübingen (Württemberg) into a family of Lutheran theologians. After his father's death in 1601, the family moved to Tübingen where he received his M.A. degree in 1605. He had to leave the university in 1606 after allegations that he had associated with prostitutes. After several journeys and several tutoring positions, he was ordained as a Lutheran pastor on 25 February 1614, assumed a position as assistant pastor in Vaihingen, and married in 1615. In 1620, he received the position of pastor in Calw near Stuttgart. Twice, in 1618 and again in 1634, Andreae lost both his house and his library to the raging Thirty Years' War (1618–1648). In 1639, Andreae was appointed court pastor for Württemberg in Stuttgart, and in 1650 he became rector at the Lutheran school in Bebenhausen. He died in Stuttgart on 27 June 1654.

Today, Andreae is best known as author of three texts generally known as Rosicrucian Writings (1615–1616)—although Andreae's authorship of the first two is not universally accepted—and for his utopia *Christianopolis* (1619). The *Fama Fraternitatis* (1615) describes the fifteenth-century journey of the protagonist C.R. to the Middle East. Upon his return, he founds a secret society dedicated to the improvement of mankind, based on the newly acquired alchemical knowledge. The *Confessio Fraternitatis* (1615) confirms the existence of the brotherhood, explains its principles for the benefit of the learned, and affirms that it does not pose a threat to Lutheran orthodoxy or to worldly government. *Chymische Hochzeit Christiani Rosenkreutz anno 1459* (The Chemical Wedding of Christian Rosenkreutz in the Year 1459), published in 1616 but probably written in 1605–1606, is a novel that describes the tests and trials the protagonist endures during the seven days between the invitation to attend the royal wedding and his final release, his soul having been cleansed of all worldly desires. At the center of the story is the ritual execution of the royal couples and their restoration to life through alchemical procedures.

Christianopolis (1619) is the tale of a shipwrecked traveler who after moral examination is introduced into the society of Christianopolis. The city is a corporate Christian community that is preparing for a new Christian age. It is governed by an elected council according to biblical principles. Science, the discovery of the true principles of nature based on neo-Platonic natural philosophy, is a central endeavor, both to promote material well-being and to attain higher knowledge of God through theosophic reflection and chiliastic speculation. There is no money or private property, but the privacy of the family home is preserved. Progress is ensured through general education for both genders.

While the reception of the Rosicrucian texts was remarkable, it remains unclear whether Rosicrucianism ever developed beyond a literary movement during Andreae's lifetime. His work promotes alchemical and mystical thinking, but it also can be interpreted as a call for a general church reform. Andreae's highly individual and antirationalist spirituality, which resists integration into the emerging absolutist state, was further developed by Pietists like Philipp Spener and August Hermann Francke and even influenced enlightened thinkers like Christian Thomasius and Johann Gottfried Herder.

Bibliography: S. Beeler, *The Invisible College*, 1991; J. Montgomery, *Cross and Crucible*, 1973; F. Yates, *The Rosicrucian Enlightenment*, 1972.

Peter Hess

ANDREWES, LANCELOT (1555–1626). Andrewes was a brilliant English prelate, preacher, and theologian who served three monarchs and who shared *William Laud's notorious Arminianism, thereby incurring the wrath of Puritans.

He was born in London in 1555, the son of Thomas and Joan Andrewes. Showing unusual ability as a child, he attended the Coopers' Free School, the Merchant Taylors' School, and Pembroke Hall, Cambridge, obtaining a B.A. (1575), M.A. (1578), appointment as catechist (1578), ordination as deacon (1580), B.D. (1585), D.D. (c.1588), and mastership (1589–1605), as well as an M.A. (1581) from Jesus College, Oxford. After serving as chaplain to the Earl of Huntingdon, President of the Council of the North, he obtained the living at St. Giles, Cripplegate, London, and prebends at St. Paul's and Southwell in 1589 with the assistance of Sir Francis Walsingham. He was chaplain to Archbishop John Whitgift and Queen Elizabeth, though he rejected the bishopric of Salisbury in 1596 and that of Ely in 1599 because he was unwilling to alienate their revenues. However, he did accept a prebend (1597) and deanery (1601) at Westminster Abbey. Under *James I, he served as bishop of Chichester (1605–1609), Ely (1609–1619), and Winchester (1619–1626); Lord Almoner (1605–1619); Dean of the Chapel Royal (1619–1626); and as a privy councilor in England (from 1609) and Scotland (from 1617). He died on 26 September 1626, only a year and a half into *Charles I's reign; however, his views exercised lasting influence on the new king.

Andrewes was a good administrator, a generous benefactor, and an opponent of corruption who avoided politics unless it directly involved the Church. Like Richard Hooker, he defended the Anglican theological *via media*, rejecting both transubstantiation and predestination. He certainly was no crypto-Catholic, but his fondness for the "high church" ceremony later exemplified by Laud and his Arminian belief in free will was bound to antagonize even moderate Calvinists. Not surprisingly, his support for James I at the Hampton Court Conference in 1604 helped to alienate the king's Puritan critics. However, he was a fierce opponent of simony and other abuses that first animated Protestant reformers. His skill as a linguist (Latin, Greek, Hebrew, and fifteen modern languages) and his extensive knowledge of patristic literature explain his deep involvement in production of the Authorized Version of the Bible (1607–1611), which not only has incalculable religious significance but has profoundly influenced the development of the modern English language. His knowledge of philosophy was such that even *Sir Francis Bacon sought his counsel. Andrewes also was a brilliant preacher, though his sermons do not read as well today as they are likely to have done in his own time. His writings are voluminous, though few appeared during his lifetime. Most famous are *Tortura Torti* (1609), a reply to *Cardinal Bellarmine's attack on the oath of allegiance instituted after the Gunpowder Plot, and *Preces Privatae*, a posthumously published collection of prayers.

Bibliography: N. Lossky, *Lancelot Andrewes the Preacher (1555–1626): The Origins of the Mystical Theology of the Church of England*, 1991.

William B. Robison

ANNE (1665–1714). Anne, queen of Great Britain and Ireland from 1702 until her death, was the second daughter of *James II. A woman of limited political skills, unprepossessing appearance, and phlegmatic personality, her life was marked by a happy though unpopular marriage to Prince George of Denmark, an agonizing series of seventeen pregnancies but no surviving children by the time of her ascendancy at age thirty-seven, and constant political infighting in her court, which she despised. Her sister *Mary, wife of *William III, was likewise a political irritant to her; her best confidante was Sarah, the Duchess of Marlborough, with whom she eventually had an acrimonious falling-out. Anne's most ardent cause was the advancement of the Church of England, and she took little or no interest in the art, music, or theater of her day.

England's most notable foreign achievement under Anne's reign was its successful role in blunting the ambitions of *Louis XIV in the War of the Spanish Succcesson (1702–1713). John Churchill, First Duke of Marlborough, had achieved signal victories at the battles of Blenheim (1704), Ramillies (1706), Oudenarde (1708), and Malplaquet (1709), making the provisions of the concluding Treaty of Utrecht (1713) favorable to broad English influence on the continent. Notably, King Louis accepted that the monarchies of Spain and France would not be joined and that a Protestant succession was ensured in England. England gained Newfoundland, Nova Scotia, and the Hudson's Bay territory, as well as Gibraltar.

At home, the factionalism between Whigs and Tories gave Anne little peace. Though Tories had approved of her coronation for her staunch support of the Anglican church, and Whigs for her recognition of the legislative authority of Parliament, she personally found it very hard to stay above party wrangling. Though inclined toward the pro-church Tories, she felt the influence of the Whigs, who were enthusiastic supporters of Marlborough's campaigns abroad and whose policies his wife Sarah vigorously promoted to Anne. Yet the queen also tried to keep at arm's length the so-called "Junto" of Whig leaders: Somers,

Halifax, Wharton, Sunderland, and Walpole. The Tory minister Robert Harley had also sought to sway Anne through his cousin Abigail Masham, a tactic that enraged Sarah Churchill. Anne's minister Sidney Godolphin threatened that he and Marlborough would resign unless the queen removed Harley from office, which she did, later severing her friendship with Sarah as well. Despite such protracted infighting and personal rancors, Anne could point to the passage of the Act of Union in 1707 as a major achievement, an action that united England and Scotland governmentally, monetarily, and legislatively. Her husband's death in October 1708 saddened the queen deeply, and she had to endure for six more years without his loving support until her own death on 1 August 1714.

Bibliography: E. Gregg, *Queen Anne*, 1984, rpt. 2001; G. Trevelyan, *England Under Queen Anne*, 3 vols., 1930–1934.

Christopher Baker

ARBUTHNOT, JOHN (1667–1735). John Arbuthnot was the son of a Scotch clergyman, but early demonstrated an abiding interest in science, mathematics, and medicine. He studied at Aberdeen and earned his doctor's degree at St. Andrews in 1696. He settled in London, where he gave lessons in mathematics, and in 1697 published an essay entitled "An Examination of Dr. Woodward's Account of the Deluge," revealing a keen sense of critical inquiry and initiating a lifelong pattern of distinguished publishing in the sciences and medicine. In 1704, he was elected a fellow of the Royal Society and a year later was appointed personal physician to *Queen Anne.

By 1712, Dr. Arbuthnot had become a favorite in the new Tory administration under Robert Harley. He forged a close friendship with Jonathan Swift, whose *Journal to Stella* provides abundant testimony to Arbuthnot's wit and humanity. Swift tells us that Arbuthnot was the author, in 1712, of *The Art of Political Lying* and *The History of John Bull*, both lively and satirical attacks upon Whig excesses. Soon Swift and Arbuthnot became intimately associated with the younger Tory wits—Alexander Pope, *John Gay, and *Thomas Parnell—and together with them formed the "Scriblerus Club" to promote a sustained satiric attack on pedantry and "the abuses of human learning in every branch." The project lapsed after Anne's death in 1714 and Swift's return to Ireland, but Arbuthnot and Pope kept the scheme alive. *The Memoirs of Martinus Scriblerus*, first published with Pope's works in 1741, is mainly the work of Arbuthnot and provides the best examples of his satire.

Dr. Arbuthnot was in attendance upon the queen at the time of her death, and in his distinguished career he was friend and physician to a catalog of luminaries that included Prince George of Denmark, Queen Caroline, Chesterfield, Pulteney, Lady Suffolk, and the writers *Congreve and *Prior, as well as *Gay and Pope. Pope called him "Friend to my life, which did not you prolong, / The world had wanted many an idle song," and honored him in *Epistle to Dr. Arbuthnot* (1735). Though Arbuthnot's spirits darkened after the death of his son

Charles in 1731, Swift remarked that Arbuthnot "hath every quality in the world that can make a man amiable and useful. . . . The doctor has more wit than we all have, and his humanity is equal to his wit." After writing affectionate farewell letters to Pope and Swift, Arbuthnot died in Hampstead on 27 February 1735.

Bibliography: G. Aiken, ed., *The Life and Works of John Arbuthnot*, 1892; A. Bower and R. Erickson, eds., *The History of John Bull*, 1976.

Dale G. Priest

ARMINIUS, JACOB (1560–1609).

A student of Peter Ramus in Leiden and of Theodore Beza in Geneva, Arminius became preacher at the Reformed Church in Amsterdam in 1588. The time was one of confusion within the Reformed Church, with debates about predestination engaging the leading thinkers, and factions surrounding these leaders. The debates would continue until the Synod of Dordrecht (1618–1619). Arminius would develop a stance that set him outside of, indeed in opposition to, normative Calvinism, though the origins of his thought are squarely within the Reformed tradition.

Brought under the tutelage of the mathematician Rudolf Snell, Arminius entered Marburg University in 1575 but shortly thereafter enrolled in the newly established university at Leiden. The beneficiary of a church endowment in Amsterdam, Arminius relocated to Geneva in 1582, where he came under the influence of Theodore Beza (1519–1605), successor to Calvin but proponent of a doctrine of predestination different from Calvin's. Although his most original work is found in other topics, Beza is best known for his supralapsarian view of predestination, according to which the divine decree determined persons' salvation and damnation before the fall from grace. Arminius absorbed Beza's teaching and in 1588 was ordained in the Reformed Church in Amsterdam.

Amid his pastoral duties, Arminius engaged in a number of theological exchanges with such figures as Francis Junius, a professor at Leiden, and William Perkins, the English Puritan. On Junius's death in 1602, Arminius became a member of the theological faculty at Leiden, prolific and controversial. It was during his time at Leiden that he developed his own theology. Arminianism, as his doctrine became known, has its roots in Arminius's reservations about the biblical basis of the predestinarian articles of the Belgic Confession (1561) and the Heidelberg Catechism (1562). According to Arminius, grace is available to all but not irresistible; hence persons may by their own will obstruct their own salvation. God grants grace to those whose divinely foreknown piety and perseverance will merit salvation. The freedom that this doctrine accorded to the human will brought a century of controversy after Arminius's death, and the passivity of the will for which later Calvinism became known originates in part with these controversies.

Bibliography: C. Bangs, *Arminius: A Study in the Dutch Reformation*, 1985; R. Muller, "God, Predestination, and the Integrity of the Created Order: A Note on Patterns in

Arminius' Theology," in *Later Calvinism: International Perspectives*, ed. J. Fred Graham (1994), 431–446.

Ralph Keen

ARNAULD, ANTOINE (1612–1694). Arnauld (known as "the Great Arnauld" to contemporaries) became the foremost advocate of Jansenism after the death of its founder *Cornelius Jansen. An outstanding theology student at the Sorbonne, he was influenced by *Jean Duvergier de Hauranne to adopt Jansenist doctrines. In 1643, he published *De la fréquente Communion*, which emphasized the importance of a proper attitude in the communicant if the sacrament were to be efficacious. This apparent diminution of the importance of communion *ex opere operato* ("by nature of its own operation," or regardless of the one administering or receiving it) aroused so much protest that Arnauld disappeared from public view for two decades, though he continued to pen defenses of Jansenism. Protests came to a head in 1656 when he was expelled from the Sorbonne; not even *Pascal's *Provincial Letters*, written expressly in his defense, could save him. In 1668, however, Pope Clement IX lifted persecution with his "Peace of the Church," and *Louis XIV welcomed Arnauld with wide popular acclaim. He then embarked upon *La Perpétuité de la foi catholique touchant l'eucharistie* (1669–1674), coauthored with *Pierre Nicole. But in 1679, the Jansenist movement flared again, driving Arnauld to Brussels, where he spent the remainder of his life as a tireless polemicist against Jesuits, Calvinists, and *Malebranche, with whom he debated the relationship of theology and metaphysics. Arnauld was the first significant theologian to accept *Descartes' philosophy.

Bibliography: A. Sedgwick, *Jansenism in Seventeenth Century France*, 1977.

Christopher Baker

AUBREY, JOHN (1626–1697). The antiquarian and biographer John Aubrey was born in 1626 in the northern Wiltshire parish of Kington. The inactivity of a solitary and sickly childhood, which led him to prefer drawing and painting to more active play, no doubt encouraged the intense powers of observation and the interest in forms and customs of the countryside so evident in his adult writings. His natural curiosity was nurtured by an excellent early education, first by the tutor who had previously taught *Thomas Hobbes and then at Blandford, in what he considered the finest boarding school in western England, where he outshone his classmates in scholarship. In 1642, the seventeen-year-old scholar enrolled in Trinity College, Oxford, but within a few months the outbreak of civil war forced him to return home. On his return to Oxford the following year, he realized that the hostilities were endangering the ruins of an ancient priory at Osney, and he commissioned drawings of them by a local artist. This was the first recorded preservationist action by a man who would become his age's most diligent antiquarian.

Driven from Oxford again in 1643, Aubrey, after several years in Wiltshire, came to London to enroll in the Middle Temple, though without any real interest in studying law. On a trip home in 1648, he stumbled upon the ancient megalithic site of Avebury that, unlike Stonehenge, had not caught the attention of scholars or the public. He speculated that the great stones had been erected by the Druids (a view no longer accepted) and later toured the ruins with *King Charles II. Despite a rapid decline in his fortunes after he inherited his father's legally entangled estate in 1652, Aubrey around this time became an active member of the circle of investigators in all branches of knowledge that would become in 1662 the Royal Society, of which he was elected a fellow in 1663. He devoted the remainder of his life to collecting and recording details of the English countryside, rural folklore, and the lives of prominent Englishmen. He published only one book in his lifetime, but the vast manuscript collection he left at Oxford's Bodleian library became an invaluable resource for historians.

In our century, Aubrey's name is associated almost exclusively with his *Brief Lives*, as his collection of over four hundred biographical portraits has come to be known. Unpublished in Aubrey's lifetime, the selected and expurgated editions that have since appeared have provided some of our most intimate knowledge about a host of personalities in seventeenth-century England. These were composed between 1669 and 1696 and consist of somewhat randomly organized but often lively remembrances of the appearances, accomplishments, and eccentricities of men, and a few women, in all walks of life, from the great philosopher Thomas Hobbes, who receives the longest treatment, to a number of people of whom nothing else is known.

If Aubrey is known in our time as the first serious English biographer, he was earlier known in other capacities. The eighteenth century admired his *Miscellanies*. The only book he published during his lifetime, this is an entertaining set of stories about ghosts and other occult apparitions reported throughout England. The nineteenth century celebrated his two *Wiltshire Collections*, which recorded the natural history and folklore of his home country. Aubrey's other works include *The Natural History and Antiquities of the County of Surrey*, a treatise on the proper way of educating a young gentleman, the vast and wide-ranging *Monumenta Britannica or a Miscellanie of British Antiquities*, a comedy entitled *The Country Revell*, and another folkloric compendium called *The Remaines of Gentilism and Judaism*.

Bibliography: A. Powell, *John Aubrey and His Friends*, 1988.

Patrick Cook

B

BACON, FRANCIS (1561–1626). The English philosopher, statesman, essayist, and collector of scientific facts was born in London on 22 January 1561. His father, Sir Nicholas Bacon, was the Lord Keeper of the Great Seal under Queen Elizabeth. His mother, Anne Cooke, the second wife of Sir Nicholas, was herself learned in Latin and Greek, the daughter of Sir Anthony Cooke, tutor to Edward VI, and the sister-in-law of Sir William Cecil, who would become Lord Burghley. Nicholas had six children at the time of his second marriage and with Anne fathered two more, Anthony and his younger brother Francis. The two young brothers, Francis being only twelve years old, entered Trinity College, Cambridge, together in 1573. Three years later, his father arranged for him to accompany a diplomatic mission to France.

Bacon returned to England after the sudden death of his father in 1579 to learn that he would receive only a tiny inheritance. The need for money to live in the manner in which he had come to expect would be a constant theme throughout his life. In need of a profession that would provide an adequate income, he enrolled in 1576 to study law at Gray's Inn. He was admitted to the bar in 1582 and elected to the House of Commons in 1584, representing Melcombe. Very soon he began appealing to his uncle, Lord Burghley, for assistance in his political career. In a letter of 1592 to Burghley, he summarized his ambitions as "moderate civil ends" but "vast contemplative ends," as he had taken all knowledge to be his "province."

Bacon spoke in Parliament in 1593 to oppose a series of subsidies to the queen, thus falling out of the royal favor. At the same time, the office of Attorney General became vacant. The Earl of Essex, who had become his close friend and was then a favorite at court, supported Bacon's candidacy for this position. The position went to Edward Coke instead, leaving Bacon only the relatively minor post of Solicitor General. In 1599, Essex himself fell out of favor and by 1601 was plotting against the Crown. The queen ordered an investigation and Essex was brought to trial. Bacon appeared as prosecutor, and despite his former friendship, attacked Essex vigorously in court. Essex was convicted of treason and executed. As Essex had considerable following, the queen insisted that Bacon publish an account of Essex's treason and Bacon complied.

The preferment that had eluded Bacon under Queen Elizabeth was granted him under *James I. Bacon was knighted by James in 1603, named Attorney General in 1613, and Lord Keeper in 1616. He was named Lord Chancellor in 1618 and Baron Verulam in 1618. He was made Viscount St. Albans in 1621 but in that same year was tried and convicted of accepting bribes and imprisoned for a time. The king granted him a pardon but on the condition that he never appear in Court or in Parliament again. Bacon would spend his last years seeking a restoration to court and a financial subsidy from the king.

Deeply in debt, in 1606, Bacon had married the much younger Alice Barn-ham, the daughter of an alderman and stepdaughter of Sir John Pakington, coheir with Alice to the estate of her father. For the wedding, paid for by Pakington, bride and groom were both most extravagantly attired. Bacon's affection did not last. Alice, not supportive of his political ambitions, was unsympathetic after his fall from grace. She would remarry soon after his death.

Bacon first appeared on the literary scene with the publication of a book of ten essays published in 1597, which he would republish in enlarged editions twice. His pamphlet on the Essex affair appeared in 1601. In 1603, he published a work on the union of England and Scotland, and in 1604, a pamphlet defending his own role in the prosecution of Essex.

Bacon's status as a philosopher and spokesman for experimental science is based principally on three of his published works. The first was *On the proficience and advancement of learning, divine and human*, first appearing in 1605. In this work, addressed to the king, Bacon criticized the scholastic philosophy then predominant in the universities and made the case for impartial investigation of nature. Bacon's next major philosophical work was the *Instauratio Magna*, or *Novum Organum*, published in 1620. The title suggests that a new organon (set of principles for scientific investigation) should replace that of Aristotle. It is here that Bacon recommends the exhaustive collection of observations as the basis for drawing conclusions and asserts the primacy of inductive logic over deductive. Bacon argues that true knowledge could be obtained only through the most vigorous study of nature, and that to do this properly, he must leave behind the various "idols" that arise in human society and in human nature as obstacles to the pursuit of truth. Bacon's one work of fantasy is the *New Atlantis*, included, unfinished, in a volume of his *Natural History* published in 1626. In this work, Bacon describes an idyllic kingdom in which an institution, "Solomon's House," has been set up for the systematic exploration of nature. This parable would have a major influence on the founders of the first scientific societies.

Bacon contracted his final illness while conducting an experiment. He was traveling by coach on the way to dinner with the king's physician when it occurred to him that refrigeration might provide a means of preserving meat.

He had the coach stop, bought a hen, and stuffed it with snow. As he resumed his journey, he became suddenly ill and sought shelter at the house of the Earl of Arundel. He died two days later on Easter Sunday, 9 April 1626, and was buried near his mother at the Church of St. Michael in St. Albans. In his will he treated his servants generously but left his wife only the minimum required by law. His intended generosity to the universities could not be realized because he died as he had lived, with substantial debt.

Bacon's reception by his contemporaries and the succeeding generation was mixed. *Joseph Glanvill saw in "Solomon's House" a blueprint for the Royal Society. *Leibniz praised Bacon as a philosopher superior to *Descartes, and Kant described him as one of the greatest of physicists. Critics, on the other hand, attacked his materialism, his avoidance of metaphysics, his lack of personal loyalty, his questionable financial dealings, and his limited actual contact with the greater scientific lights of his time. As an essayist, Bacon has been credited with having a major impact on the English prose writers of his time. Whereas English writers of the sixteenth century strove for elegance of expression, prose of the seventeenth century placed greater emphasis on clarity of expression, so that even the beauty of literary expression would not hamper the reader's understanding of the writer's ideas. Such a change, actually a return to the style advocated by some of Bacon's favorite classical writers, reflected the rising importance of scientists as writers and the growing literacy and wealth of the artisan class upon which scientists and governments would increasingly come to depend.

Bibliography: R. Adolph, *The Rise of Modern Prose Style*, 1968; F. Anderson, *Francis Bacon, His Career and his Thought*, 1962; L. Jardine and A. Stewart, *Hostage to Fortune: The Troubled Life of Francis Bacon*, 1999; P. Rossi, *Francis Bacon, from Magic to Science*, 1968.

Donald R. Franceschetti

BAFFIN, WILLIAM (1584–1622). Little is known of William Baffin before 1613, when he entered into service with the Muscovy Company. In 1615, he was commissioned by the company to find the Northwest Passage previously sought by John Davis (1587) and *Henry Hudson (1607). During a voyage under the command of Captain Robert Bylot, formerly a pilot under Hudson, Baffin examined Hudson Strait and passed into the Foxe Basin. Sighting a huge land mass to the north and east, Baffin erroneously believed that the bay was fully enclosed and that the passage did not lie in this direction. In 1821, James Parry, though more successful in finding a passage through Foxe Basin, confirmed Baffin's 1615 navigational observations and named Baffin Island in his honor. In 1616, Baffin examined Davis Strait, where he named various sounds and islands in honor of his English patrons. He traveled northwest in Baffin Bay to 77°45', a northerly latitude that would not be surpassed for another 200 years.

Now believing the Northwest Passage more easily penetrable from the Pacific coast, Baffin gained employment with the East India Company. He mapped the coast of Persia and the Red Sea from 1616 to 1622, but did not get farther northeast than Surat. In 1622, the company agreed to aid the Shah of Persia in driving the Portuguese out of Ormuz, and Baffin was killed during an altercation. He widowed a wife but had no children. Baffin's explorations were published incorrectly by *Samuel Purchas in 1625. His observations were considered dubious until they were proved remarkably accurate in the nineteenth century. Baffin was a skillful seaman, intrepid explorer, and a man of science, whose knowledge of nautical astronomy was highly advanced—he is said to have taken the first seabound lunar observation of longitude, which enabled his deep penetrations of northeastern Canada.

Bibliography: C. Markham, *The Voyages of William Baffin 1612–22*, 1881.

Ken MacMillan

BASILE, GIOVANNI BATTISTA (1575–1632). Born in Naples, author Giovanni Basile worked briefly as a soldier for Venice, Naples, and Calabria; he then returned to his native city and fulfilled various minor governmental positions and wrote extensively. His three sisters were singers, one of whom became immensely popular. Basile wrote madrigals for them to sing, sonnets, occasional poems for members of the court, and court entertainments, such as the musical tragedy *Venere addolorata (Sorrowing Venus)*. Basile is best known for his *Lo Cunto de li Cunti, overo lo Trattenimiento de Peccerille (The Tale of Tales, or Entertainment for Little Ones)*, known more recently as the *Pentameron*. The title and framing of this work derive from Boccaccio's *Decameron*: A Moorish slave girl gains her rightful place as a princess and marries her prince; they are then entertained by ten women who tell a story each day, for a total of fifty, with the last one told by the princess. The tales are nearly all of folk origin and include such well-known favorites as "Cinderella" and "Beauty and the Beast." Basile's work influenced *Charles Perrault and the brothers Grimm. Basile's *Pentameron* is written in the Neapolitan dialect, which at the time began to flower as a literary form and was important as a form of regional pride. His work helped confirm that dialect was no less an effective expressive tool than the Tuscan language and that translation into the Tuscan (which was common) was not a necessity. Italian literature of Tuscan origin was somewhat exotic in Naples as it was in other parts of Italy at the time. Basile often used rhythms, alliterations, rhymes, and puns in dialect that are difficult to translate; many translations of his work are inaccurate. The work is more than a mere collection of fairy tales; Benedetto Croce considered this the central literary contribution of the Italian Seicento.

Bibliography: B. Croce, *Saggi sulla letteratura Italiana del seicento*, 1962; J. Smarr, ed., *Italian Renaissance Tales*, 1983.

Rosi Prieto Gilday

BAYLE, PIERRE (1647–1706). Born to a Protestant family in rural southwest France, Pierre Bayle flourished intellectually under the tutelage of his father, a pastor. At the late age of nineteen, Bayle entered a Protestant *collège* (the French equivalent of middle school) in Puylaurens before continuing his education with the Jesuits at Toulouse. Influenced by one of his Jesuit teachers, Bayle converted to Catholicism for almost seventeen months. Renouncing his conversion in 1670, Bayle was branded a lapsed Catholic and forced to flee the country to Geneva, where he served as preceptor to a wealthy family. After secretly returning to France, Bayle taught philosophy at the Reformed (Protestant) Academy of Sedan from 1675–1681. There he met Pierre Jurieu, who taught theology and mentored Bayle in his career as a professor and philosopher, and who would become his detractor and bitter enemy in the 1690s.

In 1681, *Louis XIV ordered the closing of the Reformed Academy of Sedan, one of a series of measures designed to make life in France increasingly difficult for Huguenots. Bayle once again left France, this time permanently. He escaped to Rotterdam, where an influential town father had secured Bayle and Jurieu teaching positions at the École Illustre, an institution founded to provide employment to Protestant intellectuals in exile as a result of the declining political situation in France. His teaching responsibilities being minimal, Bayle was able to devote himself almost exclusively to his burgeoning literary career. His first work to attract attention, the *Lettre sur la comète* (1682), was such a resounding success that an expanded second edition was published a year later as *Pensées diverses sur la comète* (1683). Later, in response to the Revocation of the Edict of Nantes in 1685 and the death of his brother, a pastor imprisoned by French authorities, Bayle wrote a stinging pamphlet entitled *Ce que c'est que la France toute catholique sous le règne de Louis le Grand* (1686). In the same year he published his *Commentaire philosophique* (1686), calling for toleration for Muslims, Protestants, Jews, Catholics, and even atheists, an unprecedented gesture for the period.

Bayle's primary occupation during this time, however, was as editor of the *Nouvelles de la République des Lettres*, a scholarly journal appearing monthly from May 1684 to February 1687. Bayle's work as a journalist prepared him well for his most famous undertaking, the *Dictionnaire historique et critique*, first published in 1697, with a second edition appearing in 1702. The fragmentary and unsystematic structure of the text underscores one of its primary themes: that knowledge is in a constant process of revision and claims to the truth should be regarded with suspicion. In this way, Bayle follows in the footsteps of Montaigne and paves the way for Voltaire and the Encyclopedists. Indeed, many critics cite Bayle as the father of the Enlightenment and the archetype of the modern public intellectual.

Bibliography: L. Gossman, "Marginal Writing," in D. Hollier, *A New History of French Literature*, 1989; E. Labrousse, *Pierre Bayle*, vols. 1 & 2, 1963 & 1996.

Patricia Armstrong

BEAUMONT, FRANCIS (?1584–1616). Francis Beaumont gained fame primarily for dramas written in collaboration with *John Fletcher, with whom his reputation remains linked. Born in Leicestershire, he intended to follow his learned grandfather, father, and brother in becoming a barrister. Although he studied at Oxford for two years, departing in 1598 and later admitted to the Inner Temple in 1600, he abandoned his plans of a career in law. He also apparently did not adopt the family's traditional Catholic faith, an adherence that led to the impounding of much of their estate's wealth for some years.

Beaumont's writing career included a few occasional pieces and some poetry, but it produced mostly popular dramas. Plays solely credited to him include *Almacis and Hermaphroditus* (anon 1602), *Madon* (lost), and *The Woman Hater* (1607). Authorship of *The Knight of the Burning Pestle* (1607) remains open to question, with some attributing it to Beaumont alone, others labeling it a collaboration with Fletcher. Along with Fletcher, Beaumont professed to be a disciple of *Ben Jonson who wrote "To Francis Beaumont" in honor of his protégé. Some sources emphasize the devotion of the playwrights to their craft as they shared sparse living quarters close to the theaters.

Beaumont and Fletcher produced a number of popular tragicomedies. The first few included *The Faithful Shepherdess* (1608–1609), a failure, but their *Cupid's Revenge* (1610–1612) exhibits many of the characteristic plot features expected from their collaboration. These include a mixture of realistic with mythic-type characters set in an exotic locale and a distant time. Following their *The Scornful Lady* (1608–1610), the trilogy of plays representing what most critics consider their best work, *Philaster* (1609), *The Maid's Tragedy* (1610–1611), and *A King and No King* (1611) all received popular acclaim. *Philaster* represents a particular triumph as their first drama enacted by the prestigious King's Men, a venue that almost guaranteed its success. Additional collaborative works include *The Captain* (1609–1612) and *The Noble Gentleman* (prob. 1611). Many of their dramas reflected unflatteringly on the ignoble actions of princes through themes including incest, corruptive power, mistreatment of the common man, tyrannicide, a princely propensity for blood sports, and the divine rights of inheritance. One explanation of why the pair could publish such politically dangerous works rests upon *King James I's unpredictable, if not tolerant, behavior. The performance of such dramas by the King's Men suggests their openness to the presentation of plays that turned on daring political statement. One character in *Philaster* dares to state that a king deserves obedience only when his actions prove just, and the king himself admits to finite powers. Contemporary performances of their work centers on *The Knight of the Burning Pestle* (1607), a play some consider least representative of the collaboration. Wildly popular and influential in their own day, they later ranked well behind most other well-known Elizabethan-Jacobean dramatists in critical importance.

Beaumont's special strength can be seen in his plot construction and emotional expression. His collaborative works remain notable for a superior plot symmetry featuring sexual titillation, vacuous rhetoric, and reversals of fortune, ingredients foreshadowing those found in Restoration comedies emphasizing themes of lust and greed. Following his 1613 marriage to an heiress, Beaumont retired from the stage to his Kent estate, although that same year he wrote a masque in honor of Princess Elizabeth's wedding to Frederick V. Following a debilitating stroke, he died in 1616. He was honored by burial in the area later known as "Poet's Corner" in Westminster Abbey, although his particular place remains unmarked.

Bibliography: P. Finkelpearl, *Court and Country Politics in the Plays of Beaumont and Fletcher*, 1990.

Virginia Brackett

BEHN, APHRA (ca. 1640–1689). Aphra Behn was the first English woman to earn a living as a writer, but little is known of her early background. She may have been born to a Johnson family in Kent, England, or was possibly an illegitimate daughter of Lord Willoughby of Parham.

Sometime between 1658 and 1663, she traveled to the British colony of Surinam, South America, where her father, who died during the voyage, was to hold a governor's appointment. She remained there a year, possibly longer. Upon returning to England, she married a merchant named Behn, thought to have died in the plague shortly thereafter. A supporter of the crown, she was enlisted by *Charles II around 1666 to spy on British antiroyalists exiled in the Netherlands, a service for which she was never paid. After a period of poverty, including debtor's prison, she began writing plays, which proved financially profitable. *The Forced Marriage* (1670) was her first stage production, followed by a prolific output of drama, fiction, and poetry over the next eighteen years. She was friends with leading literati of the period, including *Dryden, *Etherege, *Otway, and *Tate, and was well-read in Fontenelle, *Boileau, *La Rochefoucauld, and other European intellectuals.

Behn achieved literary acclaim principally for her morally shocking portrayals of colonial slavery, female transgression, and male hedonism. *Oroonoko* (1688), her most famous work and the first antislavery novel published in England, introduced the genre of abolitionist fiction into the British and American literary canon. Based partly on her experiences in Surinam, the story exposes the slave trade's crimes against humanity. Oroonoko, an African prince raised in the royal court of Coromantien, Ghana, is kidnapped by slave traders and sold into bondage in Surinam where he is reunited with his wife, Imoinda, now also a slave. After organizing an unsuccessful slave revolt, Oroonoko kills Imoinda and their unborn child to free them from the horrors of living in captivity under a master's whip. Colonial authorities execute Oroonoko by the horrendous method of dismemberment and incineration. Behn's magnification of Oroonoko's suffering and the inhumanity of colonial rule helped to incite the outrage against slavery that was gaining momentum under the Quakers and other antiauthoritarian groups of the period.

Behn earned infamous notoriety for casting virtuous heroines in transgressive roles. *The History of the Nun* (1689), in which an innocent heroine becomes a murderess, represents that theme by challenging the patriarchal idea of feminine passivity. The beautiful young Belinda runs away from a convent to elope with Henault. Disgrace and poverty follow them, and Henault is reportedly killed in war. For financial security, Belinda marries Villenoys, who enjoys hunting and business trips more than domestic life. Surprisingly, Henault returns from war and, to resolve the problem of bigamy, Belinda murders both husbands. Although Belinda is justifiably executed for double homicide, Behn makes her a sympathetic character entangled in problems beyond her control. Like many of Behn's female characters, Belinda illustrates the plight of women in a male-dominated society.

Behn's most successful play, *The Rover* (Part 1, 1677; Part 2, 1681), portrays male hedonism associated with the aristocracy. The play dramatizes a gang of drunken, pleasure-seeking Cavaliers banished from England to Naples and Madrid during the interregnum. The most self-indulgent of the bunch is Willmore, the Rover, whose insatiable appetite for women reminded contemporary audiences of Behn's friend *John Wilmot, earl of Rochester, paragon of Restoration libertinism. The play is comic, but it hints at the sobering consequences of spending one's life in pursuit of carnal pleasure. The *Hobbesian theory that human nature inherently inclines toward corruption, as shown here, runs through much of Behn's literary canon.

Bibliography: H. Hunter, ed., *Reading Aphra Behn: History, Theory, and Criticism*, 1993.

James Norton

BELLARMINE, CARDINAL ROBERT (1542–1621). Robert Bellarmine was born into a noble family of Montepulciano, Tuscany, in 1542. He entered the Jesuit order in 1560 and studied philosophy at the Collegio Romano. He completed his Thomistic theological education in Padua and later in Louvain, a Counter Reformation outpost that, between 1550 and 1600, saw continuous clashes between Catholics and Protestants. In Louvain, Bellarmine became an expert in religious controversies and a successful preacher. Because of his talents and experience, in 1576 Bellarmine was appointed professor of theology at the Roman College. He extended his lecture material into *Disputationes de Controversiis Christiane Fidei* (1586–1593), a multivolume work that presents and criticizes fairly both Catholic and Protestant arguments. This work proved very influential, becoming the standard reference for centuries. *De Controversiis* included Bellarmine's opinion about papal temporal power, which he determined to be indirect and contingent upon the needs of the pope's spiritual leadership. As Jesuit and churchman, Bellarmine held prestigious positions: In 1588 he

became spiritual director of the Collegio Romano, rector in 1592, provincial of Naples's Jesuit province, personal theologian to Clement VIII in 1597, and was made Cardinal of Capua in 1599. During this time, he also served as a papal representative in France, with *King James I of England and with the Republic of Venice; he acted also as chief editor for the revised—and henceforth official—edition of the Vulgate. Finally, in 1605, Bellarmine became chief papal adviser on theological matters and a permanent official of the Holy Office. Among his many writings, ranging from a Hebrew grammar to a compilation of patristic works, several manuals of catechism became standard references for centuries.

Bellarmine is also famous for his involvement in the *Galileo affair of 1616. Bellarmine himself was an admirer of Galileo and his telescopic discoveries. Yet Bellarmine admonished Galileo by replying to Paolo Antonio Foscarini—a friar who had submitted a scriptural defense of Copernicanism—that the authority to interpret the Bible rested only with theologians and that such action would be considered only when demonstrative proof supporting Copernicanism was shown. In 1616, the Holy Office addressed heliocentric theory, finding it heretical. Its decree condemned Copernicanism, and the pope asked Bellarmine to deliver a verbal admonition to Galileo. Although no formal record of Bellarmine's interview with Galileo remains, the event became important in Galileo's 1633 trial, which Bellarmine did not live to witness. Bellarmine was canonized in 1930 and declared a Doctor of the Church in 1931.

Bibliography: J. Brodrick, S.J., *Robert Bellarmine: Saint and Scholar*, 1961; P. Godman, *The Saint as Censor: Robert Bellarmine between Inquisition and Index*, 2000.

Renzo Baldasso

BERNINI, GIAN LORENZO (1598–1680). Under the patronage of eight popes, Bernini gave supreme Baroque expression to the Catholic faith during the Counter Reformation. Born in Naples in December 1598, Bernini was the son of a Florentine sculptor who began working in Rome for Pope Paul V in 1605–1606. At the age of eight or so, Bernini emerged as a child prodigy in his father Pietro's workshop. Soon he was copying the Hellenistic sculpture and High Renaissance paintings in the Vatican. Already an independent sculptor in his adolescence, Bernini wanted his medium to compete with painting in capturing realistic detail. Influenced by such painters as Raphael, *Caravaggio, Annibale Carracci, and the contemporary *Guido Reni, he was a pictorial sculptor determined to create three-dimensional pictures through marble. Bernini created his first life-size group for Cardinal Scipione Borghese, the pope's nephew. From this rather traditional, static composition of Aeneas fleeing Troy (1618), the young artist quickly grew in the direction of his mature genius. The 1620 *Neptune and Triton* established a new relationship between a free-standing figure and the space beneath him through a dynamically poised trident. For his next three sculptures, again for Borghese—*Pluto and Prosperina* (1621–1622),

Apollo and Daphne (1622–1625) and *David* (1623–1624)—Bernini chose a climactic moment of psychological drama as well as physical action. Contorting facial expressions and bodies, endowing skin and drapery with tactile sensuousness, making hair and features seem to move, and differentiating textures for colorist effects, he revolutionized sculpture. By incising and ridging marble so that its uneven surface could catch light in a dramatic alternation of illumination and shadow, he gave sculpture its own version of painterly chiaroscuro.

In 1623, Cardinal Maffeo Barberini, Bernini's long-term patron and friend, was elected *Pope Urban VIII. He immediately designated Bernini the official artist to realize his pontificate's Counter Reformation ambition of impressing the public through art addressing both the emotions and the senses. A devout Catholic who attended church daily and took communion at least once a week, Bernini was the ideal genius to propagandize for the papacy. Immediately, Bernini embarked on the papal project to decorate St. Peter's Basilica. For the monument over St. Peter's tomb, Bernini designed and executed a four-story bronze canopy or "baldacchino." Its voluted roof, topped by an orb and cross, rests upon four spiraling columns surmounted by angels. This combination of architecture and sculpture, proportioned to bridge the church's vast interior and human dimensions, surging dynamically and organically, was the monument that announced the birth of the Baroque to the world.

During the decade spent on the baldachin (1623–1633), Bernini took other projects and was so in demand that he could not complete all his own commissions—especially after being appointed architect for St. Peter's and for the Palazzo Barberini. Thus he organized a studio of assistants whose combined efforts achieved a high level of artistic integrity. Bernini's unprecedented virtuosity for giving marble lifelike energy characterized his portrait busts, tomb monuments, and fountains as well as his church work. His bust of Cardinal Scipione Borghese (1632), for instance, captured his former patron with glinting eyes and parted lips, as if in mid-conversation with a viewer. The figure of Urban VIII that Bernini carved for his friend's tomb (1628–1647) was similarly vital and also innovatively combined marble with bronze. In his fountain designs, especially the Triton fountain in the Piazza Barberini (1642–1643), dynamic statuary interacted with water flow to brilliant effect.

With Urban's death in 1644, Bernini lost some of his absolute power as Rome's unchallenged artistic designer. The succeeding Pope, *Innocent X, preferred the sculptor Alessandro Algardi and the architect *Francesco Borromini. Bernini continued as architect for St. Peter's but faced disgrace in 1646: A bell tower constructed according to his 1637 designs had caused the cathedral's facade to crack, stopping construction in 1641. Now, under the new pope, the tower was torn down to make way for other artists' work. Even with this failure, Bernini's reputation suffered little. Indeed, in 1648, he replaced Borromini as the architect-sculptor of Piazza Navona's Four Rivers Fountain (1648–1651), whose engineering won over the pope. In 1655, Bernini received a magnificent welcome from *King Louis XIV, who invited him to Paris to design the new royal residence, the Louvre. On this one long absence from Rome, Bernini alienated the French with his superiority about Italian art but carved a fine, linear bust of their absolutist monarch.

In later years, Bernini's religious masterpieces included work on St. Peter's surroundings. He transformed the space in front of the cathedral into a piazza shaped by two colonnades, which he conceived of as the arms of the church stretching out to embrace its believers (1656–1657). Then he redecorated the Ponte Sant'Angelo (1667–1671), the only bridge over the Tiber River leading directly to the Vatican, with ten larger-than-life angels, two of which he carved himself. Bernini died at the age of eighty-one in the city he had made resplendent. Although his Baroque legacy—intricate grandeur, dramatic use of space, light and mass, sensuous treatment of textures—met with neoclassical disdain from the mid-eighteenth century through the 1880s, his immense influence on his age has always been unquestionable.

Bibliography: G. Bauer, ed., *Bernini in Perspective*, 1976; H. Hibbard, *Bernini*, 1965.

Margaret Goscilo

BERNOULLI, JACQUES (1654–1705). Jacques, also known as Jakob, Bernoulli, who with his two brothers, four nephews, and two great nephews formed a unique mathematical dynasty, was born in Basle, Switzerland, into a family of merchants. His grandfather, also named Jacques, had left Antwerp in 1583 to escape the religious persecution of Protestants. After a stay in Frankfurt, he established the family in Switzerland in 1622, marrying into a well-established Basle family and building up a prosperous apothecary business, which he passed on to his only son Nikolaus, the father of Jacques, and Jacques's two brothers, Nikolaus and Johan.

Nikolaus Bernoulli attempted to direct the education of his sons so that they would qualify for recognized professions. Jacques received the Master of Arts degree in philosophy in 1671 and a licentiate in mathematics in 1676, while his younger brother, Johann, received his licentiate in medicine in 1690. Both developed an interest in mathematics, however, with Jacques initially tutoring his younger brother. In 1676, Jacques began work as a tutor in Geneva and then moved to France where he studied the philosophy and scientific thought of *Descartes. On a second journey to the Netherlands and England, he met *Robert Boyle and *Robert Hooke. In 1683, Jacques returned to begin teaching mathematics at the University of Basle, being named Professor at 1687. Jacques married Judith Stupanus, the daughter of a wealthy pharmacist, in 1684. Jacques and his brother Johann pioneered the application of calculus, in the form published by *Gottfried Wilhelm Leibniz in 1684. The two brothers did not remain friendly, however, as Jacques resented Johann's accomplishments. The competition between the two would lead to the solution of some of the fundamental problems in the calculus of variations, that is, the determination of curves in space that satisfy some overall condition. In 1687, Leibniz had posed the problem of the curve of constant descent, or the shape of the wire on which a sliding bead will oscillate with unvarying period regardless of the amplitude of its motion. This problem was first solved by the Dutch physicist *Christiaan Huygens, inventor of the pendulum clock, in 1687 and then by a different method by Leibniz two years later. Jacques Bernoulli presented his own solution in 1690 and then posed the problem of the shape of the catenary, a chain hanging under its own weight. Liebniz himself solved this problem, and his solution, along with those of Huygens and Johann Bernoulli, was published in 1691. Johann, who had become professor of mathematics at the University of Groningen in 1695 then posed the problem of the brachistochrone, or the curve along which a particle, sliding under its own weight, would move most quickly between any two points. Bernoulli, Leibniz, and others furnished the solution to this problem.

Bernoulli made numerous contributions in other areas of mathematics. He is now known particularly for his work on infinite series and the theory of probability. Bernoulli died of tuberculosis on 16 August 1705. His posthumous and incomplete publication, the *Ars Conjectandi*, on probability theory is considered by many to be his most original work.

Bibliography: E. Bell, *Men of Mathematics*, 1937; C. Boyer, *A History of Mathematics*, 1968; J. Hofmann, "Jakob (Jacques) Bernoulli," *Dictionary of Scientific Biography*, 1973.

Donald R. Franceschetti

BÉRULLE, CARDINAL PIERRE DE (1575–1629).

French cardinal and originator of a tradition of mysticism, Bérulle was ordained a priest in his twenties (1599) and founded the Oratory, a congregation dedicated to sacerdotal training, in 1611. Bérulle also became the guardian of the Carmelite nuns in Paris, having brought members of that order from Spain in 1604. As counselor to the nobility, and especially as cardinal (from 1627), he was engaged at the highest level with French domestic and diplomatic affairs. His enduring influence, however, lies in his spiritual work. His religious sentiment is fully expressed in various "Discourses," essays on divine attributes and actions, in which a Christological emphasis is evident throughout. Many are addressed to Christ and thus serve as prayers and meditations rather than as analytical discussions of the sort common in the age of Montaigne. Devotion to Mary is another prominent theme, not unexpectedly in an order of religious women. Scholars have seen Bérulle's work as a polemic against the Renaissance and especially the Reformation. A reaction to the latter is evident, though the mystical tendencies of Bérulle's writings reveal debts to the Renaissance revival of Platonism. Through his texts and the force of his personality he would exert substantial influence on the course of Catholic piety in France for the rest of the seventeenth century.

Bibliography: M. Dupuy, *Bérulle: Une spiritualité de l'adoration*, 1964; F. Preckler, *Bérulle Aujourd'hui 1575–1975*, 1978; W. Thompson, ed., *Bérulle and the French School: Selected Writings*, 1989.

Ralph Keen

BLOW, JOHN (1648–1708). Despite his own voluminous compositions, John Blow is perhaps better known for his more illustrious pupil, *Henry Purcell. Blow was born probably at North Collingham in Nottinghamshire and became a chorister in the Chapel Royal in 1660, composing several anthems by 1663. Six years later, he became organist of Westminster Abbey, and in 1674 was named Master of the Children of the Chapel Royal, the same year he married Elizabeth Braddock. Blow's career was linked to the court; having been given the honorary doctor of music degree by the Dean of Canterbury, he was named in 1685 a private musician to *James II, at whose coronation he was present. In 1679, he vacated his post at Westminster, which was taken over by Purcell, but he was reappointed upon Purcell's death in 1695. Blow's only extant stage composition is the *Masque for the Entertainment of the King: Venus and Adonis* (between 1680 and 1687) with Venus acted by Mary Davies, a mistress of *Charles II. He wrote two anthems for the coronation of James II and one for the opening of St. Paul's Cathedral. In 1700, he published *Amphion Anglicus*, a song collection dedicated to Princess, later Queen, *Anne. In all, he wrote over 100 anthems and thirteen church services, as well as numerous other compositions. Blow died at his home in Westminster on 1 October 1708.

Bibliography: C. Dearnley, *English Church Music 1650–1750*, 1970; W. Shaw, *The Services of John Blow*, 1988.

Christopher Baker

BODLEY, THOMAS (1545–1612). Thomas Bodley, founder of the Bodleian Library at Oxford, was born in 1545. He fled England with his parents during the reign of Mary Tudor, and like many other English Protestants during the Marian persecution, the Bodleys sought refuge in Geneva. There Thomas attended the lectures of John Calvin and Theodore Beza, among others. Upon the death of Mary in 1558, the Bodleys returned to England where Thomas continued his education at Magdalen College, Oxford. In 1576, he left England to travel for four years on the continent, and from 1588 to 1596, he performed diplomatic duties as the English resident at The Hague.

Upon returning to England, Bodley set about restoring and promoting the library of his alma mater, a task that occupied him for the rest of his life. This work began with the donation of Bodley's own sizable collection of books, which he had received as a gift from Robert Devereaux, the Earl of Essex and Queen Elizabeth's favorite. Devereaux had acquired the collection during an otherwise unsuccessful, state-funded raiding expedition on the Spanish port of Cadiz. The property of Bishop Jerome Osorio, the collection ranked as one of the largest in Europe. The year 1602 marked the formal reopening of the library at Oxford. In 1610, Bodley reached an agreement with the Stationer's Company to provide the library with copies of every book the company published. Thus, the Bodleian became the first deposit library in England. Bodley's dedication to rebuilding the Oxford Library was immense, and he supported its growth at his own expense. Records reveal that in 1611 he sold his own household plate to help fund the completion of the library's last building. He remained instrumental in the library's development until his death in 1612.

Bibliography: E. Miner, "Sir Thomas Bodley Revisited," *The Princeton University Library Chronicle* 38 (1977): 134–145; W. Wooden, "Sir Thomas Bodley's Life of Himself (1609) and the Epideictic Strategies of Encomia," *Studies in Philology* 83 (1986): 62–75.

Charles Pastoor

BOEHME, JAKOB (1575–1624). Boehme (also spelled "Böhme" or "Behmen") was a Lutheran mystic who, though accused of heresy by the rector of his native town, gained a wide and influential following; his adherents in England ("Behmenists") later joined the Quakers, whose emphasis upon an "inner light" resembled somewhat Boehme's conception of personal revelation. He became a cobbler in 1589 (a profession he gave up in 1612), married ten years later, and began experiencing mystical insights that formed the basis of his writings. His first work, *Morgenröte im Aufgang oder Aurora* (*The Beginning of Dawn or Aurora*) was published in 1612. Six years later, he started writing a variety of theological tracts that appeared eventually in 1623 as *Der Weg zu Christo* (*The Way to Christ*). Opposition from the Lutheran pastor in Görlitz in Silesia where he lived finally drove him to Dresden in 1624, but he returned to Görlitz and died there that same year on 17 November.

The bulk of his other treatises appeared posthumously: *Die Drei Prinzipien göttlichen Wesens* (*The Three Principles of Divine Being*) on the nature of the godhead; *Signatura Rerum* outlining his cosmology; *Mysterium Magnum*, which explicated Genesis; and *Von Christi Testamentum* on the nature of baptism and the eucharist. Boehme's complex theology draws upon the abstruse ideas of astrology, alchemy, and Paracelsus. God for him constituted the *Urgrund* or *Ungrund* (foundation) of existence, a fundamental fusion of all opposites and contraries of being, toward which the individual seeks to progress. His ideas, at times both pantheistic and dualistic, had a strong influence on later German Romantics, such as Hegel, and on the Cambridge Platonists; *Newton also studied him closely.

Bibliography: J. Stoudt, *Sunrise to Eternity: A Study in Jakob Boehme's Life*, 1957.

Christopher Baker

BOILEAU-DESPRÉAUX, NICOLAS (1636–1711). French critic and poet, legislator, and a model for French neoclassicism, Boileau was born in Paris and educated at the colleges of Harcourt and Beauvais. He studied law and for a while practiced it. The death of his father allowed him to renounce the legal profession, and his considerable income enabled him to devote himself entirely to literature. He took the name of Despréaux after the name of an estate his father owned. In 1660, he wrote his first satire, and in 1663 published his first collection of poems. Boileau does not belong among the greatest of French poets because the prosaic quality of his poems does not rise above mediocrity. His real abilities were yet to be exploited in his satirical poems, which is poetry used as a medium of criticism, where reason is the principal ingredient.

In 1666, Boileau published a collection of satirical poems that were already celebrated in the court salons. He was soon a great favorite of the king, and around 1670 *Louis XIV granted him an annual pension. His campaign against bad taste and in favor of simplicity in literature made him many enemies but in compensation won him the friendship of the greatest writers such as *Molière, *Racine, and *La Fontaine. In 1674, he published his major work, *L'Art Poétique*. Ten years later, the French Academy, under pressure of Louis XIV, opened its doors to him.

Boileau had great influence on French literature, as both poet and critic. Boileau, in his satires (begun in 1660), epistles (begun in 1669), as well as in his *L'Art Poétique*, established himself as the foremost critic of his day. He dictated the principles of French classical literature and was called the "Lawgiver of Parnassus." He and other French critics saw imitation of the literature of antiquity as the only possible guarantee of excellence, but despite a talent for judging contemporaries, his criteria were limited by current aesthetic doctrines. In *The Art of Poetry*, the culmination of his literary criticism, Boileau gives final expression to the classical theories and practices that had been gradually gaining ground since the Renaissance.

Nicolas Boileau's works consist of twelve satires, twelve epistles, *The Art of Poetry* (*L'Art Poétique*), *The Lectern* (*Le Lutrin*), a few miscellaneous poems and epigrams, and a few short critiques and letters, as well as a translation of Longinus. The largest part of his poetry is written in imitation of the Latin poet Horace. His satire is witty but malevolent and very personal. His critical judgments attributed an excessive importance to reason to the neglect of the feelings and imagination; however, with one exception, his critical opinions were sound and justified. The exception was *Corneille, the great tragic writer, already advanced in years when Boileau directed his satires against him.

In *The Lectern* (1673, 1683), a mock-heroic poem that was later used as a model for Alexander Pope's *The Rape of the Lock*, the French theorist produced a masterpiece of comic writing in the classical manner. But it was *The Art of Poetry*, an imitation of Horace's *Ars Poetica*, that brought him to the height of his reputation. *The Art of Poetry* is a four-canto poem written in epigrammatic Alexandrine lines of polished and elegant verse. Boileau explains that the poet must find his ideas in nature and must follow the principle of reason. Reason will teach the poet perfect elocution, variety and nobility of vocabulary, and true versification. In the Third Canto, devoted to tragedy, comedy, and the epic, Boileau confirmed the long-established unities of "one place, one day, one dramatic fact" as the principles of composition for those genres. Boileau was the first important critic of his age, he considered himself the guardian of classicism, and his influence has been lasting in France for almost two centuries. *John Dryden, Alexander Pope, and *Joseph Addison regarded him as the postclassical canon of literary judgment.

Bibliography: E. Dilworth, ed., *Selected Criticism*, 1965; S. Menant, ed., *Boileau Despréaux*, 1998; J. White, Jr., *Nicolas Boileau*, 1969.

Angela Morales

BORELLI, GIOVANNI ALFONSO (1608–1679). Borelli was born in Naples and studied at the University of Pisa, where he became a professor in 1656, after first having taught mathematics at the University of Messina in 1649. He also taught at Rome and then at the Florentine Accademia del Cimento from 1657 to 1667. In 1667, he again began teaching at Messina, but political factors forced him to leave seven years later. For most of the rest of his life, he served the household of ex-Queen Christina of Sweden in Rome. His most famous work is *De motu animalium* (*On the Movement of Animals*), published in the year of his death. Following Galileo's mechanical principles of natural motion, Borelli undertook to explain animal movement as the result of the lever-action of bones and muscles. Though concerned primarily with human motion, his discussion also included references to birds, fish, and insects. However, Borelli recognized that muscle contraction itself was the result of a chemical process, not merely a mechanical one, though he had only an imperfect understanding of the nature of that process. His astronomical publications included *Del movmento della cometa apparsa il mese di decembre 1664*, the first scientific suggestion that comets followed a parabolic course through space, and *Theorica mediceorum planetarum ex causis physis deducta* (1666), which discussed the satellites of Jupiter. While at the Accademia del Cimento (and in the company of *Torricelli, *Steno, and *Cassini), Borelli also studied the phenomenon of capillary action, noting that liquids rose more quickly in wet rather than dry tubes and to a height inversely proportional to the tube's diameter. He also discovered that wet objects floating on a liquid attracted each other. Borelli (with Viviani) performed experiments as well on the velocity of sound, arriving at a speed of 1,077 feet per second, although he reasoned incorrectly that this speed remained unaffected by wind. He died in a monastery on 31 December 1679.

Bibliography: A. Wolf, *A History of Science, Technology and Philosophy in the 16th and 17th Centuries*, 1935, rpt. 1968.

Christopher Baker

BORROMINI, FRANCESCO (1599–1667). After *Bernini, Borromini ranks as one of the foremost exemplars of Italian Baroque architecture. While he was still a young stonemason, his drawings and designs drew attention, and in 1624 he began work under Bernini, becoming his chief assistant in 1629. That he contributed significantly to the great baldacchino (altar canopy) in St. Peter's Basilica is certain, but the relationship of his work on it to Bernini's is unclear. His own important commissions included, among others, the church of San Carlo alle Quattro Fontane (1634–1644) in Rome, with its strikingly concave bays on the facade and a complex interior design, which rejected traditional Renaissance and classical principles. The facade of his Oratory of San Fillippo Neri (1637–1642) also employs a grand curve in the upper storey, while the dome of his San Ivo della Sapienza (1642–1660) in Rome combines a triangular design with convex and concave indentations. His largest commission was the interior renovation of San Giovanni in Laterno (1646–1650). He also fashioned such secular structures as the Piazza Navone (1653–1657) and a design for the unfinished Palazzo Carpegna. Focusing his career on architecture alone to the exclusion of painting or sculpture, Borromini was praised by his clients yet also denounced by many contemporaries for his daringly curvilinear designs, which broke with sixteenth-century standards of taste, but he influenced subsequent German Baroque architecture. Evoking strong loyalty from his workers, brilliant and courtly yet melancholic and introspective, he committed suicide, dying on 2 August 1667.

Bibliography: A. Blunt, *Borromini*, 1979; R. Wittkower, *Art and Architecture in Italy, 1600–1750*, 1958, rev. 1980.

Christopher Baker

BOSSUET, JACQUES-BÉNIGNE (1627–1704). An unsurpassed master of the French Classical sermon, Bossuet was also an important controversialist, historian, and advocate of religious reform. He was born at Dijon, where he studied at the Jesuit Collège des Godrans. His father had been appointed to the Parlement of Metz in 1638 and obtained for his son a canonicate of that city's cathedral chapter; although Bossuet was then too young to hold such an office according to the decrees of the Council of Trent, such irregularities were not uncommon in early seventeenth-century France. He began to preach while studying at Paris's Collège de Navarre, where he obtained a doctorate in theology in 1652. Bossuet had prepared himself for the priesthood at Paris under the direction of *Vincent de Paul and became a disciple of the devoted missionary. Bossuet participated in Vincent de Paul's charitable endeavors, taking a leading role in the Lazarists' 1658 mission to Metz, a confessionally divided city in which Bossuet had become involved in theological debate with the Protestant Paul Ferry.

Bossuet obtained his first post at the royal court through the intercession of the queen mother, Anne of Austria, and thus became closely associated with the "dévot" faction, which sought to promote charitable works, improve morality, and enhance the power of the clergy. He therefore did not initially win much favor with *King Louis XIV, who seems not to have found Bossuet's erudite and aggressively moralizing sermons to his liking. Bossuet nonetheless later secured the esteem of the king through his important role in the conversion of prominent Protestants, the most famous of whom was the highly successful soldier Henri de la Tour d'Auvergne, Comte de Turenne. Bossuet was rewarded with a bishopric in 1669, and in 1670 was appointed to the prestigious but not publicly prominent post of preceptor to the Dauphin. In 1671, he was elected to the Académie Française.

The balance of Bossuet's career was dominated by his controversies with Protestants and other Catholics. In 1678, he counseled the suppression of the French oratorian Richard Simon's treatise on the inadequacies of the Bible as a unique source of theological guidance. Simon had sought to demonstrate the weakness of the Protestant "sola Scriptura" principle, but Bossuet, long since engaged in verbal combat with "libertines," saw in his work a threat to one of the foundations of the Catholic church's authority. Thus, Bossuet's *Discourse on Universal History* (1681) asserted that the narratives presented in the Old Testament were literally true. This universal history also sought to strengthen the Gallican position by emphasizing the role in fostering the Christianity of Constantine and Clovis, and positioning the Bourbon kings of France as the successors of these lay champions of the Church; the French Christian owed unquestioning loyalty to Louis XIV. The vogue of mysticism that France experienced in the last decade of the seventeenth century also aroused Bossuet's fear of antinomianism, and his 1697 condemnation of the *Maxims of the Saints* of his former friend and fellow Catholic *François de Salignac de la Mothe Fénelon was wanting in the courtesy and restraint of his earlier writings against Protestantism.

Bossuet's best sermons are masterpieces of the genre. While respecting the Ciceronian structural constraints of classical pulpit oratory, Bossuet expresses himself in an elegant, economical style that is enlivened by vivid and arresting imagery. These sermons point repeatedly to the obligation of the rich and comfortable, including the king, to assist the sick and destitute, while avoiding the sinful behavior their wealth makes possible; the transitory nature of human existence is also a recurrent theme.

Bibliography: T. Goyet, *L'humanisme de Bossuet*, 2 vols., 1965; O. Ranum, ed., *Discourse on Universal History*, trans. E. Forster, 1976; J. Truchet, *La prédication de Bossuet*, 2 vols., 1960.

Matthew Koch

BOULAINVILLIERS, HENRI DE (1658–1722). Born into an ancient noble family, Henri de Boulainvilliers was educated at the exclusive academy at Juilly. After financial circumstances cut short his military career in the 1680s, he turned his attentions to raising his children and began writing to educate them. Boulainvilliers' interests were wide ranging. He particularly liked astrology, but it was as a historian that he became known to his contemporaries and is remembered today.

Boulainvilliers' first historical writings were genealogical. They included a history of his family prefaced by a dissertation on the "true and old nobility" of France. As with most of Boulainvilliers' works, neither was published during his lifetime, but the latter appeared posthumously. Many of Boulainvilliers' later works were politically motivated. The last years of *Louis XIV's reign witnessed growing opposition to his policies, which led to calls for reform, particularly from the circle surrounding his grandson the Duc de Bourgogne. Boulainvilliers was connected to this circle and wrote for them prior to Bourgogne's death in 1712. Louis XIV's death in 1715 brought the Duc d'Orléans to power as regent, on whose behalf Boulainvilliers also wrote.

Boulainvilliers' writings can often appear contradictory. Nonetheless, certain key ideas and theories run through them. Central was his advocacy of the thèse nobiliaire against Bourbon absolutism, which earned him the label of aristocratic reactionary. Boulainvilliers traced the origins of the French nobility back to the Franks. He described how the Franks had conquered Gaul and reduced the Gallo-Romans to serfdom. Equal among themselves, the Franks initially established aristocratic government and only later appointed a king, who was forced to defer to the decisions of the "nation" gathered in an assembly. Though Charlemagne had reinforced these origins by instituting French feudalism, his successors had gradually eroded this system, the government had become despotic, and the French nobility decadent. Royal favor and personal wealth replaced birth, virtue, and the pursuit of glory as the ideals of elite society. Pessimistic about the prospects for the future, Boulainvilliers hoped for the rehabilitation of feudalism and urged Orléans to bring it about and to recall the Estates-General. Boulainvilliers died in Paris on 23 January 1722, but his writings continued to influence others, including Montesquieu, and contributed to the development of political culture in eighteenth-century France.

Bibliography: V. Buranelli, "The Historical and Political Thought of Boulainvilliers," *Journal of the History of Ideas* 18 (1957): 475–494; H. Ellis, *Boulainvilliers and the French Monarchy*, 1988; P. Leffler, "French Historians and the Challenge to Louis XIV's Absolutism," *French Historical Studies* 14 (1985–1986): 1–22.

Rachel Hammersley

BOYCE, WILLIAM (1711–1779). Boyce is regarded as one of the foremost English composers of his day, second only to Handel. He is known for his symphonies, masques, operas, and church music. Boyce was also a renowned organist and musical editor. Born in London in 1711, Boyce, the son of a cabinetmaker, began his musical career as a chorister at St. Paul's Cathedral. He studied vocals under the direction of Charles King and musical composition under Maurice Greene, the organist of Oxford Chapel ca. 1722. A gradual, yet progressive impairment in hearing did not diminish Boyce's musical appreciation or studies, and by 1736 he had secured a position as organist at the Wren Church of St. Michael's. It was at this time that Boyce composed the music for the oratorio *David's Lamentation over Saul and Jonathan*. In 1740, Boyce wrote the operetta *Peleus and Thetis*. He married in the summer of 1748. His son received fame as a double-bass player in the London orchestras. In 1749, Boyce's ode *O be Joyful* earned him the degree of Doctor of Music from the University of Cambridge. Boyce was appointed Master of the King's music in 1755. In 1759, he composed one of his most famous songs, "Heart of Oak," for David Garrick's pantomime *Harlequin's Invasion*. By 1768, further deterioration in his hearing forced Boyce to resign his position as organist at St. Michael's. Between 1760 and 1768, he edited the now famous collection *Cathedral Music*—the first compilation of English church music, by various English composers, to be printed after the Restoration and the first to be printed in score. The three-volume collection remained in use for almost 150 years. Boyce died 7 February 1779 and was buried under the dome of St. Paul's Cathedral.

Bibliography: R. McIntosh, *The Dramatic Music of William Boyce*, 1979; J. van Nice, *Souls of the Righteous: Two Anthems for the Georgian Court*, 1970.

David M. Rosen

BOYLE, ROBERT (1627–1691). Hailed early in the eighteenth century as the very personification of the Royal Society and as the man who put *Francis Bacon's program for the scientific investigation of nature into practice, Robert Boyle was the most famous scientist or "natural philosopher" of his time. Most of the prolific stream of publications that he produced from 1660 to 1691 appeared in both English and Latin, spreading the word of his experimental and theoretical breakthroughs throughout Europe. Like many early scientists before our age of specialization, he made contributions to the understanding of natural phenomena in a number of fields, but he was first and foremost the father of modern chemistry. Methodologically, he is important for his vigorous championing of experimentation over inherited authority to support hypotheses. With his own convincing experiments, he effectively refuted authoritative beliefs about matter—the Aristotelian notion that it is composed of the four elements of earth, air, water, and fire, and the more recent theory of Paracelsus that all bodies are made of the "principles" of salt, sulphur, and mercury—replacing such ideas with the modern notion of elements as the indivisible components of chemical compounds. In so doing, he laid the foundation for the great eighteenth-century advances in chemistry. Today, his name is invoked whenever chemistry students learn of Boyle's Law, which states that the volume of a gas is inversely proportional to the pressure applied to it, at constant temperature.

Born in 1627 at Lismore Castle in the Irish province of Munster, Boyle was the son of the first Earl of Cork, the owner of immense estates and reputedly the richest man in Great Britain. Lord Cork immediately perceived his seventh son's aptitude for study and procured distinguished tutors to teach him French and Latin. At the age of eight, he was sent to prestigious Eton where, as he recalls in an autobiographical fragment, he found himself so consumed by a passion for knowledge that his teachers had to force him to give up reading for play. In 1638, he was sent with his brother and tutor to Geneva. There for twenty-one months the Boyle brothers adhered to a rigorous schedule of study and sport. In Geneva too Robert Boyle first became enamored of mathematics and experienced a religious crisis prompted by the awesome power of a thunderstorm. From this point on, his life was one of unwavering Christian piety, and he became a diligent student of theology and an erudite biblical scholar able to read the scriptures in all of their original languages. From Geneva, the Boyles proceeded to Italy. The death of *Galileo during their stay in Florence induced Robert to study that great scientist's works on mechanics and astronomy. Financial problems delayed their return to England, and by the time of their arrival in 1644 their father had died, and England was in the throes of civil war.

Establishing himself at his father's former residence in Dorsetshire, but also frequenting London to participate in discussions of the new philosophy in what he called the "Invisible College," a group of advanced thinkers that was the forerunner of the Royal Society, Boyle divided his time among chemical research, the writing of moral essays, and the study of literature. The next stage of his career began when he moved to Oxford in 1654, by then a stronghold of the new philosophy where members of the Invisible College were holding gatherings. His first six years at Oxford were a kind of apprenticeship; he studied, built instruments and laboratories, and experimented without publishing. Taking the remarkably adept *Robert Hooke as his assistant and commissioning him to build a new air pump in order to experiment with vacuums, Boyle at this time conducted a momentous series of experiments on the physical properties of gases. He demonstrated, among other things, that air had weight and elasticity, that it was a mixture of substances rather than one of the four Aristotelian elements, and that one of its constituents was essential for combustion and respiration. These findings, together with his role as a founding member of the Royal Society, made Boyle quite famous. With the publication of his most popular work, *The Sceptical Chemist* (1661), which also captured the popular imagination, he came to be seen as the prototypical scientific virtuoso. The ten books produced at Oxford and the twenty papers placed in the Royal Society *Transactions* before 1666 represent one of the most impressive examples of industriousness in the history of science.

Boyle left Oxford for London in 1668 to facilitate his participation in the Royal Society activities. At the house of his sister, Lady Ranelagh, he entertained a constantly growing number of visitors from England, the continent, and the colonies eager to meet the renowned virtuoso. The popularity he derived from both the sheer importance of his work and the accessible style of all his writings can also be gauged by the frequency with which he and the Royal

Society were the targets of satirists, including *Samuel Butler and Jonathan Swift. By 1673, Boyle was suffering from paralysis. From this point until his death in 1691, his scientific publications were largely elaborations of earlier works. He also continued to write and lecture in promotion of Christian piety and tolerance and vigorously rejected painful charges that his philosophy of mechanistic empiricism and his advocacy of an atomistic or "corpuscular" theory of matter encouraged disbelief in God.

Bibliography: R. Hooykaas, *Robert Boyle: A Study in Science and Christian Belief,* 1997; L. More, *The Life and Works of the Honourable Robert Boyle,* 1944.

Patrick Cook

BRADFORD, WILLIAM (1590–1657). Respected governor of the Plymouth colony for over thirty years, Bradford was born in Austerfield, England, in approximately 1590 to Yorkshire yeoman parents, William Bradford Sr. and Alice Hanson Bradford. His father died while William was an infant, leaving the family a rather comfortable inheritance. His mother remarried in 1593, and most of Bradford's upbringing is attributed to his paternal grandparents and various uncles. He received no formal university education, but was apprenticed as a farmer instead. The sermons of a local Nonconformist minister, Richard Clyfton, inspired Bradford's spiritual awakening and, as a result, in 1606 while still in his adolescence Bradford became a devout follower of the Puritan denomination known as Separatists.

The Separatist movement, under the tutelage of men like William Brewster and minister John Robinson, consisted of a small group of "simple souls" who were unequivocally dedicated to Puritan doctrine. Considered to be the "purest" of the Puritans, the Separatists had no delusions about reforming the Church of England from within. Consequently, Separatists were perpetually engaged in bitter exegetical, political, ecclesiastical, and doctrinal conflicts with church authorities. Separation from the Church of England at that time, however, was considered an act of treason, so in 1609, Bradford joined the already exiled Separatists in Holland where he sustained himself working as a weaver. Under Bradford's leadership, the Separatist group soon migrated to New England aboard the Mayflower.

At approximately 100 feet and 180 tons, the diminutive Mayflower sailed out of Southampton, England, in September 1620 carrying 102 passengers, bound for the Virginia territory, but rough seas forced her to land elsewhere. In mid-December 1620, some 66 days and almost 2,800 miles after departure, the ship landed in Plymouth, Massachusetts. Bradford's first wife, Dorothy, died on the voyage, only two weeks prior to reaching their destination. It was reported that "Mrs. Dorothy Bradford accidentally fell off the Mayflower into the cold waters of Cape Cod, and drowned." (Bradford would later remarry, this time a woman from a prominent English family named Alice Carpenter Southworth.)

Shortly after his arrival in April 1621, upon the death of John Carver, Bradford was elected governor of the fledgling Plymouth colony. The position of governor constituted such responsibilities as magistrate and council, superintendent of agriculture, commerce, and land use management. He faithfully served the gubernatorial office for the next thirty of his remaining thirty-six years. He humbly performed the duties of civil servant for some eighteen years before accepting any salary, and then he would only accept £20 annually. Bradford performed his civic duties with such magnanimity that his name quickly became synonymous with the ideologies that the colony stood for. He was one of the men responsible for drafting the Mayflower Compact. Bradford and company composed the Compact during the ship's crossing in 1620. Considered to be the precursor to the bill of (civil) rights, the Mayflower Compact served as a "covenant" wrote Bradford, whereby "[we] combine ourselves together into a civil body politick." Bradford spent nearly twenty years documenting his *History of Plymouth Plantation* (c.1650). The original manuscript, once thought lost, was later found in London, and the *History* was not published until 1856. It is written in the rhetoric of a traditional Puritan pulpit discourse. Bradford's *History* is a comprehensive, pragmatic, yet sensitive theological account of the founding and functioning of the New England colony. Because of his inherent civic and moral discretion, the fastidiousness of the *History of Plymouth Plantation*, the progressiveness of the Mayflower Compact, and his intelligence, resolution, and strength of character, Bradford is considered to be the father of both the Plymouth colony and its dogma.

Bibliography: B. Smith, *Bradford of Plymouth*, 1951.

David M. Rosen

BROWNE, SIR THOMAS (1605–1682). Browne, with *Robert Burton and *John Bunyan, one of the greatest English prose writers of the seventeenth century, was born in London on 19 October 1605. After attending Winchester school, he earned a B.A. from Oxford in 1626 and an M.A. in 1629. Deciding to become a physician, he left Oxford (which had no medical curriculum) and studied at the universities of Montpellier, Padua, and Leiden, where he finally received his medical degree. He practiced for a time in Yorkshire and Oxford before settling in Norwich, where he married; his practice grew, and he became the physician of *Joseph Hall. He received a knighthood in 1671 from *Charles II for his long support of the monarchy. Browne corresponded with such notables as *John Evelyn and *John Aubrey, but for reasons still unclear he never became a member of the Royal Society. Though he took part in a witchcraft trial in 1664, the allegation that it was his testimony that secured judgment against the two female defendants is unfounded. He died on his seventy-fifth birthday.

Before the age of thirty, Browne wrote *Religio Medici* (*The Religion of a Doctor*); it was read in manuscript by his acquaintances, eventually being published in London in 1642 without his permission (an approved publication followed in 1643). The book was widely popular and was translated into Latin, Dutch, and French. It is noteworthy for its ornate yet fluid prose, Browne's wide-ranging curiosity and perceptive reflection, and its attitude of broad tolerance of divergent views of religion, philosophy, and science—a rare stance in Browne's era of sharply fought doctrinal battles. In 1646, he published *Pseudodoxia Epidemica*, or *Enquiries Into Very Many Received Tenents, and Commonly Presumed Truths*. Also known as *Vulgar Errors*, it discusses a variety of commonly held superstitions and beliefs (such as that salamanders could live in fire) with a dispassionate, "scientific" eye that echoes *Francis Bacon's emphasis upon empirical evidence.

In 1658, two essays appeared in one volume, *Hydriotaphia, Urne-Buriall, or a Discourse of the Sepulchral Urnes Lately Found in Norfolk* and *The Garden of Cyrus or the Quincunciall, Lozenge, or Network Plantantions of the Ancients*. Though the recently discovered burial urns Browne refers to were in fact Saxon rather than Roman as he thought, this error does not diminish the grand beauty of his Latinate meditations on human mortality. *The Garden of Cyrus*, in a typically seventeenth-century outlook that sought correspondences among all parts of the physical and spiritual worlds, studies the ways in which patterns of fives (the quincunx) can be found in nature. Toward the end of his life, Browne also wrote an account of the death of his friend Robert Loveday, *A Letter to a Friend* (1690). Browne is important not only for his prose, which succeeds in being at once imposing, intimate, and inventive, but also for his largeness of spirit, his refusal to rush to judgment in complex issues, and his absorbing love of meditative thought, which led him, as he said in *Religio Medici*, to "pursue my reason to an O altitudo."

Bibliography: F. Huntley, *Sir Thomas Browne: A Biographical and Critical Study*, 1962; J. Post, *Sir Thomas Browne*, 1987.

Christopher Baker

BRUEGHEL, JAN (THE ELDER) (1568–1625).

Jan Brueghel, called "Velvet Brueghel," was the most prominent among the artist children and grandchildren of the Flemish painter Pieter Brueghel the elder. Jan was born in Brussels and never knew his famous father, who died shortly after Jan's birth. Both Jan and his older brother Pieter probably received their early artistic training from their grandmother, the accomplished miniaturist Mayaken Verhulst. In the late 1590s, Jan traveled to Italy where he met the influential Cardinal Federico Borromeo who became a lifelong benefactor. Jan established a lucrative career in the artistic center of Antwerp and was highly regarded by his most renowned Flemish contemporary, *Peter Paul Rubens, with whom he collaborated.

Jan specialized in cabinet pictures, including small-scale paintings of still lifes and landscapes with figures. Such works as his *Bouquet with Irises* were prized for their refined, velvety brushwork and attention to detail. Jan also played a key role in the Ambrosiana, an art academy and library founded by Cardinal Borromeo of Milan, which promoted reform within the Catholic Church. Borromeo argued that religious art should be natural, and to this end Jan developed a new genre of devotional imagery, which became widely popular: The Madonna and Child in a Garland of Flowers.

Jan died, along with three of his children, from a cholera epidemic. His sons Jan the younger and Ambrosius became painters, thus proliferating the Brueghelian dynasty of artists. Jan the elder's popular reception was profound, and he was regarded, along with Rubens, as one of the two most prominent painters in Antwerp.

Bibliography: *Jan Breughel the Elder*, exhibit catalogue, Brod Gallery, London, 1979; P. Jones, *Federico Borromeo and the Ambrosiana: Art Patronage and Reform in Seventeenth-Century Milan*, 1993.

Pamela Merrill Brekka

BRUEGHEL, PIETER (THE YOUNGER) (1564–1637). Also known as "Hell Brueghel," Pieter the younger was the son of the famous Pieter Bruegel, patriarch of the Bruegel family of artists in Brussels. Pieter the younger established a successful career copying his father's widely popular paintings. These copies are historically significant as they have provided evidence of lost originals by Pieter the elder. Although technically inferior to his father's, Pieter's paintings were in high demand and sold cheaply as known copies of Pieter the elder's most popular compositions. His atelier produced sixty or more versions of Pieter the elder's *Winter Landscape with Skaters and a Bird Trap*. Characteristic of Pieter the elder's best known paintings, the work includes a bird's-eye view of peasant activities set in an atmospheric landscape. Pieter the younger had a thriving workshop with many assistants, but never achieved the wealth or acclaim of his brother Jan Brueghel the elder.

Bibliography: W. Liedtke, *Flemish Paintings in the Metropolitan Museum of Art*, 2 vols., 1984; J. Turner, ed., *The Dictionary of Art*, 34 vols., 1996.

Pamela Merrill Brekka

BULL, JOHN (1562/63–1628). John Bull, English composer and keyboard virtuoso, was a choirboy at Hereford Cathedral from 1573 to 1574, when he presumably joined the Children of the Chapel Royal. In December 1582, Bull became organist at Hereford Cathedral, and in January 1583, choirmaster. He was inducted in 1586 as a Gentleman of the Chapel Royal. Bull received the B. Mus. (1586) and D. Mus. (1592) degrees from Oxford, and in 1597 he became the first Professor of Music at Gresham College, London, at the queen's recommendation. In 1601, Bull applied for a leave of absence, possibly for European travel, but by 1603 he was in England because his name was listed with the Gentlemen of the Chapel Royal at Elizabeth I's funeral. On 29 June, Bull resumed his professorship at Gresham College, but in 1607 he left, at the

College's request, to marry the woman he had gotten with child. Bull survived through organ building and performing as a musician for Prince Henry. In 1612, after the prince's death, Bull became music master to Princess Elizabeth. She and her betrothed, Frederick, Elector Palatine, were the dedicatees of *Parthenia* (c. 1612), the first printed virginal music, containing pieces by Bull. In 1613, Bull was charged with adultery; in August 1614, he escaped to the South Netherlands, where Archduke Albert enlisted him as organist at the Brussels court. In 1617, he became organist at Antwerp Cathedral, at which post he stayed until his death, on 15 March 1628. Bull is best known as a composer of keyboard music, but except for pieces published in *Parthenia*, it survives only in manuscript. Many of his organ works are contrapuntal, based upon plainchant (e.g., *Miserere*). For virginals, Bull composed much dance music, like the Quadran pavans and galliards, and variations, like those on the popular tune "Walsingham." Little of Bull's sacred music remains.

Bibliography: P. Chappell, *A Portrait of John Bull*, 1970; R. Rasch, "The Messaus-Bull Codex," London: British Library, Additional Manuscript 23.623; *Revue belge de musicologie* 50 (1996): 93–127.

Susan Treacy

BUNYAN, JOHN (1628–1688). John Bunyan was born at Elstow near Bedford, England, where he received a rudimentary education and worked as a tinker with his father, Thomas. After serving in the parliamentary army from 1644 to 1646, he married and received two books from his wife's dowry, Arthur Dent's *Plain Man's Pathway to Heaven* (1601) and Lewis Bayly's *Practice of Piety* (1612), which influenced his early development as a Puritan evangelist. Much of Bunyan's theology reflects Dent's belief that everyone, regardless of social status, inherits original sin, and Bayly's belief that any sinner can receive God's grace through repentance.

Bunyan's early theological treatises, such as *Some Gospel Truths Opened* (1656), were aimed at Quakers, particularly Edward Burrough. Religious arguments between Bunyan and Burrough mostly involved Bunyan's literalist position that God speaks to believers directly through the Bible's inspired words, against Burrough's nonliteralist position that God speaks to believers metaphysically through an inner light medium. Bunyan also attacked the Latitudinarians and the Ranters, who he felt were socially dangerous heretics. He was especially hostile to Ranter beliefs that sin exists only in the mind and that God could not have created the material universe from nonexistent material. Bunyan cynically personified Ranter opinions in such allegorical figures as Badman in *The Life and Death of Mr. Badman* (1680) and Carnal Security in *The Holy War* (1682).

Bunyan joined a nonconformist congregation in 1653 that had taken over St. John's Church in Bedfordshire. He was appointed deacon there in 1657, but in the first year of the Restoration, he was arrested and sentenced to twelve years in prison for preaching without an Anglican ordination license, a violation of *Charles II's Uniformity Acts. He was released in 1672, and in the same year he was elected head pastor of the St. John's Puritan assembly, then holding secret religious services in a barn. The government arrested Bunyan again in 1677 for unlawful preaching, but released him six months later.

During his twelve-year prison term, Bunyan wrote religious books, the most famous of which was *Grace Abounding to the Chief of Sinners* (1666), his spiritual autobiography. Its theology builds on Dent's and Bayly's theories of sin and grace, and on St. Paul's belief in Christ's power to save even the worst of sinners. The autobiography falls generally into three parts: a preconversion period of youthful folly and sin, a period of spiritual awakening and conversion through grace, and a postconversion period of struggles against sin. Rooting his narrative in biblical passages, Bunyan shows that sinners need not appeal to ecclesiastical mediators to shed sin and receive God's favor through grace. His other prison works include *I Will Pray* (1662), a justification for using extempore prayer instead of the formal liturgy of the Book of Common Prayer; *Christian Behavior* (1663), a conduct book showing how to live a spiritual life without the intervention of church clerics and rituals; and *The Holy City* (1665), a millenarian allegory of Christ's second coming.

Bunyan's greatest work is *The Pilgrim's Progress* (1678). It is a dream allegory written in a provincial idiom that, antithetical to the pedantic rhetoric of scholastic discourse, became the stylistic hallmark of Puritan prose. Like the simple parables of Jesus, Bunyan's imagery is drawn from everyday life to communicate religious principles to a wide audience. The allegory relates the travails of Christian who, terrified by the damnation he reads about in the Bible, abandons his family to embark on a pilgrimage from the world here and now, called the City of Destruction, to a future heavenly world, called the Celestial City. Along the way, Christian encounters such menacing forces as Apollyon and Giant Despair who threaten his soul, but he also meets virtuous companions like Faithful and Evangelist who help him follow the righteous path. The allegory epitomizes the Puritan belief that spiritual salvation can be found through a personal quest for God.

Bibliography: C. Hill, *A Tinker and a Poor Man: John Bunyan and His Church 1628–1688*, 1989.

James Norton

BURNET, GILBERT (1643–1715). Born in Edinburgh, Scotland, Gilbert Burnet held various clerical offices beginning in 1661. He was professor of divinity at Glasgow University in 1669, where he espoused liberal principles that gained him the resentment of Presbyterians and Anglicans. Instead of accepting a bishopric in Edinburgh, Burnet went to London in 1673, and his support of the Anglican church quickly put him into favor with *Charles II. About 1674, Burnet began writing his respected *History of the Reformation of*

the Church of England (1679–1714). By 1680, Burnet had sided with the
*Shaftesbury Whigs, who wished the king's Catholic brother, James, Duke of
York, to be excluded from succession. In 1683, Burnet came to the defense of
his Whig friends following the Rye House Plot, an abortive attempt against the
persons of Charles and James. This, in addition to other factors, resulted in
Burnet's removal from court favor.

Upon the accession of *James II (1685), Burnet left England, traveling to
France and Rome and then to Holland, where he became an advisor to *William,
Prince of Orange. He returned to England with William during the Revolution
of 1688, was appointed Bishop of Salisbury, and remained a trusted advisor to
*William III and *Mary during their reigns. Burnet's most important work, *A
History of My Own Times*, was published, as he desired, posthumously (1723).
Although biased against James II and generally supportive of Whig liberalism,
the collection is a reasonably well-balanced contemporary source offering valuable information on the constitutional crises that form the basis of seventeenth-
century English political thought.

Bibliography: M. Delorme, "Gilbert Burnet: Bishop and Historian," *History Today* 29
(1979): 594–602.

Ken MacMillan

BURNET, THOMAS (?1635–1715). Thomas Burnet, later Master of Charterhouse, was born in Yorkshire about 1635 and was educated at Clare Hall and
Christ's College, Cambridge. Burnet was an exceptional student and scholar,
and through the intervention of the Duke of Ormonde, he was appointed as
Master of Charterhouse in 1685. Burnet was interested in theology and the
emerging new science; he considered the reconciliation of the old and the new
and speculated upon new directions for scholars. He was not sympathetic to
Rome and maneuvered around the plans of *James II to restore papal influence.
Burnet was considered a stylist in his day, and his two most significant works
were *Telluris Theoria Sacra, orbis nostri orginem et mutations generales quas
aut jam sublit aut olim subiturus est completens* (2 vols., 1681) and *Archaeo-
logiae Philosophicae sive doctrina antiqua de rerum originibus* (1692). In the
first book, Burnet argued that the earth was similar to an egg that had been
altered by the great flood; he recognized natural change but argued that it was
in the context of a Divine plan. In the second book, Burnet advanced a liberal
interpretation of Genesis in an attempt to understand human nature and the
struggle between good and evil. Many Anglican bishops questioned Burnet's
orthodoxy and took issue with some of his conjectures. Burnet died at Chare-
terhouse on 27 September 1715.

Bibliography: M. Pasini, *Thomas Burnet una storia del mondo tra ragione, mito e riv-
elazione*, 1981.

William T. Walker

BURTON, ROBERT (1577–1640). Robert Burton was born in Leicestershire and educated at Oxford, first at Brasenose College and then at Christ Church. Although he was diligent and intellectually gifted and was said to make considerable progress in learning logic and philosophy, he took the unusually long period of nine years to earn his B.A., which he received in 1602. While a student at Brasenose, he probably consulted the astrologer-physician Simon Forman, who shared many interests with Burton, if he is indeed the Robert Burton recorded among Forman's patients. Some have inferred from his protracted education and the treatment by Forman that he suffered from melancholic illness (what we would call depression) from early on in life. If so, he coped with his malady sufficiently to earn his M.A. in 1605 and his Bachelor of Divinity in 1614, to tutor college students and work as a market inspector, to serve as vicar of a local church from 1616 and as Christ Church librarian from 1626, to acquire one of the largest and most varied personal libraries in his time, and to become one of the most fantastic figures in English literary history.

Like most humanists of his age, Burton wrote works in Latin, including a number of very conventional commemorative poems and a satirical play entitled *Philosophaster*, an energetic satire of university life. Burton's claim to fame, however, is *The Anatomy of Melancholy, What it is. With all the Kindes, Causes, Symptomes, Prognostickes, and Several Cures of It. In Three Maine Partitions with their several Sections, Members, and Subsections. Philosophically, Medicinally, Historically, Opened and Cut Up*, or more simply, as it soon came to be called by a wide and admiring readership, "Burton's Melancholy." Written in English but filled with untranslated Latin quotations from a huge variety of authorities on an equally huge variety of subjects, the book was enormously long when it first appeared in 1621, but it continued to grow in subsequent editions as its author's steady accumulation of facts and opinions helped him to realize the title's encyclopedic imperative of "all the kinds."

The *Anatomy of Melancholy* is a book that eludes classification. It is a treatise on Renaissance psychology, reporting what Burton's contemporaries, relying largely on an understanding that mental and physical health were produced through balancing the body's four "humors," believed about causes, symptoms, and cures. It is also a work of philosophy that combines humanistic affirmation, deep skepticism, and a clear-sighted recognition that to be human, alas, is to be melancholy. It is as well a therapeutic text, both for the reader and the author, who recommends vigorous exercise of the mind as the best medicine for melancholy, and who clearly practices what he preaches. Perhaps, despite its enormity, it is best classified as what Montaigne called an essay, that is, an exploratory foray lacking the pretension of arriving at a definitive conclusion, for in a world fascinated by the new proliferation of printed texts, Burton seems always aware that more opinions will be encountered to balance and qualify those he already possesses.

An early biographer tells us that Burton predicted his own death astrologically and died precisely on schedule, prompting suspicions that he committed suicide rather than be proved mistaken in his calculations. In any event, his *Anatomy* long outlived him. In the eighteenth century, its popularity declined, but its fantastic wit, fond antiquarianism, and thoroughly unclassical style appealed to the Romantics, who initiated a nineteenth-century revival of interest. Today, it is treasured as a window into a vast range of early seventeenth-century ideas, and into the heart, mind, and soul of its strange and appealing author.

Bibliography: R. Fox, *The Tangled Chain: The Structure of Disorder in the Anatomy of Melancholy*, 1976; M. O'Connell, *Robert Burton*, 1986.

Patrick Cook

BUTLER, SAMUEL (1612–1680). Little is known of satirist Butler's life. He was born in Strensham, south of Worcester, to a lower gentry family of "slender fortune." Butler may have gone to the King's School in Worcester and certainly was enough of a scholar to be the beneficiary of the bulk of his father's library. He never matriculated at Oxford or Cambridge, but served as a secretary for a series of gentry and noble families and studied painting with Samuel Cooper. During the interregnum, Butler probably attended the Inns of Court and may have served as a royalist pamphleteer. In 1663, at the age of fifty, Butler achieved fame with the publication of *Hudibras*, the first best-seller of the Restoration and *Charles II's favorite book. Although Charles and his chief minister *Hyde took great pleasure in the mock-epic, Hyde commissioning a portrait of the satirist, the poet had to wait fifteen years before he would receive any reward, thereby becoming an exemplum of the fickle nature of royal bounty. After 1663, Butler continued to write and publish, in particular the second and third parts of *Hudibras* (1664 and 1678), and served as a secretary to the Duke of Buckingham in 1673. Butler died a relatively poor man in 1680.

Butler's fame rests soundly on *Hudibras*. The mock-epic recalls the adventures of a Presbyterian Justice of the Peace, Sir Hudibras, during the interregnum. Accompanied by his squire, Ralpho, a member of the Independent church, Hudibras succeeds, in spite of himself, in suppressing a bear baiting (although the audience of the baiting soon gained their revenge) and fails to woo a wealthy widow. *Hudibras* is somewhat off-putting to modern readers: It switches continuously from doggerel to classical verse, it is filled with long-winded satires of Renaissance learning, and expresses a sense of humor based on a conception of society as inherently stratified. Yet just such elements made this poem so endearing to Butler's contemporaries.

The unifying theme of *Hudibras* is a satire of those things which try to be what they are not. The satire runs from the main character, to Presbyterianism, the interregnum social order, false logic, and the very verse of the poem itself. Hudibras, for instance, strives to attain heroic qualities, but he not only lacks the valor, skill, and refinements of chivalry, he also cannot escape, nor even seem to realize the need to escape, his petty bourgeois values. When Hudibras prepares to suppress the bear baiters, enemies extremely unsuitable for a chivalric quest, he dons all the accoutrements of a knight but uses them more to store food than to fight, his breeches lined "with many a piece of ammunition, bread and cheese." His dagger "would scrape trenchers or break bread." The appearance of such mundane values where knightly ones should exist also comes to the fore in Hudibras's stinginess, lack of valor, and the pettiness of his quests and opponents. Yet what makes this mock-epic so amusing is that Hudibras remains blissfully hypocritical, using lengthy and false logic to prove his cowardice bravery, his lack of dignity chivalry. The protagonist employs the same sort of logic to defend Presbyterianism in his debates with Ralpho. Thus, Butler tars Presbyterianism as resting on the sort of logic that would render Hudibras's shameful actions heroic. This message is further reinforced by Butler's setting the poem during the interregnum, when, Butler implies, the acceptance of such false logic, the triumph of Presbyterianism, created a social order where Hudibras and his ilk became knights of the realm. Like Hudibras himself, the verse strives for nobility. It is written in octosyllabic rhyming couplets (a style later known as Hudibrastics), but it cannot escape its true doggerel nature. Absurdly obscure words, classical allusions, and high-flying verse sit cheek by jowl with slang, bad rhymes, and references to the most plebeian of matters. The verse, like Hudibras, tries to be what it is not and in so doing renders itself ridiculous.

Learning again became a target in *The Elephant in the Moon*, which was published after Butler's death. In this poem, members of the Royal Society, while observing the moon through a telescope, see on it what they believe to be an elephant. Soon after, they detect whole armies massing for battle. The footboys, using common sense, soon discover that the elephant and the armies were simply a mouse and gnats that had "gotten / in the hollow tube." Yet unwilling to renege on their discovery, after voting the scientists declare, "elephants in the moon / though we have discoverd none / is easily made manifest." Butler wrote a number of other satires, most notably *To the memory of the most renow'd Duval*, prose works, and a set of characters and observations. Most of this work, however, appeared in print only after his death.

Bibliography: R. Nevo, *The Dial of Virtue: A Study of Poems on Affairs of State in the Seventeenth Century*, 1963; G. Wasserman, *Samuel Hudibras Butler*, 1989.

Brian Weiser

BUXTEHUDE, DIETRICH (1637–1707).

Dietrich Buxtehude was born in Holstein, which was then under Danish rule. He presumably studied organ and possibly composition with his father, who was a professional organist. In 1668, Buxtehude was elected as successor to the famous organist Franz Tunder at St. Mary's Church in Lübeck, one of the best and most lucrative church music positions in Germany. According to custom in such successions, Buxtehude was obliged to marry Tunder's daughter.

In 1673, Buxtehude initiated his celebrated musical services, the "Abend-musiken," the evening performances, which were to become the pride of Lü-beck. These services occurred annually on the five Sunday evenings before Christmas and consisted of concerted pieces of sacred music for chorus and orchestra as well as organ performances given by Buxtehude himself. Such musical evenings were to continue throughout the eighteenth century and into the nineteenth.

Upon the approach of his own retirement, Buxtehude wished to continue the practice whereby one of his daughters would marry his successor. Handel traveled to Lübeck in 1703 with the intention of securing Buxtehude's post, but the nuptial clause, which would have compelled marriage to one of Buxtehude's daughters, deterred him from further negotiations. Two years later, Bach traveled 200 miles on foot to hear Buxtehude perform. Like Handel, Bach also declined the marriage as a means of obtaining the Lübeck position.

Buxtehude's strength as a creative musician resides in his free organ compositions not based upon chorale melodies. In the treatment of chorales on the organ, he was not equal to his contemporary colleague *Johann Pachelbel, but Buxtehude's innovations with respect to instrumental composition were profoundly influential on later composers. His multifaceted personality—organist, composer, and conductor—matches the broad stylistic range of his music. With his uninhibited individuality and virtuosic approach, Buxtehude was perhaps the most progressive composer of the seventeenth century.

Bibliography: K. Snyder, *Dietrich Buxtehude: Organist in Lübeck*, 1987; G. Webber, *North German Church Music in the Age of Buxtehude*, 1996.

James P. Fairleigh

BYRD, WILLIAM (ca.1540–1623).

Composer William Byrd was an innovator and master of musical form. His more than 500 compositions include secular songs, sacred works in both Latin and English, consort music, and a large body of keyboard works. Born in London, he received his early training at the Chapel Royal as a composition and keyboard pupil of Thomas Tallis. Later, the contemporary Italian composer Ferrabosco provided important foreign and contrapuntal influence.

Byrd left London in 1563 to take up the post of Organist and Master of the Choristers at Lincoln Cathedral. At Lincoln, Byrd produced a large amount of music for Anglican worship. His works became central to the English cathedral liturgy. His Latin motets also date from this period. He returned to London to become Gentleman of the Chapel Royal, sharing the organist position with Tallis from 1572. There Byrd traveled in powerful circles; the Earl of Worcester and the Petre family were his patrons. He also associated with the important poets of the late sixteenth century, whose work provided him with many song texts.

As a Catholic, Byrd was deeply involved in the resistance to the Anglican Church. He responded to the 1581 Jesuit executions on Tyburn Hill with a song

setting, *Why do I use my paper, ink and pen?*, followed by the prolific composition of Latin motets with texts that could be interpreted as expressing the
plight of English Catholics. He even published three masses. He and his family
were accused and fined for recusancy, but perhaps because of his favor with
Queen Elizabeth, his situation was somewhat alleviated.

Byrd and Tallis had, in 1575, received one of the first licenses to print music
in England, dedicating their first publication to Queen Elizabeth. But it was not
until after the death of Tallis that Byrd led the way in earnest for music publishing in England, with his popular *Psalmes, Sonets, and Songs* (1588), followed by the *Cantiones sacrae*. Byrd's keyboard works appeared in the
manuscript *My Ladye Nevells Booke* (1591), and in the collection *Parthenia* (ca.
1612). Along with numerous variations on popular songs, the paired pavane and
galliards are notable among his keyboard output. In 1594, he moved to Stondon
Massey, Essex, where he spent the rest of his life. A number of his *Gradualia*
were composed for the secret masses held at the nearby Petre estate. Byrd was
a major influence on English musical style, especially with regard to the verse
anthem and his development of keyboard music. *Thomas Morley, Thomas
Tomkins, Peter Philips and *John Bull all referred to Byrd as their master.

Bibliography: J. Harley, *William Byrd: Gentleman of the Chapel Royal*, 1997.

Joyce Lindorff

BYRD, WILLIAM (THE ELDER) (1652–1704).

William Byrd the elder
was the founder of the Byrd political and economic dynasty in Virginia, and
the father of William Byrd II, whose diaries provide an in-depth portrait of
Virginia colonial society. The elder Byrd was born in Britain, but his mother's
brother, a landowner in Virginia, persuaded the young man to come to America;
in 1670, he bequeathed all his holdings to his nephew. At one stroke, Byrd
became a wealthy landowner with an enviable social position. He greatly improved the estate and eventually found himself as the chief Indian trader in the
colony. He built a plantation house on the James River at Westover and made
an advantageous marriage to the daughter of a royalist Cavalier officer. Their
first child, a son named William after his father, was born in 1674; three daughters followed, and all the children were sent to England for their education.

Byrd increased his wealth, both by functioning as his neighbors' agent for
tobacco export and also by his participation in the importation of indentured
servants and slaves to the Virginia colony; he later became the principal owner
of a major slaving ship. Through his position as a state councillor, Byrd patented
virgin land for the colony and then had it repatented and assigned to either
himself or his friends. In addition to his business activities, Byrd took an active
role in colonial politics; as a young man, he served in the Virginia House of
Burgesses and was a militia commander. By 1680, he had become a member
of the colony's highest court, the Council of State, and eventually became its
president. He also took part in the committee that built the College of William

and Mary and was named as auditor and receiver general of the colony by royal appointment.

His success enabled him to build a second home, Belvedere, but he did not spend much time on the leisure activities and visiting that preoccupied the lives of his friends and neighbors. Though he maintained a sociable presence in the colony, business continued to be the focus of his life until his death in 1704. He left his son a rich legacy, both in land and in social position, a legacy that William Byrd II built on, making the Byrds one of the foremost Virginia families.

Bibliography: C. Dowdey, *The Virginia Dynasties*, 1969; L. Wright and M. Tinling, eds., *William Byrd, The London Diary (1717–1721) and Other Writings*, 1958.

Connie S. Evans

C

CADILLAC, ANTOINE DE LA MOTHE (ca. 1658–1730). Cadillac was born sometime between 1657 and 1661, probably in Toulouse, France, where his father belonged to the council of the city parliament. Apparently, Cadillac served in the French army on the continent, but in 1683 he is found at Port Royal. Four years later, on 25 June 1687, he married Marie Therese Guyon in Quebec; they had thirteen children. In 1688, Cadillac received a land grant, Mount Desert Island, not far from what is now Bar Harbor, Maine. The English incursions of 1690 and 1691 cost him his landhold; he returned to Quebec and an appointment in the colonial army. From 1694 to 1697, Cadillac commanded the post at Mackinac and, on his return to Quebec, proposed the establishment of a French fort at the lower end of Lake Huron. Permission was granted and Detroit was established in 1701. Socially, Cadillac took families to his colony, and economically, at least in the first years, the Company of the Colony received the profits from the fur trade. In 1705, *King Louis XIV assigned the revenue of the fur trade to Cadillac himself. Until his appointment as governor of Louisiana in 1710, this arrangement continued. In 1712, Cadillac took up his post in Louisiana and in 1717 returned to France where he died on 20 October 1730, having been recalled from his post for the dissension within his colonial administration. Although Cadillac succeeded in establishing the stronghold that virtually guaranteed the long-sought safe passage along the Great Lakes, he did not win the admiration of all. His fur trade was regulated, but sheer distance and the desire for profit made regulation difficult to enforce. Furthermore, Cadillac was among those who paid the coureurs de bois in brandy, and the mission trading posts soon became sites of drunkenness, so much so that in 1702 the Jesuit Carheil not only complained to the governor but also warned against the rhetoric and machinations of Cadillac, whom Carheil did not expect to desist from either practice. By the end of Cadillac's tenure in the Great Lakes, the danger to the missions of the early explorers lay more in trade and commerce than in wild weather and unknown languages.

Bibliography: A. Laut, *Cadillac, Knight Errant of the Wilderness*, 1931.

Martha Oberle

CALDERÓN DE LA BARCA, PEDRO (1600–1681). Born in Madrid in 1600, Calderón studied with the Jesuits and attended the universities of Alcalá and Salamanca. In his youth, he was a man of action as he fought in the Catalonian war of 1637. Upon *Lope de Vega's death in 1635, Calderón was recognized as his literary successor and was appointed to the court in Madrid by Felipe IV. There, Calderón played an important role in scripting the many court entertainments. Calderón never married, but he did have a son whom he passed as his nephew. In 1651, he was ordained a priest and retired from the world, although he continued to write religious and secular plays.

Calderón's contributions to literature were mainly his 120 plays; 80 *autos sacramentales*, or one-act allegorical plays; and various shorter pieces. He also wrote lyric poetry. The social, moral, and theological ideologies of his time became apparent in his dramatic production of secular, religious, and philosophical works. Calderón followed and improved upon the historical and legendary themes established by Lope as seen in the Calderonian version of *El alcalde de Zalamea* (1642?). Unlike Lope's plays, Calderón's exhibit a preference for ideological concerns over action and the frequent employment of characters as symbols of abstract concepts. Calderón is especially noted for his numerous dramas of the Spanish honor code and for the perfection of the "cape and sword" style. The honor plays had the theme of marital infidelity rather than that of individual, personal integrity. The Spanish honor code embraced the social convention of the right and duty of male family members to cleanse and silence any female sexual impropriety. *El médico de su honra* (1635) illustrates the resolution of death at the mere suspicion of female infidelity. In contrast to the extreme measures of the honor plays, the cape and sword plays had happier endings. There was usually a marriage after a series of fast-paced, complicated, amorous intrigues, such as those found in *La dama duende* (1629). The Baroque philosophical idea of life as disillusionment also permeated his plays. Life was depicted as an ephemeral phenomenon enacted in a world of false appearances that led to inevitable decay and destruction, with only virtue remaining intact. Calderón championed the theological ideas of the Jesuit teachings, especially that of the dominance of free will over predestination as seen in his best-known play *La vida es sueño* (1635). Segismundo, the prince and heir to the throne, has been imprisoned since birth by his father King Basilio because of portentous signs that were interpreted as a threat of regicide. When Segismundo is a young man, the king decides to test the prince's attitude and behavior by releasing him from prison. The ruse is to drug the prince and put him in the palace for observation. When Segismundo awakens from the sleeping potion, he proves himself to be undisciplined and arrogant and, therefore, is taken back to his cell. The prince must learn the value of reasoning to resolve his emotional conflicts and rightfully inherit the throne. His recollection of his time in the palace is interpreted as a dream in which there is a revolt during which the prince is set free. Once again, Segismundo encounters the king in the palace. Instead of killing his father, the prince bows to the king's authority. Segismundo has learned to control his actions and will now inherit the throne.

Calderón's plays manifest a Baroque style in their complex and intricate form as well as in language. To the traditional allegorical play, Calderón married the theological ideas with poetic forms and intricate scenery. It was not until Calderón's rich adaptations of the allegorical plays that the genre received its definitive form. Calderón's *autos sacramentales* are among the best ever written and were performed in the open air in public squares especially during the Corpus Christi festival. Calderón remained champion of the Spanish national theater until the neoclassical influence rejected his works. However, German Romanticists rediscovered Calderón and revived interest in the Spanish national theater of the seventeenth century.

Bibliography: F. De Armas et. al., eds., *Critical Perspectives on Calderón de la Barca*, 1981; M. McGaha, ed., *Approaches to the Theater of Calderón*, 1982.

M. Ellen Blossman

CAMDEN, WILLIAM (1551–1623). English antiquarian William Camden is a founder of modern historical methodology, primarily noted for his critical approach of collating contemporary records with antiquities. Born in London on 2 May 1551, Camden attended Christ's Hospital and St. Paul's Church. He was admitted to but did not attend Magdalen College, Oxford. Camden earned a B.A. in 1573 from Christ's Church after earlier attending Broadgates Hall. In 1575, he was appointed second master of Westminster School, where *Ben Jonson was among his students, and headmaster in 1593. In the interim, Camden was awarded the prebend of Ilfracombe in 1589. By the initiative of Queen Elizabeth and with the support of Fulke Greville, Camden was named Clarenceux king-of-arms in 1597.

To collect data for his most important work, *Britannia* (1586), Camden traveled the English countryside. This work was written in Latin, its purpose, Camden wrote, "to restore Britain to antiquity and antiquity to Britain." Camden made way for new kinds of antiquarian activity and anticipated a methodology of modern historicism by using literary texts to re-create the past, considering the natural topography of England in relation to Latin literary remains. Camden was the first historian to consider surnames and to acknowledge the importance of ancient languages in the study of place-names. He is credited with reviving the term *Anglo-Saxon* as a designation for the earliest period of Old English and to the West Germanic tribes generally, the term having fallen into disuse following the Norman Conquest. He also uncovered the existence of Romano-British coins. Camden's influence in writing *Britannia* is evidenced by his role as a founding member, along with the young Robert Cotton and others, of the Society of Antiquaries (1585). The activities of the society were coterminous with the publication of *Britannia*, and, indeed, involved with the work as a model for historical documentation and methodology.

In 1605, Camden published *Remaines*, a collection of medieval chronicles and commonplace texts. In 1615, he published the first volume of *Annales Rerum Anglicarum et Hibemicarum Regnante Elizabetha* (*Annals of the Affairs of England and Ireland During the Reign of Queen Elizabeth*). The second volume, which dealt with the period after 1588, was published posthumously in 1627. Although Camden was criticized for his portrayal of Mary, Queen of Scots, which perhaps he altered at the instigation of *James I, most later accounts of the reign of Elizabeth I are based upon this work.

The terms *antiquarian* and *historian* were less distinct during Camden's life than today. Though Camden's rival Ralph Brooke saw Camden's work as unconventional, most of his contemporaries acknowledged him as both historian and antiquarian. Before his death, Camden retired to Chislehurst, Kent, and, solidifying his role as an exceptional historian, founded a chair of history at the University of Oxford. Himself the successor of John Leland, who is credited with beginning the antiquarian movement in England, Camden influenced historical and antiquarian activity in the late seventeenth century. In 1838, the Camden Society, which later merged with the historical society (1897), published many of Camden's historical documents. In fact, this flourish of activity is sometimes referred to as the "Age of Camden."

Bibliography: T. Kendrick, *British Antiquity*, 1950; F. Levy, "The Making of Camden's Britannia," *Bibliotheque d'Humanisme et Renaissance* 26 (1964): 70–97; H. Trevor-Roper, *Queen Elizabeth's First Historian*, 1971.

Carol Jamison

CAMPANELLA, TOMMASO (1568–1639). Tommaso Campanella was born in Stilo, Calabria, in 1568. He entered the Dominican order in 1582; while studying theology in 1588 at the Studium in Cosenza, he became a follower of Telesio's natural philosophy after reading his *De rerum natura*, defending in print Telesio against a Jesuit critic. Seeking intellectual freedom, Campanella moved to Naples in December 1589, becoming tutor to the del Tufo family. Of his many writings produced there, the only to survive is the *Philosophia sensibus demonstrata*, published in 1591; in this work he presented his empirical approach to the reform of natural philosophy. In May 1592, after a Dominican court convicted him for his disobedience and support for Telesio, Campanella fled to Rome and later Florence, Bologna, and Padua, where he befriended among others *Galileo, *Sarpi, and Della Porta; his interests in politics and political philosophy originated during this journey.

Arrested by the Inquisition in January 1594, imprisoned and tortured, Campanella abjured publicly. While incarcerated, he wrote copiously; though many of his papers were lost, his *Compendium de rerum natura* composed during this period was eventually published in 1617. Rehabilitated, he was sent back to

Calabria where he organized a failed revolution against the Spanish rule. In the subsequent trial, Campanella saved his life by faking insanity but remained incarcerated for the next twenty-seven years in Naples. While in prison, Campanella wrote numerous works; most notably, *Città del Sole*, his famous political utopia, in 1602; in 1616, after hearing of the condemnation of the heliocentric theory, he wrote the *Apologia pro Galileo* defending the *libertas philosophandi* and a new relationship between religion and science.

In May 1626, through the action of *Pope Urban VIII, for whom Campanella provided political advice as an astrologer, he was moved to Rome and confined under house arrest for three more years. There, he wrote copiously and published his *Astrologicorum libri VIII*, among other works. Exasperated by his numerous enemies and the political life of Rome, in October 1634 Campanella escaped to France, where he interacted with *Mersenne, Peiresc and *Gassendi. A supporter of both *Richelieu and Louis XIII, Campanella was awarded a pension. He died in February 1639. Much of Campanella's voluminous writings on astrology, magic, natural philosophy, Neoplatonism, and political theory do not survive because of his tumultuous life and his radical ideas; however, Campanella was a prominent intellectual of his time, interested in reforming its intellectual, religious, and political institutions.

Bibliography: L. Firpo, "Campanella, Tommaso," *Dizionario biografico degli Italiani*, vol. 17 (1974): 372–401; J. Headley, *Tommaso Campanella and the Transformation of the World*, 1997.

Renzo Baldasso

CAMPION, THOMAS (1567–1620). Poet, author of masques, composer of music, theorist, and physician, multitalented Thomas Campion embodied the sentiment of the "Renaissance man." He is best known, however, for his outstanding ability to wed poetry with music, particularly airs for solo voice to the accompaniment of a lute. Campion was born in London on 12 February 1567. His father died when he was nine, leaving Campion, his mother, his sister Rose, and a stepsister, Mary Triggs. Campion's mother married Augustine Steward, her third husband, within a year of Campion's father's demise. In 1580, his mother also died, and Campion's stepfather quickly remarried. Thus, Campion found himself the foster child of Steward and Steward's new wife, Ann Sisley.

Campion's foster parents enrolled him in the University of Cambridge where Campion remained until 1584. In April 1586, he was admitted to Gray's Inn but was not called to the bar. However, he may have been involved in plays and masques, which were frequently performed nearby, and he likely developed other contacts that enhanced his budding reputation as poet and musician. He collaborated with Francis Davison, for example, in a masque performed before the queen on Christmas 1594.

Campion's first publication, five sets of anonymous verses, appeared in a 1591 edition of Sidney's *Astrophel and Stella*. Evidence suggests that in the same year Campion interrupted his studies to fight in the campaign of the Earl of Essex to help Henri IV of France against Spanish invaders in Britanny. The year 1595 marks the publication of Campion's *Poemata*, a collection of epigrams and elegies, which established his reputation as a Latin poet. In satiric fashion, he used Latin pseudonyms to mask the names of real people, including Barnabe Barnes, for whose poetry Campion had apparent contempt. From 1594 to 1602, Campion developed the art form for which he is best known: poetry accompanied by lute. During this time, he probably established a friendship with the composer Philip Rosseter. The two collaborated on *A Booke of Ayres* published in 1601. In 1602, Campion published his theories on rhyme in *Observations in the Art of English Poesie*.

In 1606, Campion wrote prefatory verses for Barnabe Barnes, whom he had earlier satirized. In this publication, Campion is first referred to as "Doctor in Physik." Where Campion received his medical degree remains a mystery, but it is conjectured that he studied medicine at the University of Caen, which may have attracted him because of its annual poetry contest, pageantry, and music. Campion began his career as a physician in London and published little in the immediate years following, though he did produce and publish in 1607 a masque for the marriage of Lord Hayes.

After the plague receded, Campion's poetic output increased. The endorsement of *William Camden helped establish his reputation and put him in demand by the court. In 1613, he published *Songs of Mourning* in memory of the death of Prince Henry and wrote and published a masque for Princess Elizabeth's wedding and another masque for the wedding of the Earl of Somerset to Frances Howard. Shortly thereafter, he published *Two Bookes of Ayres*. He is described on the title pages of these works as a physician, indicating that he did not relinquish the field of medicine.

Though linked to a scandal involving the death of Sir Thomas Overbury, Campion was later cleared. In 1617, he published *The Third and Fourth Booke of Ayres* dedicated to his patient and patron Sir Thomas Monson, and in 1618, published the words, but probably not the music, to *Ayres that were sung and played at Brougham Castle*. Also in the last years of his life, he wrote the theoretical treatise *A New Way of Making Fowre Parts in Counterpoint*. Resuming his earlier interest in Latin poems, he produced a volume of Latin epigrams and elegies in 1619 before dying on 20 March 1620.

In the centuries after Campion's death, lute songs lost popularity and remained in obscurity until the late nineteenth century. Campion's reputation reemerged with the Romantic Movement, though he was initially appreciated more for his poetic content than music. His reputation as a musical theorist, however, remained solid as evidenced by the repeated publication of *New Way of Making Fowre Parts in Counterpoint*. With Edmund Fellowes's publication in 1920 of Campion's songbooks with music, Campion's complete artistry, the ability to meld music with words, was recovered.

Bibliography: E. Lowberry, T. Salter, and A. Young, *Thomas Campion: Poet, Composer, Physician*, 1970.

Carol Jamison

CARAVAGGIO, MICHELANGELO MERISI DA (1571–1610). Born Michelangelo Merisi in autumn 1571, Caravaggio became famous under the name of the Lombardy farming community near Milan where he grew up and where his father served as steward to Francesco Sforza, Marchese di Caravaggio. On 6 April 1584, young Caravaggio was apprenticed to the Milanese painter Simone Peterzano, whose occasional realist depictions of peasants later influenced him. Between 1588 and 1592, Caravaggio moved to Rome and worked to establish himself while living in poverty and bohemian freedom. Around 1593, he worked on still lifes in the studio of Guiseppe Cesari (d'Arpino), Pope Clement VIII's favorite painter. Shortly afterward, his connections with an art dealer named Valentino attracted Cardinal del Monte's patronage. No doubt thanks to the cardinal's influence, in 1597 he earned a first public commission that proved decisive for his career: decorating the Contarelli Chapel in the Church of San Luigi dei Francesi.

By 1597, Caravaggio had already painted two-score canvases whose verisimilitude ignored contemporary academy standards. But with his two large lateral Contarelli paintings, the *Calling of St. Matthew* and the *Martyrdom of St. Matthew*, he not only moved from secular to religious subjects but also challenged tradition more sensationally than ever. He created two physically immediate, unidealized evocations of the saint in group scenes whose flat, unidentifiable backgrounds and extreme chiaroscuro intensified symbolic meaning as well as emotional impact. Baroque painting owes a considerable debt to his innovative tenebrism, that is, use of lighting contrasts to dramatize compositions. For the altarpiece, Caravaggio painted a bald, homely figure with bare feet apparently protruding from the frame; when the church rejected this naturalistic version, he substituted a more conventional *Inspiration of St. Matthew*. Meanwhile, however, he continued his subversive preference for depicting prosaic-looking saints in crowded, extremely lit, and foreshortened compositions with the *Crucifixion of St. Peter* and the *Calling of St. Paul*, painted around the same time for Rome's Cerasi Chapel.

Caravaggio's revolutionary artistic approach matched the growing lawlessness in his private life. Even while gaining fame as an important innovator, he began to express violent tendencies that led to repeated arrests and culminated in 1606 in his killing of Ranuccio Tomassoni over a tennis game in the Campo Marzo. To escape justice, Caravaggio fled to Naples, then to Malta in 1607–1608— painting all along. Despite receiving the Order of Malta, he soon needed to flee Maltese authorities too, and by 1609 he had returned to Naples, hoping for the pope's pardon. The amnesty came three days too late: In July 1610, Caravaggio, badly hurt in an ambush and sick (probably with pneumonia), died in Port'Ercole on board a ship bound for Rome.

Bibliography: H. Hibbard, *Caravaggio*, 1983; C. Puglisi, *Caravaggio*, 1996.

Margaret Goscilo

CAREW, RICHARD (1555–1620). English antiquarian and poet Richard Carew was born into a major landholding family of Cornwall. He became a "gentleman commoner" at Christ Church, Oxford, at age eleven, eventually joining the circle of *William Camden, Richard Hakluyt, and Philip Sidney. After inheriting his family estates, Carew filled a number of Cornwall offices, including justice of the peace, high sheriff, and member of Parliament. He also served as deputy-lieutenant under Sir Walter Ralegh, lord-lieutenant of the county.

Self-taught in French, German, Greek, Italian, and Spanish, Carew's first two publications (both 1594) were translations from Italian: five cantos from Tasso and a version of *The Examination of Men's Wit*, a comparison of male and female abilities by John Huarte, originally published in Spanish. Carew claimed English was equal in literary quality to Latin in "An Epistle concerning the excellencies of the English tongue," included in the second edition of Camden's *Remains* (1605). Carew's foremost work was his *Survey of Cornwall*, begun about 1589 and published in 1602. Rather than relying solely on genealogical accounts of great families, this antiquarian collection also incorporated a wealth of topographical information. Carew's *Survey* became a model for English county histories of the seventeenth century, its influence extending to such works as *John Aubrey's *Naturall Historie of Wiltshire* (1685). Carew, who died at his family home in November 1620, was the addressee of Henry Spelman's "Epistle on Tithes," and is considered on a par with *John Selden in *Ben Jonson's "Execration upon Vulcan."

Bibliography: M. Brayshay, ed., *Topographical Writers in South-West England*, 1996.

Michael R. Hutcheson

CAREW, THOMAS (?1594–1640). Thomas Carew was born in June 1594 or 1595. His father, Matthew Carew, was a lawyer in Chancery and was knighted by *James I in 1603. Carew entered Merton College in Oxford in 1608 and was intended to follow his father's profession of law. He was therefore admitted to the Middle Temple in August of 1612, but much to his father's distress, did not seem to devote very much of his time to legal pursuits. Having entered the service of Sir Dudley Carleton, he traveled to Venice, where Carleton was ambassador. He remained in Carleton's service when the latter changed his post to The Hague; however, he was dismissed for writing unfavorably about his employer and his wife, and was sent back to England with the suggestion that he might be happier employed elsewhere. In 1619, after a period of unsuccessfully seeking employment, Carew finally went to Paris with *Sir Edward Herbert, ambassador to France, and from this point on his fortunes looked up.

He was very successful in *Charles I's court, holding the titles of Gentleman of the Privy Chamber Extraordinary and Sewer in Ordinary to the King. Legend has it that he owed these positions and this favor to Henrietta Maria, whose reputation with her husband he had saved with some fast thinking, but there is no concrete evidence of this. He continued writing poetry during this time and was asked to write a masque, *Coelum Brittanicum*, in 1634, which was published later that year. In 1639, he accompanied Charles on his military expedition to Scotland. Carew died on 23 March 1640, probably of syphilis, which had plagued him on and off for years. It was said that his father was his worst enemy, and indeed Sir Matthew made some remarks to the effect that Carew had deserved his early death; they probably never reconciled. His collected poems were published upon his death and a second edition followed in 1642.

Carew was one of *Ben Jonson's friends and proteges—the "Sons of Ben"—and a friend of *John Donne. Two of his most famous poems are written about Donne and Jonson: *An Elegy on the Death of Dr. Donne, Dean of St. Paul's*, and *To Ben Jonson: Upon Occasion of His Ode of Defiance Annexed to His Play of The New Inn*. The latter reveals both Carew's admiration of Jonson and his criticism of Jonson's "itch of praise." *Coelum Brittanicum* was recognized even at the time for its mastery of the masque genre, balancing royal compliment, political philosophy, and poetry appropriate for setting to music and visual expression in *Inigo Jones's designs. However, it is for his erotic poetry that Carew was and still is most noted, especially *A Rapture*, a poem that earned him (along with his personal behavior) the reputation of a libertine in his own day and among succeeding generations of critics. In addition to its almost embarrassingly detailed portrait of sexual activity, the poem has political overtones as well in its references to enclosure laws and legal rights of husbands. These political and protofeminist overtones are consistent in Carew's erotic poetry, especially his insistence upon women's freedom and access to sexual pleasure, a protofeminism echoed in most of the writers of Henrietta Maria's literary circle.

Bibliography: L. Martz, *The Wit of Love: Donne, Carew, Crashaw, Marvell*, 1969; L. Sadler, *Thomas Carew*, 1979; E. Selig, *The Flourishing Wreath: A Study of Thomas Carew's Poetry*, 1958.

Melissa D. Aaron

CARISSIMI, GIACOMO (1605–1674). Most noted for his oratorios, motets, and cantatas, Carissimi was a major seventeenth-century Italian composer and a teacher of *Alessandro Scarlatti. As a boy, he sang in the Tivoli cathedral choir; after positions in Tivoli and Assisi, in 1629 Carissimi became maestro di cappella at the Collegium Germanicum in Rome. This Jesuit institution was a focal point of the Counter Reformation, and Carissimi's predecessors included *Tomás Luis de Victoria. Though offered other prestigious positions—for example, maestro di cappella at St. Mark's, Venice, after *Monteverdi's death—

for the rest of his life, Carissimi devoted himself to the musical training of the seminarians and the overall supervision of choristers and music at San Apollinare, the college's parish. Carissimi took holy orders and was ordained a priest in 1637. In 1656, he was honored by Queen Christina of Sweden with the title maestro di cappella del concerto di camera. Although Carissimi collaborated with the Congregation of the Oratory in music for the Oratorio del Ss Crocifisso, thus playing a part in the development of the oratorio as musical genre, there is no evidence about which specific works of Carissimi might have been performed there. Carissimi died on 12 January 1674 and was interred at San Apollinare.

Bibliography: G. Dixon, *Carissimi*, 1986; A. Jones, *The Motets of Carissimi*, 1982.

Susan Treacy

CARRIERA, ROSALBA (1675–1757). Carriera was a portraitist who rose from humble origins to become one of the most successful artists of the rococo. Few artists were as innovative or influential as Carriera. She introduced ivory as a support for miniatures and popularized pastels as a medium for portraits. The latter made her an instant celebrity in European courts, and she received constant artistocratic patronage.

Her career probably began as a lacemaker in her native Venice, and when the industry declined she turned to decorating ivory snuff boxes, which were tourist souvenirs. She also made miniatures, which were highly popular at the time. In 1720, she traveled to Paris where she was inundated with requests for pastel portraits. Her success was immediate: She was made a member of the Royal Academy, an honorary privilege rarely extended to women or foreigners. Her customers included the King of Poland (who owned the largest collection of her works) and the Holy Roman Emperor, Charles VI.

The appeal of her portraits at the time can be easily understood. She created flattering images of her sitters without losing their individuality or making them appear stilted. Moreover, the pastel technique is much faster than traditional oil paintings, which allowed shorter sessions for her sitters and promptly finished products. Speed was highly desired in portaitists at a time when such items often served the purpose that a photograph does today. The faces of the sitters in her pastel portraits are softly detailed, while their hair and clothing are more spontaneously rendered. Carriera's images of the French, German, Polish, and English gentry are intimate and charming, with a sense of spontaneity and grace. Her portrait of the boyish Louis XV shows a simultaneous mixture of aristocratic disdain and childlike innocence. The pastel painters Maurice Quentin de la Tour and Jean-Baptiste Perroneau are among the inheritors of her invention.

Bibliography: A. Harris and L. Nochlin, *Women Artists: 1550–1950*, 1976; A. Slatkin, *Women Artists in History: From Antiquity to the Present*, 2001.

Rosi Prieto Gilday

CASAUBON, ISAAC (1559–1614). A devout but undogmatic Protestant of an irenical turn of mind reputed to be the most accomplished western European Hellenist of his era, Casaubon applied his erudition effectively to the confessional disputes of the early seventeenth century. The son of Reformed refugees from France, Casaubon was born in Geneva, and his youth was much disrupted by the contemporaneous French civil wars. He returned to Geneva in 1578 to begin his studies at the city's Académie. In 1581, Franciscus Portus, the Académie's Greek professor, died after having named Casaubon as his most suitable successor. Casaubon's fame as a Hellenist grew because of his editions and commentaries as well as his work at the press of Henri Estienne II, who became his father-in-law in 1586. Casaubon published an edition of Strabo and annotations on the New Testament in 1587, a Greek and Latin Aristotle in 1590, and in 1592 an edition of Theophrastus's *Moral Characters*, which stimulated interest in this author. He left Geneva to take a post at Montpellier in 1596, but was soon called to Paris, where he was much admired by *Jacques-Auguste de Thou and his learned friends. Perhaps because he feared being coerced into conversion, he lingered in Lyon while seeing his commentaries on Athenæus through a press there, but was summoned in 1600 to depart for the capital by King Henry IV because the latter wanted him to participate in a religious disputation. The subject of discussion was a treatise on the Eucharist of the Protestant Philippe de Mornay, seigneur du Plessis-Marly ("Duplessis-Mornay"); Casaubon won the enduring suspicion of many French Protestants by siding with Mornay's Catholic opponents regarding the interpretation of the citations that were the focus of the debate.

Casaubon's major publication during the decade he resided in Paris was his 1609 edition of *Polybius*. On the 1610 assassination of Henry IV, Casaubon fled to England, fearful of what the future held in store for him in the French capital without this powerful patron. The most significant fruit of Casaubon's labors in England was his attack on Cesare Baronio's ecclesiastical history; he published the first installment of a refutation of the cardinal in 1614, but died shortly thereafter, leaving the project unfinished. The lengthy tome that Casaubon produced was nonetheless an important intervention in contemporaneous learned controversies, calling into question the authenticity of the sibylline prophecies and the reflections of "Hermes Trismegistus" by demonstrating that they contained evidence of having been composed long after the era commonly attributed to them. He again disappointed many Protestant scholars, however, by noting that the attitude of Reformed divines toward Christian antiquity was in many respects as flawed by their confessional presuppositions as the work of their Catholic counterparts.

Bibliography: A. Grafton, *Defenders of the Text*, 1991; F. Laplanche, "A propos d'Isaac Casaubon," *History of European Ideas* 9 (1988): 405–422; M. Pattison, *Isaac Casaubon 1559–1614*, 2nd ed., 1892.

Matthew Koch

CASSINI, GIOVANNI DOMENICO (1625–1712). Cassini, the patriarch of three generations of astronomers, was born on 8 June 1625 near Nice. He received a Jesuit education in Genoa and became professor of astronomy at the University of Bologna in 1650. After making a reputation for himself in Italy as a civil engineer, in 1667 he was appointed head of the Paris astronomical observatory, taking French citizenship in 1673. Cassini discovered four moons of Saturn between 1671 and 1684, and he observed the separation between the rings of Saturn, which is now known as the "Cassini Division." He accurately posited that the rings were actually bands of even smaller satellites, and he was among the first to observe Mars's polar caps and compare them with Earth's icecaps. He worked with the French astronomer Jean Richer to determine the distance between Mars and Earth and also estimated the distance from the Earth to the sun, arriving at a figure of approximately 87,000,000 miles, a difference of just over six percent from the actual mean distance of 92,827,000 miles. This conclusion was published in his posthumous *Divers Ouvrages d'Astronomie* (1731). Cassini later disputed the shape of the earth with *Newton, asserting that its polar radius was greater than its equatorial radius. However, French expeditions to Scandinavia and Peru to measure the meridians showed that Newton's conception of the Earth as an oblate, or flattened, spheroid was correct. Cassini's geographical and cartographical measurements were heavily supported by *Louis XIV, who was vitally interested in the military uses of highly accurate maps. Cassini's son, Jacques, drawing in part upon his father's work, was able in 1789 to finally prepare one of the finest topographical maps of France, measuring thirty-three by thirty-four feet.

Bibliography: L. Jardine, *Ingenious Pursuits: Building the Scientific Revolution*, 1999; A. Wolf, *A History of Science, Technology and Philosophy in the 16th and 17th Centuries*, 1935, rpt. 1968.

Christopher Baker

CAVALIERI, BONAVENTURA (1598–1647). Bonaventura Cavalieri was born ca.1598 in Milan to a noble but not affluent family. He entered the Jesuit order after childhood and received a thorough classical education in Milan. In 1617, Cardinal Federico Borromeo, having noted Cavalieri's extraordinary intelligence, sent him to Pisa and introduced him to *Galileo, of whom he become a lifelong friend and patron. Cavalieri mastered ancient geometry under the supervision of Benedetto Castelli, Galileo's pupil and a professor of mathematics in Pisa. In 1619, he pursued unsuccessfully the then-vacant chair of mathematics at the university of Bologna. During the 1620s, while continuing his mathematical investigations, publishing several treatises, and competing for an academic position (at Pisa, Rome, and Parma), Cavalieri held various Jesuit administrative positions in Lodi and Parma. In 1629, he had completed his most important work, *Geometria indivisibilibus continuorum* (1634), and, finally, he obtained an interim appointment to the mathematics lectureship at Bologna. In the subsequent years, he published extensively, both in Italian and Latin, on topics ranging from logarithmic-trigonometric tables to optics and astrology. Among his most important titles are *Specchio ustorio* (1632), *Directorium generale uranometricum* (1632), *Compendio delle regole dei triangoli* (1638), *Centuria id varii problemi* (1638), *Nuova pratica astrologica* (1639), and *Trigonometria plana, er sphaerica, linearis et logarithmica* (1643). Through this incredible intellectual production, he not only secured his academic position until his death (1647), but also gained international fame.

His major contribution to early modern science is his method of indivisibles, the earliest method of integration upon which *Newton and *Leibniz developed calculus. The method of indivisibles calculates areas (or volumes) of irregular figures by comparison to a known one, comparing the respective indivisibles that compose the figures—lines for surfaces, planes for solids. Thus, by analyzing planes and solids though their basic geometrical constituents, he furnished the basis of infinitesimal calculus. He and his followers refined and applied this method to problems of quadratures and center of gravity.

Bibliography: A. De Ferrari, "Cavalieri, Bonaventura," *Diozionario Biografico degli Italiani*, vol. 22 (1979), 654–659; E. Giusti, *Bonaventura Cavalieri and the Theory of Indivisibles*, 1980.

Renzo Baldasso

CAVENDISH, MARGARET (1623–1673). Due largely to her negative portrayal in the diary of *Samuel Pepys, for the first three centuries following her death the prevailing image of Margaret Lucas Cavendish, the first Duchess of Newcastle, was that of a notable eccentric whose flamboyant dress drew stares, whose fanciful writings drew accusations of mental imbalance, and whose ventures into scientific speculation drew charges of folly and impudence. She was also known as the second wife of the gallant Marquis of Newcastle, a model Cavalier aristocrat who commanded the royal forces in the north during the English Civil War, and her chief claim to fame was the admiring biography of her husband that she published in 1667. Today, in the light of feminist reevaluation of women writers, the image of Margaret Cavendish is much more complex. She is seen as a pioneering thinker and writer remarkable for the range of her interests and for the boldness she showed in entering traditionally male areas of endeavor.

Margaret Lucas was born in 1623 in Colchester, Essex, into an affluent family of Royalist allegiance. Her father died when she was two years old, but Margaret later recalled that her childhood had been idyllic. Like most women at the time, she received little education; as a result, she admitted later in life that she never mastered spelling and grammar. Nevertheless, Margaret showed an early interest in writing, producing an abundance of "baby books." In 1642, after the Lucas home was attacked by a Roundhead mob, the family removed to the Royalist haven of Oxford, where Margaret the following year became maid of honor to Queen Henrietta Maria. When the queen fled to her native France following the defeat of the king's forces in 1644, Margaret accompanied her. In Paris she met and married the widower Newcastle, thirty years her senior and, she found, "a pattern for all gentlemen." They resided in France and Holland for fifteen years, mostly on credit. Newcastle was a dabbler in various sciences, while his brother, with whom Margaret formed a deep friendship, was a more serious scholar well versed in mathematics, and the household hosted a number of leading philosophers, including *Descartes and *Hobbes. In this atmosphere, Margaret's interest in natural philosophy was piqued. In 1651, she began an eighteen-month visit to London to recover lost family property. It was during this visit that she also completed her first two books, *Poems, and Fancies* and *Philosophical Fancies*, both of which were published in London in 1653. The latter is an exposition in both prose and verse of Cavendish's own version of an atomistic theory of nature. The former, a varied collection that would be revised and reprinted in 1664 and 1668, represents her major work of poetry. Some of its poems examine such scientific issues as the circulation of the blood and the multiplicity of worlds. Others feature very fancifully developed analogies, such as "Man's Head Similized to the Globe of the World." Perhaps the most interesting pieces for modern readers are two narrative poems on hunting, in which she develops a powerful sympathy for the doomed hare and stag, provoking thoughts uncharacteristic of the period concerning humanity's relation to nature.

On her return to the continent, Margaret continued writing, taking up both new projects and revising and enlarging old. Noteworthy productions of this period include a number of plays: *The World's Olio* (1655), one of the most random set of observations ever published; *Philosophical and Physical Opinions* (1655), in which she rejects her earlier atomism; a collection of true and "feigned" stories entitled *Nature's Pictures Drawn by Fancies Pencil to the Life* (1656), which includes her masterpiece of short fiction, "The Contract"; a collection of orations; and a collection of "sociable letters," which reveal her comic sense and her keen eye for social foibles.

With the restoration of the Stuart monarchy in 1660, Margaret Cavendish returned to England and settled into a comfortable life at her husband's country estate, where she composed the biography of her husband and undertook the systematic study of contemporary works of natural philosophy. *Philosophical Letters* (1664) contains the fruits of this study, including arguments against positions by Hobbes and Descartes and an impassioned call, in response to both writers, to appreciate the "intelligence" of animals. Responding to *Robert Hooke's celebrated *Micrographia*, in *Observations upon Experimental Philosophy* (1666), Cavendish announced her preference for rational speculation over "deluding glasses and experiments" and defended her belief that higher matter was rational, sensible, and self-moving. Attached as an appendix to *Observations* is *The Description of a New Blazing World*, the work for which she is best known to modern readers. An extravagant blend of utopia, science fiction, romance, and philosophical speculation, *Blazing World* recounts an abducted young girl's transportation to a world where she becomes an all-powerful empress directing her subjects' investigations of the mysteries of the universe. Margaret's last work was *Grounds of Natural Philosophy* (1668). In the years remaining until her death in 1673, she devoted herself to revising her works.

Bibliography: M. Cavendish, *The Blazing World and Other Writings*, 1992; D. Grant, *Margaret the First*, 1957.

Patrick Cook

CERVANTES, MIGUEL DE (1547–1616). Miguel de Cervantes Saavedra was born in Alcalá de Henares, Spain, although he never attended the university there. Much of his childhood education was derived from travels with his surgeon father. However, he did study for some time in Madrid with the humanist educator Juan López de Hoyos. Through Cervantes' writings, it is known that he was very familiar with Renaissance doctrines and the principal classical and modern authors of Spain and Italy. From 1559 to 1570, Cervantes traveled to Italy in the service of Giulio de Acquavivas. On 7 October 1571, he fought at the naval battle of Lepanto, where he was wounded in the chest and his left hand was maimed. In 1573, he participated in the expedition to Tunis. Upon returning to Spain in 1575, he was captured by Algerian pirates and taken as a slave to Algiers. Finally, he was ransomed by Trinitarian priests in 1580. He married Catalina Palacios Salazar, but their union produced neither issue nor happiness. In 1587, Cervantes was appointed as a purchasing officer in charge of collecting provisions for the Spanish Armada. After the defeat of the Armada, Cervantes lived obscurely in Seville where he tried in vain for an appointment in Spanish America. By 1597, Cervantes was imprisoned in Seville because he had deposited money from the Royal Treasury with a banker who declared bankruptcy. Cervantes was again imprisoned in Seville for debt in 1602. He moved to Valladolid in 1604. With the success of the *Quixote*, Cervantes moved to Madrid in 1606. While the first part of his life was marked by heroic deeds, the second part was full of economic hardships and intense literary creation. Cervantes died in Madrid on 23 April 1616.

The vocation of a poet accompanied Cervantes throughout his life. He wrote many types of poetry, perhaps his best being in his plays and novels where he reveals his artistic inclination toward beautiful, idealized, pastoral scenes and other Renaissance-style popular themes. The long, allegorical poem *Journey to Parnassus* (1614) is of historic interest for his critique of contemporary poets as well as the defense of his own literary creations. As a dramatist, he figures as one of the principal playwrights before Lope de Vega imposed a new national style of action and intrigue. Cervantes' drama emphasizes conflicts of passion and character portrayal. From his earlier period only a few plays remain, one of which is *The Siege of Numancia*. The dramas from his later period in *Eight Dramas and Eight Interludes* (1615) were not staged in his lifetime. Cervantes' theatrical technique is more successful in his comic interludes because he skillfully employs movement and psychological observation with satiric intentions. In the *Interludes (Entremeses)*, he shows an ironic sensibility for human weakness and a keen understanding of motives of human conduct.

Most of all, Cervantes is remembered as a novelist. Cervantes created two types of novels that were copied in the next century: the romantic adventure novel and the novel of customs. His first novel was *La Galatea* (1587), a pastoral romance. Cervantes also began the short novel in Spain that was already so popular in Italy. In 1613, *Exemplary Tales* was published, consisting of twelve short novels that reflect his moral and poetic interpretation of life. *Don Quixote* (Part I, 1605 and Part II, 1615) is said to be the most accomplished novel of the period. The novel testifies to the aspirations and failures of the Spain of its time. In it, Cervantes manages to portray human nature in all its facets, from its greatness to its pettiness in a variety of surroundings. At the beginning of the novel, the defining character traits of Don Quixote and Sancho Panza are idealism and pragmatism, respectively. However, the traits are intertwined by the end of the novel. Don Quixote's generous and noble intentions are continually dashed against the harsh realities of Spain. During Cervantes' lifetime, the *Quixote* reached sixteen editions; its first part (1605) was so successful that it was imitated by the pseudonymous Avellaneda with a "false" sequel in 1614. Cervantes wrote his second part in 1615 as a critical response to Avellaneda's sequel.

The novel *Persiles and Segismundo* (1617) was published posthumously. It is a byzantine novel of fantasy and travel to exotic countries in which two lovers of high birth are finally united. As a true Renaissance figure, Cervantes cultivated the genres of his time: dramatic plays, comic interludes, novels, short novels, and poetry. However, his literary genus was never adequately compensated monetarily nor recognized by his literary peers. In spite of all his personal and professional adversities, he never became bitter or pessimistic. A spirit of understanding, compassion, and optimism triumphed in his literary works.

Bibliography: A. Castro, *El pensamiento de Cervantes*, 1972; M. Riquer, *Aproximación al Quijote*, 1967.

M. Ellen Blossman

CHAMPLAIN, SAMUEL DE (1567?–1635). Champlain was both an explorer and administrator; his family came from Brouage, France, a noted harbor city. As an explorer, his first voyage to the New World, from 1599 to 1601, was to Mexico and the West Indies in the service of the Spanish king. Champlain's journal of that voyage, with his maps and drawings, was not published until the Hakluyt Society did so, in English, in the nineteenth century. Remarkably, in this early journal, Champlain suggests constructing a canal across the Isthmus of Panama.

In 1603, Champlain was once again in the New World. In the service of France, he explored the St. Lawrence River, and his reports of this journey earned him royal patronage. In 1604, Henry IV sent Champlain, along with the Huguenot Pierre de Gua, sieur de Monts, to New France with a patent for the fur trade between the fortieth and forty-sixth degrees of latitude, with a company of Catholics and Huguenots, accompanied by both a priest and a minister. They established the St. Croix colony in New Brunswick, and Champlain explored Port Royal on the west coast of Nova Scotia in 1606. Remaining in Canada until 1607, Champlain ranged from Canso, Nova Scotia, to present-day Woods Hole, Massachusetts. When, in 1607, de Monts' patent was canceled, Champlain returned to France. By summer 1608, Champlain was back in Canada and exploring the Saguenay and St. Lawrence Rivers, for de Monts could regain and retain his patent if he founded a colony in the interior. Serving as de Monts' lieutenant, Champlain established the colony of Quebec in 1608. With the death of Henry IV in 1608, de Monts, ruined financially, sold his proprietary rights in Canada to the Jesuits. Champlain remained in New France; in 1609, he discovered the lake later named for him, lying at the intersection of New York, Vermont, and Quebec. In 1612, he returned to France and was appointed governor of Quebec. From 1629 until 1632, Quebec was in English hands and Champlain himself was exiled to England, where he worked on the third edition of his *Voyages de la Nouvelle France* (1632); with the return of Canada to France, Champlain once again became governor. He died in Quebec on Christmas Day 1635 and was buried there.

An explorer himself, Champlain fostered the efforts of others. In 1634, he sent *Nicolet up the Great Lakes in search of a way to the west. Although Nicolet did not find what he was looking for, he did map his route and so provided guidance for the later voyage of *Joliet and *Marquette. As entrepreneur, Champlain was part of the Company of New France, a merchant group he had helped convince *Richelieu to establish, that financed the voyages to the New World. The company was determined to keep its interest in New France alive, funneling funds away from the losses incurred by English depredation and toward the maintenance of the colony at Quebec. Members of the large general group of 100 shareholders put up 3,000 livres each, while an inner circle contributed an additional 800 livres each; to this group went monopoly of the peltries trade. However, that same group was held responsible for the safety and well-being of the colony, and those expenses were the first attended to. The Jesuits noted Champlain's piety and knowledge. He built chapels, caused the Angelus to be rung at appropriate hours, and set aside time for daily prayer and meditation for the members of his household. History constituted the morning's household reading, with saints' lives read in the evening.

In his early explorations, Champlain found the Algonquin in control of the St. Lawrence Valley, whereas Jacques Cartier, in 1535, had found the Iroquois there. In 1609, Champlain attacked and defeated the Iroquois who had begun to move back toward the St. Lawrence. Champlain was victorious, but the Iroquois developed a lasting hatred for the French and so closed Lake Erie to French trade and exploration. Although Champlain fought the Iroquois, he respected his Indian allies. At feasts, he offered them sagamite, their food. Although he used an interpreter, Champlain went to parlay and to council. He kept his word, and the Indians knew it. In 1632, with three English ships at Tadoussac and an English bark farther upriver, eighteen Indian canoes approached Quebec. Champlain's effort to keep the Indians allied with the French succeeded because one of the chiefs reminded his group that the French, under Champlain, had improved Indian life, that the French leader had always spoken the truth, and that the French, but not the English, would teach the tribes to write and, in so doing, would keep the Indian stories and ways alive for the tribes' children.

Bibliography: C. Heidenreich, *Explorations and Mapping of Samuel de Champlain, 1603–1632*, 1976; S. Morrison, *Samuel de Champlain, Father of New France*, 1972.

Martha Oberle

CHAPMAN, GEORGE (1559?–1634). Poet, playwright, and translator, George Chapman was born in Hitchin, Hertfordshire, although the date of birth is uncertain. *Anthony à Wood gives a birth year of 1557, but the portrait of Chapman in his 1616 edition of Homer's works bears the inscription "Georgius Chapmannus . . . Aeta: LVII.MDCXVI." If the inscription is accurate, he was born in 1559. Wood's interest in Chapman suggests that the latter was educated at Oxford, although he did not take a degree there. Some authorities surmise that Chapman studied Latin and Greek for several years at both Oxford and Cambridge.

Chapman was about thirty-five when he appeared on the London literary scene with the publication in 1594 of *The Shadow of Night*, a nondramatic poem that includes references, possibly autobiographical, to a military campaign in the Netherlands with Sir Francis Vere. The poem has also been linked to an allusion to "the school of night" in *Love's Labor's Lost* (c.1594), in which *Shakespeare may be satirizing *Ralegh's infamous society of skeptics and free thinkers, active in the early 1590s, that included Matthew Roydon (to whom *The Shadow of Night* is dedicated), *Thomas Harriot, Henry Percy, Christopher Marlowe, and Chapman himself. A similar speculation connects Chapman to the "rival poet" of Shakespeare's sonnets—the "he" of Sonnet 86, for instance, whose "compeers by night" do not frighten the speaker of the sonnet. In 1595, Chapman produced a collection of four popular poems based loosely on Ovid, including *Ovid's Banquet of Sense* and a pastoral called *The Amorous Contention of Phillis and Flora*. The poet claimed *Phillis and Flora* was a translation of a Latin original written by a friar in 1400, but the true origin remains conjectural. In 1598, Chapman completed Marlowe's unfinished *Hero and Leander*—possibly at Marlowe's behest—and published the continuation with the second edition of Marlowe's original fragment.

Chapman probably began writing for the stage in about 1595. The earliest reference to his plays is found in Philip Henslowe's records for the Rose Theater, where the opening of the comedy *The Blind Beggar of Alexandria* is noted for 12 February 1596. The play suited the popular taste and was quite profitable for Henslowe. Further evidence from Henslowe's records suggests Chapman wrote over the next two years several successful plays no longer extant; in 1598 Francis Meres in his *Wit's Treasury* mentions Chapman as one of the best contemporary writers of comedies and tragedies. Six other comedies written after 1598 are extant, most notably *An Humorous Day's Mirth* (1599) and *All Fools* (1605). Chapman also collaborated with *Jonson and *Marston on the comedy *Eastward Hoe* in 1605, and the playwrights were briefly imprisoned for alleged anti-Scottish satire in the play. Chapman's reputation for tragedy resides mainly in the great popular success of *Bussy D'Ambois* (1607), although *Dryden criticized it in 1681 as mostly empty bombast and "false poetry." At least four other Chapman tragedies played to large and enthusiastic audiences. *The Conspiracy of Charles, Duke of Byron* especially aroused great interest in 1606 when the play offended the French ambassador, and the authorities forbade further performances. When the players defied the order, three members of the company were arrested, but Chapman escaped. The objectionable passages had been removed by the time the play was printed—together with a second part, *The Tragedy of Byron*—in 1608. Chapman's hyperbolic dramatic style was famously admired by his contemporary *John Webster and later by Charles Lamb.

Today Chapman is best known for his translations of Homer, an achievement universally publicized in Keats's majestic sonnet *On First Looking into Chapman's Homer* (1816). The translations of the *Iliad* and *Odyssey* were an eighteen-year undertaking, beginning with the initial installment, *Seven Books of the Iliades of Homer, Prince of Poets*, in 1598. Two other partial versions of the *Iliad* appeared while Chapman experimented with format and meter before the publication in 1611 of the complete *Iliad* in rhymed, fourteen-syllable lines. The *Odyssey*—in rhymed, ten-syllable lines—followed in 1614. Finally, the two translations, in twenty-four books each, were united in one folio volume and published as *The Whole Works of Homer, Prince of Poets* in 1616. Chapman's translation is far from perfect, and his knowledge of Greek has been disparaged. Still, the work remains one of the great Elizabethan achievements, and Pope would acknowledge his debt to his predecessor's accomplishment. Coleridge, praising the poetry of the translation from a romantic viewpoint, described Chapman's Homer as "truly as original as the *Faerie Queene*."

*Wood, in his *Athenae Oxonienses* (1692), characterizes Chapman as "a person of most reverend aspect, religious and temperate, qualities rarely meeting in a poet"—a description seriously at odds with his usual reputation as a flamboyant atheist. Whatever the truth, Chapman was given a Christian burial in the churchyard of St. Giles-in-the-Fields Parish on 14 May 1634. *Inigo Jones designed the monument erected in memory of the poet.

Bibliography: A. Nicoll, ed., *Chapman's Homer*, 2 vols., 1956; A. Nicoll, ed., *Chapman's Homer: The Iliad* rpt. with new preface by Gary Wills, 1998; T. Parrott, ed., *The Plays of George Chapman*, 1961; C. Spivack, *George Chapman*, 1967.

Dale G. Priest

CHARLES I (1600–1649). King of England from 1625 until 1649, Charles I was born on 19 November 1600, the son of *James I and Anne of Denmark. His father's plan to marry him to the Infanta Maria was an effort to gain Spanish assistance in ensuring that Charles's brother-in-law Frederick V would not lose his title to the Palatinate as a result of a defeat in the Thirty Years' War. Much to the delight of the English populace, who feared closer ties to a Catholic power, the marriage plan failed. But it had succeeded in displaying the fatal indifference of Charles and his father to the pulse of popular sentiment. He later married Henrietta Maria, daughter of French King Henry IV.

Charles involved England in a variety of continental struggles (such as his support of the French Huguenots against *Richelieu), which did not reflect a tightly coordinated foreign policy, but which did exact a toll on the country's treasury. His increasingly stringent demands, including repeated calls for funds (as with his disliked policy of ship-money, taxes levied to pay for warships), led Parliament to issue the Petition of Right (1628), which demanded that he not impose taxes without their consent, compel forced loans, impose martial law in peacetime, imprison subjects without trial, or lodge troops in civilian homes. He signed the petition more out of political expediency than conviction.

The king's seeming favoritism toward Catholics (not to mention his marriage to a Catholic) continued to anger English Puritans, already irritated by his father's religious policies. His high church Anglican practices, vigorously pursued by his Archbishop of Canterbury, *William Laud, together with his leaning toward the theology of *Arminius, drove ardent Calvinists into close alliance with the Parliamentary constitutionalists; together they would become the nucleus of his opposition in the civil war. His firmly absolutist adherence to a divine-right theory of monarchy led him to govern without Parliament for eleven years (1629–1640).

Open resistance to his rule surfaced first in Scotland, leading to his comment that might sum up his whole theory of government: "I mean to be obeyed." When the Scots finally revolted in 1639, having pledged to defend Presbyterianism, Charles convened the Short Parliament in 1640 to seek money to quash the revolt. But Parliament demanded concessions and Charles dismissed them in three weeks, ordering clergy to periodically make public assertions of his divine right. Unable to contend alone against the powerful Scots, he reconvened the Long Parliament on 3 November 1640 (it was not formally recessed until 1653), which compelled him to grant fuller liberties to it during the following year. A year later, in 1642, a revolt of Irish Catholics led Parliament to tell Charles he could cope with these rebels only through its approved ministers and he had to seek its approval before levying further taxes. It put forth a Grand Remonstrance in 1641, summing up its political and religious grievances against the king. By January, 1642, Charles had had enough of Parliamentary obstinacy. With several hundred troops, he entered Parliament to arrest five of its members, but they had escaped. The king then left London, both sides began to raise armies, and that summer civil war broke out.

Charles's supporters, the Cavaliers, were Anglican nobility of the north and west of England. His opponents, the Roundheads, were Puritans of London and the southeast organized by *Oliver Cromwell into the New Model Army. This army defeated the king's troops at Marsten Moor (1644) and Naseby (1645), and Charles was captured by the Scots in 1646. Extremists gained control of the Parliament and purged ninety-six of its more moderate members. The resulting "Rump Parliament" voted to dismantle the monarchy, the Church of England, and the House of Lords, and to rid themselves of Charles himself, who was beheaded in January 1649.

Personally devoted to his family, noble and saintly at his execution, Charles gained a reputation as an Anglican martyr, an image abetted by the posthumous (but ghostwritten) account of his difficulties in *Eikon Basilike: The True Portraiture of His Sacred Majesty in his Solitudes and Sufferings*. He was not responsible for the growing power of the Calvinist merchant class in Parliament who despised his Anglican policies and royalist claims, but his failure to recognize that his absolutist convictions had become anachronistic accelerated England's plunge into civil war.

Bibliography: C. Carlton, *Charles I, the Personal Monarch*, 2nd. ed., 1995; P. Gregg, *King Charles I*, 1981.

Christopher Baker

CHARLES II (1630–1685). Charles II, eldest son of *Charles I, became king of England in 1666 and reigned for twenty-five years. He largely reversed the absolutism practiced by his father by working more closely with Parliament, which had gained a stronger sense of its legislative power during the Puritan Protectorate. Two years after his father's execution, he was crowned king of Scotland and later fought against *Cromwell at Worcester, where he was defeated. He fled first to France and then to Germany and Spain. After Cromwell's death, the Protectorate dissolved, and Parliament accepted Charles as king in 1660.

Working with his legislature, Charles rescinded laws passed under Cromwell. His new Parliament, decidedly more Anglican in makeup, took steps to bring greater consistency to English religious practice without being cruelly repressive. Many of those implicated in his father's death were exonerated; the Corporation Act (1661) required all judges to be members of the Church of England; the

Act of Uniformity (1662) required it for clergy and teachers, and the Test Act (1673), for officeholders. In practice, however, non-Anglicans (called "Dissenters" or "Nonconformists") could get by with merely an occasional participation in an Anglican communion. But anti-Catholic feeling was as strong as opposition to strict Calvinism. This posed a problem, for Charles had leanings toward the Roman faith, and his brother James, later *James II, was an ardent Catholic. Charles in fact had made a secret pact with *Louis XIV of France to support Louis's wars and to become a Catholic eventually in exchange for annual subsidies from Louis; Charles later converted on his deathbed. Many saw Catholic plots for a government takeover everywhere, especially Titus Oates, who condemned an alleged such plot in 1678. This "Popish Plot" led to the execution of thirty-five people named by Oates as conspirators, though the king knew them to be innocent. In 1681, Parliament debated an Exclusion Bill, which would have barred Catholic James from the succession. Its advocates acquired the name "Whigs" (a Scottish variation of "wigs"); its opponents were termed "Tories" (Gaelic for "robbers"). The Whigs were defeated, and Charles died shortly thereafter on 2 February 1685.

Charles had put down a Scottish revolt in 1666 and fought an expensive and finally indecisive war with the Dutch from 1665 to 1667 (his navy was ably reorganized by his Naval Secretary, *Samuel Pepys), but he did not seek a major expansion of English interests on the continent. By nature amiable and pleasure-loving (one of his many mistresses was the actress *Nell Gwyn), his court gained notoriety for its comparative moral laxity. He loved the theater, which had re-opened after the fall of the Protectorate, and his known patronage of the stage benefited such playwrights as *Dryden and *Wycherely. Charles's reign ended with a more balanced recognition of authority between king and Parliament and the return of the Church of England to the center of the nation's religious life, although Anglicans now had to accept the lively growth of a number of vigorous dissenting sects.

Bibliography: G. Davis, *The Restoration of Charles II, 1658–1689*, 1955; R. Hutton, *Charles the Second, King of England, Scotland and Ireland*, 1989.

Christopher Baker

CHARPENTIER, MARC-ANTOINE (?1645-50–1704).

Little is known about the childhood of French composer Charpentier, a younger contemporary of *Lully, but possibly between 1662 and 1667, Charpentier studied in Rome with *Carissimi. Sébastien de Brossard, an early biographer, credits him with bringing back to France copies of Carissimi's motets and oratorios; certainly Charpentier's initial compositions reflect exposure to mid-century Italian style. Although Charpentier never held an official post under *Louis XIV, he had an illustrious Parisian career. In 1672, after Lully's hoarding of royal privileges and patents had ended his musical partnership with *Molière, the playwright invited Charpentier to compose for his troupe, a collaboration that continued until about 1686. Also, until her death in 1688, he composed sacred and secular works for the Duchesse de Guise, among the wealthiest of musical patrons. The 1680s were a fruitful decade for Charpentier, with employment as music director for the grand Dauphin, for whom he composed both sacred and secular music, including the court entertainment, *Les plaisirs de Versailles*. In 1683, four new sous-maître positions at the chappelle royale were created; Charpentier auditioned but disqualified himself because of illness. Because of his employment with the Dauphin, however, the king gratefully pensioned him. As music master of the Jesuit church, Charpentier composed much liturgical music and several sacred operas, such as *David et Jonathas* (1688), sung between the acts of dramas produced at Parisian Jesuit colleges. For the Opéra, Charpentier composed only one tragédie lyrique, *Médée* (libretto by Thomas Corneille), which premiered in 1693. In 1698, Charpentier attained the most coveted church music position outside of the royal chapel, that of maître de musique at Sainte-Chapelle. Here he produced much imposing sacred music until his death on 24 February 1704. Few works of Charpentier were published during his lifetime, but most of his autograph manuscripts, now at the Bibliothèque Nationale, survive.

Bibliography: C. Cessac, tr. E. Glasow, *Marc-Antoine Charpentier*, 1995; H. Hitchcock, *Marc-Antoine Charpentier*, 1990.

Susan Treacy

CHILLINGWORTH, WILLIAM (1602–1644). Protestant theologian William Chillingworth was born in Oxford in 1602. His father was such an influential citizen that *William Laud, then a fellow of St. John's, Oxford, served as godfather. A precocious intellect, Chillingworth took his B.A. at age eighteen and was made a fellow of his college at age twenty-six. It was this intellect that made him a favorite of Sir Lucius Cary, second Viscount Falkland, who maintained a college of sorts at his country estate of Great Tew, not far from Oxford, where Chillingworth composed his only major work, *The Religion of Protestants a Safe Way of Salvation* (1638). The book was notable as the work of a keen dialectical controversialist, one who had renounced Protestantism in favor of Catholicism a few years earlier and had studied at Douay. Understood to be from a reclaimed Anglican with insider knowledge of Jesuit beliefs, yet a godson of the Archbishop of Canterbury, and having benefited from the revising influences of divinity professors and even the Vice-Chancellor of Oxford, the book inflamed most people at the time, even if it came to be, after the Restoration and into the eighteenth century, increasingly popular.

     The reason for the discrepancy in early and later responses to *The Religion of Protestants* is, in part, its brand of shorn reasonableness, something that lends itself to empirical epistemology more than to late medieval or Reformation theology. *The Religion of Protestants* takes the form of a response, which best suits the dialectical habit of its author, and as a response it redefines the terms and conditions of the debate. Blending scholastic attitudes about the subordinate though useful relation of reason to revelation with classical ideals of reason and modern desires for materiality, Chillingworth's book argues from the standpoint of "right reason, grounded on divine revelation and common notions, written by God in the hearts of all men." It was this that enraged Catholics and Puritans alike, both finding too much self-reliance and too little deference to God in the method. Edward Knott, to whom Chillingworth is replying, responded vehemently from the Catholic standpoint with *Christianity Maintained: or, a Discovery of Sundry Doctrines Tending to the Overthrow of the Christian Religion* (1638), and Francis Cheynell represented the vehement Puritan response in his elaborated obituary of Chillingworth designed to denounce his person and thereby his method, *Chillingworthi Novissima; or the Sickness, Heresy, Death, and Burial of William Chillingworth* (1644). By contrast to these early reactions, later ones, such as that of Thomas Birch in his 1742 edition and introductory biography, valued the aloof toleration and insistent rationality of the ethos implied by *The Religion of Protestants*. Kings and scientists, early and late, valued Chillingworth; devoted Catholics and Puritans, early and late, despised him.

Chillingworth died on 30 January 1644 after fighting in the royalist army in 1643, and he was buried in Chichester Cathedral. He was a delicate man, which made him exceptionally vulnerable to the rigors of battle, but he was also a man committed to a cause he felt would provide the most toleration possible for intellectual inquiry.

Bibliography: R. Orr, *Reason and Authority: The Thought of William Chillingworth*, 1967; R. Simpson, ed., *The Lady Falkland: Her Life*, 1861.

Jesse G. Swan

CHURRIGUERA, JOSÉ DE (1664–1723). Churriguera, the most prominent member of a family of Spanish architects, sculptors, and designers, was trained by his father as a maker of retables (altar panels). Many of his large, ornate altarpieces survive, such as those in the church of San Esteban in Salamanca, the Sagrario in Segovia Cathedral, and the church of the Orden de Calatrava in Madrid. Churriguera also constructed the flamboyant catafalque (1689; destroyed) for Queen Marie Louise (1662–1689). Churriguera's architectural career began in 1690. While living in Salamanca, he created the chapel of the Colegio Mayor at Oviedo and two wings of the cloister of the Colegio Mayor at Cuenca. His most important architectural contribution is his design for the town and palace of Nuevo Baztán (1709–1713), commissioned by the wealthy banker Juan de Goyeneche. It is an ambitious and highly original urban scheme built to house a community of glassworkers, the workshops and dwellings of which are organized around Goyeneche's mansion and a church. Its plan, organized on deliberately broken axes, breaks from the usual Renaissance gridiron and avoids the monotony of symmetry. The palace he began in 1724 for Goyeneche in Madrid was left incomplete at his death; its neoclassical facade is a later addition. The family lends its name to the Churrigueresque style, which is the culmination of Spanish Baroque architecture. It can be described as effusive and animated, with abundant application of ornament. Strictly speaking, the family used this style only for their retables, as their architectural designs are more restrained; it is only their followers, such as Pedro de Ribera, who apply it to architecture. The Churrigueresque was heavily criticized in the second half of the eighteenth century by leading figures of the Spanish Enlightenment for its lack of Italianate classicism. This style is often seen in Spanish colonial architecture of the Americas.

Bibliography: A. Ceballos, *Los Churriguera*, 1971; G. Kubler, "Arquitectura de los siglos XVII y XVII," *Ars hispaniae* 14 (1957).

Rosi Prieto Gilday

CLEMENT XI, POPE (1649–1721). Clement XI (pope from 23 November 1700 to 19 March 1721) was born Giovanni Albani and was trained in classics, law, and theology. Beginning at the age of twenty-eight he held a variety of church posts, was ordained a priest in September 1700, and was elected pope two months later at the age of fifty-one. He fared poorly in the political maneuverings of his era. In the War of the Spanish Succession, he was compelled to acknowledge Philip of Anjou as ruler, only to relinquish that choice in favor of the Austrian Hapsburg Archduke Charles after Austrian troops invaded Rome in 1709. In the Treaty of Utrecht (1713), he was made to give up papal rights in Sicily, Sardinia, and Parma. Like the popes of the late seventeenth century, he also had to grapple with the spread of *Cornelius Jansen's doctrines, which he condemned in the papal bulls *Vineam Domini* (1705) and, more notably, in *Unigenitus* (1713), which, though debated by the French for the next three decades, defined the church's lasting position on Jansenism. He took a strong interest in the church's missions in north Germany and the Philippines, but his most significant decision was to restrict the incorporation of indigenous Chinese customs and Confucian concepts in the Jesuit missions there. He also greatly expanded the Vatican library through the acquisition of manuscripts collected by J.S. Assemani.

Bibliography: L. Pastor, *The History of the Popes from the Close of the Middle Ages*, vol. 33; 40 vols., 1891–1954.

Christopher Baker

COLBERT, JEAN-BAPTISTE (1619–1683). Jean-Baptiste Colbert represents well the esthetics based on common sense, the conservative ideas, and the virtue of savings illustrated in the rise of the bourgeoisie during the seventeenth century. Son of a merchant, Colbert revealed his talents of organization when he first worked for Fouquet, the superintendent of finances for *Louis XIV. He soon denounced Fouquet's malversation to the king. Fouquet's arrest constituted the first political act of the young king who appointed Colbert superintendent of finances in 1661. A highly centralized administrative structure held the country in its grip; the office holders had to give way to commissaires, and the "sovereign" courts became merely higher courts. The effects of centralization were reinforced in turn by the complete linguistic unification of the kingdom and a system of privileges tightened by Colbert; they granted the exclusive rights of publication to printers/booksellers. At the same time, Colbert created industries financed by the state and developed trade and exports. Protecting French commerce, he developed Brest harbor, which was fortified by Vauban in 1683, and rebuilt the navy.

The prosperity resulting from his economic policies financed the sumptuous life of Louis XIV in Versailles and his numerous wars in Europe, which were to increase the French influence all around the world. Seeing the importance of the arts for national prestige, Colbert worked with many artists he considered to be key intellectual figures: Jean Chapelain in literature, *Le Brun in the plastic arts, and was himself in charge of writing the history of the King. He fostered academies and libraries and elicited royal propaganda by giving pensions to selected artists. He was appointed controller general of finances in 1665 and introduced the system of five great tax forms. In 1680, an order signed by Louis XIV and Colbert under the king's private seal directed that the two French theatrical troupes in Paris merge, becoming the only group authorized to perform "comedies" (i.e. plays) in Paris. The establishment of this monopoly meant privilege for actors and guaranteed exclusive rights to the most prestigious texts and roles in the French theater. Colbert died in 1683. He was responsible for a vigorous and often unpopular program of economic and administrative reforms, but his twenty years of administration coincided with the most brilliant years of Louis XIV.

Bibliography: B. Barret-Kriegel, *Les historiens et la monarchie*, 1988.

Martine Sauret

COMENIUS, JOHN AMOS (1592–1670). Born 28 March 1592 in Nivnice, Moravia (now the Czech Republic), Moravian clergyman (the last of the Moravian bishops) Jan Ámos Komenský (Czech) advocated sweeping educational reforms. He sanctioned teaching in the vernacular rather than Latin as well as teaching language through conversational methods. Comenius's parents died when he was twelve and as a result he spent the remainder of his youth in boarding school, until he entered the University of Heidelberg, Germany, where he trained for the ministry. In 1618, he and other Protestant leaders were forced to escape from Ferdinand II's Thirty Years' War, and by 1628, Comenius had settled in Leszno, Poland. It was here that he began in earnest to advocate for an educational system that taught the nation's youth both their indigenous culture and European culture. He advocated "natural" teaching methods that would increase learning speed and increase student retention by making learning environments amiable as well as academically comprehensive. He also understood the benefit in relating education to everyday life experiences. Comenius advocated a universal educational system, one which offered equal opportunities to women. His *Didactica magna* (1628–1632, *The Great Didactic*) contains a dissertation of these pedagogical principles. Comenius's other significant works include *Janua Linguarum* (1631, *The Gate of Tongues Unlocked*) and one of the earliest illustrated books designed especially for children, *Orbis Sensualium pictus* (1658, *The Visible World*).

Bibliography: L. Howard, *John Comenius: The Labyrinth of the World and the Paradise of the Heart*, 1998; J. Wolfe, *Learning from the Past: Historical Voices in Early Childhood Education*, 2000.

David M. Rosen

CONGREVE, WILLIAM (1670–1729). Though born near Leeds, the man who would arguably become the best—or at least best remembered—comic dramatist of the English Restoration was raised in Ireland from an early age owing to his father's military posting there. He attended school in Kilkenny from 1681 until 1685 and then Trinity College, Dublin, from 1686 until 1688, where he befriended Jonathan Swift. In 1690, he entered the Middle Temple in London but did not take a degree. His only novel, *Incognita* (1692), prompted Samuel Johnson to later say that he "would rather praise it than read it." Congreve fared better with comedy, premiering *The Old Bachelor* in 1693 with the help of *John Dryden. The play was an immediate success, unlike *The Double Dealer*, which flopped in October of the same year. But in April 1695, *Love for Love* brought Congreve more acclaim than anything he had yet written. In 1698, he issued *Amendments of Mr. Collier's False and Imperfect Citations*, a rebuttal of Jeremy Collier's *Short View of the Immorality and Profaness of the English Stage* published that year. Lashing out at the apparent license taken by such dramatists as Dryden, *Wycherley, and Congreve, Collier castigated what he saw as their profanity and negative portrayals of the clergy (Collier was a nonjuring Anglican Bishop). Congreve's own work was actually far less vituperative than Wycherley's, a fact lost on Collier. In 1700, he finished the play for which he is best known today, *The Way of the World*. He was manager of the Queen's Theatre in 1705, and the next year wrote a prose *Discourse on the Pindaric Ode*, issued with his *Pindaric Ode to the Queen*. In later years, Congreve suffered from gout, blinding cataracts, and obesity. He wrote little and lived on a government appointment, the Secretaryship of Jamaica. After a carriage accident while traveling to Bath, he died on 19 January 1729.

The Way of the World, a comparative failure in its own day (the legend that Congreve actually harangued its unappreciative audience is probably just that), has since become the representative Restoration comedy for most modern theatergoers. Though it employs characters and themes typical of other contemporary comedies (legacy hunting, social competition between predatory rakes and conniving society women, and sexual intrigue—all of which had enraged

Collier), its personalities are in fact more refined than many plays of a generation earlier, and it can probably stand as the best English counterpart to the work of *Molière. Congreve's men are less offensive than previous stage rakes, and his women have a feisty wit that keeps them "modern." Above all, Congreve's dialogue displays an intellectual inventiveness and a polished yet colloquial satiric vitality (as in the "proviso scene" between Mirabell and Millamant in *The Way of the World*), which holds up Restoration behavior more often to mocking scrutiny than for uncritical humor. There would be little to compare with this play until Oscar Wilde's *The Importance of Being Earnest*, two centuries later.

Bibliography: N. Holland, *The First Modern Comedies: The Significance of Etheredge, Wycherley, and Congreve*, 1959; M. Novak, *William Congreve*, 1971.

Christopher Baker

CORELLI, ARCANGELO (1653–1713). Arcangelo Corelli was thirteen years old when he traveled to Bologna to receive his first lessons on the violin. He entered the Accademia Filharmonica four years later, where he continued violin studies with Giovanni Benvenuti. After moving to Rome sometime during the early 1670s, he obtained a number of positions as a violinist at various metropolitan churches and theaters. Rome remained Corelli's chief residence to the end of his life, except for visits to Modena and Naples in 1689 and 1702, respectively. After 1680, he acquired a commanding reputation as composer as well as performer and became a favorite in the highest circles of Roman society. Cardinal Pietro Ottoboni, an enthusiastic lover of music, was his close friend and patron. Corelli lived in the cardinal's palace with brief intermissions up to the day of his death, conducting weekly concerts, which were considered to be among the most important events in Roman musical life.

Corelli's fame was not limited to Rome or even Italy. Young talents from virtually every country came to benefit from his instruction. His compositions were published throughout western Europe as well as in Italy, and illustrious foreigners visiting Rome rarely failed to pay him homage. When Queen Christina of Sweden went to live there, Corelli conducted in her palace the performances of an orchestra of 150 musicians. The King of Naples repeatedly tried to induce him to resettle in his capital and made him several lucrative offers, all of which, however, were declined by Corelli, who was unwilling to leave his position in Rome, where he was universally loved and admired. Corelli has a double claim to prominence in music history—as an outstanding performer who was the founder of modern violin technique and as a great composer who shaped the chamber sonata and the concerto grosso of the late Baroque era.

Bibliography: P. Allsop, *Arcangelo Corelli: "New Orpheus of Our Times,"* 1999; M. Pincherle, *Corelli: His Life and His Music*, rev. ed., 1956.

James P. Fairleigh

CORNEILLE, PIERRE (1606–1684). A prolific French dramatist who wrote more than thirty plays and dominated theater during the first half of the seventeenth century, Pierre Corneille was born in Rouen, Normandy on 6 June 1606, the eldest of the seven children of Pierre Corneille, lawyer, and Marthe le Pesant. He was educated at the Jesuit school, where he excelled in writing Latin verse. After becoming a lawyer in 1624, Corneille pled only one case, perhaps because of his timidity, his stutter, and his preference for the theater. In 1629, like his father before him, Corneille became king's advocate for water and forests, as well as king's advocate at the port of Rouen's Admiralty court; he occupied these positions and faithfully executed his duties there for the next twenty-one years. In that same year, however, his first comedy *Mélite* was performed by Mondory's traveling troupe in Rouen and then again, probably in early 1630, in Paris, where it was an instant success. Thus began twenty-two years of a rather surprising dual career for this provincial bourgeois who was an insecure perfectionist and, above all, a devout family man.

Between 1629 and 1636, Corneille wrote eight plays: one tragedy, *Médée*, one melodrama, *Clitandre*, and six comedies. Corneille transformed the comic genre: Instead of resorting to the ridiculous comportment of buffoon characters to make his audiences laugh, Corneille used the realistic conversations, lively spirit, and amorous adventures of well-mannered young people to create humor. During this period, *Cardinal Richelieu invited him to join *les cinq auteurs*, five authors who collaborated in the writing of plays based on Richelieu's ideas. Corneille received a pension for his participation, but broke with the group after a short time.

In 1637, Louis XIII ennobled both Corneille and his father. In that same year, Corneille presented *Le Cid*, one of the most controversial and influential plays in the history of French literature. In it, Corneille introduced what would become his trademark theme: the conflict between love and duty. Although *Le Cid* established his reputation as a creative dramatist and was a popular success, it was criticized, primarily by his rivals, as implausible and morally defective. *Le Cid* violates the unities—according to which a play should be limited to one action that occurs in one place within the time frame of only one day—and, more shocking still, it violates the rules of propriety, when the king decides that Chimène must marry her lover Rodrigue (*le Cid*), who killed her father in a duel. Because of the bitter controversy (the *Querelle du Cid*) that ensued, Richelieu asked the *Académie française*, which he had founded in 1634, to arbitrate. In 1638, the *Académie* rendered their verdict: Although they noted that the play had a certain charm, they agreed with Corneille's critics that it had violated the unities and rules of propriety. Corneille presented no new plays until 1640.

When his father died in 1639, Corneille became the legal guardian of his two youngest siblings, Marthe and Thomas. In 1641, Corneille married Marie de Lampérière. The couple set up house in Rouen, where the first of their seven children was born the following year. A devoted husband and father, Corneille also enjoyed a life-long close relationship with Thomas, who later became a playwright like his brother, married Marie's sister, and lived in an adjoining house.

Between 1640 and 1642, Corneille presented the three other plays that, along with *Le Cid*, comprise his classic tetralogy and are considered his greatest work: *Horace, Cinna ou la clémence d'Auguste*, and *Polyeucte martyr*. In 1647, after three attempts, Corneille was finally admitted to the *Académie française*. Although he continued to write plays and moved to Paris in 1660 in order to be closer to the theatrical world, Corneille saw his popularity and influence in his later years eclipsed by that of a new playwright, *Jean Racine. He died in Paris in 1684 in relative obscurity. Nevertheless, Racine praised his verses and *Molière referred to him as his master. Corneille's plays, which depicted protagonists of heroic proportions and exalted individual freedom and the triumph of human will over personal interests, continue to inspire and entertain theatergoers more than 300 years after his death.

Bibliography: C. Abraham, *Pierre Corneille*, 1970; C. Carlin, *Pierre Corneille Revisited*, 1998.

Leslie A. Sconduto

CORYATE, THOMAS (?1577–1617). A native of Odcombe in Somersetshire, Thomas Coryate found fame through his literary accounts of his wide-ranging travels on foot. While the records of his arrival at Winchester College suggest that he was born in 1579, Coryate speaks of himself as two years older, writing that he learned more in his European travels of 1608 than he had in "all the days of [his] life before in England which contayned two and thirty years." After study at Oxford, those English years consisted of unspecified service in Prince Henry's household. In 1608, Coryate undertook the journey from Odcombe to Venice, and his experiences along the way became the basis of his work *Coryate's Crudities Containing his Observations of France, Amiens, Paris, Fountaine Beleau, Nevers, Lyons, Savoy, Italy, Turin, Milan, Cremona, Mantua, Padua, an the Most Glorious, Peerlesse and Mayden Citie of Venice*. Even with the prince's blessing, however, the 200,000-word manuscript failed to find a publisher until *Ben Jonson, *John Donne, and other well-known writers composed mocking testimonial verses to Coryate, which were grouped together as the "Panegyricke Verses" and affixed to the *Crudities*' initial 1611 edition. Lambasting Coryate became so fashionable that more such verses appeared later that year under the title of *Coryate's Crambe*, as did *The Odcombian Banquet*, a pirated version of the *Crudities*.

In 1612, Coryate left England once again, this time with his sights set on exploring the East. His travels brought him as far as Surat in India, where he succumbed in 1617 to chronic flux, which became lethal when combined, according to legend, with too much sack. Coryate left no manuscript per se of his final journey, yet his letters and journals provide a valuable—and lively—record of early English interactions with and impressions of Indian culture, just as his *Crudities* had offered English audiences an entertaining view of the more familiar, but still exotic, cities of Italy.

Bibliography: J. Bate, "The Elizabethans in Italy," *Travel and Drama in Shakespeare's Time*, ed. Jean-Pierre Maquerlot, 1996; M. Strachan, *The Life and Adventures of Thomas Coryate*, 1962.

Hillary Nunn

COTTON, JOHN (1584–1652). Born in Derbyshire, Cotton emigrated to the Massachusetts Bay Colony, becoming one of the foremost Puritan spokesmen of colonial New England. He earned a B.A. (1603) and an M.A. (1606) at Emmanuel College, Cambridge, and was ordained in 1610. He married Elizabeth Horrocks three years later, developing a thriving ministry in Lincolnshire, where he also was censured several times for failing to conform to Anglican liturgical practices. His first wife died in 1631, and the next year he married Sarah Hawkredd Story, a widow; one of their six children, Maria, later married *Increase Mather. Experiencing increasing Anglican criticism for his nonconformity, Cotton sailed for Boston on 13 July 1633; en route his wife bore their first child, appropriately named Seaborn Cotton.

Cotton soon became more widely known through the many collections of his published sermons, which began appearing about a decade after his arrival, such as *The Way of Life* (1641), *A Brief Exposition of the Whole Book of Canticles* (1642), and *Christ the Fountain of Life* (1651), among others. However, despite the fame he was acquiring in Boston, he chose to support the antinomians in the theological controversy of that name from 1636 to 1638, most notably defending *Anne Hutchinson. Advocating the importance of grace over the role of good works and the primacy of the believer's immediate faith in God regardless of the efficacy of the ordained clergy, Cotton was soon as castigated in the New World as he had been in the old. He gradually disengaged himself from the antinomians, however, and by the 1640s had reemerged as a force in the colonial Puritan church. His second major controversy arose when he became embroiled in debate with *Roger Williams (who had been banished from the colony in 1635) over the treatment of those who dissented from Puritan orthodoxy. Williams, in *The Bloudy Tenent of Persecution* (1644), advocated tolerance of dissent, whereas Cotton, in *The Bloudy Tenent, Washed and Made White in the Bloud of the Lamb* (1647), argued for the necessity of punishing the transgressors. Williams answered in the debate's final tract, *The Bloody Tenent Yet More Bloody* (1652).

Cotton went on to pen works describing colonial Puritan practices and theology (e.g., *The Grounds and Ends of the Baptisme of the Children of the Faithful [1647]*), and he was a key translator of the *Bay Psalm Book* (1640), the first book published in America. He died two days before Christmas, 1652, having exerted a profound shaping influence upon the character of New England Puritanism.

Bibliography: I. Polishook, *Roger Williams, John Cotton, and Religious Freedom: A Controversy in New and Old England*, 1967; W. Stoever, *"A Faire and Easie Way to Heaven": Covenant Theology and Antinomianism in Early Massachusetts*, 1978; T. Toulouse, *The Art of Prophesying: New England Sermons and the Shaping of Belief*, 1987.

Christopher Baker

COUPERIN, FRANÇOIS (1668–1733). The most illustrious member of a notable musical family (from whom he was distinguished by the term "le Grand"), François was the son of Charles Couperin (1638–1679), organist of the Parisian church of St. Gervais. Charles had inherited the post upon the early death of his brother Louis (1626–1661)—the second most illustrious Couperin—but after only seven years, Charles died. As François was only ten, he could not yet inherit the post, so Michel-Richard de Lalande (1657–1726) was named interim organist until 1685, when François turned eighteen, the requisite age.

In 1690, Couperin brought out his first publication, *Pièces d'orgue*, containing two organ masses. Although he never again published any organ music, Couperin worked as an organist all his life. In 1693, he succeeded his old teacher, Jacques Thomelin, as royal organist. This appointment was an entrée to others, especially as harpsichord teacher to members of the royalty and nobility. Now Couperin found himself assuming other royal assignments—composer of sacred music for the royal chapel and of chamber music, and deputy for royal harpsichordist Jean-Baptiste-Henri D'Anglebert, whose eyesight was failing. In 1696, Couperin bought a title of nobility, and around 1702 he was honored with the title of Chevalier de l'Ordre de Latran. In 1713, Couperin published his first book of harpsichord pieces. He also obtained a twenty-year printing license, and the next work to emerge was *L'art de toucher le clavecin* (1716, 2nd ed. 1717), his handbook on playing the harpsichord. Probably also in 1717, Couperin's second *livre de pièces de clavecin* was published, as well as the first three of the *Leçons de ténèbres*, his masterpiece of liturgical music. The third and fourth books of harpsichord music appeared in 1722 and 1730, with the *Concerts royaux* attached as the second part of Book Three.

Couperin was a devotee of Italian music, and his mission for chamber music was to blend the best traits of both French and Italian styles. This quest began with his earliest trio sonatas, *La pucelle, La visionnaire*, and *L'astrée*. Inspired after hearing some sonatas of *Arcangelo Corelli on a concert series, Couperin composed *La pucelle*, which he later claimed was not only his first sonata, but also the first sonata ever composed in France. In 1724, Couperin paid homage to Corelli in his trio sonata *Parnasse, ou L'apothéose de Corelli*, the final work in his publication *Les goûts réünis*, or *Nouveaux concerts*. Lastly, in *Les nations* (1726) Couperin included the three early sonatas, renamed *La françoise*,

L'espagnole, and *La piemontoise*, plus another earlier work (*L'Impériale*); each of these Italianate sonatas was joined to an *ordre*, or suite, of dances in the French style. His health had been troubling him for some time, so in 1723 Couperin secured his cousin Nicolas Couperin as his assistant and eventual successor at St. Gervais. As Couperin's health worsened, he parceled out his appointments. His court harpsichordist position went to his daughter Marguerite-Antoinette, and Guillaume Marchand succeeded him in the royal chapel. Couperin finally died on 11 Sept 1733.

Bibliography: W. Mellers, *François Couperin and the French Classical Tradition*, 1950, 2nd rev. ed., 1986; D. Tunley, *Couperin*, 1982.

Susan Treacy

COWLEY, ABRAHAM (1618–1667). Ironically, Samuel Johnson's important remarks on English metaphysical poetry appeared in his 1779 biographical essay on Cowley, who by then had completely lost the wide popularity he enjoyed during the later seventeenth century and whose work is now seen as more rational than, and far less typical of, the works of *Donne or *Herbert. Cowley was born in London and attended Cambridge, where he befriended *Crashaw (on whose death he composed a memorial poem). A royalist, he sought refuge with the court at Oxford in 1643, then assisted the queen in France as a secretary. He returned to England in 1654 and was jailed the next year, but after his release he earned a medical degree from Oxford in 1657. He briefly served the queen again in France during 1659 and 1660, thereafter retiring to his country estate at Chertsey and dying there on 28 July 1667.

Cowley is important for marking a transition in English poetry from the metaphysical style to the later Augustan mode of the eighteenth century; both tendencies are present in his verse, as well as the marked influence of Spenser in his youthful *Poeticall Blossoms* (1633). His metaphysical poetry was published in *The Miscellanies* (1656) and *The Mistress* (1647 and 1656), but it is tamer than the colloquial, forceful works written by members of "the School of Donne" a generation earlier. He deserves credit for raising the poetic couplet to new prominence as a verse form in his unfinished epic *The Davideis* (1656), and his *Pindarique Odes* (1656) reveal an innovative rendering of Pindar's style that drew attention but also puzzlement for its uninhibited complexity.

Cowley's prose essays also represent a step toward the succeeding age in their polished yet conversational sentences, which anticipate *Dryden and *Joseph Addison. His topics remind one of *Bacon's essays and Montaigne's, though he is too sociable an author to truly indulge in the former's pithy brevity or the latter's provocative introspection. He not only comments upon the pleasures of moderation in his country retirement, but touches as well on medical and horticultural topics, interests that earned him a nomination for membership into the Royal Society, though he was never inducted. His interest in medicine is also reflected in his elegy on the death of *William Harvey.

Bibliography: R. Hinman, *Abraham Cowley's World of Order*, 1960; D. Trotter, *The Poetry of Abraham Cowley*, 1979.

Christopher Baker

CRASHAW, RICHARD (?1612–1649). Known for his intense religious sentiments, Richard Crashaw's verse is often said to typify metaphysical poetry. Though his ecstatic poems are often compared with those of *Vaughan and *Herbert, Crashaw's style is modeled upon the continental Baroque. His main influence was the Italian poet *Marino. From his youth, Crashaw was exposed to extreme religious views. His father was a zealous Puritan who officiated at the death of Mary, Queen of Scots. Radically departing from his father's faith, Crashaw later converted to Catholicism. His father died when Crashaw was fourteen, leaving him in the care of two lawyers, Sir Henry Yelverton, one of his father's patrons, and Sir Randolph Crew. In 1629, Crashaw entered Charterhouse as scholar, probably under the influence of Yelverton. Here, he studied classical orators and poets and learned to imitate their styles. He concentrated upon Greek, Latin, and rhetoric, and probably composed a number of epigrams as part of his religious lessons.

In 1631, Crashaw entered Pembroke College of Cambridge, which was Arminian in theology and High Church in ritual. Later that same year, he was elected a Greek Scholar. Crashaw may have entered Pembroke with a reputation as a poet, for his verses appeared in the second edition of Bishop *Lancelot Andrewes's sermons, published in 1631, and he contributed to all three volumes of occasional Latin verse published annually by the college. In fact, all of his secular poetry, mostly occasional and none autobiographical, may have been written at Pembroke. Also while at Pembroke, Crashaw began to frequent Little Gidding, a Laudian religious community founded by George Herbert's friend, Nicholas Ferrar. Crashaw apparently modeled his ascetic life on the rituals of this community.

In 1634, Crashaw received his B.A. and published *Epigrammata Sacra*, a compilation of religious Latin verses. In 1635, he was elected Fellow of Peterhouse, which was noted for its elaborately refurbished chapel and practice of Roman rituals. By 1639, he had taken Holy Orders and was curate of Little St. Mary's. With the controversy leading to the civil war mounting, Crashaw left Peterhouse in January 1644, and sought exile in Leyden. His conversion to Catholicism may have occurred during his return to England, possibly at Oxford in 1645. Soon thereafter, he took up residence in Paris. Through the influence of *Abraham Cowley, Crashaw made his way into the circle of Queen Henrietta Maria. In 1646, *Steps to the Temple: Sacred Poems, with other Delights of the Muses* was published, including both religious and secular poems in Latin and English. Also in this year, Crashaw moved to Rome and entered the service of Cardinal Pallotta, who was revered by the pope. In 1649, Crashaw was moved, for his own protection, to a minor post at Loreto, having angered his fellows by reporting to the cardinal their misbehavior. Renowned for its sacred shrines,

Loreto was a fitting place for the devout Crashaw to spend his last few months. He died of a fever on 21 August 1649.

Bibliography: W. Austin, *Richard Crashaw: A Study in Baroque Sensibility*, 1967.

Carol Jamison

CROMWELL, OLIVER (1599–1658). Oliver Cromwell's greatest mission was to rid England of the monarchy, which he regarded as an oppressive form of government, and replace it with a more republican system that would ensure protection against absolutism. However, his own personality and the course of events in the English civil war thrust him into the role of Puritan autocrat. Cromwell was born on 25 April 1599 and later entered Cambridge but never took a degree. In 1620, he married Elizabeth Bourchier, to whom he remained devoted. He represented his birthplace, Huntingdon, in Parliament when the Petition of Right was passed against *Charles I in 1628. He farmed for a time, then moved to London in 1646. He also represented Cambridge in the Short and Long Parliaments and gained influence in a variety of legislative responsibilities, being known for his outspoken comments (it was said at his death that his personality was "exceeding fyery").

After Charles I attempted to arrest five Parliamentarians in 1642, Cromwell actively began defending Puritan interests, and he fought royalist troops at the battle of Edge Hill. Becoming a capable military commander himself, he defeated the king's forces at Grantham, Gainsborough, and Winceby, and at the more important battle of Marston Moor (1644), where his cavalry gained the name "Ironsides." He was instrumental in forming the New Model Army, a highly disciplined force composed of men from a variety of locales rather than a single geographical place. This army defeated the king's forces at Naseby in 1645, and Charles surrendered to the Scots the next year.

After the end of this so-called "first civil war," Presbyterians in Parliament were willing to cooperate with the defeated king, a position Cromwell did not agree with; they also favored dissolving the New Model Army and retiring Cromwell himself, a stance that led to the army's occupation of London in 1647. Cromwell was also at this time engaged in vigorous debate with the extremist group called the Levellers, disagreeing with their calls for expansion of the voting franchise. These issues were unresolved when Charles agreed to permit Presbyterianism in Scotland in exchange for Scottish military support. Thus in 1648, the "second civil war" began, and ended, with Cromwell defeating the king's army at Preston Pans in Scotland. After 140 members of Parliament were expelled, the resulting smaller "Rump Parliament" tried and executed the king in 1649, established England as a republican Commonwealth, and declared Cromwell military commander in Ireland. He mercilessly suppressed the Irish rebels, and his return to England in May 1650 was the subject of *Andrew Marvell's poem "An Horatian Ode Upon Cromwell's Return From Ireland." In 1651, he also defeated a large Scottish force in Worcester.

Cromwell's military victories strengthened his influence in England as his disputes with Parliament worsened. Its policies appeared to him too liberal and self-interested; he and the army dissolved the legislature in 1653, and from then until 1660 Cromwell governed England as Lord Protector, the Commonwealth replaced by his Protectorate. But numerous disagreements developed with the Parliament of the Protectorate (for example, he refused to grant it authority over the army) and he dissolved it in 1655. Parliament reconvened the next year, but once again it, Cromwell, and the army failed to function harmoniously, and Cromwell disbanded it in 1658, the year of his death.

Cromwell embodied the English experiment in Puritan government. Though often hot-tempered, he was more tolerant of religious diversity than the more radical Levellers and Diggers, but he defended the authority of the army and thus aggravated many constitutional Parliamentarians. Whether viewed from the perspective of the royalist *Edward Hyde, Earl of Clarendon, as "a brave bad man," or from that of more romantic advocates of Puritan republicanism, such as Thomas Carlyle in the nineteenth century, as a "strong true giant of a man," Cromwell was clearly instrumental in helping end England's era of royal absolutism.

Bibliography: C. Hill, *God's Englishman: Oliver Cromwell and the English Revolution*, 1970; J. Morrill, *Oliver Cromwell and the English Revolution*, 1990; J. Morrill, *The Nature of the English Revolution: Essays*, 1993.

Christopher Baker

CUDWORTH, RALPH (1617–1688). Cudworth, the most systematic philosopher of the Cambridge Platonists, was born in Somersetshire in 1617. A Puritan, he entered Emmanuel College in 1632, earned a master's degree in 1639, and served as a fellow and tutor from 1639 to 1644. With the Puritan victory, he was appointed master of Clare College and professor of Hebrew in 1645. He earned a bachelor of divinity degree in 1646 defending Cambridge Platonist *Benjamin Whichcote's view that good and evil are eternal, immutable, and grounded in reason. After serving as rector of North Cadbury, Somerset, from 1650 to 1654, he became master of Christ's College where *Henry More, another Cambridge Platonist, was a fellow. He married and in 1658 had a daughter Damaris, who as Lady Masham became a good friend of *John Locke.

Most of Cudworth's writings were theological, such as *A Discourse concerning the True Nature of the Lord's Supper* (1642). In 1647, he gave a sermon before the House of Commons; with a bitter dispute raging about church discipline and an increasing Puritan presence in the government, Cudworth preached on a passage from 1 John that states that, in keeping the Ten Commandments, one would know God. By espousing views popular with no particular party, Cudworth thus advocated tolerance. His major philosophical work, *The True Intellectual System of the Universe* (1678) attacked two forms of atheism: materialism, represented by *Thomas Hobbes, and hylozoism, the idea that spirit is part of matter. Attracted by the atomism of the scientific revolution, he strove to incorporate it into a dualism not of mind and body, but rather of passivity and activity. Activity accounted for all change and motion and, in addition to the actions of the mind, included "spiritual plastic powers," which explained the activity of organisms. This idea appealed to philosophical biologists such as *John Ray. In 1731, some of Cudworth's manuscripts in the British Museum were published as *A Treatise Concerning Eternal and Immutable Morality*. Basing his ideas on Plato's *Euthyphro*, he argued that good and evil are so by their nature, not by command. This view, however, can be interpreted as a secular ethics. Additional manuscripts were published as *A Treatise of Free Will* in 1838.

After the Restoration, Cudworth's enemies unsuccessfully attempted to remove him from his university post at Christ's College, where he remained until he died in 1688.

Bibliography: J. Passmore, *Ralph Cudworth: An Interpretation*, 1951.

Kristen L. Zacharias

CUYP, AELBERT (1620–1691). An important Dutch landscape painter of the seventeeth century, Cuyp is known mainly for his later manner in which he portrayed native topography in golden yellow sunlight. He never visited Italy, restricting his travels to the Netherlands, where he sketched sites along the Rhine from 1651 to 1652. A lifelong resident of Dordrecht, Cuyp appears to have retired from painting after his 1658 marriage to the well-off Cornelia Boschman. He subsequently held offices in the Reformed Church and the High Court of South Holland, which accords with early biographer Arnold Houbraken's attestation of his irreproachable character. He died in Dordrecht in 1691.

Trained by his father Jacob in landscape painting, Cuyp initially imitated the low diagonals, loose light brushwork, and monochrome palette of Jan van Goyen's landscapes. This changed dramatically, however, when he became familiar with *Claude Lorrain's style through the landscapes of Utrecht painter Jan Both, who had returned from Rome in the mid-1640s. Cuyp applied Claude's warm tone and silhouette effects to such native themes as the *Gathering of the Fleet at Dordrecht* (c.1655) and the *Valckhof of Nijmegen* (c.1652), as well as pastoral and river scenes, and portraits. Cuyp seems to ennoble his cattle, posed in profile in warm, glowing sunlight (*River Landscape with Cows*, 1645–1650). Along with Jacob van Ruisdael, Cuyp is regarded as a key painter of the "classical" phase of Dutch seventeenth-century landscape painting.

Bibliography: A. Chong, *Aelbert Cuyp and the Meaning of Landscape* (Dissertation), 1992; *Aelbert Cuyp en zijn familie: Schilders te Dordrecht* (Exhibition catalogue), 1977; A. Houbraken, *De Groote Schouburg, I*, 1718–1721; S. Reiss, *Aelbert Cuyp*, 1975.

Lloyd DeWitt

CYRANO DE BERGERAC, SAVINIEN DE (1619–1655). French dramatist, satirist, and libertine whose writings reflected the spirit of his times and whose flamboyant life inspired a popular legend, Savinien de Cyrano de Bergerac was born in Paris to Abel de Cyrano, lawyer, and Espérance Bellanger. After completing his education at the Collège de Beauvais in Paris, Cyrano became an officer in the Compagnie des Gardes, where he earned his reputation as a swordsman. Wounded at the siege of Mouzon in 1639 and again at the siege of Arras in 1641, Cyrano ended his military career, returned to Paris, and began to study philosophy, possibly under *Gassendi. In 1645 or 1646, he composed *Le Pédant joué*, a satire on which *Molière based two scenes of *Les Fourberies de Scapin*. Cyrano then began to write the two works for which he is best known: *L'Autre monde ou L'Histoire comique des états et empires de la lune* and *L'Histoire comique des états et empires du soleil*; the first describes an imaginary trip to the moon, the second a trip to the sun. These fantasies, which were published posthumously in 1656 and 1662, influenced both Jonathan Swift's *Gulliver's Travels* and Voltaire's *Micromégas*. During the period of the Fronde (1648–1652), an aristocratic revolt against Mazarin and despotism, Cyrano wrote eight political pamphlets, at first in support of the revolt and then in support of Mazarin. Seeking protection, Cyrano entered the service of the Duc d'Arpajon in 1652. The following year, after the performance of his controversial play *La Mort d'Agrippine*, Cyrano was accused of atheism and blasphemy. In 1654, he was hit by a falling beam as he entered Arpajon's house. Cyrano died of this injury fourteen months later on 28 July 1655.

Bibliography: E. Harth, *Cyrano de Bergerac and the Polemics of Modernity*, 1970.

Leslie A. Sconduto

D

DANIEL, SAMUEL (?1562–1619). Samuel Daniel was born in Somerset and educated at Magdalen Hall, Oxford. After having traveled through Italy, Daniel entered the household of Mary Sidney Herbert, Countess of Pembroke (sister of Sir Philip Sidney), as tutor to the young William Herbert. He would also serve as tutor to Anne Clifford, daughter of the Countess of Cumberland. While the Pembrokes and Cliffords were Daniel's main patrons, he also benefited from the patronage of Lucy, Countess of Bedford; Charles Blount; Lord Mountjoy; and Edward Seymour, Earl of Hertford.

Daniel's literary career began in 1591 when twenty-eight of his sonnets were published without his consent in a pirated edition of Sidney's *Astrophil and Stella*. One year later, Daniel issued an authorized version of his sonnet sequence *Delia*, which included a prose dedication to the Countess of Pembroke. Another edition of *Delia*, containing four new sonnets as well as his long narrative poem *The Complaint of Rosamond*, appeared in 1592. A third edition of the sonnet sequence, along with an augmented *Complaint*, was published in 1594, the same year in which Daniel published his closet drama *Cleopatra*. The year 1594 also marked the appearance of the first four books of Daniel's ever-evolving but never completed *The Civil Wars* (which went through five editions between 1594 and 1605), a sober historical epic in the fashion of Lucan, chronicling the ill-effects of the War of the Roses on the English nation. Daniel would later try his hand in prose history, with the first part of his *Historie of England* published in 1612, the second in 1618. The first edition of Daniel's works, entitled *The Poetical Essays*, appeared in 1601; his folio *Works* appeared in 1602, and another folio volume, with Daniel's portrait on the frontispiece, appeared in 1623. The prefatory material of these various editions of Daniel's poetry offers a fascinating instance of early modern authorial self-fashioning.

Following the death of Queen Elizabeth, Daniel sought to establish himself in the court of *King James I and VI. In 1603, he presented James with his "A Panegyricke Congratulatorie"; he also wrote the Christmas masque of 1604, *The Vision of the Twelve Goddesses*. As his "Panegyricke" attests, Daniel hoped to use the office of poet as a means to counsel the newly crowned king. Given the decadent and absolutist nature of the Jacobean court, and given Daniel's refusal to voice mere flattery, it is not surprising that he never settled into the unofficial position of court laureate. In fact, Daniel incurred the court's wrath with the publication of his play *Philotas* (1605), which many at court viewed as a sympathetic allegorical portrayal of the rebellious Robert Devereux, Earl of Essex. Daniel did, however, become the licenser to the Childern of the Queen's Revels in 1604, and he continued to write masques for Queen Anne, for whom he served as a groom of the privy chamber.

That Daniel took poetry seriously is most evident in his *Musophilus* (1599), dedicated to Fulke Greville, and, even more so, in his *Defence of Ryme* (1603), a response to Thomas Campion's *Observations in the Art of English Poesie* (1602). Like Philip Sidney and Edmund Spenser before him, Daniel was committed to the idea that poetry should perform crucial moral and didactic work. Fittingly, his poetry is marked by a plain style that well suits its moral seriousness.

Bibliography: W. Godshalk, "Recent Studies in Samuel Daniel (1975–1990)," *English Literary Renaissance* 24 (1994): 489–502; J. Rees, *Samuel Daniel*, 1964.

Christopher Ivic

DAVENANT, SIR WILLIAM (1606–1668). Sir William Davenant was born in Oxford, the second son of John D'Avenant, a merchant and broker of wines, the proprietor of a hostelry, and, according to *Anthony a Wood, "an admirer of plays and play-makers, especially *Shakespeare," who served as the boy's godfather. Having studied privately with Edward Sylvester, Davenant entered Lincoln College around 1620–1621. After a short stay at the college, he served as a page to Frances, first duchess of Richmond, and as clerk to the statesman and poet Fulke Greville, Lord Brooke. He joined the first expedition of the Duke of Buckingham in the 1627 siege of the Isle of Rhé, then migrated to court after Brooke was murdered in 1628.

His first two plays—the unperformed *Albovine* (1629) and *The Cruel Brother* (1630), staged by the King's Servants at Blackfriars—were revenge tragedies. Davenant followed these with two tragicomedies. He achieved his first real success, however, with *The Wits* (1633), a tightly crafted comedy. He then produced a series of works that were meant to please Queen Henrietta Maria: the tragicomedy *Love and Honour* (1634), the comedy *The Platonic Lovers* (1636), and the masques *The Temple of Love* (1635) and *The Triumphs of the Prince D'Amour* (1637). The year 1638 saw the appearance of his first collection of poems and the production of two more plays, *The Unfortunate Lovers* and *The Fair Favorite*. Davenant's favor at court is suggested by his appointment as what amounted to poet laureate and his receipt of a royal patent to operate a theater. He wrote all the masques mounted at court after *Coelum Britanicum* (1634), but his finest was his last, *Salmacida Spolia* (1640), in which *Charles I played Philogenes.

The Scottish revolt against episcopacy in 1639 drew Davenant into the struggle that would culminate in Charles I's execution: He marched north with the King's troops the day after receiving his patent for a theater. In the 1640s, he was implicated in a plot to turn the army against Parliament, fought with the Royalists under the Lord General, William Cavendish, and engaged in gunrunning. He was captured en route to Maryland in 1650, where he was supposed to assume the position of lieutenant-governor. According to some contemporaries, *John Milton interceded on his behalf with the Commonwealth, a favor that Davenant is said to have returned after the Restoration.

The importance of Davenant's *Discourse upon Gondibert* and *Thomas Hobbes's *Answer* as documents in the history of literary criticism have obscured the merits of *Gondibert* itself (1650), which is the most significant epic to appear between *The Faerie Queene* and *Paradise Lost* and one whose stanza was adopted by *John Dryden and later poets for public verse. During the Protectorate, Davenant returned to England from the exiled court to stage works that prepared the ground for Restoration opera; the most successful was *The Siege of Rhodes*. With the return of *Charles II, he managed and directed one of London's two monopoly theaters, which, using casts that now included actresses, mounted some fifty productions: tragedies and comedies, heroic dramas and burlesques, operas, and farces. He directed the first two plays of *George Etherege, a comedy by Dryden, and new works by Sir Samuel Tuke, Lord Orrery, and himself. But he also revived older plays from the English and Continental traditions. More than anyone else, he perpetuated William Shakespeare in performance. Some of his productions, like *Hamlet*, hewed reasonably close to the original, while others, like his conflation of *Measure for Measure* with *Much Ado About Nothing* and his operatic adaptations of *Macbeth* and *The Tempest* (a collaboration with Dryden), did not. Davenant wrote only two new plays after the Restoration, but the improvements in theatrical technique that he introduced—movable scenery, the proscenium stage, and the more systematic training of actors—had a lasting influence on theatrical history. Upon his death in 1668, Davenant's third wife oversaw the completion of the Dorset Garden Theater and, for the next twenty-two years, distinguished herself as the first Englishwoman to manage a theater.

To follow Davenant's career is to trace the main currents of English poetry and drama at mid-century, for he wrote everything from irreverent lyrics to philosophical poems that evince, in Douglas Bush's words, "a massive dignity and sobriety of thought," from court masques to operas staged in commercial theaters. His quick fancy persuaded a few contemporaries like Sir John Suckling and *Sir John Denham to identify him as the heir of *John Donne, but in the metrical discipline and clarity of some of his best verse we can also detect the acknowledged influence of *Ben Jonson. If Davenant assimilated the main traditions of Jacobean verse and drama, his experimentation with the heroic couplet and with such hybrid genres as mock epic, burlesque, heroic drama, and opera also made him a herald of Augustanism.

Bibliography: P. Bordinat and S. Blaydes, *Sir William Davenant*, 1981; M. Emond, *Rare Sir William Davenant*, 1987.

Blair Hoxby

DEKKER, THOMAS (ca. 1570–1632). Thomas Dekker was one of the most active and collaborative dramatists during the Elizabethan and Jacobean reigns; he was also one of the most prolific pamphlet writers of the time. Oddly, however, the record of his personal life remains sparse. The assumption that he was born in London remains just that, an assumption, and this is about as close as we get to the man outside of additional speculation we can draw from reading into his plays. But there are two exceptions. We know that Dekker spent many years in debtor's prison, unable to make what he earned through his writing pay for what he spent. Beginning at least as early as 1599, Dekker was in and out of prison. At one point, he spent six straight years, from 1613 to 1619, in King's Bench Prison. In addition, we know that his debts did not keep others from wanting to work with him.

Based on the number of times other playwrights both worked with him and wrote of him in their own plays and prose, it is clear that nearly every playwright in London of the time knew of Dekker. Although he wrote what would become his most famous play, *The Shoemaker's Holiday* (1600), on his own, Dekker collaborated with others on most occasions. In 1604 and then in 1605, for example, he paired with *John Webster to write *Westward Hoe* and *Northward Hoe*, respectively. In 1604, he also wrote *The Honest Whore* with *Thomas Middleton. Each of these plays was extremely popular, but soon after, he turned for several years from writing plays to writing pamphlets. One is tempted to ascribe the switch of genre and intention to the great plague that visited London in 1603. Dekker wrote many plague pamphlets thereafter from 1603 to 1625 when the last great plague struck within his lifetime. Prior to that, however, in 1609, he wrote the most popular of his pamphlets, *The Guls Horn-book*. In it, instead of warning the citizens of London to see the plague as a sign that they must amend bad behavior, Dekker turned to depicting with affection and with some degree of ridicule London characters. Of particular value to students of theater history are his comments on how Londoners conducted themselves in the playhouses. The next year, he returned to writing of characters in drama. In 1611, he collaborated once again with Middleton to produce the famous character Moll in *The Roaring Girl* (1611). In 1621, he joined with two other famous playwrights, *John Ford and William Rowley, to write *The Witch of Edmonton*. Another well-known work entirely his own is the completion of the work done two decades earlier with *The Honest Whore, Part 2* (1630). His collaborations and individual productions kept him employed—though not out of debt—until the end of his life. Records indicate that Thomas Dekker, singly and in collaboration, produced roughly fifty plays; only fifteen to twenty survive that can be ascribed to him.

We also know that Dekker worked again in collaboration upon a different kind of entertainment altogether: He helped to create the street entertainment that welcomed *James I into London in 1603. In each of his productions, including the pamphlets, Dekker made a few things about his interests quite clear. He made great efforts to depict with affection, humor, and colloquial speech what life was like for the average early modern London citizen. He enjoyed rewarding the common man in plays such as *Shoemaker's Holiday*, in which the local and friendly shoemaker becomes the Mayor of London. Finally, he earned the love and loathing of his fellow playwrights, either collaborating with them or presenting them in satire, as he did in 1601, mocking Ben Jonson in his play *Satiro-mastix*. Although we know nearly nothing of the man's personality or habits or family, Dekker performed prolifically amid the most productive writers of his day.

Bibliography: K. McLuskie, *Dekker and Heywood: Professional Dramatists*, 1994; J. Twyning, *London Dispossessed: Literature and Social Space in the Early Modern City*, 1998.

Rebecca Totaro

DENHAM, SIR JOHN (1615–1669). John Denham was born in Dublin in 1615, son of Sir John Denham and his second wife, Eleanor, while his father was lord chief justice of the King's Bench in Ireland. The family moved to England when Denham was two. He matriculated at Trinity College, Oxford, in 1631; was examined for his B.A. three years later; married Anne Cotton; and began studying law at Lincoln's Inn.

He first rose to public notice in 1641, when, after having been admitted to the bar, he served as a witness for the defense in the trial of Thomas Wentworth, first Earl of Strafford. While he sided with Strafford, his poem on the trial shows that, as we might expect of a son whose father had opposed the Crown in John Hamden's Ship Money Case (1638), he understood Parliament's fear of arbitrary power. During the civil wars, he sided with the Royalists, serving variously as a soldier, informant, and fund raiser on the Continent. After the Restoration, he succeeded *Inigo Jones as Surveyor of the Works, in which capacity he oversaw the construction of Burlington House and Greenwich Palace. He was made Knight of the Bath and elected to Parliament. These honors were tarnished by his second wife's affair with the Duke of York, which rumor blamed for his fit of madness in 1666. Denham returned to sanity and to Parliament after the death of his wife in 1667. He was buried in the Poet's Corner of Westminster Abbey in 1669.

Denham is chiefly remembered for two works, *The Sophy* and *Cooper's Hill*, both of which appeared in 1642. The former is a tragedy set in motion when the aging King Abbas, acting on the evil counsel of his adviser Haly, orders that his son Mirza be blinded and thrown in prison. Bent on revenge, Mirza contemplates killing his own daughter, whom the old king loves, but he decides that love is the "nobler passion." In the play's resolution, the gullible king, his naive son, and the corrupt Haly all pay for their vices with their lives. Although the title page of the tragedy claims that it was acted at Blackfriars, no other evidence corroborates the claim.

Cooper's Hill is the first example of "local poetry" in English. From atop the hill, the poet surveys St. Paul's, the city of London, Windsor Castle, and Saint Anne's Hill, while obliquely introducing what Samuel Johnson calls "historical retrospection or incidental meditation," much of it concerning the balance of power between the king and his subjects. The edition of 1642 includes a hunt in which the king shoots a stag that, abandoned by all, dies a hero. This reflection on the death of Strafford is, in the 1655 edition, revised and expanded into a meditation on the execution of *Charles I, who now becomes the hunted stag. In this edition, the poet condenses the other sections of the poem and universalizes its themes. He also adds his address to the Thames, announcing what would become an ideal of Augustine versification:

> O could I flow like thee, and make thy stream
> My great example, as it is my theme!
> Though deep, yet clear; though gentle, yet not dull;
> Strong without rage, without o'erflowing full.

*John Dryden and Alexander Pope looked back on Denham as a founder of Augustine poetry and particularly admired the "strength" of his lines. Besides his two best known works, Denham wrote a few worthy lyric poems, including "On Mr. Abraham Cowley his Death," several political satires, and translations, particularly of Virgil, that placed literary felicity before literal fidelity and thus provided a model for such later translators as Dryden and Pope.

Bibliography: B. O'Hehir, *Harmony from Discords: A Life of Sir John Denham*, 1968; B. O'Hehir, *Expans'd Hieroglyphicks: A Critical Edition of Sir John Denham's* Coopers Hill, 1969; E. Wasserman, *The Subtler Language*, 1959.

Blair Hoxby

DENNIS, JOHN (1657–1734). Poet, dramatist, and critic, Dennis received his B.A. from Caius College, Cambridge, in 1679 and his M.A. from Trinity Hall in 1683. After completing his education, Dennis mixed with many of the notable literary figures of his day, including *Congreve, *Dryden, and *Wycherly at Will's Coffee House, where his knowledge of literature earned their respect and esteem. Between 1692 and 1714, Dennis wrote a number of "Pindaric Odes," which are generally unremarkable, despite their faithful adherence to the form. As a dramatist, Dennis fared no better. His first play, *A Plot and no Plot*, was acted in 1697 with little acclaim. Subsequent dramatic efforts such as *Rinaldo and Arminda* (1699), *Iphigenia* (1700), and *Apius and Virginia* (1705) were received no better, although *Liberty Asserted* (1704) had modest success. Dennis was able to offset his lack of achievement as a poet and dramatist with his considerable skill as a literary critic, which he displayed in *The*

Advancement and Reformation of Poetry (1701), and *The Grounds for Criticism in Poetry* (1704).

Dennis based much of his critical theory in the neoclassical tradition and believed that all worthy literature could be divided into types (tragedy, comedy, epic, satire, and the ode), each of which was designed to have a specific effect on its audience. Also integral to Dennis's literary theory was his assertion that the art of poetry must be conscribed within the boundaries of rules, lest it represent nothing but mere whimsy. (Dennis did, however, make some allowance for the transgression of poetic rules by those in possession of superior poetic genius like *Milton, arguing that such individuals do not break poetic rules with their poetry, but instead transcend them.) Dennis labored under a negative reputation throughout much of his later career. His critical ideas were often considered antiquated by the London literati in the age of *Addison and *Steele, and his caustic and occasionally mean-spirited application of his critical skills landed him in literary skirmishes with Alexander Pope (among others), who lampooned Dennis in his *Essay on Criticism*. Despite his flaws, however, Dennis is still a modestly important figure in the history of Milton criticism and among contemporary theorists of the sublime.

Bibliography: N. Hooker, ed., *The Critical Works of John Dennis*, 2 vols., 1939–1943.

Walter H. Keithley

DESARGUES, GIRARD (1591–1661). This French geometer born in Lyon was much admired by *René Descartes and *Blaise Pascal. Desargues's interest in the technique of perspectival representation led to an original understanding of conic sections and made him a leading thinker in projective geometry. In his *Brouillon Project d'une atteinte aux événements des rencontres du cône avec un plan* (1639), he considered the various intersections of a cone by a plane (ellipse, parabola, and hyperbola) to be images of a circle seen from various points of view. Through involution, the sections are conveniently projected onto a plane and then more easily comprehended. The perspective theorem also resulted from his research. Desargues's achievement highlights the conceptual importance of visuality in the scientific revolution: In projective geometry, parallel lines are considered to meet toward infinity, for example.

Alongside this theoretical drive, he remained committed to help painters, architects, and stonecutters by providing easy techniques for their work. His ideas became available to a still larger public when expounded in the work of Abraham Bosse (1602–1676) and Philippe de La Hire (1640–1718). An engraver and teacher of perspective, Bosse defended his master during a bitter controversy in which Desargues was accused of incompetence. This, together with the fact that Desargues's treatises were written in an obscure manner, still does not wholly explain why his work in projective geometry did not have a follow-up before Gaspard Monge. More was to be made of another innovation in mathematics that captured the interest of scientists, the algebraization of geometry by *Descartes.

Bibliography: J. Field and J. Gray, *The Geometrical Work of Girard Desargues*, 1986; R. Taton, *L'Oeuvre mathématique de Desargues*, 1951.

Jean-Vincent Blanchard

DESCARTES, RENÉ (1596–1650). From his early education at the Jesuit school of La Flèche, Descartes wrestled with the presumption of verisimilitude and tradition as criteria in the pursuit of knowledge and dedicated himself to defining an original path to certainty. His comprehensive philosophical system was elaborated for a larger part in the Low Countries while maintaining close contact with the French learned community. Descartes once wrote that philosophy is a tree whose roots are metaphysics, the trunk physics, and the branches sciences such as medicine and ethics. But even if a metaphysical foundation is necessary to guarantee truth, and evidence points to the fact that this metaphysics was constituted early, Descartes' first major writing, *Regulae ad directionem ingenii* (elaborated from 1620 to 1628 but published posthumously), was chiefly concerned with intellectual procedure and method. After dismissing the Aristotelian syllogism as a device only useful to rhetorically explain findings, he insisted on the need to rely constantly on the clear and constant intuition of certainty; this intuition is to be maintained through the reduction of a problem to simple, evident elements, and the following inferences. Descartes applied these findings to physics and biology in a work, *Le Monde, L'Homme*, that he prudently refrained from publishing after hearing of *Galileo's condemnation (1633), since it put forward the Italian's heliocentric stance. Descartes' physics is mechanistic insofar as it rejects the qualitative method of Aristotle's followers and seeks to understand things on account of their size and motion. Matter is defined by extension. But in spite of these innovations Cartesian physics was considered a failure, especially after *Isaac Newton's intervention. Descartes may have professed to follow the procedures of mathematics to obtain reliable truths, but his interpretation of phenomena does not fundamentally rely on measurement. Furthermore, his use of surprisingly poetic images as hypotheses verified by observable facts contradicts the need to establish a single intuitive inference from simple elements to phenomena. Ironically, Descartes' most successful achievements in sciences are in the field of mathematics. He found the sinus law that explains light's refraction and profoundly changed the practice of geometry by fostering its algebraization.

The cornerstone of Descartes' innovation is the subjective perspective of his philosophy. He may have borrowed many elements from the thought of his contemporaries, but this essential aspect of his system represents a definitive break with tradition. After the rejection of predetermined logical procedure in the *Regulae, Le Monde* brought him closer to the radical affirmation of subjectivity. Against a fundamental principle of Aristotelianism, he held that sensory perception provides no ontological representation of the world, but only a coded image of it (there is no "laughter" in things that tickle). The world, in essence, is completely objectified and its understanding is valid from one's personal point of view.

Descartes' first publication was a set of three treatises (*La Dioptrique, Les Météores, La Géométrie*) preceded by the famous *Discours de la méthode* (1637). Its willingness to break with tradition is signaled not only by the content of the texts but also by the fact that they were written in French instead of Latin. A new public was courted when Descartes declared that good sense was the trait that humans have most in common. The argument participated in his main philosophical thrust: Persons must claim for themselves the freedom that was enunciated in the famous utterance: "I think, therefore I am." This is why the discovery of truth in the *Discours* is told as an autobiography, from the illusions of childhood and through revelation of an admirable science, which happened, the reader is told, one evening while the thinker was sitting next to a stove. The French language of the text also suggests that it can be understood in the social and political context of absolutism and social behavior. In the *Discours*, Descartes explained how he found the first of the evident truths that he needed to build a valid chain of inferences. He encountered it within the very process of doubting that mental representations can be true: To dismiss that one is thinking is impossible. Moreover, doubting necessarily implies a proof of God, since human finite nature is only conceivable on the backdrop of infinity. We may understand God, but certainly not comprehend the infinite idea of God. Later, in what is arguably one of the most important works in the history of philosophy, *Meditationes de prima philosophia* (1641), he fully expounded his system for a public more knowledgeable in philosophy and theology. Thomist realism, for which there is an ontological continuity between object and subject through sensory perceptions, is given a final and radical blow. The objective world is now comprehended under the notion of *res extensa*, whereas the thinking subjective substance is the *res cogitans*. All things of matter can be explained by mechanistic physics. This theory, among many consequences, led Descartes to argue that animals were machines, since they had no souls. Descartes' theory of error is to be found among the many clarifications brought by the *Meditationes*: Free will, by which human beings are in the likeness of God, is infinite, whereas our intellectual capacity is limited. Error is the consequence of freedom's nonabstention in the face of an uncertain choice. Some have doubted the sincerity of Descartes' ultimate metaphysical argument on the existence of God, viewing it a safeguard against the all-too-real possibility of religious persecution. But this appears unlikely. It should be noted that the text was accompanied by objections formulated by men such as *Marin Mersenne, *Thomas Hobbes, and *Pierre Gassendi, along with Descartes' responses.

Descartes' ethics, put forward in *Les Passions de l'Âme* (1649), is grounded in the mechanistic account of the body. Thought is altered by the body's passions and must strive, in a fashion reminiscent of Stoicism, to scientifically understand these phenomena in order to contemplate them with detachment and formulate moral principles. That this knowledge is not readily available does not prevent one from acting in the freedom of its best capacity. The goal of the free mind is to attain generosity by which it esteems oneself and recognizes the freedom of others. In an enigmatic statement, to be found in a letter to Princess Elizabeth of Bohemia (28 June 1645), Descartes declared that the explanation of the body and soul unity (a third substance brought about by the existence of sensations and passions) could not be pursued and that it was better to *practice* it in conversations with other generous souls. This gesture reminds us that Descartes' project defines as much a way of life as a path to philosophical truth.

Bibliography: F. Alquié, *La Découverte métaphysique de l'homme chez Descartes*, 1950; J.-M. Beyssade, *La Philosophie première de Descartes*, 1978; J.-P. Cavaillé, *Descartes, La Fable du Monde*, 1992; D. Garber, *Descartes' Metaphysical Physics*, 1992; S. Gaukroger, *Descartes: An Intellectual Biography*, 1995.

Jean-Vincent Blanchard

DOMENICHINO (DOMENICO ZAMPIERI) (1581–1641). After about the age of thirty-three, Domenico Zampieri was known as Domenichino; he became the most well-known proponent of classicism in seventeenth-century art. The son of a prosperous cobbler, he entered the studio of Annibale Carraci around 1595, who promoted him as a rival to *Guido Reni. Having copied some of Carraci's work in his early years and completed the *Liberation of St. Peter from Prison* (1604), he turned primarily to the frescoes and altarpieces for which he became famous. Between 1604 and 1605, he completed three frescoes for Cardinal Agucchi at the church of St. Onofrio on scenes from the *Life of St. Jerome* and in 1609 finished ceiling frescoes on the *Life of Diana* for the Giustiniani palace. His canvas of *The Last Communion of St. Jerome* was unveiled in 1614, and between 1616 and 1618 he finished, with his assistants, frescoes of Apollo and Diana for Cardinal Aldobrandini at the Frascati Villa.

From 1617 to 1625, Domenichino finished three important altarpieces in Bologna: the *Madonna of the Rosary*, the *Martyrdom of St. Agnes*, and the *Martyrdom of St. Peter*. In the last of these, he revisits an altarpiece of Titian on the same theme, but reverses the image, a practice he used on subjects taken from other artists as well. His most significant secular fresco done between 1621 and 1631 was the design of *Truth Disclosed by Time* (ca. 1622) for the ceiling of the Palazzo Costaguti. The most notable project of Domenichino's career was the frescoed pendentives and apse for the church of St. Andrea della Valle in Rome. Giovanni Lanfranco had also sought this commission, and their two styles illustrated the differences between the "classical" and the "full" Baroque.

Having aroused an intense jealousy among many of his contemporaries, in his last decade Domenichino feared for his life. His final years in Naples were productive but anxious, and his death on 6 April 1641 may have resulted from poisoning. His devotion to the art of Raphael and Annibale Carraci made him a vital link in the classical tradition, and his landscapes in this vein had a decisive influence upon the works of *Claude Lorraine and *Nicolaus Poussin.

Bibliography: J. Pope-Hennessy, *The Drawings of Domenichino in the Collection of His Majesty the King at Windsor Castle*, 1948; R. Spear, *Domenichino*, 2 vols., 1982.

Christopher Baker

DONNE, JOHN (1572–1631). Esteemed clergyman and shaper of seventeenth-century metaphysical poetry, John Donne was the third of six children of devoutly Catholic parents. Donne's mother, Elizabeth Heywood Donne, daughter of Tudor playwright John Heywood, counted among her relatives Sir Thomas More and Tudor playwright John Rastell, both Catholic martyrs. Donne's father, prosperous London ironmonger John Donne, died in 1576, and Elizabeth Donne married London physician John Syminges. Initially tutored at home, Donne attended Hart Hall, Oxford (1584–87), and Cambridge, (1587–89) but as a Catholic was denied degrees. After studying law at Tavies Inn and Lincoln's Inn (1592–95), he traveled abroad (1595), then joined Ralegh's and Essex's campaigns at Cadiz (1596) and the Azores Islands (1597).

In 1597, Donne became secretary to Lord Keeper of the Great Seal Sir Thomas Egerton and, by late 1601, served briefly in Parliament for Brackley, Northampton—he served later for Taunton, Somerset (1614). When he revealed in February 1602 that he had eloped the previous December with Sir George More's sixteen-year-old daughter Anne, More withheld Anne's dowry and had Donne fired. Marriage validation from the Anglican archbishop came in April, but reconciliation with Sir George took longer. The couple lived with family and friends in Pyford (1602–1604), Mitcham (1605–1609), and London (1610–1612) and, from 1603–1617, had twelve children, seven of whom—Constance, John, George, Lucy, Bridget, Margaret, and Elizabeth—lived to adulthood. Donne sought employment wherever he could and increasingly drew upon his literary abilities to attract the favor of influential people. His lifelong friendship with Magdalen Herbert (later Lady Danvers) and her young sons Edward and *George began while Donne's family resided at Mitcham. She particularly encouraged Donne's writing of religious poems and prose, which he often shared with her. Similarly, Lucy, Countess of Bedford, bestowed patronage on Donne, frequently including him among guests at Twickenham Park, standing godmother to his daughter Lucy, and encouraging his secular poems. Sir Robert Drury, another patron, provided the Donnes' Drury House apartment in London and employed Donne as secretary on a continental trip in 1611–1612.

Although the precise date of Donne's Anglican conversion remains unknown, he wrote anti-Catholic tracts for Thomas Morton (1605–1607), argued in *Pseudo-Martyr* (1610) that Catholics should take the Oath of Allegiance, and satirized Ignatius Loyola and his Jesuit Order in *Ignatius His Conclave* (1611). Despite encouragement from Morton, *King James, and others, Donne resisted taking Anglican orders until all other avenues to advancement were closed. His ordination by Bishop John King in 1615 was quickly rewarded with an honorary Doctor of Divinity from Oxford, vicarages at Keystone and at Sevenoaks in Kent, and chaplaincy at Lincoln's Inn. After serving as Viscount Doncaster's chaplain on a diplomatic mission to Germany (1619–1620), he received lifetime appointment as Dean of St. Paul's in London (1621), additional vicarages at Blunham in Bedfordshire (1622), and St. Dunstan's-in-the-West in London (1624), and positions on the Canterbury Prerogative Court Commission (1628) and Lambeth Palace Commission (1629).

Donne's personal life appears suffused with grief and ill health. The Donnes lost a stillborn child (1612), their infant son Nicholas (1613), and both Francis and Mary, ages seven and three (1614). Shortly after delivering another stillborn child in 1617, Anne Donne died and was buried at St. Clement Danes. Donne found refuge, if not consolation, in his work, and thirteen-year-old Constance supervised the household and her six siblings. In 1623, Donne barely survived a second serious illness—he had suffered another some fifteen years earlier—and in 1625 removed his household to Magdalen Danvers' residence at Chelsea until an outbreak of plague in London subsided. Two years later, his nineteen-year-old daughter Lucy died, and by late summer 1630, his own health failed. He preached his last sermon before *King Charles in London in February 1631 and died on 31 March. Of his burial place in St. Paul's Cathedral, only the marble effigy of Donne in his funeral shroud, which he had commissioned and for which he had posed during his final days, survived London's Great Fire of 1666.

Donne's authorship of secular and religious poetry spans his lifetime. Arguably the best of his religious poems—*Holy Sonnets*; *Goodfriday, 1613: Riding Westward*; and *Hymn to God My God, In My Sickness*—are poignant meditations on events pertinent to the Christian life, and his *La Corona* is an intricate sonnet cycle on the life of Christ. His secular poems include satires, verse epistles, elegies, and a variety of occasional verse. Of his several so-called songs and sonnets, such poems as *The Indifferent*, *The Sun Rising*, *The Canonization*, *A Valediction: Forbidding Mourning*, *The Good Morrow*, and *A Nocturnal upon St. Lucie's Day* are particularly representative. Except for *The First and Second Anniversaries* and his *Elegy on Prince Henry* (1613), Donne's poetry remained unpublished until the collection entitled *Poems* (1633). His poetic method employs brilliantly unconventional techniques to startling and provocative effect: deliberately broken, often elliptic syntax; ironic inversions of conventional poetic settings, language, and themes; images and conceits fashioned from extraordinarily disparate components—such as lovers likened to a drafting compass or a sinner's heart to the gate of a besieged city; and a dramatic speaker as likely to be motivated by moral indifference ("I can love any, so she be not true"), righteous indignation ("For Godsake hold your tongue, and let me love"), or subdued astonishment ("I wonder by my troth, what thou, and I / did, till we loved") as by confident salvation ("Death be not proud") or certain depravity ("Oh my black Soule!"). Amounting to full rebellion against Elizabethan poetic conventions, Donne's methods influenced a number of younger seventeenth-century poets: George Herbert, *Richard Crashaw, *Henry Vaughan, and others in England; and in America, *Edward Taylor. But Donne had contemporary detractors: *Ben Jonson disparaged his rough meters, and William Drummond complained of his "metaphysical" concepts. Restoration and eighteenth-century critics applied the term *metaphysical* to whatever qualities they considered objectionable in these poems—*John Dryden, its excessive philosophy, and Samuel Johnson, its *discordia concors* (discordant unity). Thus Donne's method and followers came to be called metaphysical, its pejorative connotation prevailing until the early twentieth century.

In addition to *Pseudo-Martyr* and *Ignatius His Conclave*, Donne's long prose works include *Essays in Divinity*, *Biathanatos*, a defense of suicide unpublished until 1646, and *Devotions upon Emergent Occasions* (1624), meditations occasioned by his 1623 illness. Donne excelled in the pulpit, his popularity rivaled only by *Bishop Lancelot Andrewes, and several of his sermons were published and reprinted during his lifetime. Despite brief exegetical forays into Latin, the sermons remain accessible and powerful, due in part to their blend of styles carefully tailored to varying audiences and topics. The best of them, such as *Deaths Duell*, contain an aesthetic appeal—a keenness of intellect, precision of language, and sharpness of image—absent in those of Donne's contemporaries.

Bibliography: R. Bald, *John Donne: A Life*, 1970; C. Coffin, ed., *The Complete Poetry and Selected Prose of John Donne*, 1994; T.D. Pasquale, *Literature and Sacrament: The Sacred and the Secular in John Donne*, 1999.

Sallye Sheppeard

DOWLAND, JOHN (1563–1626). *Semper Dowland Semper Dolens*, a pavane by Dowland, offers an apt commentary on this English lutenist-composer. Though cited in Fuller's *Worthies* as "a cheerful person . . . passing his days in lawful merriment," Dowland's own account of his life indicates that he suffered from melancholy and perceived himself as having been slighted in his career. His melancholic temperament is evident, for instance, in the preponderance of serious lute songs (e.g., *In darknesse let mee dwell*) over light and dancelike ayres like *Fine knacks for ladies*. Biographical details are scarce for Dowland's early life. From 1580 to 1584 he was servant to Sir Henry Cobham, ambassador to France, where he converted to Catholicism. In 1588, he received the B.Mus. from Christ Church, Oxford. During the 1590s, Dowland's preeminence began to be recognized, as witnessed by the encomiums of *Thomas Campion and others. Dowland was disappointed, however, in 1594, when he was rejected for a vacancy among the queen's lutenists. Feeling resentful, Dowland embarked for the continent, where he was cordially received in Italy and Germany; he returned home in late 1596 or early 1597. Later that year, his successful *First Booke of Songs* was published. Dowland again went abroad, and by 1598 was lutenist for Christian IV of Denmark. While there, Dowland's second and third songbooks appeared; during this time he traveled between England and Denmark, permanently returning home in 1606. In 1612 he became one of *James I's "musicians for the lutes," a position Dowland kept until his death in 1626. Dowland and others arranged his lute music for various instruments, and his most famous work, *Lachrimae*, inspired countless compositions. Most of his lute music is stylistically derived from vocal polyphony, but much also from dance and popular music. Dowland's consort music also shows mastery, but his lute songs display his highest genius.

Bibliography: D. Leech-Wilkinson, "My Lady's Tears: A Pair of Songs by John Dowland," *Early Music* 19 (1991):227–233; D. Poulton, *John Dowland*, 2nd ed., 1982.

Susan Treacy

DRAYTON, MICHAEL (1563–1631). Born in Warwickshire, Michael Drayton had a long and productive literary career, one that saw him working within various genres: eclogues, odes, pastorals, satires, sonnets, and a host of historical poems. Between 1597 and 1602, Drayton also wrote for the stage, having a hand in twenty-four plays, many of them now lost. After the death of his first patron, Sir Henry Goodyere of Polesworth, Drayton received the patronage of Lucy, countess of Bedford, though his relations with the Countess went sour around 1597. Drayton was buried alongside Chaucer and Spenser in Westminster Abbey.

Drayton's literary career spanned the reigns of three monarchs: Queen Elizabeth I, *King James VI and I, and *King Charles. It was under Elizabeth and James that Drayton composed and revised the bulk of his poetry, and it is crucial to approach much of his work as that of a high Elizabethan writing within and against Stuart absolutism. Similarly, it is important to view Drayton's poetry as Spenserian in terms of not only its literary form but also its political content. If Drayton's early work—for example, *Idea The Shepheards Garland* (1593)—is indebted to the archaic diction of Edmund Spenser's *The Shepheardes Calender*, then his later work shares Spenser's intense interest in national history. This is particularly evident in his *Mortimeriados* (1596—revised as *The Barons Warres* in 1603), his *Englands Heroicall Epistles* (1597), and in his own *Faerie Queene*, the monumental 30,000-line poem *Poly-Olbion* (1612, 1622).

It is for his *Poly-Olbion* that Drayton is best known. A county-by-county description of England and Wales, organized into thirty "songs" written in Alexandrine couplets, *Poly-Olbion* is at once a chorographic and cartographic paean to the rivers, mountains, and forests that dominate the English and Welsh landscape. Accompanying the eighteen "songs" of the first edition were antiquarian annotations by *John Selden (which were dropped from the second edition). The regional maps, engraved by William Hole, include lively renditions of allegorical figures of hunters, shepherds, husbandmen, and water nymphs. The poem's dedication to the land seems to suggest political disengagement on Drayton's behalf; however, the fact that the poem has little to say about cities and courts is crucial to understanding Drayton's politics. In the wake of James's accession to the throne, Drayton published the congratulatory poem *To the Majestie of King James* (1603) with the expectation that James would bestow favor upon him. The new king never reciprocated. That James is totally absent from *Poly-Olbion* (he is omitted from the poem's list of great British rulers) is a reflection of the poet's rejection of the crown in favor of the country. Moreover, Drayton's epic was dedicated to, and included an engraved portrait of, Prince Henry, who was often celebrated by anticourtly poets as the champion of militant Protestant politics. Not surprisingly, Drayton's *Poemes Lyrick and pastorall* (1606) praise, among other poets, *Samuel Daniel, who, too, fell out of favor with the Jacobean court.

Bibliography: J. Brink, *Michael Drayton Revisited*, 1990; R. Helgerson, *Forms of Nationhood: The Elizabethan Writing of England*, 1992.

Christopher Ivic

DRYDEN, JOHN (1631–1700). John Dryden was born in Northamptonshire, the son of Erasmus Dryden and his wife Mary. He studied at the Westminster School, then matriculated at Trinity College, Cambridge, where he took his degree in 1654. Although in his youth he served as a Latin secretary for the Protectorate, he died a Catholic Jacobite. Dryden's first poem, "Upon the Death of the Lord Hastings" (1649), supports his claim that *Abraham Cowley was the "Darling of my youth," but by the time of his "Heroique Stanzas" (1659), he preferred the quatrains of *Sir William Davenant's *Gondibert* (1651). In his poems on the restoration, he uses the heroic couplets that he continued to perfect for the rest of his life. These poems also show him assembling the stock of allusions and myths, and cultivating the public voice that we associate with Augustine poetry. For models, he turned not only to Virgil but to contemporaries like *Edmund Waller and *Sir John Denham.

Although his marriage in 1663 to Elizabeth Howard brought him into contact with the Duke of York and other courtiers, Dryden depended on the stage for his living. His first play, a humors comedy called *The Wild Gallant* (1664), was a failure. He would eventually coauthor a successful farce—*Sir Martin Mar-All* (1667)—and write a great comedy that blended sexual humor, farce, and political satire—*Amphitrion* (1690)—but he found tragicomedy and tragedy more amenable to his talents. After the tragicomedy *The Rival Ladies* (1664), he coauthored the Restoration's first fully rhymed heroic tragedy, *The Indian Queen* (1664) with his brother-in-law Sir Robert Howard. He followed with the sequel *The Indian Emperor* (1665).

The year between 1667 and 1668 marked a turning point in his career. The signal service that he rendered the Stuarts with his publication of *Annus Mirabilis* (1667) made him a natural choice for poet laureate. At the same time, he found repeated success on stage. Not only did he write a tragicomedy that was a favorite with the king (*Secret Love*), he also collaborated with Davenant on a semioperatic adaptation of *The Tempest* that helped promote the use of spectacle and music on stage and placed him in a position to sign an unprecedented agreement with the King's Company to supply three plays a year in return for a share in the company comparable to that held by the leading actors. Although he had already supported rhymed drama in the dedication of *The Rival Ladies*, he expanded his remarks in his first extended work of criticism, *Essay of Dramatick Poesie* (1668). Written as a dialogue among four characters, the work shows Dryden assessing the works of such precursors as *William Shakespeare and *Ben Jonson, and trying to define a native literary tradition, but its discussion of rhyme and the unities of time, place, and action on stage provoked the sharpest contemporary reaction. Dryden believed that by borrowing some French conventions, his own generation of poets could achieve a heightened mimesis— "Nature wrought up to an higher pitch"—that would enable them to differentiate themselves from the geniuses of a prior age.

He defended and expanded his position in subsequent works that drew him into a fierce critical controversy, and he also attempted to put his theory into action in *Tyrannic Love* (1669) and *The Conquest of Granada* (1670–1671), the butt of the Duke of Buckingham's burlesque *The Rehearsal* (1671). Being ridiculed on stage as the writer Bays did not debilitate Dryden, for he wrote his greatest tragicomedy that year (*Marriage A-la-Mode*) and produced his most mature rhymed tragedy four years later (*Aureng-Zebe*). It was also around this time that Dryden wrote *MacFlecknoe*, a great mock heroic poem aimed at his literary rival, Thomas Shadwell. In the prologue to *Aureng-Zebe*, Dryden confessed that he had grown "weary" of rhyme, feeling that passion was "too fierce to be in Fetters bound" and that "Nature flies him like Enchanted Ground." He yielded the "foremost Honours" to the "less polish'd" age of Shakespeare. His next tragedy, *All for Love* (1677), was a revision of *Antony and Cleopatra* that, while written in blank verse and showing a deep engagement with Shakespeare's play, also shows the influence of *John Milton's enjambed verse and of French dramatic form.

The Exclusion Crisis prompted his next great work, *Absalom and Achitophel* (1681). Dryden expressed a lifelong ambition to write an epic, famously exclaiming "that Poet has cut us all out" after reading *Paradise Lost*, but it is in this poem, in which the events of the crisis are retold through the "throne narrative" of 2 Samuel and in which the "Majesty of the Heroique"—much of it Miltonic—is "finely mix'd with the Venom" of satire, that he comes closest to writing an original epic. The same events persuaded him to write the Juvenalian satire *The Medal* (1682) and (with Nathaniel Lee) a political play aimed at the duke of Monmouth, *The Duke of Guise* (1682). In these years, he also published *Religio Laici* and a coauthored translation of *Nicolas Boileau-Despreaux's *Art of Poetry*. Verse essays like these prompted Samuel Johnson to claim that Dryden was the first "who joined argument with poetry."

The 1680s were another period of momentous change for Dryden. The sovereign whom Dryden had served for most of his poetic life died before he could see the semiopera that was meant to celebrate his reign, *Albion and Albanius*. Dryden tried to mourn the king, and to solidify the succession of *James II, in his *Threnodia Augustalis* (1685). His decision to write a Pindaric ode for the king stemmed from a renewed interest in the sorts of complex stanzaic forms that he had once admired in Cowley and that he was now experimenting with in his translations. These experiments led to his ode on Anne Killigrew (1686) and *Alexander's Feast* (1697).

Why precisely Dryden converted to Catholicism—and publicized his choice in *The Hind and the Panther* (1687)—remains uncertain, but the move was consistent with his support of James II. His conversion made his position after 1688, when he lost his posts as poet laureate and historiographer royal, untenable. The Glorious Revolution prompted him to write his last great tragedy, *Don Sebastian* (1689), a play centrally concerned with the competing claims of loyalty and interest. His last important works for the stage were the comedy *Amphitryon* (1690) and the semiopera *King Arthur* (1691), whose score was contributed by *Henry Purcell.

With his contributions to *Ovid's Epistles* (1680), *Sylvae* (1685), and *Satires of Juvenal and Persius* (1692), he began to devote more of his energy to translating and analyzing ancient poetry. His "Discourse Concerning the Original and Progress of Satire" (1692) is his critical masterpiece. This was followed by *Aeneis* (1697) and *Fables Ancient and Modern* (1700). Dryden advocated a form of "Paraphrase" in his translations that mediated between literal "Metaphrase" and free "Imitation." He let "*Virgil* speak such *English*, as he wou'd himself have spoken, if he had been born in *England*, and in this present Age." His *Fables*, an anthology of original poems and translations, show Dryden making a final attempt to insert himself into a literary tradition that he now traced from Homer through Boccaccio and Chaucer. The poet who had thought so much about the claims of tradition and original talent, judgment and fancy, now felt thoughts "come crowding in" fast upon him. This, in Pope's view, was Dryden's "most glorious season," when his "fire, like the Sun's, shin'd clearest toward its setting."

No century since the eighteenth has shown such an appreciation of Dryden. Johnson not only named him "the father of English criticism," he claimed that no nation had ever had an author who "enriched his language with so many models." It seems appropriate to give Johnson the last word: "What was said of Rome, adorned by Augustus, may be applied . . . to English poetry embellished by Dryden . . . , 'he found it brick, and left it marble.' "

Bibliography: H. Swedenberg, ed., *Essential Articles for the Study of John Dryden*, 1966; J. Winn, *John Dryden and His World*, 1987.

Blair Hoxby

D'URFÉ, HONORÉ (1567–1625). The literary production of Honoré d'Urfé is somewhat limited, yet vast. Born in 1567 in Savoy, Honoré, like his brother

Anne, was surrounded from the outset by a typically Renaissance literary and cultural atmosphere. While he is credited with only six literary titles, one of these, the pastoral novel *L'Astrée*, is significant in both its dimension and its contribution to European literature. Published in four parts, from 1607 to 1627, with the concluding volume completed and published by Baro, d'Urfé's secretary, *L'Astrée* is important on several levels. In the two centuries following its publication, it was deemed to be a largely autobiographical account of d'Urfé's life and especially of his tumultuous relationship with Diane de Châteaumorand; however, modern scholars contend that the novel merits examination in contexts beyond that of the author's life. It is a complex yet structured work, with several distinct narrators recounting the myriad adventures of countless characters; these intrigues are regularly interrupted by various intercalated tales. Moreover, *L'Astree* has also been qualified as encyclopedic in its presentation of the Renaissance ideas with which d'Urfé was imbued. Literary and philosophical influences such as pastoral fiction and neoplatonism are readily apparent, as is a typically French Renaissance view of history in which the author depicts his idealized version of the history of fifth-century Gaul.

In addition to *L'Astrée*, d'Urfé wrote a circumstantial piece, a pastoral poem and play, an unpublished six-part epic poem, and the three-volume *Epistres Morales*. He also took an active role in political endeavors. He joined the Catholic League and served as governor of Forez, then later represented France at the Savoy court in Turin. At the time of his death, he was fighting in the War of the Valteline, during which he perished, probably of illness, at the age of 58.

Bibliography: M. Gaume, *Les Inspirations et les sources de l'œuvre d'Honoré d'Urfé*, 1977; L. Horowitz, *Honoré d'Urfé*, 1984.

Laura L. Dennis-Bay

DU VERGIER DE HAURANNE, JEAN (ST. CYRAN) (1581–1643) Born in Bayonne, Du vergier studied theology at Louvain and settled in Paris, where he distinguished himself for a learned response to the question of the circumstances under which one might give his life for his king. He retreated for years to study at his family's estate in Bayonne with his friend and colleague, *Cornelius Jansen, and then joined intellectual forces with *Richelieu against the Huguenots. Becoming in 1620 the abbot of the Benedictine monastery of Saint-Cyran, he thereafter became known by that name. Sixteen years later as head of Port-Royal, the convent which became the home of Jansenism, he developed the idea of spiritual directorship. He scrapped with the Jesuits over the questions of methods of study as well as episcopal authority. He later turned against Richelieu, who jailed him in 1638. He was released after Richelieu's death in 1642 and died in October 1643.

In the controversies with the Huguenots, St. Cyran held for careful and disciplined study of the ancients, for a Pauline-Augustinian theology with a sacramental concept of the church as the Body of Christ, and for reform of the temporal practices of the church. In the disputes with the Jesuits, he favored approaching theology in the manner of St. Augustine rather than through the dialectics of the medieval scholastic philosophers or Schoolmen. He was not persuaded that the regular clergy were superior to their secular counterparts nor did he find a worldly cleric of greater merit than a simple priest. As for Richelieu's idea that the church was to submit to civil authority, M. de St. Cyran would have none of it. He was regarded as a spiritual director of sympathy and personal piety, one who urged charity as the prime virtue. More than one hundred of his letters of direction were published in 1645 and 1647, but a comparable number went unpublished until 1962. The contributions of the Abbé to Jansenism depend on one's point of view. Certainly St. Cyran and Jansen were close friends, and it stands to reason that two gifted minds studying together contributed to the development of each. St. Cyran's part in Jansen's *Augustinus* (1640) is a matter of schoarly debate; however, St. Cyran did promulgate in France Jansen's idea of spiritual conversion. That advocacy is documented in the *Frequent Communion* (1643) of St. Cyran's pupil and follower, *Antoine Arnauld.

Bibliography: A. Sedgwick, *Jansensim in Seventeenth-Century France: Voices from the Wilderness*, 1977.

Martha Oberle

E

EARLE, JOHN (1601–1655). Tutor to Prince Charles during the exile of Royalists in France, John Earle translated from English into Latin *Eikon Basilike*, the putative meditations of *Charles I while in prison. This text helped keep the spirit of those in exile high, explaining in their own king's words his innocence, thus justifying the loyalty of his followers. Earle's own loyalty during over sixteen years of his exile in France earned him an important clergy position when *Charles II gained the throne at the Restoration in 1660; Charles II appointed him Dean of Westminster. Later, he became Bishop of Salisbury. Prior to this political prominence, the Oxford University graduate and fellow of Merton College had established himself as an important writer. Earle's *Microcosmography, or A Piece of the World Discovered in Essays and Characters* catapulted him to fame with its rendering of behavioral types, each as a distinct character. *Microcosmography* became so popular that the four separate printings in 1628 did not exhaust demand. Two expanded versions followed within the next five years, securing the work's influence. It is still valued today, in largest degree due to Earle's presentation of humans as individuals—each unique and not merely standing for a virtue or vice as was the common method behind most other early modern English "Characteries."

Bibliography: H. Osborne, ed., Introduction, *Microcosmographia or A Piece of the World Discovered in Essays and Characters*, by John Earle, 1971.

Rebecca Totaro

ELIOT, JOHN (1604–1690). Puritan apostle to the Indians of the Massachusetts Bay Colony, minister of the church at Roxbury, and author of the first book printed in North America, The Bay Psalm Book, Eliot was born in Widford, Hertfordshire, England in 1604. He earned his degree from Jesus College, Cambridge, in 1622, nine years before emigrating to Boston, Massachusetts, in 1631. From 1632 until his death, he was ordained as "teacher" (pastor) of the church of Roxbury. In 1640, along with Thomas Weld and Richard Mather,

Eliot coauthored *The Whole Book of Psalms Faithfully Translated into English Meter*, otherwise known as The Bay Psalm Book. Shortly thereafter, Eliot engaged in communal efforts to "Christianize the Indians," and by 1646 he was preaching to them in their own Algonquian tongue. He established a community and a church of "praying Indians" at Natick, Nonantun (Newton), and elsewhere. Each village had a school where Indians were taught the English language and assimilated English culture. It is reputed that, by 1674, there were some fourteen "praying Indian" villages and nearly 4,000 converts. By 1658, Eliot had completed a translation of the Bible into the Algonquian dialect. The first edition of the New Testament was printed at Cambridge and released in 1661, while the whole Bible (Old Testament of 1663, New Testament of 1662, and the metrical versions of the Psalms) was published in 1663. Many copies of Eliot's biblical translation were lost or destroyed during King Philip's Indian War of 1675–1676, and villages themselves never fully recovered from the serious devastation and persecutions the war imposed upon them. In 1677, at the age of seventy-three, Eliot began a revised edition of the entire work, which was subsequently published in 1685. Eliot died on 21 May 1690, in Roxbury, Massachusetts.

Bibliography: R. Cogley, *John Eliot's Mission to the Indians before King Philip's War*, 1999; O. Winslow, *John Eliot: Apostle to the Indians*, 1968.

David M. Rosen

EPISCOPIUS, SIMON (1583–1634). Born in Amsterdam, Episcopius's academic abilities brought him to the attention of Cornelis Bennink and, thanks to Bennink, the chance to study at Leiden, at the States College first and then, in 1603, at the theology faculty under *Arminius, whose follower and advocate Episcopius became. Episcopius signed the Remonstrance of 1610, the declaration which set forth the more liberal and less dogmatic Calvinist views of the Arminian sect and earned them the name Remonstrants. Episcopius defended the Remonstrants before the States General in 1611, and in 1612, succeeding Franciscus Gomarus, became professor of theology at Leiden. The bitter opposition of the orthodox Calvinists culminated in the Synod of Dort in 1618, which condemned, without hearing or defense, the Remonstrants' theology. At the close of the Synod in 1619, Episcopius and the other Remonstrant representatives had their offices taken from them, and all were exiled. Episcopius went to Antwerp and there published in Dutch the Remonstrants' Confession of Faith (1621). The following year he published the same work in his original Latin.

From Antwerp, Episcopius retired to France where he engaged in dialogue with the Jesuit Luke Wadding. Wadding did not draw Episcopius to Rome, if that was the Jesuit's intention. Following the death of Maurice of Orange, Episcopius returned to Holland in 1626 as preacher at the Remonstrant Church in Amsterdam. In 1634, he was appointed Rector of the Remonstrant College in Amsterdam and held this office until his death. Episcopius's great contribution was the organization and defense of the principles of Arminius. The Remonstrant position was that Christianity was more a moral power and less an intellectual system: What was necessary was the knowledge and acceptance of whatever of Christianity would cause a change in the heart and life of the individual. Free will, according to Episcopius, had been damaged but not lost in the Fall and, therefore, Man's nature, in and of itself, is capable of seeking and knowing the good. Episcopius's writings were published in a collection (1650–1665), and in a single volume (1678). In 1684, an English translation of his Confession of Faith appeared.

Bibliography: D. Nobbs, *Theocracy and Toleration*, 1938; J. Platt, *Reformed Thought and Scholasticism: The Arguments for the Existence of God in Dutch Theology, 1575– 1650*, 1982.

Martha Oberle

ESCOBAR Y MENDOZA, ANTONIO (1589–1669). A Spanish Jesuit preacher and moral theologian whose support of probabilism was ridiculed by *Blaise Pascal, Antonio Escobar was born in Valladolid, Spain. At the age of sixteen, he entered the Society of Jesus and became a distinguished scholar and a preacher of note, winning prestige and distinction for his talent among Catholic scholars. Escobar's use of probabilism in his *Manual of Cases of Conscience* provoked the French scientific and religious philosopher Blaise Pascal's bitter attack. In Pascal's *Provincial Letters* (fifth through ninth), Escobar is quoted with derision and indignation for his apparently lax moral principles that advocate the belief that the end justifies the means, although some scholars find these attacks unfounded if Escobar's words are interpreted in context.

Probabilism is a method of resolving questions of conscience by applying the premise that, when one does not know whether an action would be sinful or permissible, he may rely on a *probable opinion* for its permissibility even though a more probable opinion calls it sinful. An opinion is considered probable either if logical arguments can be cited in its favor or if recognized authorities grant it support. The *Jansenists, such as Pascal, who believed that in doubtful cases of conscience one should follow the safer solution, criticized the benignity of the Jesuit confessors as leading to a mere avoidance of sin. The question of probabilism became important during the seventeenth century, when social and economic developments, such us banking, came into disagreement with traditional moral precepts, resulting in many conflicts of conscience. Escobar y Mendoza's moral guidance was also satirized by other French authors such as *Molière (particularly in *Le Tartuffe*) and *Jean de La Fontaine. Among his supporters, Escobar's writings are recognized as classical challenges to orthodoxy.

Escobar's major works are *Sumula Casuum concientiae* (1626); *Liber Theologiae Moralis* (1644); *Examen et praxix confessionarium* (1647); *Theologia Moralis* (Lyons, 1650; Venice, 1652); *Universae Theologicae Moralis receptae* *sententiae* (1663); *De Triplici Statu Ecclesiastico* (1663); and *De Justitia et de legibus* (1663). Antonio Escobar was also a baroque poet. His two long poems are *San Ignacio* (1613) and *Historia de la Virgen Madre de Dios* (1618), the latter reworked as *Nueva Jerusalén María* (1625). The first poem is one of the many epics written around the life of founder of the Society of Jesus, Ignatius Loyola. It consists of eighteen cantos in the highly sophisticated style of the Spanish poet *Luis de Góngora. His second poem, in the same elaborated manner, comprises an allegory in which Mary, the Mother of God, is audaciously identified with a New Jerusalem whose foundations were made of jacinth, chrysoprase, and many other precious stones.

Bibliography: A. Gazier, *Blaise Pascal et Antoine Escobar*, 1912.

Angela Morales

ETHEREGE, GEORGE (ca.1634–1691). Etherege's father was a minor courtier who purchased the office of purveyor to Queen Henrietta Maria and accompanied her to France during the English Civil War. It is unknown if the playwright joined his father in France, but he certainly inherited his royalism. In 1654, he apprenticed to an attorney and became an articulated clerk in 1658. Etherege's activities until 1663 remain obscure; in that year we know he began a risqué correspondence with Lord Bathurst. In 1664, *The Comical Revenge, or Love in a Tub*, his first play, was acted by the Duke's Company and was a huge financial success.

The *Comical Revenge* was considered unique in its day and is generally believed to be the first Restoration comedy. What made *The Comical Revenge* new was partly a matter of style: "Easy Etherege," as he was known, wrote dialogue to resemble the way witty gentlemen conversed. The plot of *The Comical Revenge*, the rake or libertine protagonist being lured into marriage with a rich widow or young wealthy heroine, served as a pattern for later comedies of the era. Restoration comedy has also often been known as a comedy of manners, and Etherege certainly poked fun at those who became slaves to style—the fops—and those who had no style—boors and country folk. But perhaps the most distinguishing feature of Restoration comedy visible in this play is the contempt for the ideals so often lauded in previous Cavalier literature. *The Comical Revenge* and its successors converted ideals downward to their physical reality. A scene discussing the "love wounds" of unrequited love is immediately followed by a scene discussing the "love wounds" of syphilis. Converting ideals downward also fit in with the general concern of exploring the difference between nature and appearance, a concern that explains the ubiquity of masquerade, disguise, and deceit in Etherege's first play and its successors.

What we know of Etherege's life after the staging of *The Comical Revenge* suggests that he lived like one of the rakes in his comedies. (Indeed, one contemporary thought Sir Fopling Flutter, the title character of *The Man of Mode*, was autobiographical.) Etherege gambled and caroused with such noted wits as

*John Wilmot, earl of Rochester and Buckingham. In 1668, he wrote a second successful play, *She Wou'd if She Cou'd*. According to *Samuel Pepys, over a thousand people attended the opening, including *King Charles ll. Charles must have liked the play, because Etherege was rewarded with the position of Gentleman of the Privy Chamber in Ordinary and with the secretaryship to the ambassador to Constantinople, where he lived for two years. Upon his return, he lived a life of dissolute idleness, indicative of which was a duel (aborted because Etherege's opponent passed out from inebriation). In 1676, Etherege's most critically acclaimed play, *Sir Fopling Flutter, or the Man of Mode*, appeared on stage and Etherege received a knighthood. Among its many laudable qualities, *The Man of Mode* presented a female lead, Harriet, who matched the protagonist in terms of wit if not of ribaldry. After the success of this play, Etherege, following the lead of many a rake, married a widow, but marriage had as much success reforming him as *The Man of Mode* suggests it would have reformed his famous protagonist Dorimant; Etherege pursued gambling and lechery to the point where he accepted the ambassadorship to Ratisbon to escape his debts. In Germany, Etherege continued to chase women and play cards, although with less success and lower stakes than in England. He also wrote many letters, which have since been published. In 1687, when he discovered that his failure to mind his business was making him a minor laughingstock on the continent, he performed his duty with more diligence. Etherege gathered much information about the intentions of *William of Orange, and he warned *James II, to no avail, that his son-in-law might invade England. After the Glorious Revolution, Etherege kept his Cavalier principles, joining the Stuart court in exile until his death in 1691 or 1692.

Bibliography: F. Bracher, ed., *Letters of Sir George Etherege*, 1973; N. Holland, *The First Modern Comedies: The Significance of Etherege, Wycherley, and Congreve*, 1959.
Brian Weiser

EVELYN, JOHN (1620–1706). John Evelyn was the epitome of the English virtuoso and at the center of the social, political, and intellectual world of his day; his *Diary*, kept for eighty-five years, is considered one of the most comprehensive and historically informative documents of Restoration society. Evelyn unceasingly produced volumes on gardening, air pollution, architecture, numismatics, engraving, politics, sculpture, and participated in written debates about the relative merits of the ancients and moderns. He also was a founding member of the Royal Society, and thus part of that community of intellectuals that wished to establish a national program of scientific and technological development for the social and economic advancement of England.

Evelyn was born in 1620 in Surrey to a gentry family who made their fortune via the manufacture of gunpowder. As the English Civil War began, Evelyn prudently chose to go on an extensive "Grand Tour" of Italy and France; while in Paris, he married the daughter of Sir Richard Browne, whose household was a center for exiled English Royalists. In fact, Evelyn merits credit for helping to establish the "Grand Tour" as part of gentlemanly education. Evelyn's traveling experiences were crucial in forming his intellectual interests, particularly his fascination with humanism and classical art and architecture, and shaped his goal of reconciling the wisdom of the ancients with Restoration culture. Upon his return to England in 1652, Evelyn wrote a series of works about Renaissance architecture, city planning, painting, and sculpture to improve what he considered "backward" English tastes (*The Whole Body of Antient and Modern Architecture* [1680] and *An Idea of the Perfection of Painting* . . . [1668]). After the Great Fire of London (1666), Evelyn proposed several schemes for modeling London's city plan on the principles of Renaissance symmetry and formal balance, although ultimately these came to naught. Evelyn was also responsible for the discovery and patronage of the hitherto unknown wood sculptor *Grinling Gibbons, and for preserving his patron the Earl of Arundel's extensive collections of Greek antiquities, including the Arundel Marbles housed at Oxford. Evelyn's interest in the classical aesthetic also extended to gardening. Works like the *Sylva, or a Discourse of Forest Trees* (1664), which encouraged estate owners to plant timber for the English navy, and the *Compleat Gardner* (1693) established him as the standard English authority on the subject. His designs modeled on Roman formal gardens at Albury Park and Sayes Court, Deptford, are still extant and were influential in shaping formal English garden design until the nineteenth century.

As Joseph Levine notes, Evelyn believed that the arts could span the gap between modern sensibility and the classics by contributing to the mechanical as well as the liberal arts, to scientific knowledge as well as morality. Therefore, when the Restoration of *Charles II offered Evelyn the opportunity to participate in public affairs, he became one of the founding members of the Royal Society, contributing his botanical observations and history of trades to improve England's estate. Further, Evelyn was an appointee in 1671 to the Council for Trade and Foreign Plantations, and under *James II was made a Commissioner for the Privy Seal. A fervent advocate of the restored Anglican Church, he began an ambitious *History of Religion* in 1654, which was published posthumously. Evelyn's interest in the whole realm of learning, his extensive observations of Restoration culture in his *Diary*, and his attempts to reconcile classical learning with seventeenth-century interests in scientific, technological, and cultural progress thus made him the exemplar of the Restoration polymath.

Bibliography: E. de Beer, ed., *The Diary of John Evelyn*, 1955; M. Hunter, *The Royal Society and Its Fellows 1660–1700: The Morphology of an Early Scientific Institution*, 1994; J. Levine, *Between the Ancients and the Moderns: Baroque Culture in Restoration England*, 1999.

Anna Marie Roos

F

FABRICIUS, JOHANN ALBERT (1668–1736). Johann Albert Fabricius was born in Leipzig in 1668. He attended school in Leipzig and, from 1684, in Quedlinburg. From 1686, he studied theology, philology, and medicine at the university in Leipzig. After finishing his studies in 1688, he traveled to Hamburg in 1693 and there became librarian to Dr. Johann Friedrich Mayer. Under Mayer's mentorship, Fabricius took part in theological debates and traveled to Sweden in 1696. Mayer held appointments as pastor in Hamburg and as professor of theology in Kiel, and Fabricius was granted a doctorate of theology from Kiel in 1699. The same year, he was appointed Professor of Eloquence and Moral Philosophy at the Hamburg Academic Gymnasium. From 1708 to 1711 he was also rector of the Hamburg Johanneum.

Fabricius's works span the disciplines of theology, history, grammar, and classics. His *Bibliotheca latina* (1697), *Bibliotheca latina mediae et infimae aetatis* (1734–1736), and *Bibliotheca graeca* (1705–1707) are biobibliographies, which have made a lasting contribution to classical scholarship; reprints appear in many university libraries. In addition to his own prolific writing, Fabricius edited and translated numerous works on classical and church history, theology, classical literature, and physicotheology.

Until his death in 1736, Fabricius belonged to a series of important intellectual circles in Hamburg. Composed of historians, politicians, clergymen, poets, journalists, and academics, associations like the "Deutsch-übende Gesellschaft" and the "Patriotische Gesellschaft" dedicated themselves to improving society through education and the promotion of the vernacular language and literature. Through his pedagogical initiatives, through his many publications, whether scholarly or popular (his writings for the weekly paper *The Patriot*, for example), and through his extensive correspondence, Fabricius became an internationally recognized academic. His massive encyclopedic works provide a groundwork for classical research, while his writings on language display a humane and liberal understanding of the organic development of language and dialect.

Bibliography: Herbert Ernst Brekle, "Systemlinguistik vs. linguistischer Realismus in der Sprachbrauchdiskussion am Beispiel der Streitschriften zwischen Richey und Fabricius (1726)," *History and Historiography of Linguistics* (1990), 483–492; Jürgen Rathje, "Gelehrtenschulen—Gelehrte, Gelehrtenzirkel und Hamburgs geistiges Leben im frühen 18. Jahrhundert," *Hamburg im Zeitalter der Aufklärung* (1989), 93–121.

Janet Bertsch

FAHRENHEIT, GABRIEL DANIEL (1686–1736). A German-Dutch physicist and inventor, Fahrenheit devised the first accurate thermometer and created the temperature scale that bears his name. He was born in Danzig (now Gdansk), Poland. He was the eldest of five children; his siblings included three sisters and a brother. His father was a wealthy merchant and his mother was the daughter of a Danzig wholesaler. Fahrenheit had no formal schooling and spent the first fifteen years of his life in Danzig. When both of his parents died suddenly in 1701, his guardian sent him to Amsterdam to learn a trade. While there, the young Fahrenheit became fascinated with the making of scientific instruments, and even though the manufacture of these instruments was a small, very specialized profession in the seventeenth century, he chose this profession as his life's work.

In 1707, Fahrenheit began traveling throughout Europe in order to gain knowledge from scientists and to observe the methods of instrument manufacturers. During this time, he greatly increased his knowledge of constructing scientific instruments. In 1708, on a visit to Copenhagen, Fahrenheit met Ole Rømer, from whom he learned the craft of making thermometers. Fahrenheit also modified the temperature scale used by Rømer, which was graduated to a scale of 22.5 degrees, or three-eighths of the standard scale of 60 degrees, Rømer's boiling point for water. At first, Fahrenheit's temperature scale had the same fixed points as Rømer's. In 1714, Fahrenheit devised a scale using an alcohol thermometer with fixed points set at 7.5 degrees and 22.5 degrees. He determined the lower point by submersing his thermometer in a mixture of ice, water, and common salt. He established the higher point—also called "the blood heat" (i.e., body temperature)—by inserting the instrument under the arm of a healthy male. Fahrenheit also divided each of Rømer's degrees into four parts so that the upper reading eventually became 90 degrees and the lower reading became 30 degrees, the freezing point of water. Later, in 1717, in order to obviate the need for fractions, Fahrenheit moved his upper point (blood heat point) to 96 degrees and his lower one (ice point) to 32 degrees. In this experiment, he replaced the alcohol thermometer with a mercury-in-glass thermometer. Also in 1717, Fahrenheit began making mercury thermometers commercially. In 1724, he was elected to the Royal Society of London and, moreover, contributed five papers to *The Transactions of the Royal Society* of that year. In 1736, Fahrenheit patented a pumping device for draining the Dutch lowlands; later that year, on 16 September, he died at The Hague, in the Netherlands.

After Fahrenheit's death, his scale was again changed so that the upper reading was based on the boiling point of water (212 degrees Fahrenheit), thus changing the body temperature to 98.6 degrees Fahrenheit. However, because the two fixed points are separated by 180 degrees, the Fahrenheit scale is cumbersome, especially for scientific work. The scale is still used today in the United States by lay people and weather forecasters. In scientific work, the Celsius scale is used, in which one degree Fahrenheit is equal to 100/180 or 5/9 degree Celsius. Thus, to convert degrees Fahrenheit to degrees Celsius, or Centigrade, the following formula is used: T [Centigrade] = (T [Fahrenheit] − 32) x 5/9.

Bibliography: C. Gillespie, ed., *Dictionary of Scientific Biography*, 1970; W. Knowles, *A History of the Thermometer and Its Use in Meteorology*, 1966.

Paul C.L. Tang

FARQUHAR, GEORGE (?1667–1707). Born to an Irish Protestant clergyman probably in 1667, George Farquhar would be labeled, along with *Sir John Vanbrugh, one of the last two great representatives of the type of theater labeled Restoration drama or comedy of manners. He gained little success as a member of Dublin's Smock Alley Company, in one play clumsily drawing blood from a fellow actor. Supposedly encouraged and financially supported by Robert Wilks, Farquhar abandoned the stage in order to write.

Never coy about his materialistic motivation, Farquhar managed to place his first production, *Love and a Bottle*, in the 1698 season at London's Drury Lane Theater (The King's Company) where all of his plays except one would debut. *The Constant Couple* (1699) proved hugely successful with a record fifty-three nights of performance. While he would never match that early triumph, Farquhar immediately followed with *The Inconstant* (1702) and *The Twin Rivals* (1703). His life mirrored his art when Farquhar was later tricked into marriage to a widow he had mistakenly believed to be wealthy. Burdened by the demands of an instant family that included stepchildren, he produced *The Stage Coach* (1705) before signing on as a military recruiting officer. He abandoned that effort in 1705 to write the first of two final plays, *The Recruiting Officer* (1706), based on his army service. Happily, the play's success redeemed Farquhar's reputation, but it could not do the same for his desperate finances. Critically ill, he again acted on the advice of his old friend Wilks to write *The Beaux Strategem*. It would be produced in a new venue, The Queen's Theater in the Haymarket. Unfortunately Farquhar died in 1707 shortly before the drama's production, never knowing of its popularity.

Farquhar would later be considered a link between Restoration drama and a freer eighteenth-century dramatic sentimentality through his introduction of a comedic spirit destined to mark the later generation's drama. Although censured by Alexander Pope for his use of "pert, low dialogue," he managed to overcome the moral detachment of his age, successfully accommodating themes of morality by introducing questions of gentlemanly honor into his comedies. His creation of a wild, youthful male character, in need of domestication by a virtuous female, would quickly become a well-recognized dramatic type.

Bibliography: J. Bull, *Vanbrugh & Farquhar*, 1998; H. Ten Eyck Perry, *The Comic Spirit in Restoration Drama*, 1962.

Virginia Brackett

FELLTHAM, OWEN (1604–1667/8). Details of Owen Felltham's beginnings remain unclear, but records suggest that he was born in 1604, the second son of the gentleman Thomas Felltham in the village of Mutford in Suffolk. There is no record of his attendance at either Oxford or Cambridge, though a poem appeared under his name in a collection of works written by the schools' alumni; it is speculated that his education stemmed from tutoring and self-directed study. Felltham's best-known work, a collection of essays on subjects ranging from religion to apparel entitled *Resolves: Divine, Morall, Political* first appeared in 1623. The work's dedication to the Lady Dorothy Crane suggests that Felltham may have been employed as secretary to her father, the Right Honorable Henry Hobart, Lord Chief Justice of the Common Pleas. In the 1628 edition, Felltham added new pieces dedicated to Thomas, Baron Coventry, Lord Keeper of the Great Seal. At some point between the publication of these editions, Felltham spent three weeks in Holland and composed the humorous *A Brief Character of the Low-Countries under the States*, which circulated widely in manuscript but was not published until 1652. During the late 1620s, Felltham stayed in London, where he kept company with *Ben Jonson and his circle. He soon began serving as steward to Barnabas O'Brien, who became the Earl of Thurmond, in the early 1630s, and Felltham remained associated with the family until his death in February of 1667/68. He appears to have remained at the O'Brien manor through the civil war, revising the *Resolves* and writing poetry rather than getting involved in political debates. The *Resolves* underwent eight printings during Felltham's life; the 1661 version included Felltham's responses to the book's original critics, as well as commentary on biblical passages, many of his poems, and *A Brief Character*.

Bibliography: B. Bergquist, "Owen Felltham: A Few Biographical Facts," *Notes and Queries* 23 (1976): 233–235; T. Pebworth, *Owen Feltham*, 1976.

Hillary Nunn

FÉNELON, FRANÇOIS DE SALIGNAC DE LA MOTHE (1651–1715). Archbishop, mystic, opponent of the Jansenists, and tutor to the grandson of *Louis XIV, Fénelon was born into a large but poor family. In 1669, he matriculated at the College of St. Sulpice, was ordained around 1675, and three years later became head of the Nouvelles Catholiques, a girls' school for the education of Huguenots. This experience led to his *Traité de l'education des filles* (*Treatise on the Education of Girls*, 1687), which became an influential essay on education in the eighteenth century. In 1689, he became the tutor to the Duke of Burgundy, grandson of the king, composing for him *Télémaque* (1699), a Utopian novel criticizing oppressive absolutism and wars of aggression, and stressing a monarch's obligation to his subjects. The book, which presaged the eventual fall of the Bourbon dynasty, irritated the king and, coupled with his defense of Mme. Guyon, it contributed to his fall from influence.

Fénelon met Jeanne Guyon in 1688 and was deeply moved by her accounts of her mystical experiences and her advocacy of "quietism," a religious attitude of utter passivity and reliance upon God. Though made a member of the French Academy in 1693, his defense of Guyon's views involved him in the church's effort to suppress her in 1694. After becoming bishop of Cambrai, he signed the Articles of Issy in 1695, which condemned quietism, especially the doctrines of *Molinos. But in 1697, his *Explication des maxims des saints sur la vie intérieure* (*Explication of the Maxims of the Saints on the Interior life*) argued that Guyon's spiritual practices were not examples of a false mysticism. This angered his former mentor *Bossuet, who persuaded the king to expel Fénelon from the court. His book was attacked by the Jansenists and condemned by Pope Innocent XII in 1699. In the last eighteen years of his life, he was to leave his own diocese only once, though his work there showed him to be an exemplary bishop. The year before he died, he composed a *Letter to the French Academy* presenting his views of the state of French literature. His *Christian Perfection* (1704–1717) is an English translation of his spiritual letters, many written to *Madam de Maintenon. After a carriage mishap, he died on 7 January 1715.

Bibliography: M. de la Bedoyere, *The Archbishop and the Lady*, 1956; J. Davis, *Fénelon*, 1979.

Christopher Baker

FERMAT, PIERRE DE (1601–1665). Pierre de Fermat made seminal contributions to the mathematical areas of analytic geometry, number theory, probability, and calculus. To his name are attached the weighty titles of "The Father of Modern Number Theory" and "The Greatest Mathematician of Seventeenth-Century France." However, Fermat labored his entire life as a lawyer and parliamentarian and published only one short mathematical note during his lifetime. This "Prince of Amateurs" used personal correspondence written in leisure hours after his official duties were completed to influence the direction of mathematics and science across the continent.

Fermat, the son of a leather merchant, was born near Toulouse, France, in 1601. Little is known of his early life, but by 1631, he had received the baccalaureate in law from the University of Orleans, and in 1634 he became a councillor in the local parliament at Toulouse. In his studies, Fermat developed an interest in classical literature and joined contemporaries in attempting "restorations" of lost works of antiquity on the basis of references in extant works. While Fermat "restored" the text of *Plane Loci* by Apollonius, he made several pivotal observations that anticipated *Rene Descartes' creation of analytic geometry by several years. Unfortunately, Fermat's resulting work, *Introduction to Loci*, was not published during Fermat's lifetime and hence the scientific world was deprived of a systematic exposition that more nearly reflects the modern viewpoint on this mathematical field than does the work of Descartes.

With the exception of the calculus of *Newton and *Leibniz, no mathematical work of the seventeenth century has remained more current than Fermat's study of number theory. Fermat's investigations of this area were probably inspired by his reading of the *Arithmetica* written by Diophantus (fl. ca. 250 CE). Many of Fermat's results are communicated to us in the margins of his copy of Bachet's translation of the *Arithmetica*. Although some of Fermat's proofs of these results were transmitted in correspondence to other scholars of his day, other results were announced without proof. Much good mathematics over the next 350 years was inspired by mathematicians seeking to confirm or disprove Fermat's conjectures. A proof of Fermat's assertion that there are no positive integer solutions to $x^n + y^n = z^n$ if n is an integer greater than 2 (a result known as "Fermat's Last Theorem") was finally completed in 1995.

While involved in his study of analytic geometry, Fermat anticipated differential calculus by developing algorithms for finding maxima and minima and for finding tangent lines. In his correspondence with *Blaise Pascal on questions of games of chance, he helped establish foundations for probability theory. However, in each instance, he declined to publish his work, preferring to communicate selected successes in correspondence to friends in the scientific community. Much of this communication was directed to the Minimite friar, *Marin Mersenne, who served as the de facto communications officer for European mathematics. Fermat's reticence diminished his impact on his contemporaries but did not diminish the appreciation of his work by successive generations of mathematicians.

Bibliography: C. Boyer, *A History of Mathematics*, 1968; M. Mahoney, *The Mathematical Career of Pierre de Fermat (1601–1665)*, 1973.

                                                                    Ed R. Wheeler

FILMER, ROBERT (1588–1653). Primary theorist of patriarchalism, Sir Robert Filmer was born in 1588 into the gentry of Kent as the eldest son of Sir Edward and his wife, Elizabeth, daughter of Richard Argall. Matriculating in 1604 at Trinity College, Cambridge, Filmer left within a year for Lincoln's Inn, from whence, after eight years, he was called to the bar. Filmer married Anne Heaton, daughter of Martin Heaton, Bishop of Ely, and settled in Westminster. There he befriended the high church clergyman and historian, Peter Heylin, and no doubt socialized with his own brother, Edward, who was an esquire of the king's body. Such biographical facts, for some scholars, account for Filmer's elaborated and committed patriarchalism; certainly they are not inimical to such an ideology.

Patriarchalism is a theory of government that employs the Genesis story of Adam and Eve to justify hierarchical power relations, with a single, all-powerful male head, determined by primogeniture and called the father. All of Filmer's work, published by him, published posthumously, and circulated only in manuscript, at least draws on patriarchal theory, and most of it expounds the principles and defends them against contrary ones, such as those of the resistance theory promoted by Jesuits or the social contract theory of *Hobbes. Mostly noted today as the rebuked subject of the first of *Locke's *Two Treatises on Government*, published in 1690, Filmer's comprehensive work, *Patriarcha The Naturall Power of Kinges Defended against the Unnatural Liberty of the People By Arguments Theological, Rational, Historical, Legall*, first published in 1680 during the heat of the Exclusion Crisis to support the Tory cause, is the most extended early modern exposition of the commonly held theory of family and state organization. To understand the *Patriarcha* only through Locke's mockery of it is to misunderstand the treatise and the culture that produced it, to say nothing of what such an understanding does to the author. Other important works by Filmer include a reply to William Prynne and Sir Edward Coke, *The Free-holders Grand Inquest* (1648), *The Anarchy of a Limited or Mixed Monarchy* (1648), *The Necessity of the Absolute Power of all Kings* (1648), and *Observations Concerning the Originall of Government, upon Mr. Hobs "Leviathan", Mr. Milton against Salmasius, H. Grotius "De Jure Belli"* (1652). From the latter, Filmer's general thesis may be taken: "We maintain the natural and private dominion of Adam to be the fountain of all government and property . . . [and] that naturally the power of parents over their children never ceaseth by any separation, but only by the permission of the transcendent fatherly power of the supreme prince" (225, 228). Filmer died 26 May 1653 after suffering repeated raids on his home in East Sutton and even imprisonment by parliamentarian forces.

Bibliography: P. Laslett, "Sir Robert Filmer: The Man versus the Whig Myth," *William and Mary Quarterly* 5 (1948): 523–546; J. Sommerville, ed., *Filmer: Patriarcha and Other Writings*, 1991.

Jesse G. Swan

FINCH, ANNE, COUNTESS OF WINCHELSEA (1661–1720).

The poet Anne Kingsmill, born into an old Hampshire family, served as maid of honor in the royal court, beginning with *Charles II and then for Mary of Modena, queen of *James II in 1683. She married Colonel Heneage Finch the following year and left court service, moving to Eastwell Park, the Kent estate of the Earl of Winchelsea, which title her husband gained in 1712. Most of her poems were likely written prior to that year. She drew inspiration from this country manor and its opportunities for solitary musing on nature, as in *The Nightingale, To the Echo*, and *The Petition for an Absolute Retreat*. She published her *Miscellany Poems on Several Occasions Written by a Lady* in 1713 and gained wide fame after Wordsworth in his "Essay Supplementary to the Preface" (1815) declared her *Nocturnal Reverie* one of the only two poems between *Milton's *Paradise Lost* and Thompson's *The Seasons* that displayed "a single new image of external nature" (the other poem was Pope's *Windsor Forest*). He found her style "often admirable, chaste, tender, and vigorous." She also wrote two closet dramas, translated fables from *La Fontaine, and composed a number of other songs, lyric verses, and Pindaric odes. Her work has received renewed attention for her explicit assertions of women's creative abilities.

Bibliography: B. McGovern, *Anne Finch and Her Poetry: A Critical Biography*, 1992.

Christopher Baker

FLAMSTEED, JOHN (1647–1719). English astronomer John Flamsteed was born at Denby near Derby and left the free school there because of ill health, but by 1670 he had entered Cambridge and met *Isaac Newton. Flamsteed, an amateur observational astronomer, was appointed the first Astronomer Royal at the founding of Greenwich Observatory (1675). Contemporary Polish astronomer *Johannes Hevelius claimed that naked-eye instruments provided the reading of finer measurements in stellar positions. But Flamsteed availed himself of his own measuring instruments modified with telescopic sights to perform accurate stellar observations. Flamsteed concentrated on observing the moon and the stars with the taking of some 20,000 observations between 1676 and 1689, especially using a large sextant (angle measuring instrument of less than 90 degrees arc) accurate to ten arcseconds. His systematic observations were completed in 1705.

The growing importance of positional astronomy's accurate stellar records to the needs of astronomy and navigation was exemplified in the British furor over the use of Flamsteed's quality work. There was great pressure to publish his catalogue because of its accurate usefulness before it was completed. Though this was against Flamsteed's wishes, this first Greenwich star catalogue was edited by *Edmund Halley (1712). Flamsteed died before his completed catalogue was published in 1725 in three volumes with some 3,000 stars. His planetary data aided Newton in his monumental *Principia*. His efforts culminated a 150-year progression of instrumental improvements, which enabled new astronomical theory, particularly the validation of heliocentrism and the dissolution of a geocentric cosmos in favor of Newton's mathematicomechanical one by the late century.

Bibliography: H. Eichorn, *Astronomy of Star Positions: A Critical Investigation of Star Catalogues, the Methods of Their Construction, and Their Purpose*, 1974; J. North, *The Fontana History of Astronomy and Cosmology*, 1994.

William J. McPeak

FLEMING, PAUL (1609–1640). Fleming was born to a schoolmaster and pastor in Hartenstein, Saxony. When he was thirteen, his father sent him to the famous Thomasschule (Saint Thomas's School) in Leipzig. From 1628 until 1633, he studied medicine at the university there. During those years, he also started writing poetry in Latin and German, inspired in particular by Martin *Opitz, whose *Buch von der Deutschen Poeterey* (*Book of German Poetry*, 1624) had a profound impact on him.

In 1633, Fleming joined Adam Olearius (ca. 1599–1671) on a trade mission to Moscow. After his return, he spent a year in Reval (today Tallinn, Estonia), where he fell in love with Elsabe Niehusen. He had to leave her behind for the second stage of the mission. They left in the spring of 1636 for Moscow and went from there to Persia. In April 1639, they returned to Reval, but Elsabe had married another. He became engaged to her sister Anna instead. In 1640, he received his Doctor of Medicine degree with a dissertation on venereal disease. Shortly thereafter he died in Hamburg at age thirty-one.

Like most poets of his age, Fleming was a learned man who wrote both in Latin and German. In his poetry he adopted the conventions of his time, but he was able to develop a unique and fresh voice due to his adventurous life and experiences on his Russian and Persian journeys. His love poetry mainly follows the Petrarchan tradition, which finds its counterpart in the neostoical mood dominant in many of his poems. He also composed occasional and religious poetry. His preferred form was the ode, but he was also a master of the sonnet. Olearius posthumously published a full edition of Fleming's German poems in 1646 as *Teutsche Poemata* (*German Poems*). Today he is regarded as one of the most outstanding German poets of the seventeenth century,

Bibliography: A.J. Harper, "Paul Fleming," *Dictionary of Literary Biography* 164 (1996): 107–112; M. Sperberg-McQueen, *The German Poetry of Paul Fleming: Studies in Genre and History*, 1990.

Josef K. Glowa

FLETCHER, JOHN (1579–1625). John Fletcher is most often thought of as the other half of the playwriting team, *Beaumont and Fletcher, but he should perhaps be thought of as *Shakespeare's successor, at least in his capacity as house playwright for the King's Men. His father, Richard Fletcher, had been Bishop of Bristol, of Worcester, and finally held the title of Bishop of London, as well as Chaplain to Elizabeth. However, he lost his favor with her when he remarried in 1595 and died in debt shortly thereafter. Fletcher had entered Bene't College (Corpus Christi), Cambridge, in 1591 and very probably took his B.A. in 1595 and his M.A. in 1598. His uncle, the poet Giles Fletcher, probably supported him financially during this time.

Fletcher began by writing sophisticated comedies and pastorals for the boys' companies with his friend Francis Beaumont. *John Aubrey claimed that they were so close that they shared the same cloak, the same bed, and the same woman, but this is almost certainly Aubrian exaggeration. However, their working relationship was close until Beaumont married and retired from the theater in 1613. To this period belong the tragicomedies *Philaster* (1608–1609), *The Maid's Tragedy* (1610), and *A King and No King* (1611). *The Faithful Shepherdess*, published in 1608, is probably his first independent play, and although it did not do well initially, it was successfully revived in 1633 for Queen Henrietta Maria. He almost certainly collaborated with Shakespeare on his last two works, *Henry VIII* and *The Two Noble Kinsmen*. He also wrote a sequel of sorts to a Shakespeare play, *The Woman's Prize, or the Tamer Tamed* (1647), a continuation of *The Taming of the Shrew*. After Shakespeare and Beaumont both retired, and possibly even before this, Fletcher's chief collaborator was *Philip Massinger; he also worked with Nathaniel Field. Fletcher mastered a number of genres, including comedies such as *The Wild Goose Chase* (1621) and *Rule a Wife and Have a Wife* (1624), tragedy, and the histories such as *Bonduca* and *Valentinian*, both of which are set in ancient Britain and date between 1606 and 1612, roughly the same time at which Shakespeare was writing his own ancient Britain plays *King Lear* and *Cymbeline*. However, he is most noted for popularizing the tragicomedy and the pastoral play modeled on those of Tasso and Guarini.

Fletcher died in the especially virulent plague of 1625, having written forty-two plays, either in whole or in part. His position as chief playwright to the King's Men was taken over by Philip Massinger. Most of the plays in the 1647 folio *Comedies and Tragedies Written by Francis Beaumont and John Fletcher* were mostly or exclusively by Fletcher. His plays are the only ones other than those of *Jonson or Shakespeare to be honored in a folio edition during the early modern period, and with good reason; they formed, along with Shakespeare's plays, the staple repertory of the King's Men, the remaining members of whom signed the dedicatory letter that prefaces the edition. Until fairly recently, Fletcher criticism has centered mostly on textual issues and especially those of authorship and attribution; more recently, the political content and context of the plays, as well as the ways in which they were influenced by and shaped early Stuart culture, have received close attention.

Bibliography: W. Appleton, *Beaumont and Fletcher: A Critical Study*, 1956; P. Finkelpearl, *Court and Country Politics in the Plays of Beaumont and Fletcher*, 1990; G. McMullan, *The Politics of Unease in the Plays of John Fletcher*, 1994.

Melissa D. Aaron

FORD, JOHN (?1586–1640). Born in Devon, John Ford was likely educated at Oxford and the Middle Temple. His family name gave him the only unqualified claim to gentry among Caroline playwrights. An unabashed seeker of court patronage, Ford dedicated his *Fame's Memorial* (1606) to widow Penelope Devereaux Rich, the celebrated Stella of Sir Philip Sidney's sonnets. Early in his career, Ford purportedly authored an anonymous tract, *The Golden Mean* (1613). The tract praised the virtues of endurance and honored the Earl of Northumberland, held in the tower since 1605 for suspected involvement in the Gunpowder Plot. Ford gave an additional tract titled *A Line of Life* (1620), emphasizing the rewards of endurance, to Northumberland's son-in-law, Lord Haye, Viscount Doncaster. Ford's lost nondramatic work, *Sir Thomas Overbury's Ghost, contayneigne the history of his life and untimely death* (1615), reportedly focused on revenge through the poisoning of the knight who opposed *King James's favorite, Robert Carr, Viscount Rochester, Earl of Somerset.

Ford's career included collaboration on dramas with *Thomas Dekker and William Rowley, with whom he produced his first play, *The Witch of Edmonton* (1621). He likely worked again with Dekker on *The Welsh Ambassador* (1623) and *The Sun's Darling* (1624), a masque. His two unnotable comedies, *The Fancies, Chaste and Noble* (1636) and *The Ladies Trial* (1638), did not allow Ford to exercise the great inclination toward morbidity that would become his trademark. In his *The Lover's Melancholy* (1628), he moved closer to the genre upon which he would later establish his reputation. A great villain at last emerged in the character of D'Avolos from Ford's first real tragedy, *Love's Sacrifice* (1632).

Three subsequent works established Ford as a writer of note and have sustained critical interest. The first, *The Broken Heart* (1629), depicts love gone awry and features self-starvation on the part of one female character and death by a broken heart, following a revenge binge, by another. The second, Ford's *'Tis Pity She's a Whore* (1631), represents his most intense writing, depicting a human penchant for unrestrained violence. Probably first performed between 1625 and 1633, it features incest, adultery, betrayal and execution, all carried out in Ford's signature ceremonial manner. In one highly sensational scene, Giovanni carries onto stage the heart of his sister/lover, Annabella. Finally, in his historical chronicle play, *Perkin Warbeck* (1634), Ford closely follows fact in depicting the imposture of Warbeck as Richard, Duke of York, whom the playwright depicts as a legitimate heir to the throne.

Throughout his works, Ford emphasizes the proper role of individuals in subservience to others, whether subjects to kings, or all things female to all things male. He depicts not morality, but rather the value of a nobility gained through his characters' absolute adherence to loyalty, usually with disastrous results, in the face of slander and envy. Ford's best writing shapes characters blindly obedient to their assigned social roles, supporting a theme of ritual suffering unique in Caroline drama.

Bibliography: I. Clark, *Professional Playwrights: Massinger, Ford, Shirley & Brome*, 1992.

Virginia Brackett

FOX, GEORGE (1624–1691). Born at Fenny Drayton in Leicestershire, the son of Christopher Fox, a well-respected weaver and orthodox Puritan church-warden, and Mary Lago, George Fox became a self-proclaimed preacher and principal founder of the Quaker, or Friends, movement. He had only a rudimentary education at the tutelage of his mother and the village school before being apprenticed as a cobbler and shepherd at nearby Mancetter. Precociously pious and serious minded, he favored intensive Bible study and theological inquiry over the usual pastimes of childhood and youth. From 1640 onward, he took his quest for religious truth on the road and by 1646–1647, he had experienced what he called a series of religious "openings," moments of insight that confirmed to him certain ideas he had formed on his own and others that he had gleaned from the various Puritan, Baptist, and separatist pastors and groups he had encountered.

Although Fox and his early followers referred to themselves as Children of the Light, as Friends in the Truth, or, in time, simply as Friends, their opponents began to call them "Quakers," a coinage that although intended derisively also emphasized the singularity of their beliefs. Like his disciples, Fox suffered persecution for his beliefs, was often beaten or stoned, and over the years was arrested and jailed at Derby (1650–1651), Carlisle (1653), Leicestershire (1654), Launceston (1655–1656), Lancaster Castle (1660, 1663–1665), Scarborough Castle (1665–1666), and Worcester (1673–1674). Nevertheless, Fox and his followers remained undaunted, and Quakerism thrived in England as well as in the English colonies in Barbados, Jamaica, and North America, and in certain areas of Europe.

In 1652, Fox established a headquarters for his work at Swarthmoor Hall, manor house of the Puritan judge Thomas Fell, whose wife Margaret not only became one of Fox's most devout and industrious converts but also in 1669, nearly ten years after Thomas Fell's death, became Fox's wife. The rapid spread of Quakerism attributes in part to its ability to fuse beliefs already voiced by Puritan, Baptist, and separatist believers with those that may be called singularly Quaker, perhaps most notably the belief in salvation through the power of God's Inner Light. Fox's insistence that one may ascertain spiritual truth by subjecting one's soul to the intense scrutiny of the Inner Light led the Quakers generally to disavow social, political, educational, and religious authority, and public worship in favor of pacifism, universal brotherhood and equality, volunteer lay ministry, and silent worship. Because they eschewed oaths, titles, military service, social class, and rank and privilege, they appeared to threaten prevailing political and social structures. The Quakers' early propensity for large public meetings and aggressive proselytizing eventually gave way to the development of small private gatherings for worship and conduct of business. Fox wrote scores of tracts, letters, and pamphlets setting forth initiatives for organization and for the "convincement" of nonbelievers as well as articulating Quaker beliefs and the appropriate conduct of daily life. He is best known for his *Journal*, a narrative of the origins and development of Quakerism through 1675, with the remaining years of his life's work being filled in later with letters and other documents by his posthumous editors prior to the journal's 1695 publication. Fox died in January 1691, having spent the final years of his life in and around London.

Bibliography: W. Braithwaite, *The Beginnings of Quakerism*, 2nd ed. rev., 1955; W. Braithwaite, *The Second Period of Quakerism*, 2nd ed., 1961; N. Penny, ed., *The Journal of George Fox*, 2 vols., 1911.

Sallye Sheppeard

FRANCIS DE SALES, ST. (1567–1622). Theologian and founder of a religious order, Francis de Sales was trained by Jesuits in Paris and received a law degree from Padua in 1591; he entered public life in his native Savoy the following year and was ordained a priest in 1593. In 1597, he was sent by the pope to Geneva, where he later became bishop, charged with bringing that stronghold of Calvinism back to the Catholic church. Despite his inevitable failure at that task, he implemented reforms mandated by the Council of Trent (1545–1563), among them visitations in the district and improvements in the training of clergy. His most ambitious work, however, was establishing, with St. Jane Frances de Chantal (1572–1641), the Order of the Visitation of Mary, known as the Visitadines, in 1610. Based at Annecy, France, it is an order for contemplative women religious. The Salesian order (The Society of St. Francis de Sales) is a nineteenth-century creation, named in his honor by its founder, St. John Bosco.

The *Introduction to the Devout Life* (1609) is his most enduring text. An explication of the three ways of approaching God (purgative, illuminative, and unitive), it is addressed to a pious woman named Philothea ("God-lover") and illustrates its precepts with examples from saints' lives. Like a good confessor, de Sales counsels Philothea through the penitential process, with meditations on heaven and paradise alongside those on sin and death. The work is a case study in Counter Reformation pastoral care, an instance of ecclesiastical concern for the laity together with clerical concern to preserve the status of the priesthood. The classifications of the penitential manuals in use at that time are preserved and turned into pedagogical categories for disciplining virtue.

Bibliography: H. Bremond, *Histoire littéraire du sentiment religieux en France*, vol. 1, 1923; F. de Sales, *Introduction to the Devout Life*, trans. J. Ryan, 1972.

Ralph Keen

FREDERICK WILLIAM (1620–1688). Called the "Great Elector," Frederick William became ruler of Brandenburg and Prussia in 1640. A member of the Hohenzollern family, which with the Hapsburgs was one of the ranking powers of the Holy Roman Empire, he used an absolutist authority to rebuild his territory after the depradations of the Thirty Years' War. This involved a number of significant initiatives. He welcomed Flemings, Jews from Poland, and French Huguenots, realizing that a policy of religious toleration would attract skilled immigrants. He created an efficient standing Prussian army and navy. He succeeded in reducing the power of the Junkers (landed nobility) and in limiting the authority of the elected assemblies (or "estates"), which represented the nobles and commoners, having also cunningly arranged for Sweden and Poland to recognize him as supreme Prussian ruler. However, he also drew the Junkers into working with him for further reforms and encouraged them to serve as military officers and in the civil service. He thus crafted an absolutist government permitting a degree of cooperation that, though under his control, distinguished it from more personally centralized monarchies of the time. By the end of his rule, Prussia was one of the most efficiently run states in Europe.

Bibliography: O. Mitchell, *A Concise History of Brandenburg-Prussia*, 1980; F. Schevill, *The Great Elector*, 1947.

Christopher Baker

FRESCOBALDI, GIROLAMO (1583–1643). Frescobaldi's reputation as an organist was legendary. As a composer, he wrote highly influential keyboard and instrumental works, and to a lesser extent, vocal music. Born in Ferrara, Italy, Frescobaldi received his early training there from Luzzasco Luzzaschi, the highly regarded court organist to Duke Alfonso II d'Este. The Ferrara court had a rich musical life and was especially famous for virtuoso vocal ornamentation. The Neapolitan madrigalist Don Carlo Gesualdo, among other composers, was present at the court. At the time of the Duke's death in 1597, Frescobaldi became the organist of the Accademia della Morte of Ferrara. Guido Bentivoglio of Ferrara became patron to the young Frescobaldi, who moved to Rome around 1600. He held the post of organist at Sancta Maria in Trasevere during the early months of 1607. Bentivoglio had become an archbishop and later that year brought Frescobaldi with him to his new post in Brussels. When Frescobaldi returned to Italy after almost a year, Bentivoglio's brother Enzo, ambassador to Rome, became his patron. In 1608, Frescobaldi was appointed organist of St. Peter's, where he served for the rest of his life. But he continued to have other employers, among them the Cardinals Pietro Aldobrandini and Alessandro d'Este. In 1613, Frescobaldi married Orsola del Pino, with whom he had five children. He left Rome twice after that, first to take a position at the court of Mantua in 1615, which lasted only three months, and later to serve as court organist to the Medicis in Florence from 1628 to 1634.

Frescobaldi's compositions presented an interesting mixture of old and new thinking, including both the old polyphony and the more expressive monodic style, which supplanted it around 1600. His works were much published during his lifetime (all dates given refer to first publication). The first book of toccatas and partitas of 1615 contained a strikingly new approach to keyboard composition. Frescobaldi included instructions to the performer, indicating a desire for spontaneity, sectionality, and improvisatory spirit. He compared his toccatas with modern madrigals; they represented the new vocal aesthetic expressed on the keyboard. A second book was published in 1627. The toccatas had an enormous influence on keyboard composition and performance of the time as well as a direct impact on the later development of this genre by Johann Jakob

Froberger and Johann Sebastian Bach. On the other hand, his contrapuntally conservative ricercare and canzonas (1615) were also extremely popular during his lifetime. In the capriccios of 1624, Frescobaldi was at the peak of his compositional power. These are sectional, imitative pieces of great variety and skill. Although liturgical music is rare among Fresocobaldi's output, the three organ masses of his *Fiori musicali* (1635), also masterpieces, are his best known music today. His last keyboard works included colorful capriccios on the battle, the pastorale, and the Ruggiero aria, as well as masterful passacaglia and chaconne variations. Pieces for other instrumental combinations include *Canzonas* published in 1608, 1626, and 1634. The madrigals of 1608 were composed during his time in Brussels and demonstrate the Ferrara-based influence of Luzzaschi and Gesualdo. The two books of solo arias (1630) were written in Florence. Vocal pieces with Latin texts include the cantiones of 1616 and motets of 1618.

Bibliography: F. Hammond, *Girolamo Frescobaldi*, 1983.

Joyce Lindorff

FURETIÈRE, ANTOINE (1619–1688). A Parisian all of his life, Antoine Furetière first pursued a legal career before becoming the abbot of Chalivoy in 1662. That same year, Furetière was elected to the Académie Française, although his literary production up to that point consisted of only a few satirical verses and fables. This satirical vein is evident in the *Roman Bourgeois* (1666), a comic work that presents the habits and foibles of middle-class characters based on Furetière's acquaintances in the legal profession and his Parisian neighbors of the Place Maubert. Not as well known as the *Satires* of *Boileau or the satirical plays of *Molière, the *Roman Bourgeois* nonetheless provides insight into the world of the bourgeoisie of the day that complements the works of these authors, both friends of Furetière.

It is, however, for his work as a lexicographer that Furetière is known today, and his *Dictionnaire universel des arts et des sciences* (1690) remains an indispensable tool for scholars of seventeenth-century French literature and culture. Furetière conceived of his work as a dictionary of things, rather than of words, and so presented a variety of technical, medical, and scientific terms within its pages. As such, the *Dictionnaire universel* served as a precursor to the encyclopedic dictionaries of the eighteenth century.

At the same time that Furetière was preparing his dictionary, members of the Académie Française were compiling material for their own dictionary, a project begun in 1635. Sensing competition for its own project and judging him to be disloyal, the Académie accused Furetière of using materials compiled *by académiciens* for their dictionary and, in 1685, revoked the royal privilege granting him permission to publish his *Dictionnaire universel* and expelled him from its ranks. In response, Furetière wrote three *factum* in which he defended himself against the accusations of his former peers and claimed his dictionary to be more comprehensive in scope and more precise than the *Dictionnaire de* *l'Académie Française*, which appeared in 1694. The *Dictionnaire universel* was published posthumously in Rotterdam in 1690, with a preface by *Pierre Bayle.

Bibliography: U. Döring, *Antoine Furetière: Rezeption und Werk*, 1995; F. Gégou, *Antoine Furetière Abbé de Chalivoy*, 1962.

Patricia Armstrong

G

GABRIELI, GIOVANNI (ca.1553-56–1612). Along with his uncle, Andrea Gabrieli, Giovanni Gabrieli is noted for his monumental polychoral works, for being a paragon of the Venetian School during the late Renaissance, and for being part of a long line of outstanding composers adorning St. Mark's Basilica, Venice. Not much is known about Gabrieli's early life, but it is reasonable to assume that Giovanni studied music with his uncle. Also, like his uncle, as a youth Giovanni was employed in Munich at the court of Duke Albrecht V, whose Kapellmeister was Orlandus Lassus. In 1579, upon Duke Albrecht's death, Gabrieli possibly left, along with many other court musicians, but there is no documentation of his whereabouts until 1584, when he became temporary organist at St. Mark's, where eventually he would spend many years. In 1585, Gabrieli's position became permanent. In addition to being organist at San Marco, Gabrieli was also engaged for a part-time post as organist to a religious confraternity, the Scuola Grande di San Rocco.

Upon his uncle's death in 1586, Giovanni published two volumes of his uncle Andrea's music, titled *Concerti* (1587). From then on Giovanni would be the main composer at San Marco of ceremonial music, an important part of Venetian life. A collection of Giovanni's own large-scale works appeared in print in 1597, in the first volume of *Sacrae symphoniae*. This was the publication that spawned a number of imitations, especially in German-speaking lands, for example, *Heinrich Schütz's *Symphoniae sacrae* (1629), published in Venice. A number of the northerners, most notably Schütz, were sent by their employers to study with Gabrieli. All in all, Gabrieli provided such magnificence through his music and through hiring the best singers and instrumentalists, both for San Marco and for the Scuola di San Rocco, that *Thomas Coryat remarked of his experience at the feast of Saint Roche (San Rocco) that: "I heard the best music that ever I did in all my life . . . so good that I would willingly go a hundred miles on foot at any time to hear the like . . . so good, so delectable, so rare, so admirable, so superexcellent, that it did even ravish and stupefy all those strangers that never heard the like. . . . For mine own part I can say this, that I was for the time even rapt up with Saint Paul into the third heaven." Unlike his uncle, Gabrieli composed primarily sacred music and instrumental music, but little secular music. Following the practice of cori spezzati, instituted by Adrian Willaert and continued by Andrea Gabrieli, Giovanni's motets are principally for two or more choirs, which sing in dialogue from different locations in the church. Gabrieli died in 1612, after suffering for six years with a kidney stone. His influence on later generations of Venetian composers significantly declined after *Monteverdi became San Marco's maestro di cappella in 1613. Gabrieli's music was more influential in Germany until around 1630, when Monteverdi's music and that of his followers reached a wider audience in northern Europe.

Bibliography: D. Arnold, *Giovanni Gabrieli and the Music of the Venetian High Renaissance*, 1980; E. Kenton, *Life and Works of Giovanni Gabrieli*, 1967.

Susan Treacy

GALILEI, GALILEO (1564–1642).

Galileo was born in Pisa in 1564. His father, Vincenzo Galilei, was a musician and a proponent of progressive musical theories, who published an important treatise where he applied both mathematics and experimentation and, likely, influenced his son. Galileo received his basic education first in Pisa and then at the monastery of Villombrosa outside Florence; in 1581, he enrolled at the University of Pisa pursuing medical studies for two years. From 1583, Ostilio Ricci tutored Galileo in mathematics, who rapidly mastered the subject and, in 1586, wrote *The Little Balance*. This short treatise describes a hydrostatic scale, which, based on the Archimedean principle, is helpful for accurately measuring specific weights. Recommended by Guidobaldo del Monte, in 1589 Galileo obtained the chair of mathematics in Pisa. While teaching at the university, Galileo produced manuscript commentaries on Ptolemy's *Almagest* and Aristotle's *On the Heavens* and *On Generation and Corruption*, and on works by Tasso and Ariosto. He also composed extensive notes on motion entitling them *De motu antiquiora*; these show Galileo's indebtedness to the Jesuits' contemporary scientific work at the Collegio Romano and are a sketch for a new science of motion.

In 1592, Galileo obtained the more prestigious chair of mathematics at the university of Padua, an institution known for its faculty's intellectual freedom. Noteworthy among Galileo's Paduan writings are an anti-Aristotelian *Dialogo*—written in the Venetian dialect under the pseudonym of Cecco di Ronchetti—about the appearance of new star in the sky (1605), and *The Operations of the Geometric and Military Compass* (1606) dedicated to the young prince and future patron, Cosimo II de' Medici, whom Galileo had tutored for many summers. He also cultivated the friendship of Marina Gamba, *Paolo Sarpi, Francesco Sagredo, and Benedetto Castelli; in 1597, he had a brief epistolary exchange with *Johannes Kepler in which he declared his subscription to Copernicanism.

Although Galileo did not invent the telescope, in the winter of 1609 he pointed the new instrument (*perspicillo*) to the night sky making several astronomical discoveries published as the *Starry Messenger* (March 1610). Dedicated to his pupil, now the Grand Duke of Tuscany, the *Starry Messenger* announced four discoveries, confirmed by Kepler in April 1610 and sanctioned by the Jesuits at the Collegio Romano in May 1611. First, the moon's surface is not polished but earthlike; second, the telescope shows many more fixed stars than the eye can see; third, the Milky Way is a celestial rather than meteorological phenomenon; fourth, Jupiter has four satellites, the Medician Stars. These discoveries gained him both instant fame and the post of mathematician and philosopher at the Medici's court in Florence. As a courtier, Galileo dedicated himself to scientific research and writing, becoming also a member of the Accademia dei Lincei. Under its auspices, in March 1613 Galileo published his *Letters on the Sunspots*, but not after some considerable conflict with the religious authorities. In 1612, Galileo had also engaged in public controversy with an Aristotelian philosopher on the principles of buoyancy, publishing the *Discourse on Bodies in Water*. Written in response to the Jesuit *Christoph Scheiner's work, the *Lettere sulle macchie solari* introduced the phases of Venus as evidence to challenge Aristotelian cosmology while supporting the heliocentric worldview. Galileo drew further support from the sunspots' circulation around the sun, which suggested the sun's power to make both the sunspots and the planets rotate around itself.

Subsequently, Galileo defended Copernicanism against scriptural authority on several occasions before the 1616 condemnation, which prohibited Copernicus's work and the teaching of heliocentrism. In 1613, Galileo circulated his *Letter to Castelli*—a pupil who had publicly defended Copernicanism—arguing for compatibility between the scriptures and heliocentric cosmology. Encouraged by Monsignor Dini, in the 1615 *Letter to the Grand Duchess Christina*, Galileo explicated both his interpretation of biblical passages taken to support geocentrism and the relationship between theology and natural philosophy. By order of Pope Paul V, in February of 1616, Cardinal *Bellarmine required Galileo to cease publicly defending Copernicanism. While Galileo complied, writing to Cardinal Orsini in January 1616, he had already described his tidal theory proving the earth's motion.

In 1623, Galileo's friend Maffeo Barberini's election to the papacy as *Urban VIII, brought him relief from an investigation by the Inquisition for his atomistic views presented in the *Assayer*. Published in 1623, the *Assayer* defended the Aristotelian position that comets are sublunar phenomena against the celestial interpretation by the Jesuit Grassi. Here appeared Galileo's famous statement that "the book of nature is written in the language of mathematics."

After gaining Urban VIII's consent for considering the earth's mobility hypothetically, in 1624 Galileo began composing his *Dialogue on the Two Chief World Systems*. Though ready by 1630, the work appeared in 1632, delayed by slow delivery of the ecclesiastical permission to publish and the printing. Divided into four days, the *Dialogue*'s three characters are Salviati (Galileo), Sagredo (an educated layman), and Simplicio (an Aristotelian). In the first day, Galileo argues against the celestial/terrestrial dichotomy, a basic feature of the Aristotelian cosmos. The second day is dedicated to the refutation of the classical arguments against the earth's daily motion, while the third day discusses its annual motion around the sun. The fourth day presents Galileo's tidal theory to prove the earth's motion definitively. Since the *Dialogue* presented the earth's motion as real rather than hypothetical, the Inquisition banned it in July 1632; Galileo was tried in 1633; he abjured and was sentenced to house arrest. The trial's details remain controversial and originated the "Galileo affair."

Despite the loss of his freedom and the death of his beloved daughter Maria Celeste, in 1634 Galileo was already working on his mechanics, a subject that had occupied him throughout his career. Published in Leyden in 1638, the *Discourses and Mathematical Demonstrations on Two New Sciences Concerning Mechanics and Local Motions* retains the characters, dialogical style, and quadripartition of the *Dialogue*. Deduced from geometrical principles and arguing against Aristotelian physics, the two new sciences are the strength of materials and dynamics, which included Galileo's analysis of free fall. Although blind and in poor health, Galileo continued his scientific research. When he died on 8 January 1642, he was preparing yet another dialogue.

Bibliography: S. Drake, *Galileo at Work—His Scientific Biography*, 1978; A. Fantoli, *Galileo: For Copernicanism and for the Church*, 1994; P. Machamer, ed., *The Cambridge Companion to Galileo*, 1998.

Renzo Baldasso

GARCILASO DE LA VEGA (1539–1616). Historian and soldier, Garcilaso de la Vega was born in Cuzco in 1539 to don García Lasso de la Vega, a Spanish captain who was a cousin of the Spanish poet Garcilaso de la Vega (1501–1536), and Princess Isabel Chimpu Ocllo, a cousin of Atahualpa, last of the Inca kings. In order to draw attention to his dual lineage, Garcilaso called himself Inca Garcilaso de la Vega. Although he was born in Peru, Garcilaso lived there for only twenty years. In 1560, following the death of his father, Garcilaso traveled to Spain to complete his education. After unsuccessfully petitioning the city of Madrid for possession of land owned by his mother, Garcilaso joined the military and fought against the Moors in 1570, earning the rank of captain. Because Córdoba reminded him of Cuzco, Garcilaso settled in the Andalusian city in 1591 and remained there until his death in 1616.

Garcilaso first used the name Inca on the title page of his translation of León Hebreo's *Dialoghi d'Amore*, a sixteenth-century neo-Platonic essay on love. Garcilaso titled his translation *La traducción del indio de los tres diálogos de amor de León Hebreo (The Indian's Translation of León Hebreo's Three Dialogues About Love)*. As a child, Garcilaso learned many details about his mother's family and the Inca civilization from family members, especially his mother. These memories later proved valuable to Garcilaso as he chronicled the history of the Inca civilization. He published in 1605 the historical novel *La Florida del Inca*, an account of the expedition of the Spanish explorer Hernando de Soto (1500?–1542). Garcilaso is best known for *Comentarios reales que tratan del origen de los Incas (Royal Commentaries of the Origin of the Incas)*, a two-part history of the Inca civilization. The first part, published in 1609 and titled *Historia del Perú (History of Peru)*, deals with events leading up to the arrival of the Spanish conquerors. Garcilaso compares the Inca civilization to famous societies and cultures, including the Roman Empire. He contends that the Inca civilization embodied many Christian ideals and corrects those chroniclers who portrayed the Incas as savages. The second part, published posthumously in 1617 and titled *Historia del reino del Perú (History of the Kingdom of Peru)*, describes the Spanish colonization of Peru and the civil wars fought among the conquerors. The humanist ideas of the Spanish Renaissance are present in Garcilaso's historical accounts, which are more like novels because of his colorfully picturesque descriptions and attention to details. Garcilaso's books have genuine historical importance and are considered valuable sources of information for the study of the Inca civilization.

Bibliography: D. Castanien, *El Inca Garcilaso de la Vega*, 1969; J. Varner, *El Inca, the Life and Times of Garcilaso de la Vega*, 1968.

Michael J. McGrath

GASCOIGNE, WILLIAM (?1612–1644). William Gascoigne was born not later than 1612, son of Henry Gascoigne, Esq., in the parish of Rothwell near Leeds, Yorkshire. He lived with his father near Leeds where he developed his knowledge of astronomy. Through his letters to *Jeremiah Horrocks and William Crabtree, it is apparent that Gascoigne attended Oxford but found the lack of mathematical training discouraging. Gascoigne is generally admitted to be the inventor of the wire micrometer and of the attachment of this device to the interior of the eyepiece of a telescope. Gascoigne did not receive recognition for the invention until the publication of Adrien Auzout's (1622–1691) reports on the success of a similar device as detailed in a letter to the *Philosophical Transactions of the Royal Society*, number 21.

This action prompted Richard Townley to submit a letter to the *Philosophical Transactions*, published in number 25, explaining that he had some papers from Gascoigne that provided descriptions of the micrometer and its usage that predated the findings of Auzout. Townley relates that Gascoigne used the micrometer to calculate large distances on land and, more importantly, the moon's distance based upon the horizontal and meridian diameters. Townley provides additional information, complete with engravings, on the construction of the micrometer with his own improvements in *Philosophical Transactions*, number 29. The instrument originally consisted of either two parallel wires or small plates of metal attached to the focus of a telescope. The wires or plates could be spread apart by turning a screw until the object filled the space between them. Two scales, or indexes, measured the divisions of a foot, thus allowing for the calculation of expanse distances—the larger scale divided the foot into hundreds, and the smaller scale broke those divisions into hundreds. Townley claims that Gascoigne had a *Treatise of Opticks* ready for the press but admits to having never seen the copy. The treatise was later recovered and printed by Rigaud. In addition to the invention of the wire micrometer, Gascoigne is recognized by Sir Edward Sherburne as having invented methods for grinding glasses and as the first to use two convex lenses in a telescope. Gascoigne was killed on 2 July 1644 in the battle of Marston Moor while fighting for the Royalist army in the English Civil War.

Bibliography: *Dictionary of National Biography*, vol. 7, 47; A. Wolf, *A History of Science, Technology and Philosophy in the 16th and 17th Centuries*, vol. 1, 1935, rpt. 1968.

Michael W. Jackson

GASSENDI, PIERRE (1592–1655). Although he taught in Aix-en-Provence and Paris, the better part of the life of this French priest was centered around Digne, a southern city where he became provost of the cathedral. He exchanged ideas with the foremost scientists of his time and wrote part of the *Objections* to *Descartes' *Méditations métaphysiques* (1641). Like *Marin Mersenne, one of his closest acquaintances, Gassendi sought to make new scientific ideas compatible with religious orthodoxy. This was a perilous task considering that his rejection of scholastic method and natural philosophy was motivated by an espousal of atomism. The philosophy of Epicurus, Lucretius, and their followers was highly suspicious and considered often to be synonymous with free-thinking and dissolute morals. Gassendi therefore had to discard or provide a new interpretation for the most problematic aspect of Epicureanism: the mortality of the soul, the notion of pleasure, or the question of chance. His first writings put him close to Pyrrhonian skepticism: In *Exercitationes paradoxicae adversus Aristoteleos* (1624), he staunchly criticized the vanity of scholastic demonstrative science. Because Gassendi later had to explain natural phenomena as invisible movements and configurations of atoms, he defined the status of his statements as probabilistic, which tempered his earlier stance. He held that knowledge was founded on appearance and that, together with reason, the testimony of senses was reliable. But this procedure did not yield certainty: Conclusions do not make qualitative and essential claims about phenomena, because they are based on mere effects of things. Further proof of the probable status of scientific statements can be taken from the fact that progress in technology, as evidenced by the invention of the telescope, sometimes reveals erroneous human interpretations. Gassendi's works were collected after his death in the *Syntagma philosophicum* (1658). After Bloch, there has been much discussion about whether Gassendi's faith was real and not a cover for materialism. Against this opinion, scholars point to writings where probabilism is viewed as the mark of human frailty facing divine powers.

Bibliography: O. Bloch, *La Philosophie de Gassendi: Nominalisme, matérialisme et métaphysique*, 1971; L. Joy, *Gassendi the Atomist: Advocate of History in an Age of Science*, 1987.

Jean-Vincent Blanchard

GAY, JOHN (1685–1732). John Gay was born at Barnstaple and educated at the Free School there. He moved to London in his early twenties and earned a respectable reputation as a poet and a member of the Scriblerus Club with the publication of *The Shepherd's Week* (1714), a volume of pastoral verse. His early dramas, including *The Wife of Bath* (1713), a redoing of Chaucer, and *The What D'ya Call It* (1715), light-hearted satire on tragedy, were unsuccessful, but he achieved public acclaim for his *Fables* (1727), verse stories written for Princess Caroline's son. Although Gay was supported by patrons, his income was always meager until the enormous success of his play *The Beggar's Opera*, for which he is chiefly remembered.

The Beggar's Opera opened in 1728. In its first season, it ran for a record-breaking sixty-three straight days in London and fifty in Bristol and Bath, becoming the most performed play of its time. The production's popularity was due to Gay's creative departure from standard dramatic entertainment. Gay's innovation was to invert the conventions of Italian opera by staging thieves and whores instead of aristocratic lovers, and by using folk ballads and alehouse songs instead of florid arias. The main character, Peachum, based on the real-life criminal Jonathan Wild, employs thieves to steal goods to be resold through London warehouses. Peachum regularly betrays one of his thieves to the police to be hanged, a strategy by which he maintains good relationships with the law, earns bounty money, and keeps other gang members in fearful servitude. Peachum's daughter Polly, hopelessly in love with the hard-drinking crook and womanizer Macheath, became the play's most popular character. On the surface, *The Beggar's Opera* seems to be primarily a good-hearted spoof of European opera and English outlaws, but its underlying social message champions British nationalism through native songs and exposes the corrupt London legal system that allowed such men as Peachum to organize and control underworld crime syndicates without punishment.

Bibliography: D. Nokes, *John Gay: A Profession of Friendship*, 1995.

James Norton

GENTILESCHI, ARTEMISIA (1593–1652/3). Considered the most famous woman painter in early-modern Europe, Artemisia is the only female follower of the influential painter *Michelangelo Merisi Caravaggio. Born in Rome on 10 July 1593, she lost her mother at the age of twelve and learned painting from her father, the painter Orazio Gentileschi. Joining Orazio's workshop, she was also sent to pursue her studies with another painter, Agostino Tassi. In 1612, Tassi was accused of—as well as tried and sentenced for—repeatedly raping her. After a widely publicized rape trial, Artemisia married Pierantonio di Vincenzo Stiattesi, moved to Florence, and by 1618 (or 1623) bore four (or five) children, of whom only a daughter, Prudentia, seems to have survived. Artemisia's husband left the family in 1623; Artemisia continued to ply her trade independently—and, at times, in temporary cooperation with male painters—in Florence, Rome, Venice, and Naples. In 1636–1637, she joined her father in London and collaborated with him on a commission from Queen Henrietta Maria before returning to Naples in 1640. Although she tried to secure the patronage of Grand Duke Francesco I d'Este in Modena, she was to remain in Naples until her death in 1652/3.

Instead of portraits, which were the kind of subjects more typically commissioned from the few known successful women artists in Italy, Artemisia painted various images of biblical, historical, and mythological women—Susanna, Lucretia, Mary Magdalen, Cleopatra, Esther, Bathsheba, and Diana. While depictions of these female protagonists were popular at the time and were created also by most of her male competitors, Artemisia is viewed as interpreting them from an unprecedented feminine perspective. Many, if not all, of these figures in her works are less idealized, less sexualized, and more true to life. They atypically appear not as helpless but as strong and self-sufficient beings. Artemisia's five or six surviving paintings of the apocryphal story of Judith's beheading of Holofernes are the most characteristic of her creative identity. Two of the most innovative—one in the Uffizi, Florence, and one in the Institute of Arts, Detroit—especially reflect the influence of Caravaggio even as their dramatic conception is completely original. The former presents Judith and her maid, Abra, in the process of cutting off Holofernes's head whereas the latter depicts the two women immediately after the decapitation, about to steal away from the enemy camp.

Indicative of Artemisia's extraordinary success is the fact that she was the first woman artist to be accepted by the Academy of Design in Florence (19 July 1616). She continued to receive commissions from some of Italy's most prominent patrons: Michelangelo Buonarroti the Younger, Grand Duke Cosimo II de' Medici, Cassiano del Pozzo, and Empress Maria of Austria, and she corresponded with *Galileo Galilei.

Bibliography: W. Bissell, *Artemisia Gentileschi and the Authority of Art*, 1999; M. Garrard, *Artemisia Gentileschi: The Image of the Female Hero in Italian Baroque Art*, 1989; M. Garrard, *Artemisia Gentileschi around 1620–22: The Shaping and Reshaping of an Artistic Identity*, 2001.

                                                                        Yael Even

GERHARD, JOHANN (1582–1637).

Johann Gerhard was born into a well-to-do family in Quedlinburg, Upper Saxony, in 1582. At fifteen, he came under the influence of Johann Arndt, then a pastor in Quedlinburg, who encouraged him to study theology. After enrolling in philosophy and medicine in Wittenberg, Gerhard studied theology in Marburg and Jena. In 1606, he received his doctorate and published his famous devotional work *Meditationes sacrae ad veram pietatem excitandam*. Although only twenty-four, he was appointed superintendent in Heldburg by Duke Johann Casimir of Coburg. By 1615, Gerhard had been appointed general superintendent (bishop) in Coburg, a position which he relinquished in favor of a theology appointment at the university in Jena. As the writer of works like his nine volume *Loci theologici* (Jena, 1610–1622) and the four-volume *Confessio catholica* (Jena, 1634–1637), which sought confirmation of the teachings of the Augsburg Confession in the writings of Roman Catholic theologians, Gerhard was responsible for consolidating orthodox Lutheran doctrine.

The practical aspects of Gerhard's religious belief appear in his *Schola pietatis* (1622–1623), with its concern for moral integrity and good works, motivated by a deep and sincere faith. He occupied administrative positions as rector and dean of the theological faculty of the university. Gerhard regularly advised Lutheran rulers on matters of politics and religion, and sought to oppose the growth of absolutism during the Thirty Years' War.

Gerhard continued to preach regularly until his death in 1637. Much of his work is exegetical, and the explication and application of biblical passages is a central focus of his sermons, as of his more academic works. Like Chemnitz in the preceding generation, Gerhard appears as a champion of Lutheran thought, establishing a middle way between the extremes of Calvinism and Roman Catholicism. He is the most prominent representative of Lutheran orthodoxy during the age of confessionalism.

Bibliography: Johann Anselm Steiger, *Johann Gerhard: Studien zu Theologie und Frömmigkeit des Kirchenvaters der lutherischen Orthodoxie*, 1997; Martii Vaahoranta, "Sola Scriptura—Sed Scriptura Numquam Sola: Bibelauslegung bei Johann Gerhard und die lutherische Gruppenidentität; Versuch einer thesenhafte Analyse," *Bibelauslegung und Gruppenidentität* (1992):162–177.

Janet Bertsch

GERHARDT, PAUL (1607–1676). Gerhardt is the most widely known German Baroque poet today due to his religious hymns, which are sung in Protestant and Catholic churches alike. Born in Gräfenhainichen, Saxony, to the mayor and a clergyman's daughter, he studied theology in Wittenberg and then spent several years as a private tutor in Berlin. In 1651, he accepted a position as pastor in Mittenwalde, married, and had five children, only one of whom survived. In 1657, he returned to Berlin to become deacon at the St. Nicholas's Church (Nikolaikirche), where he enjoyed widespread popularity as a preacher and poet. There he also met the organist Johann Crüger, a gifted composer, who was instrumental in popularizing Gerhardt's poems by providing tunes for some of them.

In 1666, his life was disrupted by a confrontation with the Great Elector of

Brandenburg, who was a Calvinist and who for political reasons demanded that all Lutheran pastors sign the Edict of Toleration. It forbade them to criticize the Great Elector's faith. As a staunch Lutheran, Gerhardt refused to comply, and he was dismissed. Popular support forced authorities to reinstate him, but he chose to resign. He subsequently became a pastor in Lübben, southeast of Berlin, where he remained until his death. The Berlin incident and the early death of his children and his wife (1668) cast a cloud on the last years of his life. He wrote no more poetry after moving to Lübben. A synthesis of medieval mystical thought and Luther's theology, Gerhardt's hymns, many of which were set to music by J.S. Bach, are perhaps the finest example of the genre. His style strikes a fine balance between emotion and reason, and his emphasis on the "I" or voice of the individual foreshadows German pietism. Two collections of his hymns were published during his lifetime, *Praxis pietatis melica* (1647) and *Geistliche Andachten* (1667).

Bibliography: E. Metzger, "Paul Gerhardt," *Dictionary of Literary Biography* 164 (1996): 113–120; G. Rödding, *Paul Gerhardt*, 1981.

Josef K. Glowa

GESUALDO, CARLO (ca. 1561–1613). Gesualdo was a composer and lutenist, significant chiefly for his unique and extreme madrigal style at the end of the Renaissance tradition. He was the prince of the principality of Venosa, married Maria d'Avalos, and acquired great notoriety for murdering her and her lover, the Duke of Andria. Gesualdo retired to his estate, but in 1594, he traveled to Ferrara, a center of progressive musicmaking, and married Leonora d'Este, niece to the Duke of Ferrara, Alfonso II. In Ferrara, he met the modernists among composers, especially Luzzasco Luzzaschi, who improvised on Vicentino's chromatic arcicembalo (a type of harpsichord). Gesualdo never lost his passion for music, but in later years he became increasingly melancholy living in seclusion on his estate. His circle of musicians established after and probably inspired by his Ferrarese experience included Stella, Nenna, and Effrem.

Gesualdo's works consist of six books of five-voice madrigals, one of six-voice madrigals (published posthumously in 1626), two books of sacred vocal works, and various other compositions. However, his reputation rests on the madrigals. The first published books show a relatively conventional madrigal style, but in Book IV (1596) his chromatic style emerged, increasing in Books V and VI (1611). Gesualdo was intensely interested in text expression, his musical means very bold. He tended to choose epigrammatic poetry by poets such as Guarini and Tasso. Many texts feature oxymorons and antithetical juxtapositions ("life" and "death"), which provide the opportunity for radical contrasts in word painting achieved through contrasting textures, rhythms, diatonicism and chromaticism, consonance, and dissonance. Such stark juxtapositions carry the Renaissance madrigalism to extremes and sometimes threaten the cohesion of the work. Gesualdo's stylistic means have often been criticized or attributed to his neurotic personality, but they were clearly shaped by musical developments in his native Naples and in Ferrara. Moreover, Gesualdo's later style can be seen in the larger context of the artistic and musical movement of "Mannerism," whose representatives share some of his stylistic traits (Luzzaschi, Marenzio, and others). However, Gesualdo adhered to the multivoiced Renaissance madrigal as his basic model; thus, it was left to other composers, notably *Peri, Caccini, and *Monteverdi, to transform the genre and help usher in the early Baroque style in music.

Bibliography: S. Sadie, ed., *The New Grove Dictionary of Music and Musicians*, vol. 7, 1981; G. Watkins, *Gesualdo: The Man and His Music*, 1973.

Olga Termini

GEULINCX, ARNOLD (1624–1669). A proponent of Cartesian philosophy, Geulincx was born in Antwerp and studied theology and philosophy at Louvain. He taught there from 1646 until 1658, when he lost his post and joined the Dutch Reformed (Calvinist) church, having criticized scholasticism and monasticism. From 1663, he taught at Leiden, a vigorous center of Puritan intellectual activity, publishing works on logic and argument in 1662 and 1663, as well as his most important work *De Virtute et Primis Ejus Proprietatibus* (1665), a discussion of ethics; his complete "Ethics" was published posthumously in 1675. He also completed commentaries on *Descartes (1690, 1691), physics (1688), and metaphysics (1691). Geulincx advanced a doctrine known as "occasionalism," which asserts that it is God rather than human will that causes a specific event (Latin, *occasio*) to occur in a finite body. He agreed with Descartes that the self is a thinking entity, but proposed that the individual mind cannot be the cause of a bodily change or motion because it lacks prior knowledge of how the action is to be done. The true cause of all change and thought is God, who acts upon human reason when it is attuned to the divine will. This concept was also developed in his "Ethics," which stressed the importance of four cardinal virtues; *diligence* in following sound reason; *obedience* to divine will; *justice*, or willing what reason dictates; and *humility* before God. Geulincx's concept of occasionalism influenced *Malebranche and *Spinoza, while his emphasis upon God's sovereignty drew him also to the doctrines of *Jansen.

Bibliography: F. Cross and E. Livingstone, eds., *The Oxford Dictionary of the Christian Church*, 3rd ed., 1997; W. Doney, "Geulincx," in *The Encyclopedia of Philosophy*, ed. P. Edwards, vol. 3, 1967.

Christopher Baker

GIBBONS, GRINLING (1648–1721). Gibbons was a woodcarver of outstanding virtuosity, investing his creations with a naturalism and vitality that transformed this craft in late seventeenth-century England. He was born in Rotterdam and probably served an apprenticeship in the Netherlands before emigrating to England, where he found employment as a ship carver. His decoration of the interior of London's Dorset Garden Theatre, completed by 1671, brought him, through the painter *Peter Lely, to the attention of the architect Hugh May, who employed him at Cassiobury Park, Hertfordshire, and for the renovation of Windsor Castle, the latter project establishing Gibbons' reputation as a craftsman of consummate skill.

In these commissions and at Sudbury Hall, Derbyshire, and Badminton House, Gloucestershire, Gibbons developed an innovative, vivacious, and masterly treatment of the carved surround unmatched by any of his English contemporaries. He rejected oak, the traditional medium in this context in England, for the more pliable limewood and not only greatly enlarged the range of fruits and flowers displayed on these surrounds, but also added game birds, fish, and crustaceans to the English woodcarver's decorative vocabulary. Gibbons's naturalism is reminiscent of Dutch still life painting, then in vogue in England. His approach was congenial to the aristocrat seeking to portray himself as a country gentleman and was soon widely imitated. Gibbons's masterpiece is widely considered to be the "Cosimo Panel" (Palazzo Pitti, Florence), a gift from *Charles II to Cosimo III of Tuscany, completed in 1682. Here Gibbons transforms the conventions of the trophy genre by producing a celebration not of martial prowess, but of peace and its fruits, celebrating the two rulers as peacemakers and patrons of the arts. Gibbons's virtuosity is in abundant evidence; here he presented one of his limewood point lace cravats. Through this display of virtuosity and Gibbons's insistence on clearly identifying himself as the panel's creator, he challenged the relatively low status customarily assigned to his craft.

Gibbons later collaborated extensively with *Christopher Wren, providing decoration for a chapel added to Whitehall Palace for *James II and the royal apartments of Wren's additions to Hampton Court Palace. For St. Paul's Cathedral, London, Wren's masterpiece, Gibbons directed the adornment of the bishop's throne, organ case, and stalls, projects completed in 1698. Gibbons's contributions to the library Wren designed for Trinity College, Cambridge, involved providing a series of benefactors' coats of arms in limewood and his lively depiction of the idiosyncracies of these arms offered another stunning demonstration of his skill. From 1702, Gibbons devoted himself to working in stone, a medium in which neither his contemporaries nor later commentators found him to command noteworthy dexterity. He died at London in 1721.

Bibliography: G. Beard, *The Work of Grinling Gibbons*, 1989; D. Esterly, *Grinling Gibbons and the Art of Carving*, 1998.

Matthew Koch

GIBBONS, ORLANDO (1583–1625). English musician Orlando Gibbons is best known for vocal compositions, especially his Anglican church music. He was also the preeminent keyboardist of his time. His brothers Edward (1568–1650) and Ellis (1573–1603) were also composers; the former held positions at King's College, Cambridge, and at Exeter Cathedral. Born in Oxford, he was a chorister at King's College, Cambridge, before beginning his studies there, graduating with a Bachelor of Music degree in 1606. He became organist of the Chapel Royal, which post he held until his death. In 1619, he became virginalist to the king, in 1622 received the Doctor of Music degree from Oxford, and in 1623 was named organist of Westminster Abbey. He died at Canterbury, while awaiting the celebrations honoring *Charles I's new queen, Henrietta Maria. Gibbons's predilection for traditional polyphony showed itself in his services as well as in his many anthems. These were praised for their grandeur by no less a musician than *Thomas Morley. His keyboard works include fantasias and highly ornamented pavanes and galliards. Although Gibbons's music was widely disseminated in copies during the early seventeenth century, only a small amount of his music was published during his life. These include a collection of secular vocal pieces in 1612; *Parthenia*, a collection of keyboard music of *Byrd, *Bull, and Gibbons dating from around the same time, and nine of his trios (ca.1620), which are thought to be among the predecessers of the Baroque trio sonata.

Bibliography: J. Harley, *Orlando Gibbons and the Gibbons Family of Musicians*, 1999.
Joyce Lindorff

GILBERT, WILLIAM (1540–1603). Gilbert was a physician, scientist, and author of *De Magnete Magneticisque Corporibus et de Magno Magnete Tellure Physiologia Nova* (*On the Lodestone and Magnetic Bodies and on the Great Magnet the Earth*) (1600); the first major study of physical science published in England, it won international renown and praise from such luminaries as *Sir Francis Bacon, *Galileo Galilei, and *Johannes Kepler.

Gilbert was born in Essex on 24 May 1540, the eldest son of Hierome Gilbert, recorder of Colchester and member of a well-established family in that town. He attended St. John's College, Cambridge, where he received the B.A. (1560), M.A. (1564), and M.D. (1569), becoming a fellow (1561) and then senior fellow (1569). After establishing a successful medical practice in London in 1573, he became a fellow of the Royal College of Physicians in 1576, serving as its censor and treasurer before being elected president in 1600, the same year that he published *De Magnete*. With fellow members of that organization, he helped to begin work on the *Pharmacopeia Londinensis*, which eventually was published 1618. In 1601, Queen Elizabeth appointed him as her physician, and thereafter he resided at the royal court. At her death in 1603, she bequeathed him money to support his research. *James I retained Gilbert as royal physician, but he died on 30 November 1603. His body was returned to Colchester for burial. Having no wife or children, he left all of his books, equipment, and other supplies to the Royal College of Physicians; regrettably, all were destroyed in the Great Fire of London in 1666.

Being a bachelor scientist with a well-stocked library and laboratory at his St. Peter's Hill home, he hosted monthly meetings that were an early predecessor to the Royal Society (founded by *Charles II in 1662). Influenced by the mathematician Henry Briggs, compass maker Robert Norman, and instrument maker and theorist Edward Wright, his account of magnetism is methodical, thorough, and based on extensive empirical research, including the observations of metallurgists and hundreds of experiments with small lodestones. He accurately described the properties of magnets, recognized that the earth is itself a magnet and has a metallic core, introduced the term *electricity* to scientific discourse, understood the application of his discoveries to determining latitude (though erring in some particulars), and reportedly invented two useful navigational instruments. Rejecting Aristotle's physics and Ptolemy's celestial mechanics, Gilbert accepted a rotating earth but otherwise was torn between the heliocentric theory of Nicolas Copernicus and the geocentric theory of Tycho Brahe. Like many figures in the First Scientific Revolution, Gilbert had one foot in the future and another firmly planted in the past. Influenced by Hermeticism, he believed in a living earth and spontaneous generation of life. A collection of his papers, *De Mundo Nostro Sublunari Philosophia Nova* (*A New Philosophy of Our Sublunar World*), was published posthumously. *De Magnete* was not published in English translation until 1893.

Bibliography: A. McLean, *Humanism and the Rise of Science in Tudor England*, 1972.

William B. Robison

GIORDANO, LUCA (1634–1705).

The Baroque painter of Naples Giordano traveled to Venice, Florence, and Madrid during his career and produced altarpieces, paintings of mythology, and fresco cycles for palaces and churches throughout Europe. According to his early biographer Filippo Baldinucci, Giordano joined the Neapolitan painters' confraternity in 1665. The brilliant colors and dynamic portrayal of dramatic action characteristic of his mature work show the influence of the Venetian paintings of Paolo Veronese and the Roman paintings of Pietro da Cortona.

After visiting Florence, Giordano continued to work for the prominent Florentine patrons Andrea, Ottavio, and Lorenzo del Rosso, the Sanminiati family, and Pietro Andrea Andreini. During a later sojourn to Florence in 1682, Giordano painted frescoes for the Palazzo Medici-Riccardi, for the Riccardi, and for the Palazzo Pitti owned by the Medici. During this period, Giordano reserved an elegant classicism for secular decorations, employing a powerful Baroque style for religious art developed from the illusionistic style of Cortona and *Gian Lorenzo Bernini in Rome.

Appointed court painter by Charles II of Spain in 1694, Giordano painted frescoes of *St. Lawrence in Glory, Adored by Charles V and Philip II* and the *Foundation of the Order of Golden Fleece* for the Escorial and Buen Retiro. After Charles II died in 1700, Giordano worked for private Spanish patrons before returning to Naples in 1702. His late Neapolitan painting *Meeting of Saints Carlo Borromeo and Filippo Neri* for the church of the Gerolamini portrays popular saints of the Counter Reformation.

Although his large workshop was said to have thirty assistants, Giordano himself was renowned for his speed of execution and improvisation. Some scholars believe he wrote the *Relatione della vita di Luca Giordano pittore celebre* (1681).

Bibliography: O. Benesch, *Luca Giordano*, 1923; O. Ferrari and G. Scavizzi, *Luca Giordano*, 3 vols., 1966, rev. 1992.

Deborah H. Cibelli

GLANVILL, JOSEPH (1636–1680). Born in Plymouth and raised in a Puritan household during times of heightened sensitivity to religion, Glanvill gained an education at Cambridge and later became an ardent supporter of the Anglican Church. During his years there, he joined the Cambridge Platonists in redefining the relationship between science and religion. These men—Cambridge University theologians, philosophers, and scientists all—championed Plato as a pre-Christian epitome of the successful philosopher: a thinker who used reason to illuminate faith in things beyond reason and faith to illuminate reason applied to material reality, all with visible, social results. Specifically, Glanvill and the Cambridge Platonists added the influence of Christianity and of new scientific endeavors to Plato's philosophy. The Cambridge Platonists neither denied God a primary role in natural processes as *Hobbes and *Descartes had nor embraced the idea of "enthusiasm"—whereby God brought revelation and fervor to individuals—often invoked by Puritans in response to experimental science. Moreover, with the Platonists, Glanvill ardently turned from the scholastic philosophy of Aristotle and "The Ancients" to embrace instead "The Moderns" in advocating new approaches to science. In 1664, the Royal Society—likewise practicing these new methods—elected Glanvill a fellow. Two years later he became rector of Bath, and in 1672, he became the chaplain to *Charles II.

His influence and knowledge were clearly respectable. His position appears at the outset complicated, however, because in his most significant scientific experiments, he sought to prove the existence of spirits. Specifically, he attempted to demonstrate that, even in an age when men conducted experiments to control nature, Satan might strike in the form of a spirit. Even in a Protestant nation, Glanvill contended, the acknowledgment of evil and its many manifestations should not be thrown out with Catholicism. His method for gathering evidence to support his case did suit the Cambridge Platonists; he relied both on theology and on experimentation in the search for a key to understanding natural phenomena. In 1661, he published *The Vanity of Dogmatizing* (later reissued as *Scepsis Scientifica* in 1665), which attacks medieval scholasticism, argues for a *Baconian approach to learning, and contains the story of the Scholar Gypsy, later appropriated by Matthew Arnold for his poem of that name. Among his many publications, the most influential include *A Philosophical Endeavour Towards the Defense of the Being of Witches and Apparitions* (1666) and *Sadducismus Triumphatus: or, Full and Plain Evidence Concerning Witches and Apparitions* (1689). In the latter, Glanvill used scientific experimentation to justify his claims about the spirit world. That particular and famous work was published posthumously with a preface by fellow Cambridge Platonist and student *Henry More. Another work of equal fame is Glanvill's *Plus Ultra: or, The Progress and Advancement of Knowledge Since the Days of Aristotle* (1668), which, unlike most of Glanvill's writing, functions as a historical document. By writing it, Glanvill continued the work started by Thomas Sprat, whose *History of the Royal Society* recorded the achievements of that important group of early scientists. To study Glanvill, then, is to fathom at once religion, science, rhetoric, and witchcraft and to access the interdisciplinary nature of early modern scholarship.

Bibliography: S. Talmor, *Glanvill: The Uses and Abuses of Skepticism*, 1981.

Rebecca Totaro

GLAUBER, JOHANN RUDOLF (1604–1670).

German chemist Johann Glauber was born in Karlsstadt, his only education being an unfinished period at the Latin school there. Between about 1626 and 1632, he underwent an assimilation of chemical, alchemical, and iatrochemical experience by studying in laboratories in Paris, Basel, Salzburg (1626), and Vienna (1625/1626), where alchemical patronage was a high priority of the enigmatic Emperor Ferdinand II. These would be his principal disciplines with secondary emphases in pharmacology and metallurgy. He also discovered the mineral salts named for him. By 1635, he worked as the apothecary at the court of Giessen. Between 1636 and 1640, he moved to Frankfurt, Bonn, Amsterdam, and then returned to Giessen, perhaps as principal court apothecary (1644). Repercussions of the Thirty Years' War and his liberal religious attitudes convinced him to return to Amsterdam in 1645 until the war's end.

He returned to Germany, settling in Wertheim (1651), and began experiments in wine improvement and probably continued alchemical experiments. Living in Kitzingen the next year, he dispensed pharmaceuticals (free drugs given one hour daily), primarily antimony-based medicines, a new and different direction in pharmacology. Income came from chemical preparations and a method of tartar fabrication (1654). His humanity moved him to advocate several means of economic recovery from the Thirty Years' War, particularly, the export of wine and beer concentrate recipes (1656). Back in Amsterdam, he constructed what has been called perhaps the "most impressive laboratory in Europe." He wrote a tract (among at least thirty) for the Dutch East India Company called "Trost der Seefahrenden oder Consolatio Navigantium," containing methods for concentrating and preserving rations, medicines against scurvy, and preparation of fresh water from salt water. He experimented with artificial fertilizer (tartar chemically derived from wine) in infertile soils. Despite his alchemical beliefs,

Glauber's technological anticipations in production and selling of drugs and chemical mixtures possibly make him the world's first industrial chemist.

Bibliography: K. Gugel, *Johann Rudolf Glauber, 1604–1670: Leben und Werk*, 1955; J. van Spronsen, "Glauber grondlegger van chemische industrie," *Nederlandse Chemische Industrie* 5, 3 March 1970:3–11.

William J. McPeak

GÓNGORA Y ARGOTE, LUIS DE (1561–1627). Luis de Góngora y Argote was a lyric poet and predominant representative of *culteranismo*, a style known for its accentuated brilliance, sensory perception, and exuberant imagery. A native of Córdoba, Luis de Góngora studied in Salamanca. He returned to Córdoba where he served as prebendary at the cathedral, his two main passions being bull fights and poetry. In 1617, he was ordained and later became Chaplain of Honor to Filipe III in Madrid where he participated in the literary circles of the capital. Upon tiring of the Spanish court, he returned to Córdoba where he died in 1627 of a stroke. Góngora composed both short meters with popular themes and long meters with cultured themes. His burlesque poetry degrades reality while his ornamental poetry seeks absolute beauty as typical of the *culturanismo* style. Góngora made extensive use of hyperbation, metaphors, mythological allusions, and neologisms. Typical of the cultured style is *"Fábula de Polifemo y Galatea"* (1612), which is based on Ovid's theme of the cyclops who killed the shepherd Acis over the love of the nymph Galatea. Extensive use of colorful imagery contrasts the fury of the cyclops with the beauty of Galatea. However, not all of Góngora's poems can be classified in the *culteranismo* style. The majority of his poems are rondelets (*letrillas*) and romances, popular meters that expressed light sentiments or burlesque perspectives on the social and human conditions of the times. *"Ande yo caliente y ríase la gente"* is an example of this lighter style. In 1927, Góngora's extremely cultured style was rediscovered and thoroughly appreciated by the group of Spanish poets called the "Generation of 27."

Bibliography: D. Alonso, *Estudios y ensayos gongorinos* 1961; E. Gates, *The Metaphors of Luis de Góngora*. 1933.

M. Ellen Blossman

GRAAF, REGNIER DE (1641–1673). Graaf began his medical studies in 1660 at Utrecht and continued them at Leiden. In 1664, he published his observations on the pancreatic juice; this work was immediately translated into French and often reprinted. He received a medical degree at Angers in 1665 and became a practicing physician in Delft, where he conducted private scientific research. He held no university posts, apparently because he was a Roman Catholic. His most important contribution to science is his role in the accurate and concrete description of the anatomy and physiology of the female mammalian reproductive system. The reproductive process was vigorously debated around

1665; some at the time believed that the ovum had no function at all; others thought that an ovum was formed only on consequence of mating; another theory was that the ovum passed from ovary to uterus through uterine tubes and that its passage gave pleasurable sensations and often caused swooning in the mother. Clearly, these were merely imaginative hypotheses. Graaf diligently observed and dissected the female reproductive organs of various mammals and succeeded in isolating the ovarian vesicles with their envelopes. He was also the first to discover the morphological changes of the ovary (a term coined by Graaf) upon mating. Graaf was also the first to recognize the glandular nature of the corpus luteum, which indicates rupture of the ovarian follicles. His published observations of the mammalian reproductive organs included descriptive drawings of such accuracy that they are still considered useful today. Graaf also published a treatise on the male reproductive organs in 1668, which was often reprinted; unfortunately, it contains little in terms of original material.

Bibliography: C. Gillespie, ed., *Dictionary of Scientific Biography*, 1981; A. Meyer, *The Rise of Embryology*, 1939.

Rosi Prieto Gilday

GRACIÁN Y MORALES, BALTASAR (1601–1658). Gracián was a Spanish writer and thinker whose work has been translated widely despite his obscure style. He uses *conceptismo*, a style characterized by its extreme brevity and wit. His works were very admired by *La Rochefoucauld, Nietzsche, and, especially, Schopenhauer. What attracted them to Gracián was his deeply pessimistic yet very acute vision of the decadent Spain of the period and of the world in general. Gracián's views incarnate the Baroque *desengaño* (disillusionment). He presents the world as a deceiving place and life as a personal struggle.

His life is connected to the Jesuit order, where he studied and became an important member. He was a confessor, chaplain, and administrator, as well as a professor of theology. He published most of his works under the pseudonym of Lorenzo Gracián. He was disciplined for not seeking the order's authorization to publish and was removed from his teaching position to a lesser one, where he died at age fifty-seven.

His most famous work is the extensive allegorical novel *El Criticón* (*The Critic*, 1651–1657), which narrates the travels of Critilo—the logical man—and Andrenio—the natural man—until they reach the island of immortality. Most influential are the treatises in which he presents an ideal, virtuous individual fit for the times, such as in *El Heroe* (*The Hero*, 1637), *El Político* (*The Statesman*, 1640), *El Discreto* (*The Discrete*, 1646), and *El Oráculo Manual y Arte de la Prudencia* (*The Art of Worldly Wisdom*, 1647), a collection of near-Machiavellian aphorisms regularly reedited. He also wrote a theological work, *El Comulgatorio* (*Preparation for the Communion*), the only one signed with his real name.

Bibliography: V. Foster, *Baltasar Gracián*, 1975; M. Hafter, *Gracián and Perfection*, 1966.

Enrique Fernández

GRAUNT, JOHN (1620–1674). In 1662, Graunt, a haberdasher born in London on 24 April 1620, published *Natural and Political Observations mentioned in a following index, and made upon the bills of Mortality*, one of the earliest attempts to draw conclusions about social phenomena from quantitative data. In Graunt's day, this form of study was termed "political arithmetic"; it is now called statistics. The bills of mortality were lists of deaths published weekly and arranged by church parish and cause of death; they began to appear early in the sixteenth century and were published starting in 1563. Graunt's work, which reached a fifth edition in 1676, provided a brief history of these lists and paid special attention to the varying death rates ascribed to different diseases, questioning whether these causes may have overlapped or been misidentified. He did not advance any "statistical methods" in the modern sense of that term, but his primary achievement was in scrutinizing the validity of his data before making inferences from it. Given the data he had, his estimation of mortality rates proved remarkably accurate well into the nineteenth century. He also noted that, despite the higher mortality rates for men than for women, there still remained enough men in the population to provide husbands for women who sought them, thus ensuring human propagation, an observation that was not without its implications of a divine providence (raised a Puritan, Graunt died a Roman Catholic). His work later drew the attention of the economist Robert Malthus (1766–1834) and was developed further by his friend Sir Richard Petty. Graunt was inducted into the Royal Society in 1663, but after the Great Fire of London in 1666, he became impoverished and his name no longer appears as a society member thereafter, a fact probably influenced by his decision to become a Catholic. Graunt himself became a statistic on 18 April 1674 when he died of jaundice, leaving behind a son and daughter.

Bibliography: M. Greenwood, "Medical Statistics from Graunt to Farr," in *Studies in the History of Statistics and Probability*, ed. E. Pearson and M. Kendall, 1970.

Christopher Baker

GRECO, EL (1541–1614). El Greco was born Domenikos Theotokopoulos of Greek parentage in Candia on the island of Crete, then a Venetian possession. Very little is known about El Greco's early life. He began his professional life as a painter on Crete where he may have been trained in the late Byzantine manner of painting. El Greco's earliest known work, now lost, was a small painting with a gold background that was offered at auction in 1566. By 1567 or 1568, El Greco had traveled to Venice where he came under the influence of Titian and the Italian Mannerists. Though direct evidence is lacking, El Greco is often portrayed as having been a student of Titian or possibly the Mannerist painter Tintoretto. El Greco certainly admired and emulated Titian's use of color. However, the emotionalism and arbitrary spatial construction of his work are nearer conceptually to Byzantine and Mannerist painting.

El Greco settled in Rome in 1570 with the intention of establishing himself as a painter. He was admitted to the Roman Academy of Saint Luke, shortly thereafter setting up his own shop. While in Rome, he was exposed to the work of Michelangelo. Though El Greco was ambivalent about Michelangelo's ability as a painter, he admired Michelangelo's designs, as well as his draftsmanship, which he consulted throughout his life. While reasonably successful in Rome, El Greco never achieved the recognition he felt entitled to. In 1577, by now middle-aged, he arrived in Toledo, Spain, apparently with the intention of moving to Madrid to seek royal patronage. Toledo, however, was to be his home for the rest of his life. In addition to its importance as a manufacturing city, Toledo was a powerful religious center. The archbishop of Toledo, who held the title Primate of Spain, was among the wealthiest persons in the country. In addition, he oversaw eight bishoprics extending over a large area of central Spain.

In 1577, El Greco gained the patronage of Diego de Castillo, dean of the Toledo Cathedral chapter. El Greco was commissioned to paint a series of pictures for the altar in Santa Domingo el Antigua in Toledo. The first of these paintings, *The Disrobing of Christ* (1577–1579), provoked a conflict. The final price of the painting was set by two groups of appraisers chosen by the artist and the patron. In the case of disagreement, an arbitrator would set the final price. The respective groups of appraisers disagreed sharply over the amount El Greco should be paid. Questions were raised about El Greco's interpretation; some of the incidental figures were elevated above Christ, and the three Marys were thought to be too prominently displayed in the lower left corner of the picture. Changes were requested, which El Greco refused. In the end, he received far less than he had been originally led to expect. Worse yet, he was placed in the position of a shopkeeper rather than, as was his view, a professional person deserving of respect. *The Martyrdom of Saint Maurice* (1580–1582), a commission from Phillip II, would have established El Greco's position as a royal painter rather than a mere craftsman had it been successful, but Phillip rejected it. Perhaps El Greco's cool, dispassionate presentation that minimized the Saint Maurice's sacrifice was at odds with the clarity and immediacy demanded for Spanish Counter Reformation religious art. Pictures were expected to evoke piety, not an intellectual discourse. El Greco never achieved all he hoped for in Toledo, but he nevertheless prospered, becoming an important painter in central Spain. He found patrons who became friends. They regarded him as a learned artist and their equal, more a philosopher than a painter. By 1596, he headed a thriving workshop. Jorge Manuel, El Greco's son (born 1578), became his partner and exponent, which relieved El Greco from having to deal with financial matters. Though productive, El Greco was plagued by debt throughout his life, despite completing a number of now highly regarded works: *The Assumption of the Virgin* (1577), *Christ Driving the Money Changers from the Temple* (ca. 1571–1576), *Annunciation* (1600), and *Fray Hortensio Felix Paravicino* (1609).

El Greco left no school; other than his son, there were few followers. By the late seventeenth century, his reputation had declined as his painting came to be seen as extravagant or eccentric. Critics were puzzled at the expressive liberties El Greco took with form, scale, color, and content. They were particularly unsympathetic toward the elongated, frequently gravity-defying, figures, which were especially evident in his religious works. Yet the perceived anomalies of El Greco's style were acceptable to his contemporaries because their origins in Italian Mannerism were well understood and correctly seen as a revitalization of spirituality in painting. The rise of classicism and naturalism conspired to diminish his status, because El Greco practiced neither. By the early twentieth century, the perceived defects of El Greco's style had become the foundation of his reappraisal. The shift toward synthetic form and color in avant garde painting began with Postimpressionism in the late nineteenth century and continued in the work of Cézanne, the Expressionists, the Fauves, and ultimately the Cubists. El Greco's tendency to reconstruct the figure for expressive purposes came to be regarded as a path out of the near-photographic historic painting that passed for high art in the salons of the day. El Greco's work provided authority as well as encouragement to a generation of young painters who were determined to rescue painting from the stagnation of the academy. When Dada and Surrealism began around 1916, El Greco's painting, again, became a source of inspiration. The juxtaposition of spatial and temporal incongruities in El Greco's compositions provided a reference for the deliberate absurdities of Surrealistic art. Though El Greco's intentions were far different from those of the painters who referenced his work, they understood the power and revolutionary character of his artistic means far better than the critics who had denied him. Thus, over time, El Greco achieved the universal fame that eluded him during his lifetime.

Bibliography: J. Brown et al., *El Greco of Toledo*, 1982; L. Murray, *The Late Renaissance and Mannerism*, 1967.

John Schmidt

GREGORY, JAMES (1638–1675). Gregory, a Scots mathematician and astronomer, was born near Aberdeen in November 1638, first learning mathematics from his mother. He invented a type of reflecting telescope in 1663, five years before *Newton developed a related model of somewhat different design; his description of his design, the first practical reflecting telescope, was described in his *Optica Promota* (*The Advance of Optics*) in 1633. It is now known as the "Gregorian" telescope. Whereas Newton's telescope directed the image from the collecting mirror by means of an angled, flat mirror inside the telescope tube and then out the side of the telescope to an eyepiece, Gregory employed a perforated main collecting mirror the image from which was reflected from a secondary concave mirror through the main mirror's opening to an eyepiece. His plans for constructing a six-foot focus mirror in about 1664–1665 were foiled when it proved impossible to construct mirrors of sufficient quality. The virtue of his design was that it employed both mirrors and lenses, and it also enabled construction of a telescope shorter than the total focal lengths of its mirrors. He also developed a method of measuring the distances of stars through photometry (the measurement of the intensity of a source of light).

Gregory's work in mathematics led him to independently arrive at the first proof of the fundamental theorem of calculus prior to Newton. In about 1664, he traveled to the University of Padua and later completed *Vera Circuli et Hyperbolae Quadratura* (*The True Squaring of the Circle and of the Hyperbola*), published in 1667, one of the first discussions of the differences between convergent and divergent mathematical series, as well as commentary on other topics such as iterations and algebraic and transcendental functions. Gregory later sent a copy of this book to *Christiaan Huygens, who published a review of it in which he offered cogent challenges to some of Gregory's mathematical conclusions but also claimed to have been the first to arrive at some of Gregory's proofs, an unfounded accusation which offended Gregory. He responded to Huygens's mathematical critiques in the work *Exercitationes Geometricae* (1668). The following year saw the publication of *Geometriae Pars Universalis* (*The Universal Part of Geometry*), setting forth rules for finding the areas of curves and the volumes of their solids of revolution. This has been called the first textbook on calculus and contains the first known proof of the difference between differential and integral calculus, another discovery Gregory arrived at independently of Newton, whom he also predated in discovering the interpolation formula and the general binomial theorem.

In the summer of 1668, Gregory became a member of the Royal Society and *Charles II created for him the Regius Chair of Mathematics at the University of St. Andrews, where Gregory taught from 1669 to 1674. Among his other work there, he discovered Taylor's theorem in 1671 (some forty-five years before it was published by Brook Taylor), and he discovered the diffraction grating after observing the pattern made by a beam of light passing through a feather. After a dispute, he left St. Andrews for the University of Edinburgh in 1674, becoming the first occupant of its Chair of Mathematics. In October 1675, he suffered a stroke while observing the moons of Jupiter with his students, went blind, and died several days later at the age of thirty-six.

Bibliography: C. Gillispie, ed., *Dictionary of Scientific Biography*, 1970–1980; H. Turnbull, ed., *James Gregory: Tercentenary Memorial Volume*, 1939.

Christopher Baker

GRIMALDI, FRANCESCO MARIA (1618–1663).
Francesco Maria Grimaldi was born and died in Bologna. At fourteen, he entered the Society of Jesus, taking a doctorate in 1647 and his vows as a priest in 1651. A professor

of mathematics by 1638, Grimaldi collaborated with the Italian astronomer Giovanni Battista Riccioli, an opponent of heliocentrism. This theoretical inhibition did not prevent the two scholars from making notable contributions to astronomy. They compiled a map of the lunar surface and named physical features for prominent astronomers, a practice that continues to the present day. In addition, Riccioli and Grimaldi produced a reliable pendulum clock for measuring the acceleration of falling bodies and published a star catalogue. Grimaldi aided *Giovanni Cassini in establishing the meridian line for Bologna as well.

More important, during Grimaldi's twenty-five year career in Bologna, was his work with the nature of light, particularly his demonstration of diffraction. A very able experimenter and exceptional observer, Grimaldi described his findings and theoretical conclusions in *Physico-mathesis de lumine, coloribus, et iride*, which was published posthumously in 1665. Allowing a beam of sunlight to pass through a small opening in a screen, Grimaldi observed how the light striking a surface on the other side produced a band wider than that entering the aperture. Light was "split apart," and at the edges of the shadow created by the effect, colored fringes appeared, leading the scientist to conclude that light was bent from its course as it passed these edges. He labeled the phenomenon "diffraction." The discovery influenced Grimaldi to associate light with waves (as in a fluid) rather than with the competing particle (atomist) theory.

The significance of Grimaldi's discovery was not completely understood until the early nineteenth century when Thomas Young and Augustin Fresnel, working independently, used the principle of interference to provide fundamental proofs of the wave nature of light. Only in the twentieth century with the advent of quantum mechanics would the wave function that Grimaldi assigned to light be applicable to particles.

Bibliography: P. Dear, "Jesuit Mathematical Science and the Reconstitution of Experience in the Early Seventeenth Century," *Studies in History and Philosophy of Science* 18 (1987):133–175.

Joan Klobe Pratt

GRIMMELSHAUSEN, HANS JACOB CHRISTOFFEL VON (ca.1621–1676).

The most important German author of the seventeenth century, Grimmelshausen was born the son of a Protestant innkeeper and baker in Gelnhausen near Frankfurt. His father died early, and his mother abandoned him when she remarried. In 1634, Imperial soldiers ransacked his hometown. He escaped but was later kidnapped by Croatian and then Hessian troops, who forced the young boy to serve in the army. There he rose from stableboy to the position of a regimental secretary. After the war, he converted to Catholicism, married, and was successively an administrator for noble estates, an innkeeper, and finally the mayor of Renchen in Baden.

Though mostly self-taught, he acquired an encyclopedic range of knowledge, which he skillfully wove into complex narratives. Grimmelshausen composed his extensive body of works within only ten years. His crowning achievement was the novel *Der Abentheuerliche Simplicissimus Teutsch* (1668–1669) (*The Adventurous Simplicissimus*). This instant bestseller paints a rich panorama of life during the Thirty Years' War. A grotesque world turned topsy-turvy provides the framework for the satirical tale of Simplicius, an innocent young boy who becomes entangled in the brutality and temptations of the war. The novel traces his adventures, his ups and downs, until Simplicius finally decides to quit the world of action and become a hermit.

Modeled on the picaresque novel, especially on Albertinus's translation of *Guzmán de Alfarache* (1599), Grimmelshausen's tale may be partially autobiographical. However, recent scholarship has shown how masterfully he also drew on numerous sources for inspiration, notably on the works of such authors as Thomaso Garzoni, Charles Sorel, and Sir Philip Sidney. The success of this book prompted him to write several continuations, works that constitute a cycle known as the *Simplician Writings*. His lasting impact on German literature can be seen in the influence on moderns such as Thomas Mann, Bertolt Brecht, and Günter Grass.

Bibliography: V. Meid, *Grimmelshausen: Epoche-Werk-Wirkung*, 1984; H. Wagener, "Johann Jacob Christoffel von Grimmelshausen," *Dictionary of Literary Biography* 168 (1996):121–139.

Josef K. Glowa

GROTIUS, HUGO (1583–1645). Born in Delft, Holland, on 10 April 1583, Easter Sunday, Huigh de Groot was the son of Jan Huigh de Groot, a prominent lawyer and beer exporter, and Alida Borren van Overschie, a Catholic who would convert to Calvinism twelve years later to satisfy her son Huigh. More commonly known by the Latin rendition of his name, Hugo Grotius, Grotius was a prodigy in the liberal arts, notably philosophy, languages, and literature. His chief contribution to modern knowledge is his legal treatise *De Jure Belli ac Pacis* (*The Law of War and Peace*), published in 1625, which has made him known as the progenitor of international law.

The Law of War and Peace is distinguished by its statement of principles and complex, multilayered structure of support. Motivated by the question of what the law says about war, the treatise notes, instructively, that peace cannot be conceptualized as the antithesis to war—indeed, according to the treatise, war forms an important avenue toward peace in certain circumstances and that the law teaches toleration of many expressions of specific or local social organizations. These tenets, scholars understand, are supported by arguments drawn from Pythagoras and musical structure. The five sections of the work contrapuntally develop the themes, thereby making compelling the essential warranting assumption of Natural Law. For best appreciation of its substance and argumentative method, *The Law of War and Peace* should be read with the earlier treatise, *Mare Liberum* (*The Freedom of the High Seas*), published in 1609. The two works complement one another's cultivation of a postskeptical Stoicism that allows a materialist ethics and politics, notably developed by *Thomas Hobbes. However, the works provide the foundation of modern political and especially juridical thought by in large part resisting Machiavellian influences in their appeals to Christian ideals. Such Christian appeals sometimes are overlooked by current readers of Grotius, a situation that causes significant misunderstanding of his contribution in his historical moment and his actual influence on subsequent thought, such as on that of Hobbes.

Although most important for his contribution to modern law, Grotius was also accomplished in languages and letters. His stoic-Christian epic in his mother tongue, Dutch, *Bewijs van de waeren Godsdienst* (*Proof of True Religion*) contributed to his contemporary international renown. Translated into twelve languages, this vernacular epic may have been more influential in the seventeenth century than his work in Latin, the lingua franca of cosmopolitan humanism. Similarly, his scholarly activity in editing and annotating Greek, Latin, and Biblical texts, notably *The Seven Liberal Arts* of Martianus Capella, the *Civil War* by Lucan, and his *Annotations to the Old Testament* and *Annotations to the New Testament* contributed to his credibility both as an author and as an expert in his own time. Notwithstanding such linguistic and literary accomplishment, since Grotius's day, his neo-Latin drama has risen to prominence among his more strictly literary endeavors. Of his three neo-Latin plays, *Adamus Exul* or *Adam in Exile* is most important. Like the other two, *Christus Patiens* (*Christ's Passion*) and *Sophompaneas* (literally, "he that is all-wise," but commonly *Joseph in Egypt*), *Adam in Exile* is a Christian humanist rendition of an important biblical story, specifically, that of Genesis 3. Although its influence on *John Milton in composing *Paradise Lost* has been suggested and seems quite clearly significant, such influence has not yet been fully enunciated. Likewise, the relation between this literary work and his own and others' legal thinking and writing has been advanced but not fully developed.

Such prodigious intellectual productivity should not suggest a contemplative as opposed to an active life. Indeed, that Grotius could produce so much aesthetically worthwhile and intellectually significant work during his very engaged public life impressed even the most reserved and accomplished of scholars, such as Grotius's own teacher, the revered Joseph Justus Scaliger. After studying at the University of Leyden under Scaliger, Grotius attached himself first to Johannes Uitenbogaert, a Calvinist court preacher, and then to Johan van Oldenbarnevelt, the Land's Advocate, or prime minister. Then Grotius went on a mission to the French court where he was made honorary doctor of laws of the University of Orleans and where King Henry IV presented him with congratulations and a gold medal. Returning to Holland, Grotius rose quickly from being admitted to the bar in 1599 to being appointed attorney general and first public comptroller of the three state courts in 1607. The next year, Grotius married Marie van Reigersbergh, daughter of the mayor of Veere and Grotius's father's choice. They appear to have been happy in marriage, seeing five of their eight children to adulthood. In 1613, he went on a mission to England where his disagreements with *James I's counter-Remonstrant views hardened. Returning to Holland, Grotius received the governorship of Rotterdam. From this point, his fortunes turn. Imprisoned in 1618 with his patron, Oldenbarnevelt, Grotius escaped in 1621 with the help of his wife, who impersonated Grotius in his chamber while he escaped in what was to be a trunk of books. Moving into the service of Sweden, then, as diplomat in Paris, Grotius successfully kept Sweden's subsidy payments coming. Despite such effectual service, Queen Christina of Sweden dismissed him and tried to offer him alternative, reduced engagements. Grotius refused and hastened across the Baltic, displeased. The sea trip was plagued by storms for days, eventually wrecking the ship and overtaxing Grotius. Making his way to Rostock, Germany, he died of exhaustion shortly before midnight on 28 August 1645. For a while, his organs remained in a cathedral in Rostock, but eventually they were placed with his corpse at Nieuwe Kerk in Delft, Holland. Grotius's epitaph, composed by himself, suggests the ambivalence of a patriot who feels less loved by his country than seems just: *Grotius hic Hugo est, Batavum captivus et exul, Legatus Regni, Suecia magna, tui* ("This is Hugo Grotius, captive and exile of the Dutch, but envoy of the great Kingdom of Sweden").

Bibliography: C. Gellinek, *Hugo Grotius*, 1983; R. Tuck, "Grotius, Carneades and Hobbes," *Grotiana* n.s. 4 (1983):43–62.

Jesse G. Swan

GRYPHIUS, ANDREAS (1616–1664). Gryphius generally is thought to be the most significant poet of the German Baroque. His plays are unsurpassed in his time, and he mastered the sonnet form. Gryphius was born on 2 October 1616 in Glogau (Silesia), the son of a Lutheran pastor. The horror of the Thirty Years' War (1618–1648) and the ongoing confessional strife were formative experiences for Gryphius. His father died in 1621, the same year he entered the Latin School in Glogau. In 1634, he had the opportunity to enter the Academic High School in Danzig where he was exposed to Copernican cosmology and to *Martin Opitz's German poetry reform. A serious illness sent him home in 1636.

Gryphius went to Leiden in 1638 to study law and other subjects. In 1644, he was given the opportunity for an academic journey to France and Italy. In the fall of 1647, he returned to his war-torn native Silesia for the first time in almost ten years. In spite of several attractive offers for university teaching positions, Gryphius decided to stay in Silesia. In 1649, he married Rosine Deutschländer, and in 1650, he was appointed syndic for the Glogau estates, a position he held up to his premature and sudden death on 16 July 1664.

As syndic, he served as legal representative of the Protestant Silesian estates and had to mediate conflicts with the Catholic imperial court. This delicate political position highlights an ethical conflict that was representative for many Protestant servants of state at the time. Gryphius had to reconcile political loyalty to the imperial court, whose claim to power was irrefutable, with his service to the local Protestant nobility, with his open opposition to imperial efforts to re-Catholicize Silesia, and ultimately with his unwavering allegiance to his Lutheran faith.

Not surprisingly, we can find similar types of conflict in his tragedies. Regicide is rejected in two plays, *Leo Armenius* (1646–1647), where Leo seizes the crown with illegitimate means and finds a violent death at the hands of his successor, and *Carolus Stuardus* (1649–1650), where the violation of the divine rights of kings is deplored. The martyr drama *Catharina von Georgien* (1646–1647) pitches Catharina's Christian virtue and constancy against the barbarian tyranny of the Shah of Persia. In *Papinianus* (1657–1659), the legal scholar Papinian endures execution rather than legitimize the murderous rule of the tyrant Caracalla. *Cardenio und Celinde* (ca.1649), the only nonpolitical tragedy, sees the divine order and chaste love triumph over intrigue and lust.

Gryphius was the first major playwright in the German language; he also wrote four comedies. His tragedies are not action-driven; the focus is on the gradual development of the characters and on their response to adverse circumstances. Gryphius offers a critique of the use of force as political tool and implicitly of Machiavellianism and of the absolutist state. Gryphius is equally renowned for his poetry, most notably for his sonnets, which are mostly devotional and spiritual in character. The atrocities of the war, the *vanitas mundi* motif, the Christian virtue of constancy, and the juxtaposition of stoic acceptance of physical existence and of faith in eternal life are central themes in his writings.

Bibliography: E. Metzger and M. Metzger, *Reading Andreas Gryphius*, 1994; B. Spahr, *Andreas Gryphius*, 1993.

Peter Hess

GUERICKE, OTTO VON (1602–1686). Guericke was born into a family that had dominated Magdeburg city politics since the thirteenth century. (At his birth his family name had been spelled "Gericke," but it was officially changed in 1665 to facilitate pronunciation among French-speaking diplomats when he became ennobled as Otto von Guericke.) From 1617 to 1624, he studied at the universities in Leipzig, Helmstedt, Jena, and Leiden. In the midst of the Thirty Years' War, Guericke filled various posts for the city of Magdeburg, including serving as the official in charge of fortifications. In 1631, the imperial armies led by Tilly laid siege to the city, culminating in its devastation, burning, and plundering. Guericke was taken prisoner, but managed to purchase his freedom. After the imperial armies left in 1632, Guericke led efforts to rebuild Magdeburg, as well as other cities that had suffered similar fates. For decades, Guericke implemented his city plan, directing the reconstruction of city walls, bridges, churches, and other structures. Guericke served as mayor of Magdeburg from 1646 to 1676, and he represented the city at meetings of the Holy Roman Empire and other political bodies.

In addition to his political duties, Guericke devoted much effort to natural philosophy. His studies focused on pneumatics, particularly the nature of space and the vastness of the universe. The topic had important theological and philosophical implications: While some argued that God would not create a space with nothing to fill it, and others suggested that such space predated the Creation, *Rene Decartes insisted that space and matter were equivalent and thus denied the existence of vacuums. With increasing success, Guericke designed the pumps, vessels, and seals that helped uncover the nature of air and vacuums. In one of the most famous early modern experiments, Guericke demonstrated that two teams of eight horses working in opposite directions could not pull apart two evacuated copper hemispheres until air was readmitted into them. In addition, by weighing the spheres before and after evacuation, he was able to determine the approximate weight of air.

Guericke's fascination with air also led him to study connections between air pressure and weather. He built a four-story instrument at his Magdeburg home with a small mannikin that rose and fell in relation to barometric pressure. Guericke's interest in the attraction of celestial bodies prompted studies of electrical phenomena. In a famous demonstration using a sphere of sulfur that, when spun or rubbed became charged with static electricity, Guericke showed the forces that could keep feathers and similar objects suspended in midair. In his major published work *The New (So-called) Magdeburg Experiments of Otto von Guericke* (1672), and through scientific colleagues throughout the continent, Guericke gained a reputation as one of the leading European scientists of his day. Guericke left Magdeburg in 1681 and died in Hamburg in 1686. In all, Guericke was an important north German municipal leader who also represented a transitional period in the history of science. Like the Renaissance magus, he sought to understand connections between the terrestrial realm and the mysterious forces and fluids of the universe. Like modern scientists, though, Guericke also understood the importance of empiricism, mathematics, and the international dissemination of knowledge.

Bibliography: T. Coulson, "Otto von Guericke: A Neglected Genius," *Journal of the Franklin Institute* 236 (1943):241–264, 333–351; D. Schneider, *Otto von Guericke: Ein Leben für die Alte Stadt Magdeburg*, 1997.

Mark Finlay

GUNTER, EDMUND (1581–1626). English mathematician and inventor, Edmund Gunter was born in Hertfordshire of Welsh ancestry and matriculated at Christ Church, Oxford, in 1599. After taking the B.A. and M.A. degrees in 1615, he earned a divinity degree, was ordained, and became rector of St. George's Church in Southwark; four years later, he became professor of astronomy at Gresham college and held both posts until his death. In 1620, he published *Canon Triangulorum, or the Table of Artificial Sines and Tangents*. This volume of sines and tangents (extended to seven decimal places) also described

"Gunter's line," a logarithmic scale inscribed on a ruler for the solution of problems. This device was not, strictly speaking, the first slide rule, as it had no moving bar; it remained for *William Oughtred to join two such scales into the first slide rule. Gunter also originated the terms *cosine* and *cotangent*, and the contractions *sin* for sine and *tan* for tangent. In 1624 appeared his *Description and Use of the Sector, Crosse-staffe, Bow, Quadrant, and other Instruments*. Gunter's quadrant enabled the determination of the hour, azimuth of the sun, altitude in degrees of an object, and other problems of spherical navigation. His other inventions included Gunter's Chain, twenty-two yards in length and composed of 100 links, used in surveying (one acre equaled ten square chains). He also devised Gunter's scale, a large plane scale marked with various number lines for solving maritime navigational problems. Another navigational work was his *New Projection of the Sphere* (1623), and he also was the first to notice (around 1622–1625) that the earth's magnetic declination varies. He died on 10 December 1626.

Bibliography: B. Babcock, "Some Notes on the History and Use of Gunter's Scale," *Journal of the Oughtred Society* 3 (1994):14–20; C. Gillispie, *Dictionary of Scientific Biography* (1970–1990).

Christopher Baker

GWYN, NELL (1650–1687). Nell Gwyn, actress and prostitute, found favor with *Charles II, and became one of his most influential mistresses. An impoverished and fatherless Nell became a prostitute at the age of twelve. Her dream of appearing on stage was realized by 1663, but the combination of the 1665 plague and the 1666 fire in London meant that entertainment venues remained virtually closed until December 1666. When Nell returned to the stage, she began to take on a variety of roles and several prominent courtly lovers. She attracted the notice of the king, becoming his mistress in early 1668. Nell continued her work on the stage and gave birth to two of the king's sons by 1671, but never became an "official" mistress; however, she functioned as a valuable political hostess for the king. Nell, the only Protestant mistress of the king, was generally popular and took relatively little from her lover's coffers. By 1677, Nell had retired from the theater and hoped that her long service to the king might finally be rewarded with a title, but the king's advisor, Lord Danby, opposed the idea.

By 1678, Nell was having some financial difficulties, and after the king recovered from a serious illness in 1679, she began to lobby in earnest for a title, a quest cut short by her own illness in May 1680, and the death of her younger son in June. Nell began to concentrate less on political matters and more on charitable works, such as the founding of a hospital for veterans.

Through the next few years, Nell remained high in the king's affections, and in early January 1685, her surviving son was named Duke of St. Albans. However, Charles suffered a stroke in February, and though Nell was excluded from his bedchamber, he bade farewell to their son before dying. Nell, left to fend for herself, petitioned the new king, *James II, for financial help, which he granted. She feared for James's political future, but her own worries and a paralyzing illness in March 1687 diverted her attention elsewhere. Her deteriorating condition led her to make her will, in which she left everything to her son. She died on 14 November 1687, and her funeral was preached by a future Archbishop of Canterbury. Though Nell Gwyn's name graces many pubs today, her real legacy lies in her political acuity and her devotion to charitable works.

Bibliography: R. MacGregor-Hastie, *Nell Gwyn*, 1987; J. Wilson, *Nell Gwyn, Royal Mistress*, 1952.

Connie S. Evans

H

HALL, JOSEPH (1574–1656). Joseph Hall was bishop of Exeter and Norwich, a moral philosopher, and an innovator of a number of minor prose styles. While a student at Cambridge, he wrote the *Virgidemiarum* (1597–1602), the first English satire that effectively captured the spirit of Latin, particularly Juvenalian, satiric models. His satire was followed by *Mundus Alter et Idem* (*The World Different and the Same*, 1605), a work of fantasy travel literature that influenced Swift's *Gulliver's Travels* (1726).

After Hall was ordained in 1596, he served as rector at Hawsted Hall in Suffolk for Sir Robert and Lady Anne Drury, niece of *Sir Francis Bacon. While at Hawsted, his writings changed from those of a satirist to a moralist, as he composed practical guides to Protestant meditation such as *Meditations and Vows* (1605) that christianized Stoicism and possessed a Senecan aphoristic style. Hall also became friends with *John Donne and ultimately composed two prefatory poems to Donne's metaphysical *Anniversaries* (1613) on the death of Elizabeth Drury, daughter of his patrons. Hall subsequently served as chaplain to Prince Henry, and with *Characters of Vertues and Vices* (1608) became the first writer in English to imitate Theophrastus, an ancient Greek philosopher who composed a book of "characters," short sketches of differing character types.

After serving as a representative to the Synod of Dort and as Dean of Worcester, Hall was consecrated as Bishop of Exeter in 1627 and Bishop of Norwich in 1642. During the reign of *Charles I and the English Civil War, Hall became well-known for his religious moderation, adhering to the Elizabethan "middle way" between *William Laud's high church Anglicanism and radical Puritanism; Hall's *Episcopacy by Divine Right* (1640) was a classic defense of Anglican polity, one which led to a heated polemical exchange with *John Milton. Despite his pleas for tolerance among Christians, after Parliament passed the Ordinance of Sequestration (1643), Hall was deprived of his bishopric and his living, and moved his family to a small house in Higham. Here he continued to write and publish devotional works until his death in 1656, his satiric and devotional writings his most enduring legacy.

Bibliography: R. Applebaum, "Anti-geography," *Early Modern English Studies*, 4, 2, Special Issue 3 (September 1998): 1–17 [on Hall's *Mundus Alter et Idem*]; P. Fincham and P. Lake, "Popularity, Prelacy and Puritanism in the 1630s: Joseph Hall Explains Himself," *English Historical Review* 111, 443 (1996): 856–881; F.L. Huntley, *Bishop Joseph Hall, 1575–1656: A Biographical and Critical Study*, 1979.

Anna Marie Roos

HALLEY, EDMOND (1656–1758). Halley, whose first name is sometimes spelled "Edmund," was an astronomer, mathematician, and physicist. Using *Sir Isaac Newton's theory of mechanics, Halley was the first person to predict successfully the motion of comets. Halley was born near London, England, and educated at St. Paul's School and later at Queen's College, Oxford. From an early age, Halley displayed a great interest and aptitude in mathematics and science, and he published his first paper, on the topic of planetary orbits, at the age of twenty. This article appeared in the *Philosophical Transactions of the Royal Society* in London. Halley's interest in astronomy was greatly influenced by his friendship with the Astronomer Royal, *John Flamsteed. Together with *Robert Hooke, Halley helped Flamsteed design and construct the Royal Astronomical Observatory at Greenwich. Furthermore, Halley assisted Flamsteed in the observations leading to a complete catalogue of the stars of the Northern Hemisphere. Fascinated by this work, Halley decided to leave his studies at Oxford and complete a similar catalogue for the stars of the Southern Hemisphere. From his observational base on the island of St. Helena in the south Atlantic Ocean, Halley catalogued 341 stars and published his results in 1678, the same year in which he was elected a fellow of the Royal Society of London and was awarded an honorary M.A. degree from Oxford.

During his stay at Oxford, Halley befriended Sir Isaac Newton. Together these two scientists would solve the problems of the nature of the motion of comets. In 1682, Halley observed the bright comet that appeared over London, a comet that eventually was to bear his name. Although most people at the time feared comets as bad omens, Halley was determined to study them from a scientific perspective. Correctly, he hypothesized that comets are objects of the solar system that are governed by the same laws that govern the planets. In 1684, Halley approached Newton with a question concerning gravitation and, much to Halley's surprise, learned that Newton had found the answer to it nearly twenty years earlier. Halley then encouraged Newton to publish his work and even offered to pay for its publication. Newton accepted Halley's offer and published *Philosophiae Naturalis Principia Mathematica* (*Mathematical Principles of Natural Philosophy*) in 1729, perhaps the greatest scientific book of all time. Collaborating with Newton, Halley was able to predict the motions of comets correctly.

Although about 100 years earlier the Danish astronomer Tycho Brahe recognized the elliptical paths of comets, he did little to advance the knowledge of
their relationship to the sun and the planets. As a first step, Halley listed the
brightest comets with sufficient observations in order to plot their motions. From
this data, he hypothesized that three such comets appearing at seventy-five year
intervals—in 1531, 1607 and 1682—might actually be a single object. If true,
comets would be ordinary objects that moved around the sun rather than supernatural objects of evil. Halley applied Newton's law of gravity as the force
that moved the comets. When Halley applied Newton's formula to his observations, he was able to predict the motion of the 1682 comet. Halley's data of
1705 suggested that the comet would return in 1758, which it did, putting to
rest the geocentric view of the universe and superstitions about the nature of
comets. Today, that celestial body is known as "Halley's Comet."

Halley's scientific work was not limited to the study of comets. From 1698
to 1701, he made a series of voyages to study the Earth's magnetism; in 1716,
he predicted the 1761 and 1796 transits of Venus; and in 1718, he noted the
direct motion of the stars Sirius, Aldebaran, and Arcturus. He received much
fame for his study of comets and in 1703 was appointed Savilian Professor of
Geometry at Oxford. In 1720, he was appointed Astronomer Royal of England.

Bibliography: L. Baldwin, *Edmond Halley and His Comet*, 1985; C. Ronin, *Edmond
Halley: Genius in Eclipse*, 1969.

Paul C.L. Tang

HALS, FRANS (ca. 1582–1666). Hals lived in Haarlem, now the capital of
North Holland, from 1585 until his death in 1666. He studied with Karel van
Mandler, a Mannerist painter who had slight influence on Hals. Of more importance to Hals's artistic development was the Utrecht Caravaggisti, a group
of Dutch painters who imitated *Caravaggio's use of highly contrasting cellular
light as well as his penchant for painting pictures of revelers. By 1610, Hals
had become a member of the Guild of Saint Luke's in Haarlem; very little is
known of his early work apart from a few genre paintings and portraits, none
of which would predict his great future.

In 1616, Hals painted the *Banquet of the Officers of the Saint George Civic
Guard Company of Haarlem*, a work that established him as a painter of the
highest caliber. The painting is a portrayal of a group of common, but not
ordinary, men of the kind who help create an independent republic free from
kings and foreign overlords. A dozen men are gathered around a banquet table
in the process of sharing a meal. Some are engaged in conversation; others look
outward as if to assess a newcomer to their table. Hals's achievement of this
sense of spontaneity in what is, after all, a formal group portrait, lies at the heart
of his genius. Hals was presented with two seemingly contradictory require-
ments—the creation of a hierarchical arrangement of strong individual portraits
on the one hand combined with communal activity in the context of a unified

composition on the other. Unlike his predecessors, who generally painted a collection of individual portraits of the same size in a shallow space, Hals created a believable environment by enlarging the nearer figures slightly. Each portrait has individual characteristics and is a product of Hals's unique ability to capture the particular gestural nuances that identify each sitter's character. Hals integrated his group of portraits into a grand whole through a series of brilliant compositional devices. Three senior officers are seated at the left of the picture; about them, a gold drapery is drawn back as if to reveal the scene. To their left a young ensign stands holding the partly unfurled company banner. The banner is thrust diagonally across the middle third of the painting. The picture reaches a visual climax with three standing figures at the extreme right. The central figure holds his hat in hand gesturing toward the left side of the picture. The elliptical shape of the hat's brim reflects the size and shape of the plates on the table, which in turn lead back to the three leftmost figures.

During the period 1620–1640, Hals's mastery of painting continued to evolve. By the 1640s, he concentrated almost entirely on portraiture, his palette became darker and more somber as he also gained further insight into the human psyche. His painting, while technically undiminished, is not so broadly gestural but rather more concerned with the inner being. *The Regentesses of the Old Men's Alms House* (1664) is perhaps Hals's greatest late painting. As with much of his late work, color is minimized in favor of stark value contrast. The four regentesses, who comprise a board of governesses, are seated around a small table attended by a servant standing at the extreme right of the painting who is about to hand a note to one of the regentesses. Each woman is simply clothed, wearing a simple black dress with a white collar and cuffs as if to emphasize her sobriety. Each of the four women is similarly illuminated by a cellular light from above. The left and rightmost figures are slightly more brightly lighted, perhaps as an indication of their greater importance. Each of the four women exhibits an individual expression ranging from slight annoyance to introspection to amusement and finally careful appraisal. Because of the narrow focus of Hals's lighting scheme, each woman is isolated in the dimly lit room. The gloom is mercifully broken by a single landscape painting on the wall behind them. Hals manages to unite the women by the simple device of having a hand or sleeve overlap each of the figures successively from left to right. On the table between them, a ledger rests next to a crucifix as if to underscore the notion that their charity is a response to the concerns of this world and the next. Though Hals was not known for a profound grasp of human nature, *The Regentesses of the Old Men's Alms House* approaches the degree of insight found in the work of his contemporary, *Rembrandt. Hals had few students, though all of his sons were painters. Neither the students nor his sons distinguished themselves, perhaps because his universally acknowledged gift for bravura painting was not easily instilled in others.

In part because of the realism and directness of his painting, Hals was rediscovered in the later half of the nineteenth century with the rise of realism in literature and painting. His importance was especially notable in France, where Hals influenced Manet and Van Gogh, among many others. Hals's reputation rose steadily in the nineteenth century, until today he is ranked alongside Rembrandt as the greatest of Dutch painters; other notable works of his are *The Laughing Cavalier* (1624), *Willem van Heythuysen* (1625), *The Officers and Sergeants of the St Hadrian Civic Guard of Haarlem* (1633), and *Regents of the Old Men's Alms House* (1644).

Bibliography: S. Slive, *Dutch Painting 1600–1800*, 1995.

John Schmidt

HARRINGTON, JAMES (1611–1677). James Harrington used poetry to convey the political and social theories that eventually led to his fame, to his imitation by leaders and reformers across the seas, and to his imprisonment and irreparable health. Like Thomas More before him, Harrington used utopian fiction to suggest his concept of the best social and political program for England. Harrington's lasting fame comes in part from the claim that he was the first English writer to view the English civil war as a "revolution." What is clear is that Harrington borrowed from Plato's *Republic*, More's *Utopia*, and Aristotelian political theory to create his ideal world. *The Commonwealth of Oceana* gained publication in 1656—a decade prior to the restoration of the monarchy and a half dozen years into a republic under *Cromwell. Harrington found Cromwell's republic lacking and in order to correct it, he prescribed his own. To this end, Harrington emphasized in general the public good over individual freedom or gain. He saw the best form for such a goal in an aristocracy—with an elected "Archon" or Prince—balanced by a strong middle class. More specifically, Harrington saw the management of land and of government positions as the greatest factors in determining equality. Laws in his utopia provide the middle class with land that in England at the time belonged only to the aristocracy, so that in Oceana, "no one man or number of men, within the compass of the few or aristocracy, can come to overpower the whole people by their possessions in lands." Equally as important to maintaining equality in Oceana, government positions rotate at terms of equal intervals and are not established by either election or inheritance. Harrington was then clearly sympathetic to republicanism in theory.

At the same time, however, he supported *King Charles I and the aristocracy. Not surprisingly then, Cromwell had Harrington imprisoned upon Charles's execution and, disagreeing with his theories, had all copies of *Oceana* found with Harrington at the time of his arrest confiscated. Harrington was released shortly thereafter, but his plight did not end even a decade later when *Charles II was "restored" to the throne in 1660. Rumors spread then that Harrington had plotted against Charles II, and he was again imprisoned. There, his health declined rapidly until he was released. Harrington's ideas were never at any point silenced; in fact they grew as *Oceana*'s fame traveled across the sea. Willing and ready audiences in the American colonies and in France embraced its model of government as they each pursued democracy. The Constitution of the United States directly displays Harrington's influence in many ways, most obviously in the fact that government in Oceana is constitutional and that it upholds the system of checks and balances through the separation of executive and legislative powers into two branches.

Bibliography: W. Dickinson, *James Harrington's Republic*, 1983.

Rebecca Totaro

HARRIOT, THOMAS (ca. 1560–1621). Thomas Harriot was born in Oxford around 1560. Little is know about his early life; he graduated from Oxford in 1580. Afterward, Harriot served as the mathematics tutor of *Sir Walter Ralegh and was sent by him to Virginia as the scientist accompanying the colonizing expedition. Returning in 1586, he wrote *A Briefe and True Report on the New Found Land of Virginia* (1588). Although Harriot was a very prolific writer, the *Briefe Report* is the only tract that he saw printed; the rest of his writings (more than 10,000 folios) and voluminous correspondence remain unpublished, save for his *Artis analyticae praxis ad aequationes algebraicas resolvendas*, which was published posthumously in 1627. In spite of this situation, caused partly by his failing health during the last years of his life and partly by his pursuit of perfection, Harriot made major contributions in the fields of mathematics, applied mathematics, optics, cartography, and astronomy; generally, it is impossible to assign a specific date to his achievements. Harriot developed a comprehensive theory of equations (from linear to quintic), improved algebraic notation by introducing the inequality signs, and found rules to compute areas of spherical surfaces. In optics, he studied prismatic colors and devised a theory explaining burning glasses and the rainbow. He measured specific weights of many substances with great precision. He studied projectile motions and proved that ballistic trajectories are parabolas; he investigated collisions between balls and their resulting kinematic motions, though he lacked the modern concept of conservation of energy in perfectly elastic collisions. A convinced Copernican and an early subscriber to *Kepler's elliptical orbits, between 1610 and 1613, Harriot made many telescopic observations, focusing his attention on the sunspots, Jupiter's satellites, and the face of the Moon; he can rightly be considered a codiscoverer of the sunspots. Despite his failing health, he observed the 1618 comets. He died in 1621 after suffering from cancer for eight years.

Bibliography: J. Shirley, *Thomas Harriot: A Biography*, 1983; J. Shirley, ed., *Thomas Harriot, Renaissance Scientist*, 1974.

Renzo Baldasso

HARRIS, JOHN (?1667–1719). John Harris was born about 1666, probably in Shropshire. Harris attended Oxford where he received a B.A. in 1686 and an M.A. in 1689, at which point he took holy orders and was granted the degree of B.D. at Cambridge in 1699 and the degree of D.D. at Lambeth in 1706. Throughout his life, Harris held several positions in the Church of England, from being vicar of Ickelsham, Sussex (1690), to curate of Strod, Kent (1711). Many of these appointments were attained through the support of Harris's patron, Sir William Cowper. English parsons of the eighteenth century often devoted substantial portions of their time cultivating interests outside of their profession, and Harris found himself continually drawn to natural science. While at Oxford, Harris taught mathematics and continued to tutor his parishioners in this field throughout his lifetime. From 1698 to 1707, Harris gave public lectures on applied mathematics after being elected to the Royal Society of London in 1696.

In 1698, Harris was chosen as a Boyle lecturer and delivered a series of sermons titled "The Atheistical Objections against the Being of God, and his Attributes, fairly considered and fully refuted." These lectures argued against atheism while attempting to demonstrate the compatibility of science and the Church of England. Harris belonged to a group of liberal Anglicans who used current scientific findings, specifically the natural philosophy of *Newton, to support the supremacy of Protestant social and political order. Among his first publications in support of this cause was his 1697 defense of Dr. Woodward's theory of the deluge titled, "Remarks on some late Papers relating to the Universal Deluge, and to the Natural History of the Earth," in which Harris provided a detailed rebuttal of the attacks against Woodward's theory of the earth's creation. Woodward attempted to incorporate scripture into natural philosophy to provide a scientific explanation for such things as fossils, earthquakes, and volcanoes that would fit into the teachings of the Church of England.

Harris published a variety of books and pamphlets on a wide range of topics, but he is most notably remembered for his *Lexicon technicum: or, an universal English dictionary of the arts and sciences explaining not only the terms of art, but the arts themselves* (1704). By the third edition in 1716, Harris had compiled nearly 12,000 entries covering such diverse fields of inquiry as grammar, philosophy, technology, geometry, algebra, law, navigation, and heraldry. The strengths of the *Lexicon* are found in the entries that cover physics, chemistry, mathematics, and geology. Of particular importance is the 1710 addition of a previously unpublished essay on acids by Newton. The majority of scientific information is taken from the work of Newton, *Boyle, *Robert Hooke, and a wide range of scientists and philosophers who published in the *Philosophical Transactions of the Royal Society*. The *Lexicon* includes detailed entries for all branches of science and philosophy popular in the eighteenth century, as well as entries of lesser quality in medicine and botany. Beyond the importance of such a collection of scientific and philosophical information, the *Lexicon* also provides detailed engravings to accompany many of the entries for technological advancements and instrumentation, as well as thorough charts and graphs for entries in trigonometry and economics. The *Lexicon technicum* stands as the most influential resource for Ephraim Chambers's *Cyclopedia; or, An universal* *dictionary of arts and sciences* (1728). For all of the success of the *Lexicon*, Harris died penniless in 1719 and was buried in Norton Church at the expense of John Godfrey, Esq.

Bibliography: F. Kafker, ed., *Notable Encyclopedias of the Seventeenth and Eighteenth Centuries*, 1981.

Michael W. Jackson

HARTSOEKER, NIKOLAAS (1656–1725). A physicist and scientific instrument maker, Hartsoeker was born on 26 March 1656 in Gouda, the Netherlands. He received some education at the University of Leyden, after which he supported himself primarily as an instrument maker and secondarily in academics. In 1677, *Antoni van Leeuwenhoek described his discovery of spermatozoa. Hartsoeker, who had been grinding the tiny globular lenses used in contemporary microscopes, claimed, without merit, that he had discovered spermatozoa two years earlier. In 1678, he went to Paris with *Christiaan Huygens, who introduced him to the scientific circles there. After failing as a wine merchant, he supported himself by making optical instruments. In *Essai de dioptrique* (1694), he discussed physics and chemistry and presented a general philosophy of nature, including the idea of an infinite universe. In chapter 10 appeared the drawing of the homunculus, a tiny man inside the head of a sperm. He thus accepted animaculism—the idea that individuals are preformed in the spermatozoa and grow into adult forms. Hartsoeker did not claim that he saw the homunculus; he believed that one would see it if the outer surface of the spermatozoon were removed. Later reports on the regeneration of the crayfish claw convinced him to reject preformationism. Returning to the Netherlands in 1696, he tutored *Peter the Great, but refused an offer of a chair in physics at Saint Petersburg. He lived in Düsseldorf from 1704 until 1716 under the patronage of the Landgrave of Hesse Kassel, who secured him a position of honorary professor at the University of Heidelberg. He also served as court mathematician for the Elector Palatine, Johan Willem. Upon Willem's death, Hartsoeker returned to the Netherlands, where he died in Utrecht on 10 December 1725.

Bibliography: C. Pinto-Correia, *The Ovary of Eve: Egg and Sperm and Preformation*, 1997.

Kristen L. Zacharias

HARVARD, JOHN (1607–1638). The first major benefactor of Harvard University was born in Southwark, England, on 29 November 1607. The son of Robert Harvard, a successful butcher, and his second wife, Katharine, he attended the grammar school of Saint Savior, a Puritan church. In 1625, the plague killed Robert and five of his children. Robert left Katharine financially comfortable, and with her second husband she resolved to educate Harvard well. In 1627, he entered the Puritan Emmanuel College at Cambridge to study for the ministry. Required to subscribe to the Articles of Faith, which supported the king and the practices of the Church of England, Harvard earned a bachelor's degree in 1631 and a master's degree in 1635. He married Ann Sadler in 1636. In response to government censorship of Puritan tracts, Harvard began to collect smuggled religious literature as well as classical texts. In 1635, his mother died, leaving him money and property acquired upon the death of her second husband. Harvard sold the real estate and sailed with his wife to Massachusetts in 1637. At this time, Thomas, his only living sibling, died, leaving Harvard wealthier than most of the colonists. He settled in Charlestown and accepted a ministry at First Church. On 14 September 1638 Harvard died of tuberculosis. He had stipulated that his library of around 300 books and half his estate of approximately £1600 be given to the new college established in 1636 in neighboring New Towne (Cambridge) for the purpose of training ministers. While the college probably did not receive the full amount, these funds enabled it to begin operation. On 13 March 1638, it was officially named after Harvard. Today a seated statue of John Harvard, dedicated in 1884, stands in front of University Hall.

Bibliography: H. Shelley, *John Harvard and His Times*, 1908.

Kristen L. Zacharias

HARVEY, WILLIAM (1578–1657). Harvey discovered the circulation of the blood, publishing his insight in a work of less than eighty pages in 1628 entitled *De motu cordis et sanguinis in animalibus (On the Motion of the Heart and of Blood in Animals)*. His discovery benefited from the excellent education he received, having first entered Cambridge at the age of fifteen, then proceeding to Padua in 1599, where he studied with Fabricius ab Aquapendente (1533–1619), who had once treated *Galileo and also constructed a large observation theater for dissections. Harvey took his M.D. in 1602 and returned to England, marrying the daughter of Elizabeth I's physician and later becoming physician to *James I and *Charles I. His medical practice flourished, and he moved in the company of such figures as *Bacon, *Selden, and *Hobbes. He traveled to Europe again from 1629 to 1636 and was later with Charles I at the Battle of Edge Hill (1642). From then until 1646, he lived in Oxford then returned to London, taking up the study of animal reproduction, which resulted in his final publication *Exercitationes de generatione animalium (On the Generation of Animals*, 1651).

When Harvey began his research on circulation, the prevailing understanding was that of the early Greek physician Galen (ca. 130–200), the roots of whose concepts reached back to Aristotle. Galenic physiology held that the liver created blood, which flowed to the heart where it was transformed into a "vital spirit," a process accompanied by the "innate heat" the heart provided, a fundamental quality of life and hence of the soul. The mechanical function of the heart as a pump was not understood by Galen, nor, although the blood was known to move, was its recirculating flow recognized. Progress in understanding the circulatory system was made by the sixteenth-century anatomists Andreas Vesalius, Michael Servetus, and Realdo Colombo, all of whose works were known to Harvey. But it was he who finally grasped that blood passed from the right to left ventricles by way of the lungs (pulmonary circulation) and that the heart was in fact a muscular pump, driving the blood by its rhythmic contraction and expansion (systole and diastole). He was the first to recognize that the amount of blood pumped by the heart within thirty minutes exceeds the total volume of blood in the entire body, a phenomenon which can only be explained by the fact that the same blood circulates throughout the body, blood flowing away from the heart through the arteries and toward it through the veins. Harvey's research was a model of the emerging scientific revolution of the era. He built on the work of his predecessors in medicine, amassed data through dissection and careful analysis, and followed a clearly inductive method of inquiry, declaring "I do not profess to learn and teach Anatomy from the axioms of the Philosophers, but from Dissections and the Fabrick of Nature."

Bibliography: K. Keele, *William Harvey, the Man, the Physician and the Scientist*, 1965; G. Whitteridge, *William Harvey and the Circulation of the Blood*, 1971.

Christopher Baker

HAUKESBEE, FRANCIS (d. 1713). Haukesbee was a celebrated English electrician and experimentalist. Early in his career, Haukesbee discovered the "lateral communication of motion in air," suggesting an improvement for air pumps, one type of which still bears his name. It was also in the early stages of his career that Haukesbee experimentally determined, before the Royal Society, that water is approximately 885 times heavier than air, a measurement that has proved respectably accurate. Haukesbee was inducted into the Royal Society on 30 November 1705, a year in which he made several important chemical and electrical observations, including the fact that Mercury shaken in a glass vessel produces light and that many bodies (such as amber and glass, or glass and glass) produce light when they are rubbed together in a vacuum. In 1706, Haukesbee invented the first electric machine, employing, in his own words, "a pretty large glass cylinder, turned by a winch and rubbed by the hand," producing an electrical light that he termed "mercurial phosphorous." In 1709, Haukesbee published his *Physico-Mechanical Experiments on Various Subjects, Containing an Account of Several Surprising Phenomenon Touching Light and Electricity, Producible on the Attraction of Bodies*, which was afterward translated into French and Italian. Additionally, many of Haukesbee's observations and experiments were published in *The Philosophical Transactions* (the last posthumously in 1713). Haukesbee was probably the father of Francis Haukesbee the younger (1687–1763), who was also active in the scientific community, serving as a clerk of the Royal Society. There is no extant biography of the elder Haukesbee.

Bibliography: L. Stephen, ed., "Francis Hauksbee," *The Dictionary of National Biography*, 1921–1922.

Walter H. Keithley

HAWKSMOOR, NICHOLAS (1661–1736). Hawksmoor was one of the most important of England's Baroque architects, combining in his projects elements derived from the work of *Christopher Wren, with an extensive knowledge of ancient Roman, English Gothic, and contemporary French architecture. He was probably born at East Drayton, Nottinghamshire, and entered the service of Wren in about 1679. By 1683, he was supervising work at Winchester Palace. He aided Wren during the later phases of the replacement of London's parish churches destroyed in the fire of 1666 and was Wren's amanuensis for the last stage of the reconstruction of St. Paul's Cathedral, continuing to serve in that capacity until 1712. By 1699, he was collaborating with *John Vanbrugh, a partnership whose most noteworthy fruits are generally held to be Castle Howard, Yorkshire, and Blenheim Palace. He worked with Wren and Vanbrugh on the completion of the Royal Naval Hospital at Greenwich, where he was Clerk of Works from 1698 until shortly before his death. Hawksmoor worked more autonomously when designing churches for London's growing suburbs, financed by the Act for Building Fifty New Churches of 1711; six of the handful of projects brought to completion under the provisions of the act were designed and directed by him.

The precise extent of Hawksmoor's contribution to edifices such as Blenheim Palace and Castle Howard is difficult to gauge, but the not unreasonable surmise has been advanced that although Vanbrugh was the innovative designer, he relied on Hawksmoor's extensive practical experience and wide-ranging knowledge of the history of architecture. The mausoleum at Castle Howard, however, can be ascribed to Hawksmoor with great confidence, and here his Baroque taste for the effect of massiveness is displayed with striking virtuosity in a Doric temple that stands seventy-six feet high on an isolated hill. Hawksmoor's eclecticism and wide reading is evident in his London churches: The design of St. George's Bloomsbury was dominated by his understanding of ancient architecture, but his three Stepney churches, St. Anne's Limehouse, St. George's-in-the East, and Christ Church Spitalfields, present a unique combination of English Gothic, classical, and Baroque elements.

Bibliography: K. Downes, *Hawksmoor*, 1979; P. de la Ruffinière du Prey, *Hawksmoor's London Churches*, 2000; C. Smith, *The Building of Castle Howard*, 1990.

Matthew Koch

HEARNE, THOMAS (1678–1735). One of the most prolific English antiquaries of his time, Thomas Hearne was born in White Waltham, Berkshire, in 1678. Through the patronage of Francis Cherry, Hearne attended school at Bray and in 1695 went to Oxford, taking a B.A. (1699) and M.A. (1703). Throughout adulthood, Hearne rejected a variety of positions proffered—as missionary to Maryland, chaplain to Corpus Christi and All Souls' Colleges, and librarian of the Royal Society—to remain at Oxford's Bodleian Library, where he became second keeper in 1712. Hearne's nonjuring stance, however, cost him this position: In 1716, he was dismissed for refusing to take the Hanoverian oath of allegiance. He lived his remaining nineteen years at Edmund Hall, preferring the dictates of his nonjuring conscience to several honorary positions, including head librarian of the Bodleian.

Hearne's writings total 145 manuscript books and over forty published volumes. His early scholarship was devoted to Oxford history and to solid but unexceptional editions of classical authors, including Pliny the Younger (1705) and Livy (1708). From 1709, Hearne's scholarship turned to the English Middle Ages. Among his most influential publications were editions of the *Itinerary of John Leland* (1710–1712) and numerous medieval chronicles, including the *Textus Roffensis* (1720) and *Robert of Gloucester's Chronicle* (1724). Hearne's editions were the best available until the nineteenth century and secured his scholarly reputation. Hearne's occasionally uncritical enthusiasm for any medieval document drew the criticism of Edward Gibbon and of Alexander Pope, who portrayed Hearne as "Wormius" in Book III of the *Dunciad*.

Bibliography: D. Douglas, *English Scholars 1660–1730*, 2nd rev. ed., 1951; J. Levine, *Humanism and History: Origins of Modern English Historiography*, 1987.

Michael R. Hutcheson

HEINSIUS, DANIEL (1580–1655).

Heinsius was one of the foremost scholars of the Dutch Golden Age. Born in Ghent, his family fled the unrest in Flanders after its reconquest in the Spanish war during the 1580s, moving from Zeeland to England to Ryswick to Flushing. He entered the Francker Academy at sixteen as a law student, but then entered the University of Leiden two years later where he associated with the noted humanist J.C. Scaliger. His skill in classical languages became widely known, and he rejected numerous offers to teach outside Holland. He was successively professor of poetry and Greek and then of politics and history, then university librarian and secretary of the university senate at Leiden. There he became good friends with *Hugo Grotius; however, unlike Grotius, who was imprisoned for his support of *Jacob Arminius, Heinsius adhered to the orthodox theology of the Reformed church and gradually moved into more influential positions, becoming, for example, a reviser of the "Staten" Bible, the official scripture text authorized for use in the Dutch church.

Heinsius's publications spanned theology, literature, classics, history, and patristic studies. His editions of such classical authors as Horace, Seneca, Ovid, and Livy as well as church fathers like Clemens Alexandrinus numbered about thirty and were published by the most reputable presses of the day (he was for a time also managing editor for the Elzevir publishing firm). He published works of biblical criticism in *Aristarchus sacer* (1627) and *Exercitationes sacrae* (1629), the latter arousing the anger of *Salmasius, who accused him of favoring the political views of *John Milton. His own neo-Latin poetry was highly regarded, enjoying thirteen editions during his lifetime. His Auriacus, with Grotius's *Adamus exul*, set a new standard in neo-Latin tragedy. The poetry of Heinsius is a high point of the Dutch Baroque and was a significant influence upon the Germans *Martin Opitz and *Andreas Gryphius; he also popularized the composition of love emblems in the early years of the seventeenth century. His most important work of literary criticism was *De tragoediae consitutione* (1610, 1643), an interpretation of Aristotle's *Poetics*. Heinsius died on 25 February 1655 at The Hague.

Bibliography: B. Becker-Cantarino, *Daniel Heinsius*, 1978; P. Sellin, *Daniel Heinsius and Stuart England*, 1968.

Christopher Baker

HELMONT, JOHANNES BAPTISTA VAN (1579–1644).

Born in Brussels, Johannes Baptista Van Helmont grew up in affluence, and although he attended the university, he turned down an M.A. degree and decided to serve humanity through the study and practice of medicine. In 1599, at the age of twenty and only five years after entering training, he earned his medical degree. Helmont had contracted a disease that practitioners adhering to strictly Galenic medical practices could not cure. Their assessment of his condition, based upon assessing the balance of four bodily humors (fluids), often resulted in painful attempts to reduce the quantity of one or another through bloodletting, cupping, and purging. His own suffering from the disease and from doctors' failed attempts at a cure was said to contribute to his passion for this particular vocation; he vowed to find better remedies than those currently prescribed. Turning to the Paracelsian medical approach that utilized man-made, chemical remedies—rather than herbals to induce vomiting and leeches for draining blood—Helmont found a viable alternative. Helmont's work in this new approach was, however, condemned by the Catholic Church of which he was member, because many considered Paracelsus—founder of that approach—a magician who perverted nature through unnatural, alchemical means. From 1625 until 1642, the Church pursued one course of action after another, keeping Helmont altogether from publishing. Still, his fame spread and he became—by the majority of contemporary accounts—the first chemical theorist. He discovered carbon dioxide, distinguished gases as a class of substance distinct from solids and liquids, and was the first to employ the term *gas* in its modern scientific sense. In addition, Helmont performed one of the first experiments utilizing quantitative measurements. Growing a willow tree in a carefully weighed amount of soil, he measured the soil before and after, concluding that because the tree had "used" little soil in its growth, the tree must have taken in the water. His works were published posthumously in 1648 as *Ortus Medicinae*.

Bibliography: W. Pagel, *From Paracelsus to Van Helmont: Studies in Renaissance Medicine and Science*, 1986.

Rebecca Totaro

HERBERT, EDWARD, FIRST BARON OF CHIRBURY [OR CHERBURY] (1582–1648). Probably born in 1582, Edward Herbert of Chirbury (the "er" spelling, used in Herbert's seventeenth-century publications, is a Latinization of the original "ir" spelling of the patent of nobility) was a member of a cadet branch of the family of the Earls of Pembroke. His writings include poetry, history, an autobiography, and philosophical treatises, and he both collected and wrote musical settings for lute. He has been termed the father of English Deism for his treatise *De veritate*, which treats the rational basis of knowledge, including religious knowledge. Herbert was the eldest brother of *George Herbert the poet. He was well-educated, going to boarding school and New College, Oxford, though like most gentlemen he took no degree. In 1599, he married a cousin, Mary Herbert. He continued to study languages, including Welsh, became sheriff of Montgomeryshire in 1605, and received the Order of the Bath when *James I ascended the English throne. In about 1608, he went abroad to continue his education; while in France he met King Henri IV, *Isaac Casaubon, and de Montmorency, leader of the "Politiques." In 1610, he served as a member of Parliament for Merioneth, then went once again to Europe, serving at the siege of Juliers. By this time, he was writing poetry and appears to have exchanged verses with *John Donne and perhaps *Ben Jonson. Herbert served as ambassador to France from 1619 to 1624 but was recalled when he advised against the marriage of Prince Charles (later *Charles I) to the Catholic French Princess Henrietta Maria. He received no other appointments but was elevated to First Baron of Chirbury in 1629. During the civil war, he reluctantly sided with Parliament before dying in 1648.

Herbert has had the misfortune of being eclipsed in all his areas of literary and philosophical endeavor by other writers. He is viewed as a minor metaphysical poet; his poetic corpus (including satires, poems on platonic love, occasional verse, and a Latin philosophic poem) was published after the Restoration by his brother Sir Henry Herbert, Master of the Revels. His history of the Duke of Buckingham's failed military expedition to the Isle of Ré is regarded as an apologetic for a political benefactor; his history of King Henry VIII, while innovative in its extensive use of original documents and notable for its attempt to assess its subject impartially, often makes for dull reading. He wrote an autobiography as a family record for his descendants and thus concentrates inordinately, at least by later tastes, on matters related to family and personal honor, including his many duels, discussing only his philosophic treatise *De veritate* among his intellectual pursuits. As a composer, his taste in music was conservative, even old-fashioned. Although he corresponded with intellectuals such as *Gassendi, *Grotius, and *Campanella, and his philosophical writings were influential in the seventeenth and eighteenth centuries, his importance pales in comparison with *Bacon, *Descartes, *Hobbes, and *Locke. The central concern of his treatise *De veritate*, first published in 1624, is to find a way, free from the commitments of prior philosophic schools, among the virulent competing claims to religious truth. His chief conclusions are that the claims for exclusive truth and salvation are the work of self-serving priests and diverge from true religion, which consists in five notions found in all religions, among them the belief in a supreme deity and an afterlife. The treatise was revised and republished with two additions, *De causis errorum* and *Religio laici*, during the civil war in 1645. He supported his claims about these universal religious beliefs with an additional treatise, *De religione gentilium*, an early study in comparative religion, which *Vossius undertook to publish at Herbert's request. The most important of Herbert's philosophical ideas are summed up in the English work *Dialogue Between a Tutor and His Pupil,* probably by Herbert, though not published during his lifetime.

Bibliography: J. Butler, *Lord Herbert of Chirbury (1582–1648): An Intellectual Biography*, 1990; S. Lee, ed., *Autobiography of Edward, Lord Herbert of Cherbury*, 1906.

Sara V. Fink

HERBERT, GEORGE (1593–1633). Born 3 April 1593 in Montgomery Castle, Wales, George Herbert is considered to be one of the seventeenth century's most important devotional poets, second only in reputation to his mentor *John Donne. Although Donne's influence is evident in his poetry, Herbert's sincere meditations stand well on their own merit. His reputation as a poet is based primarily upon a collection of poems entitled *The Temple*. These poems are both conscientiously devout and secularly idiosyncratic. Herbert was the fifth son of the distinguished Richard Herbert and Magdalen Newport-Herbert, a longtime friend and devotee of John Donne. Herbert's father died in 1596. George lived with his mother, migrating from Eyton to Oxford and finally to London where their estate was established at Charing Cross. In 1608, Magdalen remarried Sir John Danvers and moved to his home in Chelsea on the banks of the Thames.

Herbert was educated first at Westminster School and then at Trinity College, Cambridge. He received his master of arts degree in 1616. He quickly rose through the ranks of scholastic hierarchy at Cambridge. By 1618, he had reached the role of university public speaker in rhetoric. In 1620, he was elected to the prestigious position of University Orator. In 1624, Herbert took leave from the university to be ordained an Anglican deacon. It is supposed that he did not return to his post at Cambridge after his leave expired. By 1625, he was well enough acquainted with *Sir Francis Bacon that the latter dedicated his *Translations of Certain Psalms* to him. Bacon thanked Herbert for his assistance in translating *The Advancement of Learning* into Latin and noted in the dedication that Herbert was the person in whom "Divinity and Poesy" met. Late in that same year, the plague ravaged London and John Donne stayed with Herbert at the Danvers home in Chelsea. Herbert's mother, Lady Danvers, died in 1627, and Herbert all but retired from public life. Donne delivered Magdalen's eulogy, and Herbert's *Memoriae Matris Sacrum* was published along with Donne's memorial sermon. Herbert then moved to the Earl of Danby's (his step-uncle) home in Wiltshire. In 1629, he married his stepfather's cousin, Jane Danvers. They lived at Baynton House, his wife's estate. Herbert was made rector of the chapel of Bemerton St. Andrew and the parish church of Fugglestone St. Peter near Salisbury in Wiltshire in 1630. In 1631, Herbert's longtime associate John Donne died.

Meanwhile, Herbert had developed a correspondence with Nicolas Ferrar (1592–1637) a deacon who, in 1625, had established a rural religious community of about thirty people at Little Gidding. Ferrar had sought Herbert's expertise on his translation of Valdesso's *Considerations*, and this collaboration lead to Herbert's "Brief Notes" on Valdesso's work. From his deathbed, Herbert sent Ferrar a collection of poems and requested that he either publish or destroy them. Herbert died of tuberculosis in March 1633. He is said to have been buried either underneath or very near to the altar in St. Andrew's Church, Bemerton. Several months after his death, Herbert's most acclaimed work, *The Temple: Sacred Poems and Private Ejaculations* was published. Nicholas Ferrar wrote the preface. Other posthumous publications followed including *Outlandish Proverbs* (1640), *Jacula Prudentum* (1651) an enlarged edition of *Outlandish Proverbs*, and *Herbert's Remains* (1652) including *The Country Parson* (or *A Priest to the Temple*). But nothing would ever match the unequivocal grace, charm, style, or popularity of *The Temple*, for Herbert was a quintessential metaphysical poet who employed all the tropes of the genre, including allegory, conceit, sensuality, analogy, and catachrestic language to express his genuine religious devotion. Ironically, Herbert is perhaps best remembered for his pattern poems, "Easter Wings" and "The Altar," in which the lines of verse form the shape of the object.

Bibliography: R. Ray, *A George Herbert Companion*, 1995; H. Wilcox, ed., *George Herbert: Sacred and profane*, 1995.

David M. Rosen

HERRICK, ROBERT (1591–1674). Robert Herrick was born in 1591, the son of Nicolas Herrick, a goldsmith in London, and his wife Julian. Herrick was apprenticed to his uncle Sir William Herrick, also a goldsmith, but in 1613, he went to Cambridge and entered St. John's College. He transferred to Trinity Hall in 1615 and took his B.A. in 1617 and his M.A. in 1620. Like many, if not most, students at Cambridge, Herrick must have been training for the church, for he was ordained a deacon on 24 April 1623 and priest on the following day, an accelerated progression through ecclesiastical orders that is relatively rare. Despite his ecclesiastical pursuits, he seems to have been a resident of London, a friend of *Ben Jonson and his literary circle, and an habitué of the Devil Tavern. Presumably his patron and employer, the Duke of Buckingham, did not mind such irregularities in his chaplain, for Herrick accompanied him on his military expeditions and was still his chaplain when the duke was assassinated in 1628.

Herrick was appointed vicar of Dean Prior in Devonshire in 1629 and installed on 29 October 1630. He was to spend most of the rest of his life there, an urbanite apparently both delighting in and chafing against being in the country. He was clearly Royalist in his sympathies and connections, and among his poems are several intended as royal entertainment or addressed to the queen, the king, and the rest of the court. His political sympathies caused him to be expelled from his vicarage at Dean Prior in 1647, and he returned to London, supervising the publication of his collection of poems, *Hesperides*, in 1648. He remained in London until the Restoration, when he successfully petitioned to be returned to his living. In 1660, he returned to Dean Prior, where he remained until his death in 1674.

Herrick's volume *Hesperides* and the subsegment of religious poetry it contains, *Noble Numbers*, are usually classed as Cavalier poetry. His royalism manifests itself not only in explicit compliment and address, but in his implied theology. His Christmas day hymn, *What Sweeter Music*, addresses the Christ Child as "our King / And Lord of all this revelling." The latter epithet echoes *Charles I's and *Archbishop Laud's support for traditional sports such as Christmas revels and May games, a theological and political stance, which lends perspective to what is assuredly one of Herrick's most famous poems, *Corinna's Going A-Maying*. Herrick is also notable for his use of the pastoral (and anti-pastoral), his frequent glosses on *carpe diem*, and his sense of humor, exemplified in his passionate *His Farewell To Sack* and the enraptured tone of its companion piece, *The Welcome to Sack*. Recent studies of Herrick have examined his poetry as Cavalier apologetic, as Laudian theology, and as argument for Anglican survivalism.

Bibliography: R. Deming, *Ceremony and Art: Robert Herrick's Poetry*, 1974; L. Marcus, *The Politics of Mirth: Jonson, Herrick, Milton, Marvell, and the Defense of Old Holiday Pastimes*, 1986; R. Rollin, *Robert Herrick*, 1992.

Melissa D. Aaron

HEVELIUS, JOHANNES (1611–1687). Hevelius was born in Danzig (Gdansk) 28 January 1611 into a wealthy brewer's family. Educated in the Netherlands and France, where he studied mathematics, astronomy and optics, Hevelius returned home to run the family business and to build a career in astronomy. The last important astronomer to base his work on unaided sight observations, Hevelius produced an atlas of the moon's surface, two books on comets, a star catalog comparable in accuracy to that of Tycho Brahe, and descriptions of astronomical instruments. Several of Hevelius's publications became standard reference works. His reviews of past research and observations, combined with his own discoveries, provided useful historical and contemporary information. *Cometographia* (1668) and *Selenographia* (1647) were state-of-the-art publications as well as showcases for his own empirical studies. Arguably, *Selenographia* was Hevelius's most influential work. The astronomer illustrated his findings with copper engravings of the moon's surface and its phases, demonstrating how that body's oscillation is viewed from earth. His names for some lunar features are retained today.

Scientists as prominent as *Robert Hooke and *John Flamsteed criticized Hevelius for his continued reliance on naked eye observations, but he certainly did not neglect instrumentation. *Machina coelestis pars prior* (1673, 1679) describes in two volumes the instruments that he built and used in his own observatory. The engravings for this work are especially noteworthy and attractive. One plate shows Hevelius and his second wife, Elizabeth Koopman, manipulating a sextant. Unusually keen eyesight may have allowed him to make better records without the still cumbersome telescopic aids, and he did use the tubeless, refracting telescopes of his day for nonpositional work. Hevelius contributed to telescopic lens technology and to development of support mechanisms for long telescopes. One telescope that he built had a spar length of 140 feet.

Johannes Hevelius died on his birthday, 28 January 1687. In his marriage to Elizabeth Koopman, Hevelius found a companion and a colleague. Koopman was a woman of means who collaborated with her husband in his work and supported their efforts with a personal fortune. After her husband's death, she edited and published his star catalog, *Prodromus astronomica* (1690), and another volume illustrating new constellations they had observed.

In addition to demonstrating the lunar libration in longitude, among Hevelius's more important contributions to astronomy were his descriptions of faculae, the exceptionally bright areas found on the sun's surface near sunspots, the discovery of four comets, a credible value for the sun's rotational period, the phases of Mercury, and his suggestion that the revolution of comets could be described in parabolic routes about the sun.

Bibliography: J. Field and F. Frank, eds., *In Renaissance and Revolution: Humanists, Scholars, Craftsmen, and Natural Philosophers in Early Modern Europe*, 1993; R. Glebocki and A. Zbierske, eds., *On the Three Hundredth Anniversary of the Death of Johannes Hevelius*, 1992.

Joan Klobe Pratt

HEYWOOD, THOMAS (ca. 1573–1641). This dramatist was one of the most productive playwrights of the Elizabethan and Jacobean theater, but is remembered for only one play, his domestic tragedy *A Woman Killed with Kindness*.

Thomas Heywood was born in Lincolnshire, the son of Reverend Robert Heywood and his wife Elizabeth. He was one of eleven children, but little else is known about his early life. He entered Emmanual College at Cambridge, a stronghold of Puritan thought, when he was sixteen or seventeen years old. The theological environment in which he was educated, along with having grown up the son of a clergyman, explains the strong moral thrust that characterizes so much of Heywood's writing. Heywood's father died in 1593, leaving him without resources to continue his education, so he left Cambridge without a degree and migrated to London to seek employment on the stage. In 1593, Heywood was hired by Philip Henslowe to work as an actor for the Admiral's Men.

Heywood's career as an actor was brief. He quickly turned his hand to writing plays and was hired in this capacity by Derby's Men in 1600. In 1601, he became an actor-sharer with Worcester's Men. This position brought Heywood into the circle of writers and performers that included Henry Chettle, *Thomas Dekker, Wentworth Smith, and *John Webster, with whom he often collaborated in the writing of some of the over 200 plays that he says he either produced himself or coauthored. In 1603, Heywood produced the play upon which his reputation rests, *A Woman Killed with Kindness*. This play, which is the best example of seventeenth-century domestic tragedy, deals with the marriage of John and Anne Frankford, Anne's adultery with the convincing gallant Wendoll, Frankford's discovery of his wife's infidelity, and her banishment from the family and eventual death. As his wife is dying, however, Frankford forgives her the sin she has committed against him with the hope that she will find forgiveness from God as well.

Heywood's play demonstrates well several elements of the domestic tragedy. First, its subject matter was taken from a scandal reported in the chapbook literature so popular in the early seventeenth century. Second, it is filled with Christian sentiment. Frankford, for example, refuses his right to avenge himself on his sinful wife and her lover by claiming he will not take the lives of two whom Christ had shed his blood to save. Third, the play demonstrates the efforts made by many of Heywood's contemporaries to fuse middle-class values with the form of classical tragedy. Finally, this play is one of the first in English drama to employ a woman as its protagonist.

With the success of *A Woman Killed with Kindness*, Heywood was secure in his position and, in the same year it was produced, married Ann Butler, with whom he had six children. With his place in the theater secure, Heywood began in 1608 to turn his hand to other literary forms. In 1609, he published *Troia Britannica*, his best work of extended poetry; in 1612, *An Apology for Actors*, a didactic work on the London theater; and in 1624, *Gunaikeion: Or, Nine Books of Various History Concerning Woman, Inscribed by the Nine Muses*, a work on the nature of women and their sufferings. In 1619, Heywood began working for Elizabeth's Men, and in 1624 joined the company sponsored by Queen Henrietta. Through the 1630s, Heywood wrote a number of city pageants and various didactic pieces, such as *Philocothonista, Or, The Drunkard, Opened, Dissected, and Anatomized* (1635) and *A Curtain Lecture* (1636), which celebrates marriage and basic domestic virtues.

Perhaps the most interesting aspect of Heywood's career is not the number of works he either wrote or coauthored, his substantial contribution to the London theater in ways other than his writing, or the importance of *A Woman Killed with Kindness*; rather, his work documents the emerging values of an increasingly influential middle class, values often rooted in Puritan sentiment. History shows that it was this very middle class, and the values it held, that would provide the foundation of Parliament's assault on the monarchy, which led to civil war in 1642. Heywood's writing does not have a political component, but it does suggest the shift in social mores that would have overwhelming political consequences.

Bibliography: F. Boas, *Thomas Heywood*, 1950; A. Clark, *Thomas Heywood*, 1931.

Gerald Morton

HILLIARD, NICHOLAS (1547–1619).

Nicholas Hilliard, noted English miniaturist, was born in Exeter. Son of a goldsmith, he was a member of the goldsmith guild but early showed a talent for portraiture and became the most renowned English painter of the Elizabethan period. Miniature portraits in water-based paints, a technique known as limning, constitute the majority of his known work; he is said to have painted some large portraits, but no such work by his hand is known for certain. He painted the first self-portrait by an Englishman, in which he appears a dashing and self-assured young man. His known sitters are of the gentry, nobility, and royalty, but many remain unidentified. Patronized by both Queen Elizabeth and *King James I, he and his workshop produced many royal portraits that were given as gifts to courtiers. He also designed the second Great Seal for Elizabeth's reign, and he was responsible for decorative work on royal charters.

Some of his most well-known work includes the mysterious full-length portrait miniature known as *Young Man Among Roses*, a portrait of a lover surrounded by flames emblematic of his enamored state, and a portrait of Queen Elizabeth known as the *Drake Jewel*, enclosed in an elaborate case also made by Hilliard and given by the queen to Sir Francis Drake; all these are now at the Victoria and Albert Museum in London. *Young Man Among Roses* is one of several impresa portraits, which match emblematic pictorial devices with a motto or "word" chosen by the sitter and having some personal significance. Hilliard also painted miniature "table" portraits, full-length portraits of the sitter in an interior or exterior scene; one of the best known of these is of George Clifford, Earl of Cumberland, in dress armor, gauntlet thrown down, with London in the distant background; it probably commemorates his appointment as Queen's Champion in 1590. Persuaded by Richard Haydocke, translator of Lornazzo's famous treatise on painting, Hilliard wrote a lively explanation of his methods. This treatise on *The Arte of Limning*, now published but existing only in manuscript during his lifetime, includes anecdotes of conversations with the queen and with Sir Philip Sidney, as well as information on his own technical innovations.

Bibliography: M. Edmond, *Hilliard and Oliver: The Lives and Works of Two Great Minaturists*, 1983; N. Hilliard, *Nicholas Hilliard's Art of Limning*, 1983.

Sara V. Fink

HOBBES, THOMAS (1588–1679). A philosopher who studied political theory, mathematics, science, and religion, Hobbes was born on 5 April 1588 when his mother went into premature labor upon learning of the approach of the Spanish Armada. In 1604, his father, a semiliterate curate, abandoned his family after physically attacking a clergyman who had sued him for libel. By that time, Hobbes was already a student at Magdalen Hall, Oxford, from which he graduated in 1608. He then became the tutor and companion of William Cavendish, thereby commencing a lifelong association with the family. While serving as Cavendish's secretary in the 1620s, Hobbes deepened his knowledge of history and improved his Latin and Greek. He translated Thucydides' *History of the Peloponnesian Wars*. Hobbes admired Thucydides' preference for the monarchy and disdain for democracy, which encouraged the use of rhetoric. Hobbes felt that the democratic rhetoric of the time was destroying England's stability.

Cavendish died, and Hobbes, in Europe with another pupil, had a geometrical epiphany, claiming, consistent with rationalism, that natural science was *a priori* and necessary and that its proofs followed the form of geometrical demonstrations. In 1630, Hobbes became the tutor of William Cavendish's son and spent 1634–1636 on the continent with him. During this period, Hobbes developed his ideas on materialism, the view that matter and motions alone could account for all phenomena, including sensation and human cognition. He became interested in optics and met with *Galileo Galilei, whose view that the secondary qualities (color, taste, and so on) were not properties of matter Hobbes later developed. The most important and enduring contact he formed at this time was with *Marin Mersenne, through whom he interacted with *René Descartes and *Pierre Gassendi.

His participation in the discussion group called the Great Tew Circle (1636–1640) stimulated certain religious views. Basing his opinion on a reading of the Bible, he wrote in 1640 that the fundamental property of Christianity was that Jesus was God. Hobbes's studies of this period produced three books: *De Cive* (1642), *De Corpore* (1655) and *De Homine* (1658). During this period, he also developed the major outlines of his political views. In *De Cive*, he asked his readers to consider what life would be like in a state without laws—Hobbes's famous state of nature. He contended that in it all people were equal and had complete liberty to claim a right to anything that they desired—an unbearable state. Hobbes thus arrived at his three laws of nature: to "seek peace," to give up the right to all things, and to keep covenants. Escape from the state of nature thus required all people to agree to forfeit their rights to everything and to transfer them to the sovereign, whom they must then obey. The only right retained was that of self-preservation, an exception that Hobbes's critics found contradictory to his belief, held from 1640 on, that the sovereign had absolute power.

Hobbes promoted these views during the reign of an increasingly unpopular *King Charles I and felt it necessary to leave England. He lived in France from 1641–1651. During this time he wrote six objections (communicated by Mersenne) to Descartes' dualism of an incorporeal thinking substance and corporeal extended substance. Against Descartes' argument that humans have an idea of God and that they have free will, Hobbes argued that God was incomprehensible and that all actions were determined and necessary. Hobbes conducted a debate on free will with William Bramwell in 1645, positing that while humans are free, their wills are not; human actions are caused by internal desires. This position raised the issue of the origin of evil and the problem of responsibility, to which Hobbes responded that even if humans do not have free will, they must be punished for breaking a law. Hobbes wrote his most famous and important work, *Leviathan*, between 1649 and 1651. Almost all its central points had appeared in earlier works: a mechanistic understanding of human existence, the essential selfishness of human nature that demands an absolute sovereign to control social conflict, and the superiority of temporal to religious authority. It contains his famous statement that in the state of nature, life is "solitary, poor, nasty, brutish and short." Included in *Leviathan* are long discussions of religious issues, such as the authorship of the books of the Bible, the interpretation of the "opening" of Adam and Eve's eyes, the characteristics of Heaven and Hell, the nature of church authority, and the corruption of the Roman Catholic Church.

After the publication of *Leviathan* (1652), Hobbes became involved in a variety of disputes on political theory, education, government, and geometry. For example, he attacked *Robert Boyle's experiment that artificially produced a vacuum and asserted that air entered the glass vessel through tiny pores. Hobbes also had unpleasant dealings with the Royal Society of London. Devoted to the Baconian ideal of experiment and empirical justification of science, it never considered Hobbes for membership, partly because of his rationalism and general disdain for experiment, and partly because of his contentious personality. Remaining intellectually active, Hobbes wrote *Behemoth* (ca. 1668), a case study of the English Civil War illustrating the dangers that would result if his principles concerning the absolute power of the monarch were not followed. He also published translations of the *Odyssey* (1675) and *Iliad* (1676). At the age of ninety, he published a work defending his physical theories (1678) in which he asserted that God was the most real substance, that he filled all space, and that, therefore, a vacuum was impossible. Hobbes died on 4 December 1679 at a residence of the Cavendishes, who arranged his funeral. Although Hobbes's detractors erroneously accused him of atheism, it must be admitted that his religious views were hardly conventional. He ultimately believed, for example, that God, angels, and spirits consisted of very fine corporeal matter; he also asserted later in life that the sovereign must be head of the church. The questions posed in his works on the nature of sensation and matter and the relationship between science and religion stimulated much future work, while the problems formulated in his political works inspired philosophers such as *John Locke.

Bibliography: A. Martinich, *Hobbes: A Biography*, 1999; T. Sorrell, ed., *The Cambridge Companion to Hobbes*, 1996.

Kristen L. Zacharias

HOFFMANNSWALDAU, CHRISTIAN HOFFMANN VON (1616–1679).

Born into a wealthy Lutheran family in Breslau (today Wroclaw, Poland), Hoffmannswaldau enjoyed a peaceful aristocratic childhood in a town that was spared the ravages of the Thirty Years' War. He attended schools in Breslau and Danzig (today Gdańsk, Poland), where he met the influential poet *Martin Opitz, who encouraged him to write poetry. In 1638—together with *Andreas Gryphius—he enrolled at the University of Leiden to study law and philology. Gryphius, however, stayed on for almost six years, whereas Hoffmannswaldau left in 1639 to embark on a long journey across Europe. He returned to Breslau in 1641, where he married and rose through various administrative positions to become the city's major in 1677.

Hoffmannswaldau's life is remarkably similar to his fellow, Silesian Gryphius. Their work differs in that Gryphius's writing focuses on the vanity of human existence, while Hoffmannswaldau's emphasizes the opposite extreme, namely, the pleasures of life through his use of the carpe diem motif. Love becomes for him the major force to overcome the bleakness and gloom of earthly life. Martin Opitz and *Giambattista Marino are considered his literary models. By emulating their works and theories, be became a poet of astonishing technical virtuosity. Though hardly any of his works were published during his lifetime, he was a major source of inspiration for many poets of the Second Silesian School, such as Daniel Caspar von Lohenstein and Johann Christoph Männling. His work was rejected by following generations, in particular scholars of the nineteenth century, for being too intellectual and sexually explicit. Only since the end of World War II have scholars come to appreciate Hoffmannswaldau as one of the important Baroque poets. Most noteworthy are his *Heldenbriefe* (*Heroic Letters*, 1664), a collection of verse epistles consisting of synoptic narratives and letters exchanged by lovers. Other significant works include sonnets, odes, elegies, and translations, notably Guarini's popular tragicomedy *Il pastor fido*.

Bibliography: F. Cohen, *The Poetry of Christian Hofmann von Hofmannswaldau: A New Reading*, 1986; M. Metzger, "Christian Hoffmann von Hoffmannswaldau," *Dictionary of Literary Biography* 168 (1996): 194–202.

Josef K. Glowa

HOLLAR, WENCESLAUS (1607–1677).

An indefatigable illustrator and topographer who worked primarily in England, Hollar produced countless drawings and over 2,500 plates in his lifetime, most of them meticulously detailed.

His wide-ranging interests are evident in his varied subject matter, which includes landscapes, portraits, mythological narratives, flora and fauna, coins and medals, and women's costumes. Many of his prints would have been bound in books and formed part of a naturalist's or antiquarian's library. Hollar's works are today important to scholars because they record seventeenth-century society with precision.

Hollar was born in Prague, where he probably received early training. His career began in 1629 when he traveled to Strasbourg and undertook commissions from print publishers. By 1635, he had visited several Dutch and German cities; his journeys produced the *Reisbuchlein* ("Little Travel Book") of 1636.

The paramount event in Hollar's life was meeting Thomas Howard, Earl of Arundel, in Cologne in 1636. Arundel's objective was to request from the emperor the restitution of lost lands on behalf of Elizabeth of Bohemia; Hollar's task was to document the endeavor graphically. Hollar then traveled to London with Arundel, one of the great art collectors of the period, where his job was to make drawings and prints of the collection. The paucity of talented etchers in England at the time ensured his position. His series of etchings from this period, *Ornatus Muliebris*, presents the attire of English women and highlights the intricacies of their hairstyles, costumes, and accessories.

This comfortable phase of his employment was short-lived. Civil war began in 1642, and Arundel's hasty departure disenfranchised him. Out of necessity, he became one of the first self-employed artists and survived by sheer productivity, especially after his move to Antwerp in 1644. Hollar returned to London in 1652, where he remained the rest of his life. Years of personal hardship and disappointment followed. Perhaps most important are Hollar's etchings of London monuments, such as St. Paul's Cathedral, destroyed in the Great Fire of 1666. This event thwarted his ambitious project for a Great Map of London, although many bird's-eye views of the city related to the project survive.

Bibliography: R. Godfrey, *Wenceslaus Hollar: A Bohemian Artist in England*, 1994; F. Grossman, *Wenceslaus Hollar, 1607–1677: Drawings, Paintings and Etchings*, 1963.

Rosi Prieto Gilday

HONTHORST, GERRIT VAN (1590–1656). Honthorst had an international reputation as an artist. Born in Utrecht, Holland, he was the son and grandson of artists, and was apprenticed to Utrecht's Abraham Bloemaert. Honthorst traveled to Rome sometime between 1610 and 1612, where he came under the influence of *Caravaggio and his Italian followers, and his fellow northerners Dirck van Baburen and Hendrick Ter Brugghen. Known in Italy as a successful painter of night scenes often illuminated by candle or lamp light, Honthorst earned the nickname "Gherardo delle Notte" (Gerard of the Night), and counted among his patrons the Marchese Vincenzo Giustiniani and the Grand Duke of Tuscany. His Italian paintings of biblical themes in the Caravaggesque manner are considered by most scholars to be his finest work. Returning to Utrecht in

1620, his painting showed the influence of chiaroscuro, the use of dramatic light and shadow to achieve effect, and he added to the painting of religious subjects scenes of revelry and music making, as well as allegorical and mythological themes, with great success. He joined the Utrecht Guild of St. Luke in 1622 and served as its dean during that decade. The Utrecht school flourished during the 1620s, and Honthorst, unlike Baburen and Ter Brugghen, survived to have a lifelong career. He traveled to England in 1628 to paint *Mercury Presenting the Liberal Arts to Apollo and Diana* at the invitation of *Charles I, and in 1637 became the court painter to The Hague, where he lived until 1652. During his later career, he abandoned his signature candlelight for the clear illumination found in the naturalistic works of Annibale Carracci and produced decorative large-scale paintings in the manner of Orazio Gentileschi (father of *Artemesia Gentileschi), Italian influences from his early years in Rome. He lived in Utrecht from 1652 until his death in 1656.

Bibliography: J. Judson, *Gerrit van Honthorst: A Discussion of His Position in Dutch Art*, 1959; D. Weller, *Sinners and Saints, Darkness and Light: Caravaggio and His Dutch and Flemish Followers*, 1998.

Mark Williamson

HOOCH, PIETER DE (1629–1684). The Dutch painter Pieter de Hooch, son of a bricklayer, was born in the thriving port city of Rotterdam. He spent the majority of his career in Amsterdam, but his most significant works are products of his Delft period (1653–1661). De Hooch is best known for his scenes of domesticity, which corresponded to contemporary Dutch moralizing literature. The family took on a special role in the Reformed Dutch Republic as the primary social and moral unit. The ideal homes, such as those depicted by de Hooch, included women and children quietly tending to household tasks. In his *Woman Nursing an Infant, with a Child Feeding a Dog*, de Hooch describes a well-ordered burgher home, in which a child mimics her mother's behavior. Unlike his Delft contemporary *Jan Vermeer, de Hooch was not predominantly concerned with the psychological or symbolic content of his paintings, but focused on the overt problems of light, perspective, and the arrangement of architectural spaces. De Hooch popularized the courtyard scene, which was composed with the same spatial elements as his domestic interiors, such as deep views into adjoining spaces and floor patterns that articulate perspective. These tidy sun-filled courtyards were an extension of the Dutch house, a kind of home set outside, which invited inspection.

Dutch interest in optical devices, such as mirrors and lenses, led to advancements in realistic effects in painting, including the representation of natural light by Delft artists. De Hooch was a virtuoso of painted light and used it in order to infuse his paintings with atmosphere and to lead the eye through his compositions. De Hooch died in Amsterdam during a period of personal and artistic decline, but his style was emulated by other artists, which is testimony of his popular reception. He is considered one of the most significant genre painters of the Dutch Golden Age.

Bibliography: P. Sutton, *Pieter de Hooch*, Complete Edition with a Catalogue Raisonné, 1980; P. Sutton, *Peter de Hooch, 1629–1684*, 1998.

Pamela Merrill Brekka

HOOKE, ROBERT (1635–1703). Among the scientific virtuosi of England's Royal Society in the seventeenth century, Robert Hooke was perhaps the most energetic and prolific. His legacy was not a theoretical revolution, as can be said of his Society colleague *Isaac Newton, but the number of discoveries that he made and anticipated and the mechanical devices that he contrived make him a unique figure in the history of early science.

The son of a parish minister, Hooke was born in 1635 in the village of Freshwater on the Isle of Wight. As a child, he was plagued by headaches, which prevented the application of his lively mind to study. Instead, he developed his mechanical skills through constructing wooden toys, such as an operable wooden clock and a model ship with small working guns. He also showed great artistic talent, astonishing a visiting painter with his self-taught ability to copy pictures. This talent led to his apprenticeship in London, following the death of his father, to the distinguished portraitist *Sir Peter Lely. At school in Westminster, he excelled in languages, music, and mathematics. In 1653, he entered Christ Church college, Oxford, where his skills and inventiveness quickly caught the attention of the university's leading experimenters. In the 1650s, he built the air pump used by *Robert Boyle in his investigations into the nature of gases, designed contrivances for human flight, discovered the relation between changes in the weather and barometric readings, and invented a number of useful devices, including several that led to improved clocks and watches. After publishing a treatise on capillary action, in 1661 he was appointed curator of experiments for the Royal Society and elected a Society fellow in 1663. For the rest of his life, he resided in apartments at London's Gresham College furnished by the society, lecturing regularly and continuing his frenetic pace of discovery.

Although he made significant contributions in every field of science, Hooke first became truly famous with the 1665 publication of *Micrographia, or some Physiological Descriptions of Minute Bodies*, which revealed to an astonished public the new world to be seen through the microscope. Readers of this remarkable volume, copiously illustrated by Hooke himself, saw for the first time such wonders as the composite eyes of the house fly, the cellular structure (Hooke coined the term *cell* to explain plant structure) of thin slices of cork, microfungi, the arrangement of particles in crystals, and a louse holding a human hair.

After fire destroyed most of London in 1666, Hooke played a major role in reconstruction, developing, for example, the engineering principles used by

*Christopher Wren to build the great dome of St. Paul's Cathedral. At the same time, his other discoveries and inventions continued apace: the bubble level, the rotation of Jupiter, the parallax method for determining stellar distances, an odometer, the universal joint, and many others. He wrote and lectured on felt making, celestial motions, the reflection and refraction of light, the physics of elasticity, meteorology, comets, memory, the origins of fossils, and earthquakes. In his last two decades, Hooke's activities were—astonishingly—more varied: he even taught himself Chinese and lectured on the nature of Chinese characters. His energy was such that he usually worked through the night, sleeping very little. In the year or so before his death in 1703, suffering from blindness and probably congestive heart failure, he could no longer maintain such a pace, so, according to his early biographer Richard Waller, he saved time by taking short naps in his clothes.

Despite his achievements, Robert Hooke has not received the recognition by historians that some less accomplished scientists have received. In part, this neglect is due to the unparalleled variety of his interests: Except for microscopy, he is not the leading figure in any of the fields to which he nevertheless contributed remarkably. In part, his contributions are also due to his personality. His friend *John Aubrey echoed popular opinion when he wrote that Hooke was not only prodigiously inventive, but also a person of great goodness and virtue. It is clear that Hooke was more concerned with increasing knowledge and improving the lot of humankind through invention than in recognition, and unlike some of his colleagues, he always acknowledged his intellectual debts. Many of his contributions also came in the form of ideas that enabled others to develop theories more systematically. It has recently been demonstrated conclusively, for example, that Hooke anticipated much of *Newton's theory of gravitation, but many of Hooke's papers on this and other subjects have been lost, leaving credit for discoveries to those who more successfully claimed it.

Bibliography: M. 'Espinasse, *Robert Hooke*, 1956; E. Tan Drake, *Restless Genius: Robert Hooke and His Earthly Thoughts*, 1996.

Patrick Cook

HOOKER, THOMAS (1586–1647). The distinguished Puritan preacher Thomas Hooker was born to yeoman parents most likely in Marfield, Leicestershire. It is believed that his father was Thomas Hooker Sr., but church and civic records prior to 1610 are so sketchy that even the Christian name of his mother has been obscured. He entered school on a scholarship, first attending Queen's College, Cambridge, then transferring to Emmanuel. He received his B.A. in 1608 and his M.A. in 1611. He was made a fellow of Emmanuel College, and between 1609 and 1618, he acquired a reputation as a renowned teacher and preacher. By 1620, Hooker was both the rector of Esher in Surrey and a sought-after lecturer at St. Mary's Church Chelmsford in Essex. However, in 1626, *William Laud, the Archbishop of Canterbury (1633–1645), renowned for his persecution of the Puritans, forced Hooker to retire due to his overt Puritan sentiments. For the next year, Hooker was living at Cuckoos farm in Little Baddow, a hamlet near Chelmsford, with his wife and four children (three daughters and one infant son). There he established a grammar school with his assistant *John Eliot, who later achieved fame as the minister of Roxbury and as Apostle to the Indians in New England. In 1630, with pressure mounting from the High Commission, Hooker fled to Holland where he spent two years as minister of an English Puritan church, first at Delft, then at Rotterdam. There he was associated with the renowned William Ames, professor of theology at the University of Franeker, Netherlands.

In approximately 1633, a group of Puritan associates from Chelmsford migrated to New England and persuaded Hooker to come over and be their pastor. Hooker accepted and was immediately ordained pastor of the church at Newtown. He moved his congregation to Hartford, Connecticut, in 1636, where he soon became one of the community's most dominating figures. Cotton Mather once remarked that Hooker "could put a king in his pocket" if he so desired. Hooker is regarded as one of the more dialectical, eloquent, and powerful Puritan preachers among all the ministers of New England. His homiletic style is that of other traditional orthodox Puritan ministers, consisting of a sophisticated blend of doctrine, reason, and application. His sermons employed verbose logic, didactic rhetoric, grammar, and figurative language to motivate his flock. Like most Puritan scholars, Hooker was prolific in his writings. Some of his more notable works include *The Souls Engrafting Into Christ* (1638), *The Saints Guide, in Three Treatises* (1645), *The Saints Dignity and Duty* (1651), *A Comment on Christ's Last Prayer in the Seventeenth of John* (1656), and *The Application of Redemption, By the effectual Work of the Word, and Spirit of Christ for the bringing home of lost Sinners to God. The first eight books* (1656), with *The Ninth and Tenth Books* of the same general title published in a separate volume in London in 1659.

Bibliography: F. Shuffelton, *Thomas Hooker 1586–1647*, 1977.

David M. Rosen

HORROCKS, JEREMIAH (1618–1641). Jeremiah Horrocks was born in approximately 1618 in Lancashire, outside Liverpool in a Puritan community. Still a teenager, he went to Cambridge, entering Emmanuel College in 1632. Although he left in 1635 without completing a degree, in Cambridge he befriended *John Wallis, *Ralph Cudworth, and John Worthington. The last introduced Horrocks to William Crabtree, who became his lifelong scientific partner. Horrocks taught himself astronomy, becoming familiar with the latest work from the continent. An advocate of heliocentric cosmology since 1637, Horrocks was an early follower of *Kepler; yet he was not afraid of correcting and deviating from the German master. Although Horrocks died in 1641, during his short scientific career he proved to be a brilliant astronomer, making important contributions to both theoretical and observational astronomy. He proposed a new dynamical theory accounting for planetary motions. Although subscribing to elliptical orbits, Horrocks parted from Kepler's celestial dynamics in believing that each celestial body emanated an attractive power—but suffered no repulsive one—and that the sun retains its central position because of its size. Horrocks also proposed a lunar theory, published posthumously in 1672, which was improved by *Newton. His observational skills matched his intellectual creativity. He proposed very accurate measures for several apparent diameters of celestial bodies, improved considerably the measure of horizontal solar parallax, corrected Kepler's *Rudolphine Tables*, and, above all, determined that Venus's solar transit would occur every eight years. His prediction and observation of Venus's transit on November 1639 formed the basis of his *Venus in sole visa* (published by *Hevelius in 1662), a short tract in which Horrocks, after presenting his observations and their cosmological and philosophical significance, reviews older cosmological theories and glorifies the Copernican tradition. The early members of the Royal Society admired Horrocks and understood the importance of his work; they published his *Opera posthuma* in 1672–1673.

Bibliography: W. Applebaum, "Horrocks, Jeremiah," *New Dictionary of National Biography* (forthcoming); Allan Chapman, "Jeremiah Horrocks, the Transit of Venus, and the 'New Astronomy' in Early Seventeenth-Century England," *Quarterly Journal of the Royal Astronomical Society* 31 (1990): 333–357.

Renzo Baldasso

HOWELL, JAMES (?1594–1666). A prolific and eclectic author, Howell wrote poetry (he was of the so-called Tribe of Ben) and political tracts, but today he is remembered for his achievements in three prose genres that experienced a great impetus in the second half of the seventeenth century. He is an extraordinary composer of imaginative familiar letters, which span a lifetime of travels, political liasons, and compensatory fantasies. Although Douglas Bush once called him the "supreme epistolizer of the age" on account of the rhetorical mastery evinced by his *Epistolae Ho-Elianae*, Howell's tendency to enliven his factual reports with self-aggrandizing fictional episodes alienated some students of Renaissance epistolography. He is the author of a popular little book called *Instructions for Forreine Travell*, and he produced many grammar books and collections of refrains and proverbs gathered during his voyages: He could justly be called the precursor of comparative paroemiology. Because these compilations, along with some of his epistles and "discourses," constitute an unmatched source of information about seventeenth-century Spanish language and culture, he has also been called the first English Hispanist. Howell's entire career corresponds to the pattern of the disaffected Stuart intellectual who repeatedly failed to obtain opportunities for promotion commensurate with his talents and education. He was probably also slighted because of his inability to stick to one potential patron long enough. His shifting political allegiances become apparent in his readiness to dedicate his latest work to whoever was in a position of power at the time of the work's completion: *Charles I, the Duke of York, *Oliver Cromwell, Prince Charles, and even Parliament.

The son of a rector at several Carmarthenshire parishes, Howell studied at Oxford's Jesus College, where he obtained the B.A. in 1613. In 1618, Sir Robert Mansell appointed him "foreign agent," charging him with the tasks of hiring skilled workers in Italy and purchasing diverse materials from Spain, France, and other countries, thus beginning a life of journeys on the continent. In 1627, he secured the position of personal secretary to Lord Scrope, and in 1632 he became a Royalist spy. In August 1642, he obtained the king's promise of reversion of the Clerkship of the Council at Nottingham, but three months later a warrant from Parliament decreed his arrest. He was summarily committed to the Fleet Prison, where he remained until 1650.

His *Instructions for Forreine Travell* (1642) went through numerous editions at regular intervals well into the eighteenth century. The first two volumes of his *Epistolae Ho-Elianae* appeared in 1645 and 1647. There is very little information about Howell's life between the years 1650 through 1660, in which appeared the second and third volumes of the *Epistolae Ho-Elianae* (1650, 1655), *Paroimiographia: Proverbs in English, Italian, French and Spanish* (1659), a work of great linguistic and ethnographic value, and *Lexicon Tetraglotton: An English-French-Italian-Spanish Dictionary* (1660). In 1661, he successfully petitioned *Charles II for the creation of the office of Royal Historian, which carried a generous stipend. In the early years of the Restoration he produced, among other books, *A New English Grammar . . . Whereunto Is Annexed a Discourse . . . Containing a Perambulation of Spain and Portugall* (1662) and *Mr. Howell's Poems upon Divers Emergent Occasions* (1664). Howell survived the Great Plague and the Great Fire of 1666, dying in London in November 1666.

Bibliography: F. Escribano, *James Howell: Un hispanista inglés del siglo XVII*, 1980; D. Woolf, "Conscience, Constancy and Ambition in the Career and Writings of James Howell," in *Public Duty and Private Conscience in Seventeenth-Century England*, 1993.

                                                          José María Rodríguez-García

HUDSON, HENRY (?–1611). Neither the birth date nor the birthplace of Henry Hudson, English explorer, is known, and no record of his life exists until 1607, the year of his first expedition. Sponsored by the Muscovy Company, Hudson set sail on 1 May, hoping to discover a polar route to Asia. By 15 August, he had reached a latitude of 80°23', bringing him nearer to the North Pole than any previous explorer. Stopped by ice, Hudson headed east, investigating the shores of Spitzbergen and Bear Island, and on his return trip discovering Jan Mayen island. His discovery of whales off Spitzbergen Island contributed enormously to the English whaling trade. Hudson returned to England on 15 September 1607. Once again exploring for the Muscovy Company,

Hudson began his second voyage the following April, his objective this time being to discover a northeast passage to the orient between Spitzbergen Island and Novaya Zemlya. Sailing with Hudson on this occasion was Robert Juet, who would serve as first mate on this and Hudson's next two expeditions. On 3 June, at a latitude of 75°29' he again found himself blocked by ice. With a discontented crew and insufficient provisions for an attempt at a northwest passage, Hudson returned home. He arrived in England on 20 August 1608.

Unable to find English backers, Hudson embarked on his third voyage on 25 March 1609, this time for the Amsterdam Chamber of the Dutch East India Company. Sailing under the Dutch flag, Hudson sought a northeast passage, via the Strait of Viagach. Once again, he found the way blocked by ice. On his own initiative, Hudson then turned west, seeking a passage through the North American continent between Labrador and Virginia. Hudson worked his way into New York Bay and found the mouth of the large river that now bears his name. Thinking he may have discovered a water route through the continent, Hudson navigated the Hudson River almost as far as Albany, where shallow water prevented further travel. He returned to England on 2 November 1609, where government officials detained him and seized his charts and journals.

Hudson's departure on 22 April 1610 marked his fourth and last voyage, with his son John as ship's boy and Juet, once again, as first mate. At the request of a group of London merchants, Hudson sought a northwest passage. By June, he had reached Hudson Strait and by August had entered the large body of water now called Hudson Bay. For the next three months, he sought a channel leading westward but was eventually trapped by ice in James Bay. Hudson and his men endured a harsh and hungry winter on shore: One member of the crew died while others suffered illness, and relations between captain and men were strained. When sailing once again became possible the following spring, Hudson wished to resume his search for the passage, but his crew, led by Robert Juet, refused and mutinied. On 23 June 1611, they set Hudson, his son, and seven other men in a shallop, never to be seen again. The ringleaders, excluding Juet, were supposedly killed in an attack by the Inuit, and Juet died later of hunger. In all, only nine members of the crew made it back to England in October 1611. They were imprisoned for mutiny but were eventually pardoned.

In addition to pioneering exploration of polar waters, Hudson vastly increased European knowledge of the geography of the New World. He charted much of the Hudson Bay and discovered the Hudson River, a vital artery in the exploration of eastern North America. In addition, he was the first European to land in what would become New Amsterdam and, eventually, New York.

Bibliography: D. Johnson, *Charting the Sea of Darkness: The Four Voyages of Henry Hudson*, 1995.

Charles Pastoor

HUET, PIERRE-DANIEL (1630–1721). Born in Caen on 8 February 1630, Pierre-Daniel Huet is known primarily as a French historian and prelate, although his interests in mathematics and the sciences are also noteworthy. The son of Daniel Huet and Isabelle Piton de Bertouville, Daniel and his two sisters were orphaned at an early age and raised by guardians. The Jesuits in Caen educated him and emphasized geometry and philosophy. The influence of Cartesian philosophy on Huet's intellectual formation was significant and inspired much of his later work. In 1652, Huet traveled with Samuel Bochart to Sweden where he was introduced to Queen Christina. Upon returning to France in 1653, he divided his time between Paris and Caen, and became acquainted with such intellectual figures as *Pierre Gassendi and Henri Montmor. Huet was instrumental in founding a scientific academy at Caen, which had the sponsorship of *Louis XIV. He corresponded regularly with both Henri Justel of the French Académie Royale and Henry Oldenburg of the Royal Society of London, relating experiments performed at Caen. Huet was personally involved with numerous dissections of birds and other animals, and did work which detailed the structures of the eye.

In 1670, Huet permanently left Caen when he was selected by Louis XIV to assist *Jacques Bossuet, tutor to the Dauphin Louis. Six years later, Huet took Holy Orders and eventually became Bishop of Soissons (1685) and Avranches (1689–1699), ultimately returning to the abbey of Fonteray near Caen. In 1689 Huet's *Censura Philosophiae Cartesianae* was published, criticizing *Descartes' methodology and his claim of philosophical originality. Huet's 1692 *Nouveaux memoires pour servir a l'histoire du cartésianisme* offered further criticisms. Throughout his life Huet remained an active scholar, compiling an impressive library of over 9,000 volumes. He died in Paris on 21 January 1721.

Bibliography: R. Hall, ed., *The Correspondence of Henry Oldenburg*, vol. 4, 1967; A. Levi, ed., *Guide to French Literature*, vol. 2, 1994.

Nicole Howard

HUTCHINSON, ANNE (ca.1591–1643). One of the founders of Rhode Island, Hutchinson was born in Alford, Lincolnshire, England, to Francis Marbury, an Anglican priest, and his wife Bridget Dryden. In 1612, she married William Hutchinson, a merchant. The couple had fourteen children. The Hutchinsons became Puritans after hearing the teachings of *John Cotton, and followed him after he emigrated to Massachusetts Bay. They arrived in 1634. At her home in Boston, Hutchinson organized weekly prayer meetings, attracting women to discuss sermons and theology. Later, the meetings grew to include numbers of men. During these sessions she expressed Antinomian views, stressing that it was not important to prepare to receive God's grace by living a moral life or performing good works. Her followers gained control of the Massachusetts government with the election of Governor Henry Vane, one of her supporters. When Puritan leader *John Winthrop returned to power in 1637, he viewed her as a heretic and political threat, and Hutchinson was placed on trial. During the proceedings, which were held away from Boston to lessen the influence of her adherents, Hutchinson defended her views admirably, but sealed her fate when she claimed direct revelation from God—an idea untenable in Puritan theology. She and her family were banished from the colony. Hutchinson was pregnant during the trial, and the child was stillborn. Winthrop called the severely deformed infant a "monstrous birth," proof of God's displeasure.

Hutchinson followed *Roger William's lead, moving with her family to Rhode Island, where they established the town of Portsmouth. Her brother-in-law, Thomas Wheelwright, and other supporters left Massachusetts as well and were instrumental in founding New Hampshire. Even while living in Rhode Island, Hutchinson continued to be hounded by the Puritan leaders of Massachusetts, who sent deputations to try to prevent her teachings. When her husband died in 1642, Hutchinson was despondent. She decided to move to Long Island, territory controlled by the Dutch, where she would avoid further persecution. Unfortunately, a year later, Indians massacred Hutchinson and her entire family, save one daughter who managed to escape. Once again, John Winthrop celebrated the event as a divine judgment against this female leader.

Bibliography: D. Hall, *The Antinomian Controversy, 1636–1638*, 1968.

Christopher E. Hendricks

HUYGENS, CHRISTIAAN (1629–1695). Huygens is famous for his work in astronomy, mathematics, physics, and technology. He had an exceptionally brilliant career that included the discovery of the rings of Saturn, the invention of the pendulum clock, and important contributions in the development of the wave theory of light. Huygens was born in The Hague, the Netherlands. His influential father, *Constantijn Huygens, directed the University of Breda and was friends with the philosopher *Rene Descartes. Until the age of fifteen, Christiaan was schooled at home, a training that included the reading of the works of Descartes, an activity that was to have an important influence on his later scientific life. In 1645, Huygens entered the University of Leiden to study law but later transferred to the University of Breda. In 1654, he published his first article on a new way to determine accurately the area of a circle.

In 1655, Huygens received an honorary doctor of laws degree from the University of Angers, and, working with his brother, developed a better way to grind lenses; this led to the construction of their first telescope. Huygens used this instrument to study Saturn and, in the process, discovered its largest moon, Titan. He also noted that Saturn's unique appearance was due to a disk around the planet in the same plane as Titan's orbit. Although he published his discovery of Saturn's moon in 1656, he did not want to publish his findings about Saturn's rings until he had a chance to study the subject further. He therefore encoded the discovery of the rings in an anagram in his paper about Saturn's moon should a dispute ever arise over who saw the rings first. He also reasoned that Saturn's rings could not be solid but must be composed of individual particles.

Huygens was also one of the first scientists to measure the rotational period of the planet Mars, which he accomplished by timing the passage of surface markings across the face of the planet. Huygens also invented a micrometer to measure the angular sizes of planets. Also in 1656, Huygens, extending the work of *Galileo, invented a weight-driven pendulum clock, which is regarded as the first truly reliable chronometer. In turn, this pendulum clock initiated a study of precision measurements of time.

In 1637, Huygens published a treatise on probability. In 1663, he was elected a fellow of the Royal Society of London and in 1665 he was appointed as the first foreign resident of the French Académie Royale des Sciences in Paris. From 1666 to 1681, Huygens lived in Paris where he studied many topics, including a description of the physical pendulum, the theory of rotational motion and centrifugal forces, and elastic collision and forces, in which he supported Descartes' ideas on forces rather than *Newton's. Indeed, Huygens' basic philosophy of nature followed the Cartesian mechanistic view.

During his stay in Paris, Huygens made one of his most significant achievements: the development of a wave theory of light, publishing it in *Traité de la lumière* in 1690. In accordance with his mechanistic view of nature, Huygens hypothesized that light was created by oscillations of some type and that light propagated itself by triggering other oscillators. These oscillators would cause yet more distant oscillators to begin oscillating, thus propagating the light as a wave. Further wave fronts, called "secondary wave fronts," would be formed from the sum of the wave fronts of earlier wave fronts. This principle, that each part of a wave front can be the center of new wave fronts, is now called "Huygens' Principle" and is still used today to describe the wave properties of light, such as diffraction, interference, and refraction. This wave concept of light seems to require that light must pass through a medium, called "the ether." In the nineteeth century, Huygens' wave theory of light, which opposed Newton's theory of light based upon the movement of corpuscles, was confirmed by James Clerk Maxwell. However, it was not until the twentieth century that the existence of the ether was rejected by the experiments of Albert Michelson and Edward Morley.

In 1682, Huygens' poor health forced him to return to The Hague. He built new telescopes and invented a new type of eyepiece that minimized color distortions. Such an eyepiece is still called a "Huygian eyepiece." In his last days, Huygens studied philosophy and speculated on the structure of the cosmos. He suggested that, with Earth being merely one of six known planets, the other planets might be similar to Earth, complete with living beings.

Bibliography: A. Bell, *Christiaan Huygens and the Development of Science in the Seventeenth Century*, 1947; J. Weaver, *The World of Physics*, 1987.

Paul C.L. Tang

HUYGENS, SIR CONSTANTIJN (1596–1687). Huygens may have been the most multitalented figure of all the Dutch notables in Holland's great seventeenth-century cultural flowering; his correspondence, numbering in the thousands of letters, records his familiarity with many of the finest minds of his day, such as *Descartes and *Corneille. He was born on 4 September 1596 at The Hague, the son of Christiaan, secretary to the state council. He studied law at Leiden, becoming an expert Latinist and fluent in seven languages as well as some native dialects. Entering the diplomatic corps, he was secretary to three Dutch princes and visited England several times, where he met and formed a close friendship with *John Donne, many of whose poems he translated. In 1618, he studied for a time at Oxford; an accomplished musician and composer, he on one occasion played the lute at the court of *James I, who knighted him in 1622. During his third visit to England, he finished his *Batava Tempe* (published 1621), a poetic praise of his birthplace displaying the style of *Marino. The next year he issued *Costelick Mal*, a satiric poem on the excesses of female fashions, dedicated to the contemporary Dutch poet Jacob Cats. Also appearing in 1622 was *Uytlandighe Herder*, his celebration of Dutch rural life. His wife, Susanna van Baerle, died in 1637, and he later wrote lovingly of their marriage in his collection of didactic poetry *Daghwerck* (1627–1638). His *Hofwijck* (1653) praises his home at The Hague. His only drama, *Trijntje Cornelis* (1653), retelling the adventures of a ship captain's wife, reveals his fondness for lower-class characters and avoids the nostalgic, pastoral moods of his earlier work. His shorter verse and epigrams display a witty intellect, as in *Sonnetten op de Feestdagen*. Huygens received the distinction of having a road from The Hague to the coastal city of Scheveningen built in his honor; his poem *Zeestraet* records his thanks. His semiautobiographical *Cluyswerck* was published in 1683. His second son, *Christiaan Huygens, became a noted astronomer and mathematician. Huygens died at the age of ninety-one on 28 March 1687 and received a state funeral on 4 April at the church of St. Jacob.

Bibliography: A. Bachrach, *Sir Constantijn Huygens and Britain, 1596–1687: A Pattern of Cultural Exchange*, 1962.

Christopher Baker

HYDE, EDWARD, EARL OF CLARENDON (1609–1674).

Advisor to both *Charles I and *Charles II, Hyde was instrumental in creating the settlement that restored the Stuarts to the throne in 1660 and wrote an acclaimed chronicle of the civil wars that defined seventeenth-century Britain.

Of gentle birth, Hyde began his career in the law. He entered Parliament for the first time in 1640, and his determination on episcopal primacy won him Charles's regard, whose champion Hyde became. By 1642, Hyde was a private advisor to the king, but Charles ignored much of his advice and chose to pursue a course that inevitably meant a showdown with Parliament. Hyde eschewed involvement with military matters and was named chancellor of the Exchequer and admitted to the Privy Council and knighted in 1643. He was determined to achieve a settlement between king and parliament, a task which grew increasingly more difficult. When war broke out, Hyde was sent to aid Charles, prince of Wales, and later followed him into exile.

In the aftermath of Charles I's execution in January 1649, Hyde acted as Charles II's chief advisor and spent the interregnum holding the royalist party together. With Cromwell's death in 1658, Hyde made contact with sources in England favorable to a restoration of the king and drew up a number of plans by which that event could be achieved. Hyde thus became the major figure in the Restoration settlement that effected the return of the king to England, authoring the Declaration of Breda, a document that parried the issue of concessions until a new Parliament could be called.

After the king was restored, Hyde enjoyed great power and control, but the secret marriage of his pregnant daughter, Anne, to *James II, Charles's brother and heir, caused him much concern. In recognition of Hyde's sterling service to the crown, Charles ennobled him in 1661 as the earl of Clarendon. Clarendon was active in imposing the Restoration settlement and took an instrumental role in establishing the Restoration church. His moderate approach was rejected by Parliament, which passed an Act of Uniformity and other legislation that placed severe restrictions on nonconformist clergy; despite Clarendon's opposition, these acts came to bear his name as the Clarendon Code.

Increasing isolation from the king made Clarendon the target of much criticism, and he was blamed for the king's barren marriage and the outbreak of the second Anglo-Dutch war. Though the war was going badly, Clarendon advised the king against calling Parliament, advice which many parliamentarians regarded as treasonous. When Parliament was called and dismissed after only four days, Clarendon was taken as the scapegoat for its frustration; many members also saw him as having siphoned off war funds for use by the court and favored courtiers, himself included.

By August 1667, Charles was convinced that Clarendon had to go, but he refused to resign and Charles was forced to order him to do so. Parliament voted charges of impeachment against him, including one for treason, and Charles prevailed on Clarendon to go into voluntary exile to avoid being arrested on the charges.

After some restless travel, Clarendon settled down and completed his works on the civil war, his autobiography, and the papacy, along with a personal vindication against the charges brought by Parliament. By 1674, he had moved to Rouen in the hopes that he might be allowed to return to England and requested the king to grant him that right. He was refused, and the disheartened earl died in December of that year, achieving in death what he could not hope for in life: his body was returned to England and interred at Westminster Abbey.

Bibliography R. Harris, *Clarendon and the English Revolution*, 1983; G. Miller, *Edward Hyde, Earl of Clarendon*, 1983.

Connie S. Evans

I-J

IBERVILLE ET D'ARDILLIÈRES, PIERRE LE MOYNE D' (1661–1706).

A typical seventeenth-century military entrepreneur who successfully coupled private enterprise and public service, Iberville also played a key role in the establishment of the colony of Louisiana. He was born at Montréal, where he received little schooling, but had become a dedicated sailor by the age of fourteen. Over the next decade, he sailed regularly along the Saint Lawrence River and to France. His first recorded participation in armed conflict came in 1686, when he and two of his brothers led a force of *voyageurs* who were largely responsible for driving the British from their James Bay posts.

Iberville remained involved intermittently in the struggle for the Hudson Bay region until 1697, persuading the crown to support the efforts of the French Canadians to control the fur trade of the region and achieving notoreity as a brave but ruthless warrior and a brilliant naval tactician. The crown granted him a monopoly of the Hudson Bay fur trade for most of the period extending from 1694 to 1700. He also served in other theaters of the War of the League of Augsburg: He was second in command of the expedition that in 1690 attacked the town now known as Schenectady, massacring sixty of its inhabitants and, along the disputed border between New England and Acadia, oversaw the surrender of Fort William Henry to the French in 1696. In 1696, Iberville was appointed commander of troops who proceeded to loot British fishing villages of the Avalon Peninsula of Newfoundland; Iberville's profits included a large quantity of cod destined for European markets.

In 1698, Iberville led an expedition that was to find the mouth of the Mississippi from the Gulf of Mexico, previous French expeditions to the region having always proceeded southward from the Great Lakes. Iberville accomplished this mission and established a fort on Biloxi Bay. Returning to France in 1699, he became a fervent advocate of the colonization of Louisiana. He believed that a French colony centered on the Gulf Coast could hinder westward expansion of British colonies and rival New France and New York for the domination of the fur trade. During his second expedition to Louisiana, he established a fort on the Mississippi and sought to win the friendship of the region's Indian peoples, discouraging them from trading with the British. A third expedition set sail from France in 1701 and resulted in the foundation of Mobile, which Iberville hoped would ultimately rival Montréal as an entrepôt of the fur trade.

Iberville's deteriorating health and the preoccupation of the crown with its European involvements brought an end to his expeditions to Louisiana; the French settlement at Mobile remained precarious. In 1706, Iberville was entrusted with a naval squadron whose objective was to harass British colonies of the Caribbean. Iberville ravaged the British settlement at Nevis, but died at Havana before he could attack other British colonies. He was thus spared participating in a long legal battle that was to find him guilty of exporting goods from France for illegal trade in the Caribbean, embezzlement, and tax evasion.

Bibliography: G. Frégault, *Iberville le conquérant*, 1944; M. Giraud, A *History of French Louisiana*, trans. J. Lambert and B. Pearce, 5 vols., 1974–1991.

Matthew Koch

INNOCENT X, POPE (1572–1655). Born Giovanni Pamphili, Innocent X received degrees in civil and canon law. Ordained in 1597, he spent twenty-five years as an auditor of the Rota, the papal court of appeals. Later nuncio to Naples and to Spain, he was elected pope at the age of seventy-two, his pontificate beginning on 15 September 1644 and ending on 1 January 1655. He objected to the anti-Catholic provisions of the Treaty of Westphalia (1648) and aided the Irish Catholics in their struggles against the Protestants. Though *Urban VIII had condemned the doctrines of *Cornelius Jansen's *Augustinus* in 1642, Jansenism gained growing influence in France, causing Innocent to convene a group of theologians before rendering a final judgment. After two years of deliberation, he condemned the Jansenists in the papal bull "Cum Occasione" (31 May 1653). He closed many monasteries in Italy that had lost members, but in the final years of his pontificate he was unduly influenced by Olimpia Maldachini, widow of his elder brother, who sought preferences from him for family members. Innocent improved the Papal States' prison system and showed great humanitarianism toward his people during the natural disasters of 1646 and 1647. His portrait by *Velazquez hangs in the Galleria Doria in Rome.

Bibliography: L. Pastor, *The History of the Popes from the Close of the Middle Ages*, vol. 30; 40 vols., 1891–1954.

Christopher Baker

INNOCENT XI, POPE (1611–1689). Born Benedetto Odescalchi, Innocent XI gained a doctorate in civil and canon law from the University of Naples in 1639. He held various church posts until 1645, when he was named cardinal, but was not ordained priest until 1650, the same year he became Bishop of Novara. In 1669, he was defeated in a bid for the papacy owing to *Louis XIV's influence, but he was later elected on 21 September 1676, remaining pope until his death on 12 August 1689. Supported by members of the Sorbonne, Louis had also asserted his right to receive the revenues of vacant church benefices; when Innocent fought this absolutist tactic, the French clergy declared the church in France to be exempt from papal authority in the Gallican Articles issued in 1682. The issue of "Gallicanism" was but one of several issues dividing this king and pope; Innocent also disagreed with Louis's rescinding of the Edict of Nantes in 1685, which resulted in the wide persecution of Huguenots. Innocent also contested the efforts of *James II to establish Roman Catholicism in England in a way that would have increased his royal authority over church issues, as spelled out in James's Declaration of Indulgence (1687). Innocent also issued pronouncements condemning slavery as well as the Quietist doctrines of *Molinos (in the papal bull "Coelestis Pastor" of 1687). Because he was believed to have had Jansenist leanings, the process for Innocent's beatification was stalled through the eighteenth and nineteenth centuries, but in 1956, this honor was granted him by Pope Pius XII.

Bibliography: L. Pastor, *The History of the Popes from the Close of the Middle Ages*, vol. 32; 40 vols., 1891–1954.

Christopher Baker

JAMES I (1566–1625). James I governed England with a more intrusively absolutist hand than had his predecessor, Elizabeth I. Born on 19 June 1566 in Edinburgh to Henry Stewart, Lord Darnley, and Mary Queen of Scots, James early on sought to position himself to ascend the English throne after Elizabeth. He did this by balancing the conflicting expectations of Protestants and Catholics, risking the prospect of being co-opted by either one. He favored the Presbyterians in 1592 by establishing that method of church polity in Scotland, but then acted to moderate Protestant influence by administering affairs through his privy council, which he could more efficiently direct.

The growing evidence of absolutist tendencies in his administration is evident in his disquisitions *The True Law of Free Monarchies* (1598) and *Basilikon Doron* (1599). Unfortunately, he underestimated the opposition of the House of Lords to his more extreme Anglican religious positions and of the increasingly Puritan House of Commons to almost every anti-Calvinist position he advocated, especially the episcopal structure and Anglican practices he had succeeded in installing in Scotland by 1621. The intensity of contemporary religious partisanship was revealed when Guy Fawkes, a Catholic radical, laid a plot to blow up the Parliament on 5 November 1605. The plot was discovered and Fawkes and his conspirators executed, but the event aggravated growing anti-Catholic sentiment. In 1621, Parliament issued the Grand Remonstrance, a complaint against James's fiscal profligacy, refusal to seek parliamentary approval for his policies, and calls for higher taxes.

Growing increasingly intransigent, James stated in 1610 that "it is sedition in his subjects to dispute what a king may do in the height of his powers." The House of Commons just as firmly insisted on its constitutional prerogatives, and in 1611, the king dissolved the Parliament which, with a brief exception, did not reconvene until 1621. A further unwise initiative was his decision to have his son and heir Charles marry the Spanish Infanta, Anne of Austria. This would likely, if successful, have further inflamed the Calvinist elements of the country as it tugged British foreign policy into the orbit of the Counter Reformation. But this "Spanish Match" failed in 1623, two years before the king's death on 27 March 1625.

Despite his instigation of a new translation of the Bible in 1611, later known as the Authorized or King James Version, James failed to ameliorate the growing religious dissension in England, while his insistence on monarchical divine right resulted in a growing tide of Calvinist parliamentarians who resented his condescension toward principles of representative government. He saddled the treasury with enormous debts, further alienating the Commons. Intelligent but pedantic, he authored *The Essays of a Prentice in the Divine Art of Poetry* (1584); a condemnation of witchcraft, *Demonologie* (1597), and one of tobacco, the *Counterblast to Tobacco* (1604); as well as other prose and verse.

Bibliography: I. Carrier, *James VI and I: King of Great Britain*, 1998; R. Lockyear, *James VI and I*, 1998; W. Patterson, *King James VI and I and the Reunion of Christendom*, 1997.

Christopher Baker

JAMES II (1633–1701). The high-handed policies of James II, who in his short reign (1685–1688) sought to reinvigorate an absolutist and Roman Catholic monarchy in England, distanced him from Whigs and Tories alike, who had come to value the more moderate administration of his predecessor, *Charles II. After the death of his father, *Charles I, James fought with some distinction against Protestant forces as a member of both the French and Spanish armies. Charles II named him Lord High Admiral after the Restoration and, with the help of able assistants such as *Samuel Pepys, he made progress in improving naval efficiency. His religious leanings were, however, growing more clearly Romanist, and his conversion to Catholicism occurred likely before the death of his first wife, Anne, in 1671. He then married Mary of Modena, an act that increased national fears of a possible Catholic king of England and fueled wider support of an Exclusion Act, which would have barred him from the throne. Passage of the Test Act (1673) compelled him to resign his admiralty position.

But the Exclusion Act did not pass, James ascended the throne, and for a time Parliament cooperated with his rule, showing restraint at his public displays of Catholicism; however, it soon became clear that he aimed to establish this faith in the kingdom. Catholics were given military commissions and university posts despite the provisions of the Test Act, and Anglican divines were expelled from their positions. In 1687, James issued a Declaration of Liberty of Conscience, a gesture of broad religious toleration which Protestants viewed as a pro-Catholic step. When his wife bore him a son in 1688, fears of a Catholic heir-apparent spread rapidly, Whigs concerned for the survival of parliamentary authority and Tories for the survival of the Anglican church. Both parties then cooperated to offer the crown to *William of Orange, the Calvinist Dutch ruler and husband of James's Protestant daughter, Mary, by his first wife.

William accepted this offer, thus enacting a transition of power known since as the "Glorious Revolution." He landed in England on 5 November 1688; when James abruptly fled to France the following month, Parliament declared the throne vacant. James attempted to regain his crown, backed by an army of 25,000 Irish Catholic supporters known as "Jacobites," but his defeat at the Battle of the Boyne in County Kildare (1 July 1690) by William's force of 35,000 disciplined English, German, and Dutch troops spelled the last time an English monarch would so rashly attempt to impose an absolutist control over the nation's religion.

Bibliography: J. Miller, *James II: A Study in Kingship*, 1978.

Christopher Baker

JANSEN, CORNELIUS OTTO (1585-1638). Founder of the Catholic group known as the Jansenists, Jansen arrived in Paris in 1604 after two years of study in Louvain. He studied with *Jean DuVergier de Hauranne from 1612 to 1617, when he became director of the College of St. Pulcheria in Louvain, which catered to Dutch theology students. His work there was castigated by the Jesuits, against whom he debated in 1626–1627. In 1628, he embarked on his most notable work, the *Augustinus* (published in 1640), a study of St. Augustine's theology (for which he read Augustine's works ten times) but also his fullest exposition of Jansenism. He became bishop of Ypres in 1636 and died on 6 May 1638.

The center of Jansenism was Port-Royal, a Cistercian abbey of nuns near Paris; in 1625, the sect moved to a new monastery in Paris, Port-Royal de Paris. Jansenism rested upon five central tenets: God's grace is irresistible; without this grace it is impossible to obey the commandments; actions are meritorious only when free of compulsion; individual free will has no influence over God's will; and Christ died for an elect group, not for all. Jansenism furthermore discounted the efficacy of the sacraments when considered apart from the attitude of their recipient. These convictions earned Jansen the enmity of the Jesuits, the church's key intellectual vanguard of the Counter Reformation, and thus lent the movement a reputation as having dangerous parallels with Calvinism. This similarity appeared strengthened by Jansen's emphasis on rigid moral and ecclesiastical discipline.

Among Jansen's defenders were *Pierre Nicole, *Antoine Arnauld, and *Blaise Pascal, whose *Provincial Letters* also attacked the Jesuits. Jansen's five propositions were condemned as heresy by *Pope Innocent X in 1653, but the movement continued to gain strength for a time. Pasquier Quesnel's *Reflexions Morales* (1693) attempted another defense and was likewise condemned by the Pope in 1713. *Louis XIV also invoked sanctions against Arnauld and Quesnel, who were exiled from France. Ardent Jansenists later settled in Holland and Tuscany, but by 1800 its influence in France was limited to isolated parishes and schools.

Bibliography: M. Escholier, *Port-Royal: The Drama of the Jansenists*, 1968.

Christopher Baker

JOHANSSON, LARS (1638–1674). Johansson, a Swedish lyric poet who wrote under the pseudonym of Lucidor, was born in Stockholm, Sweden, on 18 October 1638. His poems are considered to be some of the most powerful of the Swedish Baroque period. Orphaned at an early age, Johansson was raised by an uncle who saw to Johansson's education, and as result, young Johansson was educated both in Sweden and abroad. After he returned to Sweden, Johansson became known as a writer of funeral elegies and epithalamions (poems written to celebrate a marriage). His most personally introspective poems are drinking songs and funeral hymns—a typically Baroque convention. Baroque traits often included emotional tensions, such as those between life and death, sensuality, exuberance, whimsy, splendor of form and movement, all juxtaposed to sounds and images of the grotesque and bizarre. As such, Johansson's poems were patterned after those of the German Baroque poets, but far exceeded theirs in both intensity of feeling and power of expression. "Skulle jag sörja så vore jag tokot" ("Were I to grieve, then I were a fool") is considered his most famous lyric. His poetry was published posthumously in *Helicons blomster* (*Helicon's Flowers*, 1689). Johansson was only thirty-six years old when he died on 13 August 1674 in Stockholm, Sweden.

Bibliography: A. Preminger, "Swedish Poetry," *The New Princeton Encyclopedia of Poetry and Poetics*, 1993.

David M. Rosen

JOLIET, LOUIS (1645–1700). French explorer and, with *Jacques Marquette, discoverer of the upper Mississippi River, Joliet was baptized in Quebec in September 1645 and schooled in the seminary there at least to the level of minor orders. Joliet left the order, probably in 1667, but remained on good terms with the Jesuit missionaries. In 1669, having returned from France, Joliet was sent by Talon, Intendant of Justice and Finance in the New World, to seek the copper mines around Lake Superior. In 1672, the governor of New France, Louis Frontenac, appointed Joliet to find the great river of the Indian tales. In December 1672, Joliet made his way to the mission at St. Ignace and to Father Marquette. The following spring, with five Indian companions and two canoes, Joliet and Marquette set out down Lake Michigan, into Green Bay, to the Fox River, the route of the earlier *Jean Nicolet (who had served under *Champlain). A short portage completed, Joliet and Marquette followed the Wisconsin River to the Mississippi, the first white men to do so. Just below the entry of the Arkansas River into the Mississippi, unsure of their personal safety but very sure that the Great River emptied neither into the Atlantic nor the western ocean but into the Gulf of Mexico, they turned for home. Joliet's journal, with his maps and drawings, was lost in the overturning of his canoe en route to Quebec, but from conversations with Joliet, Father Tablon, Jesuit Superior at Quebec, reconstructed the explorer's notes. According to Tablon, Joliet clearly understood the importance of what he and Marquette had discovered: When the Iroquois finally lost control of Lake Erie, as Joliet knew was inevitable, a waterway navigable from the mouth of the St. Lawrence through the great interior of the country to the Gulf of Mexico would open. That this waterway would come into existence and that the abundance of the interior of the New World would sustain colonization and create trade was Joliet's vision. In 1679, Joliet traveled to Hudson Bay and, in 1680, he became the king's hydrographer and received grants of land in 1679, 1680 and 1697.

Bibliography: J. Delanglez, *Life and Voyages of Louis Jolliet, 1645–1700*, 1948; F. Steck, *The Jolliet-Marquette Expedition, 1673*, 1927.

Martha Oberle

JONES, INIGO (1573–1652). Jones is sometimes thought of as an architect and sometimes as the earliest of English set designers, but he should, as J. Alfred Gotch has noted, probably be thought of as the surveyor of the King's Works to *James I and to *Charles I, a job that encompassed both these skills as well as those of city planner and preservationist. Born in 1573, he was baptized on 19 July at the church of St. Bartholomew the Less in Smithfield, London. He seems to have been named after his father, a clothworker, and he had three sisters, Joan, Judith, and Mary. Little is known of his early life or education. It is assumed that he made an early trip to Italy, where he may have studied the classical architecture he was to popularize in England. However, the first account that names Jones refers to him as a picture maker. He was employed by Christian IV of Denmark, Queen Anne's brother, and in 1605 he designed the first of the masques that made his career, *The Masque of Blackness*. The following year he designed *Hymeneai*, and thereafter he was regularly employed as the designer for royal masques until the last one, *Salmacida Spolia*, performed in 1639.

Jones was employed by Henry, Prince of Wales, and designed his *Barriers* in 1610 and *The Masque of Oberon* in 1611. He was also hired by Queen Anne and designed for her the above-mentioned *Blackness*, *The Masque of Beauty*, *The Masque of Queens* (1609), and many others until her death in 1619. Most of these masques were devised in collaboration with *Ben Jonson, but in 1632 the jealousy and irritation underlying the partnership erupted and the partnership was terminated permanently. Despite Jonson's vitriolic attack on Jones in *An Expostulation with Inigo Jones*, and in the censored characterization of him as

Vitruvius Hoop in *The Tale of a Tub*, the judgment of the court seems to have been that Jonson was replaceable but that Jones was not, for thereafter masque texts were provided for him by Aurelian Townshend (*Albion's Triumph* and *Tempo Restored*, both in 1632), *Thomas Carew (*Coelum Brittanicum*, 1634), and *William Davenant (*Salmacida Spolia*). During his career as masque deviser, Jones experimented with effects of light and darkness, most noticeable, perhaps, in *Blackness* and *Luminalia* (1638), and stage technology, from the *periaktoi*, or painted scenes that turned, in his earliest works, to the complex machinery in *Salmacida Spolia*.

Jones traveled to Italy in 1613–1614 along with the Earl of Arundel and there made copious notes in his volumes of Palladio's architecture. These notes were to prove useful in his architectural career, for he was asked to design a number of buildings. The most important of these are the Queen's House at Greenwich, the new Banqueting Hall at Whitehall, built after the first burned down in 1619, and a new facade for St. Paul's Cathedral, unfortunately later destroyed in the fire of 1666. He also designed a chapel for Queen Henrietta Maria at Somerset House, first used in 1636; oddly, she sent abroad for a scene designer to erect a scene in it, instead of using Jones himself. As surveyor, Jones supervised Charles I's comprehensive city-planning projects and more than once ran afoul of local builders, as when he stopped an excavation he deemed too close to the foundations of St. Paul's. He enforced regulations on water supply, sewage, and brick making in addition to his building and masque-making projects. During the war, Jones was captured at the fall of Basing House on 14 October 1645, where his captors maliciously wrote that he had been "carried away in a blanket, having lost his cloaths." He died without issue on 21 June 1652 and left the bulk of his estate to his assistant, John Webb. Webb later published a memoir of Jones in his *The Most notable Antiquity called Stone-Heng* (1655) and *A Vindication of Stone-Heng Restored* (1665), in which he argues, based on Jones's notes, that Stonehenge is Romano-British in origin. Jones's masque designs have recently been studied, not only as the first examples of changeable scenery in Britain, but as giving visual form to the absolutist politics and Neoplatonic philosophy of the early Stuart court. Roy Strong and Stephen Orgel have produced seminal work in this field, work that succeeding critics have further elaborated and built upon.

Bibliography: S. Orgel, *The Illusion of Power: Political Theater in the English Renaissance*, 1975; S. Orgel and R. Strong, *Inigo Jones: The Theatre of the Stuart Court*, 1973; J. Summerson, *Inigo Jones*, 1966.

Melissa D. Aaron

JONSON, BEN (1572–1637). Ben Jonson is the best-known Renaissance dramatist other than William Shakespeare. Born the posthumous son of a clergyman, Jonson was raised by his mother and stepfather. He was sent to Westminster School by an unknown benefactor, where *William Camden was his master, but his stepfather, a bricklayer, pushed him into the trade shortly after Jonson's graduation. Jonson served in the military in 1591 and married Anne Lewis in 1594. About his wife Jonson had little to say, but he wrote famous epitaphs following the deaths of two of his three children.

Jonson's first association with theater, probably as an actor, can be traced to 1595 by a reference in Henslowe's diary. *Every Man in his Humour* (1598) was Jonson's first original play, and it included *William Shakespeare in the cast. In this play and in his second, *Every Man out of his Humour* (1599), Jonson developed a specialized type of comedy in which characters represent different aspects of human temperament based upon bodily humors, liquids believed since ancient times to influence personality. Jonson's classical education influenced his convictions that Elizabethan drama should conform to classical models. He frequently criticized his fellow playwrights, including Shakespeare and Marlowe, for not closely following rules of ancient drama. Particularly, he believed that drama should adhere strictly to the unities of time, place, and action. Jonson's strong views on drama offended some of his peers, whom he satirized in retaliation. Among those at odds with Jonson were *Thomas Dekker and *John Marston, with whom Jonson engaged in a "war of theaters," termed by Dekker *poetomachia*. He later mended differences with both and collaborated with each as playwright.

A complex man whose temperament and satirical writing style were often met with controversy, Jonson was imprisoned thrice, once for his involvement in writing and acting in a satire entitled *The Isle of Dogs*, again for murdering an actor (though released by claiming benefit of clergy), and finally for offending the crown with his biting satire *Eastward Ho* (1604), cowritten with *George Chapman. His classical tragedy *Sejanus* (1603), a work of which Jonson was particularly proud, both angered authorities with its heavy-handed satire and frustrated audiences who were impatient with the play's slow-moving plot. Jonson's religious affiliations, too, hampered his reputation. While imprisoned for murder, he converted to Catholicism, making him later the subject of suspicion after the Gunpowder Plot of 1605. Although Jonson's writing sometimes earned him disfavor because of his satirical style, he nonetheless wrote a number of plays that were favorably received. His four greatest comedies, *Volpone* (produced in 1606), *Epicoene, or the Silent Woman* (1609), *The Alchemist* (1610), and *Bartholomew Fair* (1614) earned him fame. In each, he creates characters whose self-destructive behavior mirrors faults that he witnessed in everyday London.

Jonson walked from London to Edinburgh in 1618–1619, where his meeting with William Drummond led to Drummond's publication *Conversations* from which originate several anecdotes attributed to Jonson, particularly concerning his views of his contemporaries. Most famous among these is his comment that "Shakespeare wanted art." Despite such seemingly virulent remarks, Jonson was not without respect for Shakespeare. He later wrote verses praising Shakespeare's first folio collection of plays and claimed that in Shakespeare "there was ever more . . . to be praised than to be pardoned."

Many of Jonson's poems are similarly directed to fellow writers or to critics who disapproved of his satirical style. His poems embody classical forms and literary types, such as the ode and epigram. Jonson was aware not only of his own close imitation of classical writers, but also of his role as progenitor of poets. He emerged as the master of the "Tribe of Ben," a group of London writers including *Robert Herrick, *Thomas Carew, and *Richard Lovelace. Jonson also discussed his reliance upon and influence over other poets in a fragmentary text entitled *Discoveries*, published after his death.

Jonson began writing masques for the court in 1611 and became poet laureate in 1616, the date of his first published folio, *Works*. Although he received a substantial pension and enjoyed success as a court poet during the reign of *King James I, Jonson was less successful after *Charles I succeeded to the throne. Failing health and an intense rivalry with court architect *Inigo Jones compounded Jonson's difficulties. His late plays, in which he experimented with Elizabethan forms, were not well received. Jonson was particularly bitter over the poor reception of *The New Inn* (1629), blaming the actors for negligence in their performance. Jonson died in 1637 in Westminster; his final play, *The Sad Shepherd*, was left unfinished. A prolific writer and a man intensely involved with his own society, Jonson is remembered as much for his opinions about his fellow poets and dramatists as his own works, the plays being infrequently adapted for modern stage.

Bibliography: I. Donaldson, ed., *Ben Jonson*, 1985; R. Miles, *Ben Jonson: His Life and Work*, 1990; D. Riggs, *Ben Jonson: A Life*, 1985.

Carol Jamison

JUNIUS, FRANCISCUS, THE YOUNGER (1589–1677).

This philologist and scholar was born in Heidelberg, Palatinate (Germany). His works were instrumental in stimulating interest in the study of both the East German languages of the Goths, known as Gothic (a language known primarily from its use in the fourth-century Bible translations of Bishop Ulfilas) and Anglo-Saxon (Old English). Junius the Younger was the son of a French Protestant theologian, Franciscus Junius (1545–1602). In 1617, Junius the Younger was ordained in the Netherlands. He relocated to England in 1620 where he eventually befriended *John Milton and became a librarian and private tutor to the family of Thomas Howard, fourteenth Earl of Arundel. Howard was a renowned patron of the arts, and it was while working for Howard that Junius began collecting ancient manuscripts. In 1651, Archbishop *James Ussher of Armagh asked Junius to edit a manuscript believed to have been written by Cædmon, the first Christian poet to write in the vernacular and also credited as being the first to introduce pagan heroic verse to Christian themes. The manuscript, first published in 1655, contains the poems *Genesis*, *Exodus*, *Daniel*, and *Christ and Satan*.

Today, the manuscript given to Junius for translation bears the names of both Cædmon (the Cædmon manuscript) and Junius (the Junius manuscript) and resides in the Bodleian Library of Oxford University. Contemporary scholars now attribute the poems to several authors, suggesting that the poems were written during various time periods. Junius's other significant contribution was his *Etymologicum Anglicanum (English Etymology*, 1743), a text said to have been used by Samuel Johnson (1709–1784) in the composition of his dictionary. Junius died on 19 November 1677 in Windsor, Berkshire, England.

Bibliography: R. Bremmer, Jr., *Franciscus Junius F.F. and His Circle*, 1998; P. Lucas, *Franciscus Junius Caedmonis Monschi Paraphrasis Poetica* (*The "Caedmon" Poems*), 2000.

David R. Rosen

K

KEN, THOMAS (1637–1711). Born at Little Berkhampstead, Hertfordshire, in 1637, Ken, later a prominent English prelate and composer of hymns, was the son of Thomas Ken of Furnivall's Inn. His maternal grandfather was John Chalkill, whom *Izaak Walton called an "acquaint and friend of Edmund Spenser." Walton himself married Ken's step-sister, Anne, in 1666. Ken began his studies at Winchester College in 1652, moved to Hart Hall, Oxford, in 1656, and a year later, given a fellowship, joined New College. He took his B.A. in 1661, was ordained in 1662, and earned his master's in 1664. *Anthony à Wood mentions Ken's singing in the musical gatherings of the university and, indeed, Ken's significance is two-fold: his music and his ministry. By 1672, Ken had returned to Winchester as fellow of the college and chaplain to the bishop. Ken spent the next two years in service to the poor, in preparing the Manual of Prayers for the scholars of Winchester College, and in composing hymns, two of which, among the best-known of all hymns in the English language, are Ken's glory: "Awake, my soul, and with the sun" and "All praise to Thee, my God, this night."

In 1674, Ken and young Walton made a trip to Rome, and ever after, Ken was solidly Anglican. In 1679, *Charles II appointed Ken as chaplain to Princess Mary, wife of *William of Orange. At the Dutch Court, Ken insisted that an unwilling relative of the prince keep his promise of marriage to an English lady; unpopular for this stand, Ken returned to England and soon became chaplain to King Charles. In 1683, once more at Winchester, Ken refused to yield his quarters to *Nell Gwyn on the occasion of the king's visit to the city. Rather than securing the enmity of Charles II, Ken gained the king's respect, and when the see of Bath and Wells fell vacant in 1684, Ken was appointed to fill it, the king reportedly remarking, "None shall have it but that little man who refused lodging to poor Nellie!" Shortly after, Ken attended Charles II on his deathbed; his ministry earned the praise of all but *Dr. Gilbert Burnet. In 1685, Ken published his *Exposition on Church Catechism, the practice of Divine Love.* In 1688, Ken, with six other bishops, refused to back *James II's reissue of the Declaration of Indulgence. During three weeks in June 1688, Ken was imprisoned, tried, and released from the tower. However, with the accession of *William, Ken, who had sworn allegiance to James, found himself unable to swear the same allegiance to William, and joined the other "non-juring bishops" in this decision; in 1691, he was superseded in his diocese by Bishop Kidder of Peterborough. In retirement, Ken lived at Longleat, the home of his friend, Lord Weymouth. Death came on 19 March 1711.

Bibliography: F. Hoyles, *The Edges of Augustinism: The Aesthetics of Spirituality in Thomas Ken, John Byrom, and William Law*, 1972.

Martha Oberle

KEPLER, JOHANNES (1571–1630). Johannes Kepler was born to a poor family in southern Germany in 1571. Distinguishing himself as a bright pupil, Kepler received a scholarship from the duke of Württemberg, which allowed him to study theology and later mathematics under Michael Maestlin, one of the earliest Copernicans, at the University of Tübingen. After completing his degree in 1594, Kepler moved to Graz as a mathematics teacher at the Protestant school. While in Graz, Kepler published the *Mysterium Cosmographicum* (1596) or *Mystery of the World*, claiming that the key to the universe's structure and harmony rests in the Platonic solids. Used only as archetypes, these five perfect polyhedra circumscribe concentrically nested spheres whose radia determine the distances among the planets, as well as explain the number and place of the planets in the universe. Subscribing to Copernicanism, Kepler placed the sun in the middle of the heavens, making it responsible for all planetary motions. The *Mysterium Cosmographicum*, the first major Copernican treatise after Copernicus, brought Kepler to the attention of Tycho Brahe, who was the court mathematician to the Emperor Rudolph II.

In the autumn of 1600, Lutheran Kepler left Catholic Graz and moved to Prague to work with Brahe, the most skilled observational astronomer of his day and originator of the geoheliocentric cosmological system. In 1601, when Brahe died, Kepler assumed his position, which allowed him basic financial security as well as time for scientific research. Most important, Brahe had accumulated several decades of observations clearly suggesting that Mars was not following a circular path around the sun. Kepler carefully studied this data, and, by 1605, he had concluded that not only Mars's path, but all the planets, followed elliptical orbits with the sun at one of the foci. These results were published in the *Astronomia Nova* (1609)—its full title being *New Astronomy Based upon Causes or Celestial Physics Treated by Means of Commentaries on the Motions of the Star Mars from the Observations of Tycho Brahe*. Containing what are known as Kepler's First and Second Law (planets move through elliptical orbits with the sun at one focus, and planets sweep equal areas in equal times within their orbital plane), this book presented a new astronomical system. While defending a Copernican universe with elliptical orbits against the scriptures and Ptolemaic cosmology, it furnished the system with a dynamic to explain planetary motions in the absence of crystalline spheres—recently discredited by Brahe's research and observations. Kepler's celestial dynamics was based on a rotating sun, which, by emanating magnetic-like rays guided the planets' movements.

Before 1609, Kepler had published several smaller works. *A Defense of Tycho against Ursus*, composed in 1600, presents Kepler's epistemological considerations about physical and geometrical astronomical hypotheses, explaining the superiority of arguments based on physical causes with observable consequences; it also offers a selective history of theoretical astronomy from the earliest times to the advances of Tycho. In *Astronomia pars opticae* (1604), Kepler discusses optical topics relevant to astronomy; in *De Stella Nova* (1606), he considers astronomical and astrological issues related to the 1604 nova. After following closely the telescopic observations announced by *Galileo's *Sidereus Nuncius*, Kepler responded with his *Dissertatio cun Nuncio Sidereo* (May 1610), lending support for all the astronomical discoveries. He replicated Galileo's observations and published the *Narratio de observatis quatuor Jovis satellitibus* (1611), writing also a thorough description of the telescope's optical principles in his *Dioptrice* (1611). The same year, his *Strena* or *On the Six-Cornered Snowflake*, a short tract on the snowflake's structure, appeared.

In January 1612, after losing his wife, his favorite son, and his patron when Rudolph II was deposed, Kepler moved to Linz as provincial mathematician. For several years, Kepler paused in his main research work, publishing only two small tracts: *De vero anno quo aeternus Dei Filius naturam in utero benedictae Virginis Assumpsit* (1613) presents astronomical evidence for redating Christ's birth to 4 B.C.; *Stereometria doliorum vinariorum* (1615) is a mathematical investigation of casks whose volume Kepler computed by considering an infinite stacking of circular disks. Finally, in 1619 came the *Harmonice mundi*, Kepler's complete theory of the cosmos's harmony, considered from the viewpoint of geometry, music, astrology, and astronomy. Trying to reinterpret elliptical orbits into a mathematical archetype, he discovered his third planetary law: The square of the period of rotation to the cube of the mean solar distance is a constant for all planets. His last two major works are an introduction to heliocentric astronomy, *Epitome astronomiae Copernicanae* (1617–1621), and the astronomical tables, *Tabulae Rudolphinae* (1627), which became both widely used and the effective vehicle for the spread of heliocentric astronomy. The impact of his interpretation of Copernicanism is further proved by demand for a second edition of the *Mysterium Cosmographicum*, which Kepler published in 1621, adding footnotes to update his earlier theories with his later results. After fleeing Catholic Linz in 1626, Kepler found a protector as astrological advisor to the Catholic general Wallenstein; he died in 1630.

Bibliography: Max Casper, *Kepler*, Trans. Doris Hellman, 1993; J. Field, *Kepler's Geometrical Cosmology*, 1988; J. Voelkel, *Johannes Kepler and the New Astronomy*, 1999.

Renzo Baldasso

KIDD, WILLIAM (ca.1645–1701). The Captain Kidd of legend ranks with Blackbeard (another figure shrouded in myth) and the purely fictional Long John Silver and Captain Hook as a pirate whom even the most casual layman recognizes; however, the real Kidd was a merchant shipowner, privateer, and pirate hunter whose brief career in piracy was relatively unsuccessful and whose execution may have had more to do with politics than his crimes.

According to tradition, Kidd was born in Greenock, Scotland, in 1645, but his life is obscure prior to 1689, when, during the War of League of Augsburg, he became privateer captain of the *Blessed William*, raiding the island of Marie Galante and otherwise harassing the French. Arriving in New York in 1691, he sided with Royalists against Leisler's Rebellion, married wealthy widow Sarah Oort, befriended Robert Livingston, acquired property on Wall Street, and became a prominent citizen. In London in 1695, he and Richard Coote, earl of Bellomont, governor of New York and Massachusetts, conspired with members of the Whig Junto—John, Baron Somers (lord chancellor); Edward Russell, earl of Orford (first lord of the Admiralty); Henry Sidney, earl of Romney; and Charles Talbot, duke of Shrewsbury—to share any loot Kidd recovered while acting on a commission to hunt for pirates in the Indian Ocean. Sailing in the *Adventure* and operating from Madagascar between 1696 and 1699, Kidd had little success, taking a mere handful of ships and only one major prize, the 400-ton Indian vessel *Quedah Merchant*. Though he claimed the captured ships carried French passports, he antagonized the Mughal emperor and thus the East India Company, which had him condemned as a pirate. The duplicitous Bellomont had him arrested in Boston in 1699 and sent to England. There the Tories tried unavailingly to use the investigation of Kidd's activities to further impeachment proceedings against the junto. Kidd, though, was convicted of piracy and murder (he had hit his gunner William Moore over the head with a bucket). On 23 May 1701 he was hanged twice after the rope broke the first time. His tarred body then hung in chains along the Thames River.

Unaccountably, Kidd has become the stuff of legend. Tales have abounded about a nonexistent buried treasure, inspiring numerous stories including Edgar Alan Poe's *The Gold Bug* and Robert Louis Stevenson's *Treasure Island*. More generally, exaggerated stories about his exploits and cruelty have contributed to the popular image of the swashbuckling pirate from "The Ballad of the Captain Kidd," which appeared shortly after his death, down to the present. The cinema has reduced Kidd's career quite literally to farce—a cheap, inaccurate production entitled *Captain Kidd* in 1945, starring Charles Laughton, who reprised the title role in 1952 in *Abbott and Costello Meet Captain Kidd*. For skeptics, it remains to be seen if an expensive epic film about Kidd slated for the early twenty-first century will get history right.

Bibliography: R. Ritchie, *Captain Kidd and the War against the Pirates*, 1986.

William B. Robison

KILLIGREW, THOMAS (1612–1683). Descendant of an old Cornish family that began royal employment during Henry VIII's reign, Thomas Killigrew was born on 7 February 1612. As the fourth son of Sir Robert and Mary Woodhouse Killigrew of London, he had little hope of significant inheritance, and in lieu of formal education beyond grammar school, he was sent to be a page in *Charles I's court, perhaps as early as 1625. Accepted into Queen Henrietta Maria's literary circle, he learned to love the theater and wrote *The Prisoners* (1635), *Claricilla* (1636), and *The Princess* (1636) for private performance at Court. In 1631, he married Cecilia Crofts, daughter of Sir John and Mary Crofts and maid of honor to the queen. She died in January 1638, not long after the birth of their only child, Harry. In 1640–1641, he wrote *The Parson's Wedding*, a bawdy comedy intended for public performance.

Killigrew fled to the continent and placed himself in the service of the exiled royal family during the interregnum. He may have written his romantic tragedy *The Pilgrim* for an English acting troupe that performed for Prince Charles in Paris in 1645. He also raised money for Prince Charles and served as groom of the bedchamber to the Duke of York at the Hague (1648–1649). Killigrew traveled to Northern Italy and Venice as ambassador for the newly crowned but still exiled *Charles II (1649). Pressured by the English Commonwealth to remove this Royalist representative from their court, the Venetians abruptly dismissed Killigrew for alleged immoral conduct (1652). Between 1649 and 1654, Killigrew wrote *Cicilia and Clorinda, or Love in Arms; Bellamira her Dream*; and *Thomaso, or the Wanderer*, extended "closet dramas" not intended for the stage. In 1655, he married Charlotte de Hesse, settled at Maestricht, and served in the House of Orange, but returned to London with his family at the Restoration. He profited from continued royal favor, including joint authorization with *Sir William Davenant to organize and manage London's acting companies and theaters in August 1660. Killigrew and Davenant established rival acting companies— the King's Men and the Duke's Men, respectively—and dominated the London stage. Killigrew moved his company from the Red Bull to Gibson's Tennis Court near Lincoln's Inn Fields in 1660, then moved again to the Theatre Royal, which he built in 1663. An initially successful manager, Killigrew began to suffer financial loses, compounded by an outbreak of plague that closed the theaters in 1665 and the burden of raising money to rebuild the Theater Royal after it burned in 1672. Although the King's Men continued to perform in borrowed quarters until the new Theatre Royal opened in 1674, Killigrew's increasing indebtedness forced him to relinquish management to his son Charles (1677). However, he continued living beyond his means for the remainder of his life. He died on 19 March 1683 and was buried beside *Ben Jonson in Westminster Abbey.

Although Killigrew wrote no plays for the Restoration theatre, his *Claricilla* and *The Parson's Wedding* were among those he produced until the appearance of the better-crafted plays commonly associated with the Restoration and eighteenth-century stage. His plays being regarded as the second-rate work of a "doodling amateur," Killigrew has been remembered not as a playwright but as a manager. The fierce competition between him and Davenant has been credited as a catalyst for the rapid development of drama and dramatic innovations of the day, with Davenant traditionally regarded as the more important and more reputable of the two. Certainly no candidate for sainthood during his lifetime— he aggressively sued for and received properties forfeited to the crown, lived lavishly with his creditors at his heels, enjoyed bawdy humor, and possessed a sharp, often caustic wit—the accuracy of his reputation as a licentious rakehell that preceded him into subsequent centuries was substantively challenged by his twentieth-century biographers.

Bibliography: A. Harbage, *Thomas Killigrew, Cavalier Dramatist 1612–83*, 1967.

Sallye Sheppeard

KING, HENRY (1592–1669). Born in Worminghall, Buckinghamshire, Henry King stood fourth-generation heir apparent in an Anglican dynasty established under Henry VIII. His father, John King, was chaplain to Elizabeth I and *James I, dean of Christ Church and vice-chancellor of Oxford, and bishop of London. Henry attended Westminster School and Oxford, earning the B.A. (1611), M.A. (1614), and B.D. and D.D. (1625). Appointed prebend of St. Pancras at St. Paul's (1616), archdeacon of Colchester (1617), and sinecure rector of Fulham (1618), he advanced no further for twenty-one years. His marriage to Anne Berkeley of Kent (1617) proved bittersweet. Anne died just months after their fifth child's birth in 1623, and only two of the children survived childhood. Whether King remarried remains unsubstantiated, but he maintained close ties with his brothers John, Philip, and William, also clergymen, and Robert, a military man, and his sisters Elizabeth Holt, Dorothy Hubert, and Anne Dutton (later Lady How) and their families, who sometimes resided with him and his children.

In 1639, King was appointed dean of Rochester, and in 1642, bishop of Chichester. Stripped of his ecclesiastical livings and personal property and obliged to flee when the civil war began shortly thereafter, King spent the eighteen-year interregnum living with various family and friends, writing, and secretly ordaining Anglican clergy. Reinstated as bishop of Chichester in 1660, he remained unwilling or unable to recast his religious and political views to suit Restoration tastes and received neither royal favor nor ecclesiastical advancement. He died at Chichester on 30 September 1669 and was buried in its cathedral eight days later.

Initially regarded a lackluster preacher, King gained considerable distinction and respect as he matured in his vocation. Only the eleven sermons comprising *An Exposition Upon the Lord's Prayer* (1628) and ten other sermons printed individually during his lifetime survive, remaining unpublished again until the last decade of the twentieth century. In rhetoric as well as theology, King spurned excess and austerity in favor of a middle style, and his sermons contribute significantly to the development of seventeenth-century prose. Like his father, he never abandoned his loyalty to the Anglican Church or the rule of absolute monarchy, his sermons urging their hierarchies as reflections of God's universal order. England's political and religious instability forced King in his 1621 Paul's Cross sermon to deny allegations of his father's deathbed conversion to Catholicism and in a sermon preached in 1662, of having himself succumbed to Puritanism.

King's close working relationship with *John Donne is well documented. King supervised the carving of Donne's funeral effigy, safeguarded his sermons, and probably assisted the 1633 publication of his *Poems*. *Ben Jonson, *Izaac Walton, Henry Wotton, and Bishop Brian Duppa are among King's other friends and acquaintances. From his early Oxford days until the interregnum, King composed lyrics and occasional poems. Enjoying an appreciative audience among contemporary poets and other readers, his poems appeared anonymously in three unauthorized collections (1657, 1664, 1700). John Wilson, musician to *Charles I, set King's sonnets and "A Penitential Hymne" to music, and King's metrical *Psalms of David* appeared in 1651 and 1654. However well received during his lifetime, King's occasional verses, elegies, and sonnets fell into obscurity after his death. Early twentieth-century critic George Saintsbury celebrated "Tell me no more" as "one of the most faultless and perfect" of all English poems, and T.S. Eliot in *The Metaphysical Poets* lauded "The Exequy" as "one of the finest poems of the age." But critics esteem his verse as more transitional, representative, and imitative if metaphysical.

Bibliography: M. Crum, ed., *The Poems of Henry King*, 1965; M. Hobbs, ed., *The Sermons of Henry King (1592–1699), Bishop of Chichester*, 1992; L. Mason, *The Life and Works of Henry King, D.D.*, 1913.

Sallye Sheppeard

KIRCHER, ATHANASIUS (1602–1680). Athanasius Kircher was born in 1602 in Geisa at the Ulster, Germany. He attended the Jesuit Gymnasium at Fulda from 1612 to 1618, entering the order in 1616. Subsequently, he studied at Paderborn, Cologne, Koblentz, and Mainz, where he was ordained a priest in 1628. Through his education he mastered mathematics, theology, philosophy, and languages. In 1628, Kircher was appointed professor of philosophy, mathematics, Hebrew, and Syriac at Würzburg. He later lectured at Avignon, where he befriended Peiresc and *Gassendi, who introduced Kircher to their intellectuals' network while encouraging him to pursue his interests in Egyptology. Although appointed professor of mathematics at the university in Vienna through Peiresc's influence, Kircher was asked by *Pope Urban VIII to teach mathematics at the Roman College, the center of Jesuit culture. Eventually,

Kircher resigned this position and dedicated his last thirty years to studying and writing. He died in 1680.

Kircher was a true polymath, an intellectual characteristic much appreciated within Baroque culture. He wrote more than forty books on a variety of topics and maintained an extensive correspondence with many European intellectuals. His interests extended from archaeology and ancient languages to science. Within science, Kircher investigated through experimentation a variety of subjects and avidly collected specimens of all sorts creating the "Kircher Museum." Among his many publications, the following are important milestones: The four-volume *Oedipus Aegyptiacus* (1652–1654) comprises the fruits of his Egyptian studies; his scientific investigations about magnetism are presented in *Magnes, sive de arte magnetica* (1641), on gnomonics in *Ars magna lucis et umbrae* (1646), on acoustics in *Musurgia universalis* (1650), and his theories about underground phenomena in *Mundus subterraneus* (1664–1665); his interest in hermeticism and mystical architecture are reflected in *Arca Noe* (1675) and *Turris Babel* (1679). Finally, *Ars magna sciendi* (1669) presents his theory for a universal method of knowing derived from Ramon Lull's combinatorial art.

Bibliography: J. Godwin, *Athanasius Kircher: a Renaissance Man and the Quest for Lost Knowledge*, 1979; C. Reilly, ed., *Athanasius Kircher S.J.: Master of a Hundred Arts, 1602–1680*, 1974.

Renzo Baldasso

KNELLER, SIR GODFREY (1646–1723).

Kneller was the dominant portrait painter of the late seventeenth and early eighteenth centuries in Britain. He was born at Lübeck and christened Gottfried Kniller. Sent by his father Zachary to the University of Leiden with a view to preparing him for a military career, Gottfried exhibited such skill as a draughtsman that he was encouraged to develop his talent as an artist. He studied with *Rembrandt van Rijn at Amsterdam, and in 1672 installed himself at Rome, fostering an enduring fascination with the work of Raphael and a predilection for a classicist approach to portraiture. Kneller arrived in England in 1676 and soon won the favor of *Charles II, which assured him a diverse aristocratic patronage. His popularity was not adversely affected by the political turmoil, which followed Charles II's death; indeed, he achieved a still greater degree of royal favor and noble patronage during the reign of *William III and *Mary II. He was appointed Principal Painter jointly with John Riley in 1688 and became sole occupant of that post on the latter's death in 1691. Kneller's Whig politics and the favor that had been lavished upon him by William virtually guaranteed the painter a cool relationship with *Queen Anne, but his popularity was now too firmly established for royal snubs to erode his fashionable clientele.

Kneller was renowned for the lifelike quality of his portraits; his Baroque naturalism was a key component of his success. Though under William his studio produced an unprecedented volume of copies of his portraits of the king as publicity for the somewhat uneasily established ruler, Kneller's best royal portraits are remarkable character studies; his impressive range can be appreciated by comparing his last portrayal of Charles II (Walker Art Gallery, Liverpool), in which the worldly-wise gaze of the aged monarch arrests the viewer's attention, to his 1698 depiction of *Peter the Great, whose youthful ambition and resolve are evident. Many of his most accomplished canvases feature fellow artists, intellectuals, and writers; particularly outstanding are his portraits of José Carreras y Coligo, his close friend Antonio Verrio, *John Evelyn, *John Dryden, *Grinling Gibbons, and his 1697 depiction of *John Locke. His naturalism is nowhere more striking than in his portrayals of soldiers, such as the 1694 portrait of Thomas Wharton and the depiction of a formidable unknown soldier standing before a burning town.

Another component of Kneller's enduring success was his ability to adapt to changing expectations of the portrait painter. Thus, as British aristocratic culture developed a more serious tone during the reign of William, Kneller obliged his patrons by providing them with the pensive "sensible" likenesses then in vogue, canvases such as his poignant 1698 portrayal of Elizabeth Villiers, Countess of Orkney, and former mistress of William. From 1697, Kneller adopted a lighter presentation and more restrained use of color, which further reinforced his classicism and foreshadowed the rococo.

Bibliography: J. Stewart, *Sir Godfrey Kneller and the English Baroque Portrait*, 1983.

Matthew Koch

L

LA BRUYÈRE, JEAN DE (1645–1696). Jean de La Bruyère is considered a French classical author and moralist. Little is known about his life. Born into a well-to-do family in Paris where he lived until his death, he never married. After studying law, he was appointed trésorier general (general treasurer) in 1673, and in 1684 became the tutor to Duke Louis de Bourbon, grandson of Condé. He managed to write during these busy years. He published his only book, *Les Caractères*, in 1688 and was elected to the Académie Française in 1693; his argumentative speech given at his reception dealing with La Querelle des Anciens et des Modernes (the polemic of the Ancients and the Moderns), increased the tensions between these two groups.

The final edition of *Les Caractères* in 1694 includes more than 1,120 sections and shows the evolution of his work. The book contains descriptions of human character ranging from a few lines to several pages in length. Four general styles are exhibited: general maxims, aphorisms in the style of *La Rochefoucauld, thoughts devoted to contemporary society, and portraits of individuals. As a moralist writer, he studied "morals" in the sense of customs or manners (*mœurs*). Before this, the term was applied to either religious writers or "bad" philosophers, while the "moralist" writers were most often simply philosophers. La Bruyère describes human foibles and draws the distinction between psychology and sociology, undertaking a critique of humankind that drew upon Montaigne's brand of humanism, the Jansenist revival of St. Augustine, and even the legacy of *Descartes.

The three categories of *Les Caractères* are grouped into sixteen chapters, in which particular subjects such as incidents, conversations dedicated to social classes such as the nobles, and the court are extremely varied in the central section "De L'homme." The book evolves from satirical views of the world to a religious conclusion in the final section, "Des esprits forts." The unity of the work is achieved from an overall vision.

The *Caractères* have become inextricably linked to the author's name and have acquired a, literally, sui generis character, to the point that the late eighteenth-century moralist Nicolas Chamfort could embrace the tradition with his title *Maximes et pensées, caractères et anecdotes* (1795). The style of La Bruyère conveys through short sentences and repetitive details all of humanity, and it has much in common with *Boileau's satires and *Molière's comedies; he tends to dehumanize his characters, while at the same time presenting a new twist on the old world. Played on stage as puppets, people seem grotesque. The result is a fascinating and vivid view of the world that is very humorous. His political point of view defending a more charitable social order would later be attractive to the *philosophes*. It is also the sign of his modernity.

Bibliography: R. Barthes, "La Bruyère," in *Critical Essays*, 1972; P. Bénichou, *Man and Ethics: Studies in French Classicism*, 1971.

Martine Sauret

LA CALPRENÈDE, GAUTIER DE COSTES (1610–1663). La Calprenède, a popular novelist who was considered the master of the heroic novel in the mid-seventeenth century, was born near Cahors around 1610. After studying in Toulouse, he joined the army and fought in the German campaign. In 1632, he moved to Paris and acquired a post of ordinary gentleman in the king's chamber. He henceforth devoted himself to literature, living in the society of the Grand Condé in Paris and Chantilly before the prince's exile after the Fronde. La Calprenède was a prolific author, penning nine tragedies and tragicomedies between 1635 and 1642, turning after 1642 definitively to the novel. His ten-volume *Cassandre*, published between 1642 and 1645, knew an enormous and enduring success, which contributed to his eventual immense wealth. Described in the precious style of the period, La Calprenède's characters exemplify generosity, gallantry, and courage; he prided himself on his novels' historical and military exactitude as well as on their lack of explicit violence or passion. The novels' historical milieus included the Persian Wars, the Roman empire, and the Merovingian dynasty, and La Calprenède engaged in a public literary dispute with *Mlle de Scudéry by blaming her novels for undermining the requisite grandeur of the heroic novel. In 1646, he married Madeleine de Lyée, a fellow writer with the pen name of Octavie. La Calprenède's twelve-volume *Cléopatre* of 1647 cemented his popular reputation. His third novel, *Faramond*, was begun in 1658 with only two volumes published before his death on 20 October 1663, the result of a fall from a horse. La Calprenède influenced subsequent writers: *Boileau included La Calprenède's Faramond in his 1666 *Dialogue des Héros des romans* and *Mme de Sévigné admitted preferring La Calprenède's novels despite their faded status by the end of the century.

Bibliography: A. Adam, *Histoire de la littérature française au XVIIᵉ siècle*, Vol. 2, 1962; M. Lever, *Le Roman français au XVIIᵉ siècle*, 1981.

Margaret Harp

LAFAYETTE, MARIE-MADELEINE (PIOCHE DE LA VERGNE), COMTESSE DE (1634–1693). French novelist who privileged the analysis of feelings over action, Marie-Madeleine Pioche de la Vergne was born in Paris to Marc Pioche de la Vergne and Isabelle Pena. After her father's death in 1649 and her mother's marriage to Renaud-René de Sévigné in 1650, Marie-Madeleine met her lifelong friend, the Marquise de Sévigné. In 1651 Marie-Madeleine became a lady-in-waiting to Anne d'Autriche and entered court life. Four years later she married Jean-François Motier, Comte de Lafayette, a man almost twice her age whom she respected but probably did not love, and moved to Auvergne with him. In 1659 Lafayette wrote *Portrait de la Marquise de Sévigné*, the only work published under her name during her lifetime. In 1661, Lafayette left her husband to return to Paris, where she resumed her social life and became friend and confidante to *Louis XIV's sister-in-law, Henriette d'Angleterre. Lafayette's first novel, *La Princesse de Montpensier* (1662), published anonymously, was based on Henriette's love for the Comte de Guiche. Lafayette held a literary salon at her home in Paris, where she received high society and writers such as *Jean de La Fontaine, *Jean-Baptiste Racine, and *Jean-Baptiste Molière, who is said to have read his *Femmes Savantes* at her salon. She was also an intimate friend of *François de La Rochefoucauld, with whom she maintained an amorous, though possibly chaste, relationship until his death in 1680. Lafayette's second novel, *Zaïde* (1670), was a collaboration with La Rochefoucauld and Jean Segrais published under the latter's name. Her masterpiece, *La Princesse de Clèves* (1678), also published under Segrais' name, is considered the prototype of the psychological novel. Her last novel, *La Contesse de Tende*, as well as her biography of Henriette d'Angleterre and her court memoirs were all published after her death.

Bibliography: S. Haig, *Madame de Lafayette*, 1970.

Leslie A. Sconduto

LA FONTAINE, JEAN DE (1621–1695).

French poet Jean de La Fontaine is certainly best known for his fables, which have been studied, recited, and staged in schools for over three centuries. Considered a French classicist, his different writings show a love for experimentation as well as a love for universal themes (love, fear, power). Born in the small town of Château-Thierry, near Paris, he remained faithful to his little provincial home throughout his life. Appointed "maître des eaux et forêts," he loved to study animals and wild life. He started to publish at the age of thirty-seven and became the first official poet of Fouquet, to whom he remained loyal after the superintendent's disgrace. His ouput was varied. He wrote numerous epigrams, epistles, poems, elegies, and discourses, in particular the *Elegie for Fouquet* written after his patron's downfall, and *Epître à Huet* in 1687. His dramatic efforts range from the adaptation of Terence's *L'Eunuque* (1654) to the unfinished tragedy *Achille*. The libretto for opera *Daphné* (1690) was rejected by *Lully, the official opera composer, and provoked a sharp polemic, *Epître à M. Niert*. His play *Astrée* (1691), based on the popular novel of *d'Urfé, his *Contes* (1674), and his letters composed of prose and verse for his wife during his travels to the Limousin, *Le songe de Vaux* (1663) and *Le Psyché de Cupidon* (1669), reveal his versatility in genres. Though these works knew various fortunes, all of these writings show a personal tone of independence and his carefree but sensitive nature.

La Fontaine started to publish his *Fables* in 1668. Originally dedicated to the king's son to present him a moral teaching in a lively way, they quickly became famous. The *Fables* are narrative poems that illustrate in a satiric way not only the seventeenth century, but humanity in general, with its passions and weaknesses. La Fontaine was inspired by many sources: Aristotle, Aesop, Phèdre, Boccaccio, Rabelais. Often seen as one of the principal authors of French classicism, La Fontaine thought originality lay more in the style than in the substance. His style is extremely original; writing in a freer verse, a real innovation for the century, he was in love with the twelve-foot Alexandrine verse line. The work is a masterpiece of dramatic concision, of rhythm, linguistic variety, and personal charm. The *Fables* feature different subjects engaging with the world of power, cruelty, and failure, presenting in each scene a distinguishable and precise lesson. They show great verbal mastery and a constant taste for experimentation and novelty. La Fontaine is considered one of France's great poets and a dedicated artist who succeeded in achieving a universal reach. As one of his fables, "L'homme et son image," asserted, La Fontaine speaks to everyone.

Bibliography: L. Van Delft, M. Gutwirth, *Un merveilleux sans éclat; La Fontaine, ou la Poésie exilée*, 1987; *Le moraliste classique; Essai de définition et de typologie*, 1982.

Martine Sauret

LA MOTHE LE VAYER, FRANÇOIS DE (1588–1672).

Son of a Parisian magistrate, François de la Mothe le Vayer inherited his father's post before abandoning it in 1625 to devote himself entirely to his studies. Two works produced in the 1630s, including the *Considérations sur l'éloquence française* (1638), in which he asserts the superiority of the Ancients over the Moderns, gained him such widespread recognition that he was elected to the Académie Française in 1639.

The following year, La Mothe le Vayer published *De l'instruction de monsieur le Dauphin* (1640), prompting *Cardinal Richelieu to name him preceptor to the future *Louis XIV, a role that he would take on, after initial opposition on the part of Anne of Austria, in 1652. Once the king began his personal reign in 1661, La Mothe le Vayer served as both an official advisor and historiographer to the king.

A friend of the freethinkers *Gassendi, *Naudé, and Diodati, La Mothe le Vayer is known above all for being a skeptic in the same vein as Montaigne and *Bayle. His *Dialogues faits à l'imitation des Anciens* (published posthumously in 1698) provide the best example of La Mothe le Vayer's skepticism. He was particularly interested in comparative historiography and applied his skepticism to the study of history in *Du peu de certitude qu'il y a dans l'histoire* (1668).

La Mothe le Vayer was quick to defend himself and his skeptical attitude against accusations of irreligion, depicting himself rather as a Christian skeptic who sought to discount the pursuit and validity of earthly knowledge and to sweep away religious superstitions as a means of providing a surer foundation for religious aspirations.

Bibliography: R. Pintard, *Le libertinage érudit dans la première moitié du XVIIe siècle*, 1943; F. Wickelgren, *La Mothe le Vayer, sa vie, son œuvre*, 1934.

Patricia Armstrong

LA ROCHEFOUCAULD, FRANÇOIS VI, DUC DE (1613–1680). Born into a 500-year-old family whose members had fought on both the Protestant and Catholic sides in the Wars of Religion, the Duke of La Rochefoucauld married at fifteen and started his military career at sixteen by fighting in Italy and Holland. In his twenties, he joined the French nobility's struggle at home to preserve their independence from the absolutist measures of Louis XIII and *Louis XIV. In 1635, La Rochefoucauld's participation in plots against Louis XIII's autocratic chief minister, *Cardinal Richelieu, led to brief imprisonment followed by exile to his estate. Next, he joined the Fronde, rebels under the Prince de Condé who mutinied against Louis XIV in 1648 through 1653. Wounded twice, La Rochefoucauld also had a passionate affair with Condé's sister, the Duchess de Longueville.

After the Fronde's failure, the disillusioned La Rochefoucauld became prominent in Parisian salon life, particularly through his friendships with the exceptionally cultured Madame de Sablé and *Madame de Lafayette. Salon life demanded a witty, intellectual dissection of society's mores and manners that raised conversation to the level of an art. More than equal to the challenge, La Rochefoucauld eventually began recording and refining his salon insights for anonymous publication. A pirated Dutch version of his manuscript led him to claim his authorship in the first official edition of *Réflexions ou sentences et maximes morales* (*Reflections or Moral Maxims*) in 1665.

The maxim, a genre just gaining acceptance in the 1660s, conformed to the classical principles of brevity, balance, and restraint. La Rochefoucauld presented 317 in his first edition, which opened with an essay on self-interest that he later condensed pithily into maxims such as "Virtues are lost in self-interest as rivers are lost in the sea." Overall, his book challenged the Stoic philosophy, dominant in the first half of his century, that will and reason do and must subdue the passions. Although some early readers found him too pessimistic and cynical about human nature, the Maximes came out in four more editions in La Rochefoucauld's lifetime. Since he kept revising and perfecting his work over almost twenty years, these texts differed considerably, with the 1666 version offering only 302 maxims, while the definitive 1678 edition totals 504.

By 1661, La Rochefoucauld had sufficiently regained the King's favor to be made a knight of Saint-Esprit; in 1662, he wrote his *Mémoires*. He died in Paris two years after supervising the fifth edition of the *Maximes*. On this book rests his stature as a master—alongside *Racine—of classical French literature.

Bibliography: D. Culpin, *La Rochefoucauld: Maximes*, 1995; W. Moore, *La Rochefoucauld: His Mind and Art*, 1969.

Margaret Goscilo

LA SALLE, JEAN-BAPTISTE DE (1651–1719). Jean-Baptiste de la Salle was born on 30 April 1651 in Reims, France. He was born into the French nobility and early pursued a vocation in the Church; he was ordained in 1678 and developed an interest in serving the poor. After several years of considering which approach to undertake, La Salle founded the Brothers of the Christian Schools in 1684 and devoted himself to the education of poor boys. This initiative was a new direction for so-called "charity schools" in France. Previously, these institutions had failed because of the lack of state support and the difficulties associated with developing and sustaining a dedicated and qualified faculty. La Salle expended his fortune on the schools and his new congregation; the "Brothers' Schools" quickly developed a reputation for quality, and tuition-paying parents sent their sons to be educated. Day schools were expanded to boarding schools and, in 1705, La Salle moved his headquarters from Paris to Rouen where he established his first school for delinquent boys.

At the same time he continued to write; La Salle's publications included *The Duties of a Christian* (1703), *The Conduct of Christian Schools* (1720), and the posthumously published *Meditations* (1730–1731). The last years of La Salle's life were directed at the stability of his congregation; he feared that upon his death, a priest would be named to lead the Christian Brothers. He averted such a development when a General Assembly of the order named a Brother as a successor two years prior to La Salle's death. He died in Rouen on 7 April 1719. In 1725, Pope Benedict XIII proclaimed that the Christian Brothers were to be a Papal Institute and enjoy the protection and support of Rome. Jean-Baptiste de la Salle was canonized a saint of the Roman Catholic Church in 1900.

Bibliography: W. Battersby, *St. John Baptist de la Salle*, 1965; E. Fitzpatrick, *La Salle: Patron of All Teachers*, 1951.

William T. Walker

LA SALLE, ROBERT CAVALIER, SIEUR DE (1643–1687). An important French explorer of North America, La Salle was born in November 1643 in Rouen. Entering the Society of Jesus as a young man, La Salle was trained in mathematics and natural sciences. Unhappy with the Jesuit life, he left the order in 1666 or 1667; all accounts speak of his subsequent and lifelong distaste for the Jesuits. In Canada by 1666, he set about learning the languages and ways of the Indians and became so expert that he earned the confidence of Frontenac, governor of New France. Knowing La Salle daring enough to risk the Iroquois in their own country and able to make his way with the other tribes, Frontenac sent him to build and command a fort at present-day Kingston, Ontario. Frontenac wished to intercept any Indian trade headed toward Albany and the English, and he also wished to move the French trading facilities closer to the tribes. In 1674, La Salle secured the seignory of Frontenac from the king. In return for the title, LaSalle had to rebuild the fort with masonry and to maintain it at his own expense. The years 1675 and 1676 found La Salle building and developing Fort Frontenac into a thriving trading post.

In 1677, La Salle was once again at court, and this time the king granted him the right to build forts throughout the interior provided that they be maintained at La Salle's expense and that he not trade with those tribes whose custom was already established at Montreal. Envisioning an organized and extensive system of forts and fur trade, La Salle built, in 1679 and above Niagara Falls, the *Griffon*, a vessel designed to carry guns as well as goods and the first ship deliberately built for trade on the lakes. La Salle's other establishments were Fort Crevecoeur, near present day Peoria, and Fort St. Louis, near Utica, Illinois. However, neither establishment succeeded; the settlers left by La Salle to maintain Fort Crevecoeur actually destroyed it, and La Barre, the governor who succeeded Frontenac, required La Salle to hand over Fort St. Louis.

In 1681, La Salle journeyed the whole length of the Mississippi River, claimed that river's valley for *King Louis XIV, and named it in his honor: Louisiana. In 1683, La Salle went to court to propose building a new fort, one at the mouth of the Mississippi River. With the king having provided men, arms, supplies, and money, the expedition sailed from La Rochelle on 24 July 1684 for the Gulf of Mexico. Somehow, La Salle managed to miss his goal, the mouth of the Mississippi, and sailed as far west as Matagorda Bay, half-way between present-day Galveston and Corpus Christi, Texas. He landed there with disastrous results. The colonists suffered and sickened in the heat, and many died. La Salle set out several times overland for the Mississippi, but always had to turn back. During his final attempt to bring aid to his colony, the explorer was murdered by his own men on the banks of the Trinity River on 19 March 1687.

La Salle qualifies as being among the most brilliant of the North American explorers. Although extremely ambitious, his idea of organizing a string of trading posts and of building ships specific to the Great Lakes trade made economic and military sense, as did his idea of fortifying the mouth of the Mississippi River. La Salle's temper and disposition are another matter; Joutel, a contemporary and loyal follower, remarked on La Salle's arrogance and pride. The explorer could not brook any questioning of his authority or judgment and could not admit a mistake. Some suggest this combination of faults led to the catastrophe of Matagorda Bay: LaSalle knew he had sailed too far and simply refused to acknowledge the mistake and turn around. In fairness to him, however, the problem of longitude had not yet been solved, and so LaSalle had no dependable means of measuring his east-west distance. La Salle's aides likewise deserve mention. Tonti, his lieutenant and author of a journal of La Salle's explorations and discoveries, was noted for his ability, fidelity, and ambition. Hennepin, the Recollet friar who frequently accompanied La Salle, wrote the journal usually considered the best authority on La Salle's first western exploration, although Hennepin's later work is marked by exaggeration, if not outright invention. He was the first to describe Niagara Falls and the upper Mississippi River.

Bibliography: E. Osler, *La Salle*, 1967; F. Parkman, *La Salle and the Discovery of the Great West*, 1889, rpt. 1968.

Martha Oberle

LA TOUR, GEORGES DE (1593–1653). The son of a baker, Georges de La Tour's artistic career was spent in the small French town where he died, Lunéville, in Lorraine. Biographical facts about him are scanty, but it has been assumed that he studied from 1605 to 1610 in the studio of Jacques Bellange in Nancy. In 1617, he married Diane Le Nerf, daughter of a nobleman; his work appears to have prospered, and he is known to have had five apprentices over the course of his career. His region of France was sorely despoiled by the Thirty Years' War, a possible reason for the fact that no commissions of his before 1641 are known. However, after a trip to Paris in 1639 he acquired the title of Peintre Ordinaire du Roi, and between 1644 and 1651, the number of his commissions increased. La Tour's oeuvre is complicated by difficulties of chronology and attribution, but about forty paintings can be assigned to him with certainty. Those completed before 1630 include the *Musicians' Quarrel, Old Man, Old Woman, Cheat with the Ace of Diamonds*, and *Cheat with the Ace of Clubs*; paintings from the last phase of his career include the *Adoration of the Shepherds, St. Sebastian Tended by St. Irene, Dice Players*, and *New-born Child*. La Tour's work is noted for its general avoidance of Baroque or Mannerist tendencies in favor of a realism in the fashion of *Carravagio. His work favors genre and religious topics, and is marked by the typical presence of a small number of figures set in simple scenes lit from a single light source that casts a stark contrast of light and shadow through the painting. His works thus often acquire a serene, contemplative atmosphere, which links them with the spiritual tendencies of the Counter Reformation. His nighttime scenes, which have brought the most critical praise, are seen in the *Repentance of Mary Magdalene* and in *Christ with St. Joseph in the Carpenter's Shop*. Though his works lost favor after his death, they have acquired increasing popularity in the twentieth century.

Bibliography: P. Conisbee, *Georges de La Tour and His World*, 1996.

Christopher Baker

LAUD, WILLIAM (1573–1645). The principle ecclesiastical advisor to *Charles I, this Archbishop of Canterbury was a major contributor to the conflicts of the English Civil War. William Laud was born on 7 October 1573, the son of a Reading clothier, William Laud. From such a modest beginning, Laud would rise to become one of seventeenth-century England's most powerful ecclesiastical figures and Charles I's principal advisor on religious matters. Laud was educated first at the Reading Free school and then St. John's College, Oxford, from which he received his B.A. in 1594 and Doctor of Divinity degree in 1608. While at St. John's, Laud cultivated the theological mindset that would make him a strong proponent of episcopacy in the English church and an energetic enemy to the various reform ideologies, which fell under the general term *Puritanism.*

Laud held numerous church positions during his rise to power. He was ordained in 1601 while still a fellow at St. John's College. In 1603, he became chaplain to the Earl of Devonshire, Charles Blount. In 1607, he was appointed vicar of Stanford in Northamptonshire and then a year later, chaplain to Richard Neile, bishop of Rochester. Neile recognized in the young Laud talents that matched his zeal and promoted his career. In 1611, Laud was elected president of St. John's and in 1616 became dean of Gloucester. By this time, the ambitious Laud had captured the attention of *James I and accompanied the king to Scotland where his custom of wearing the surplice invoked the wrath of the Scottish Presbyterians. This collision between Laud and those who saw all forms of episcopacy as being evidence of Catholic tendencies was only the first of many conflicts that cast Laud as a pawn of Rome. It was, however, the involvement of Laud, then bishop of St. David's, in the 1621 dispute over the countess of Buckingham's conversion to Catholicism that finally defined Laud's sentiments. He expressed at this time his belief that the Church of Rome was a true church, even though he rejected such components of Catholic ideology as the infallibility of the pope.

After the death of James I, Laud found a new patron in Charles I, who first appointed him dean of the Chapel Royal, then later bishop of Bath and Wells, bishop of London, and finally in 1633 archbishop of Canterbury. Charles appreciated Laud's conviction to orthodoxy and the belief that ornamentation did not detract from, but rather enhanced, religious service. Laud also shared the king's belief that bishops served a viable purpose in government. Puritans in England, and particularly in the House of Commons, rejected both beliefs, so Charles I turned to Laud to help him suppress those who would attack the crown for its religious positions by making him a judge in the Star Chamber. Using this position, as well as that of archbishop, Laud was able to promote his agenda of religious conformity and persecute dissenters who chose openly to attack both himself as well as the king and the Catholic Queen Henrietta.

Perhaps remembering his first experience with the Scottish Presbyterians, Laud encouraged Charles I's efforts to impose in 1637 a new Book of Common Prayer on the Scottish Kirk. The reaction of the Scottish leaders was to sign the National Covenant and march a formidable army into England where the king was forced to negotiate a settlement, much to his disadvantage. In 1640, in debt and facing a determined foe, Charles I convened in 1640 his first Parliament in eleven years. The House of Commons, now reconstituted and controlled by Puritans, such as John Pym, was finally in a position to pursue its own ecclesiastical and political agenda. In 1640, it impeached Thomas Wentworth, earl of Strafford, Laud's close associate at the court of Charles I. In that same year, Laud suffered the same fate and was imprisoned in the Tower. Those whom he had so rigorously persecuted during Charles's personal rule took particular pleasure in undoing most of what Laud had achieved to impose uniformity on the Church of England and to defend the rights of bishops to hold political position. Largely because of the efforts of *William Prynne, who had suffered extensively at the hands of Laud, the former archbishop was tried in 1644 and sentenced to death. He was beheaded on 10 January 1645, almost four years to the day before the king, whom he had served so faithfully if not always advisedly, would suffer the same fate.

Bibliography: E. Bourne, *The Anglicanism of William Laud*, 1947; C. Carleton, *Archbishop William Laud*, 1987; H. Trevor-Roper, *Archbishop Laud, 1573–1645*, 1940, rev. 1962.

Gerald Morton

LE BOSSU, RENÉ, ABBÉ (1631–1689). Le Bossu, a French cleric whose private life remains obscure but whose literary criticism exerted a profound influence, was born in 1631. An adolescent at the time of his religious orders, Le Bossu spent his life at the church of Sainte Geneviève de Paris. In 1694, he published a philosophical treatise, *Parallèle des principes de la Physique d'Aristote et de celle de René Des Cartes*, in which he argues that the Aristotelian method, thanks largely to the exactitude of its definitions, is more useful than *Descartes'. Little read, this volume announced the intellectual rigor found in Le Bossu's next work, *Traité du poème épique* (1675). An intensive Aristotelian application to the study of literature, which set forth exacting rules for the epic, the *Traité* knew immediate popularity throughout Europe. Five subsequent editions appeared in Paris, Amsterdam, and The Hague between 1677 and 1714.

It is the enthusiastic English response to the *Traité* that ensured Le Bossu's legacy. Pope included Le Bossu's commentaries on the *Iliad* and the *Odyssey* in his own translations of Homer, while *Dryden and *Congreve cited the *Traité* frequently. In contrast, Voltaire criticized Le Bossu's inflexible rules and ignorance of Dante, Arisoto, Spenser, and *Milton, claiming that a poem written according to Le Bossu's rules would not only be impossible but also unread. Laurence Sterne offers a biting satire of the *Traité* in his *Tristam Shandy* (1759–1767). Le Bossu argues principally that the epic poem must, first, offer a universal moral lesson and, second, be based on the rational rather than the supernatural. By studying the epic, Le Bossu analyzed the literary genre ignored by France's unrivaled neoclassical literary critic *Boileau, in his *Art poétique*

(1674). Today, the *Traité* is little read but is considered a decisive element in the late seventeeth-century writers' Quarrel of the Ancients and the Moderns.

Bibliography: S. Curran, *Le Bossu and Voltaire on the Epic*, 1970; H. Gillot, *La Querelle des Anciens et des Modernes en France*, 1968.

Margaret Harp

LE BRUN, CHARLES (1619–1690). Charles Le Brun (or Lebrun) was a talented and apparently willful child who was introduced into the studio of Simon Vouet by Chancellor Pierre Séguier. Le Brun soon left Vouet's studio, striking out on his own. In 1642, Le Brun traveled to Rome with *Nicolas Poussin, where he came under the influence of Poussin and Raphael, eventually coming to believe he was their logical inheritor. By 1645, Le Brun had returned to France, resuming a career that would make him the most highly regarded painter of the age of *Louis XIV. Le Brun was not only a talented painter but also a consummate politician who, throughout his life, cultivated powerful friends. Nor was he hesitant to use his influence to thwart his rivals; as a consequence, Le Brun was able to retain his power over the arts at the French court until near the end of his life. Another source of Le Brun's unequaled power began with the founding of the Academy of Painting and Sculpture, whose ostensive purposes were to provide training in the arts and separate them from the craft guilds and thereby elevate visual artists above the level of craftsmen. Le Brun, a founding member of the academy, became its rector, chancellor, and finally director. He was able to achieve much to elevate the status of artists, while at the same time assuring his own position.

In 1658, by virtue of his ability to organize complex decorative programs, Le Brun was charged with decorating Nicolas Fouquet's chateau; Fouquet was Louis XIV's superintendent of finance. The decorations were a great success; unfortunately, they aroused Louis's jealousy and suspicion. Fouquet was replaced with his subordinate, *Jean-Baptiste Colbert, whose good graces Le Brun had also managed to cultivate. Colbert saw in Le Brun an instrumentality through which French culture might be elevated while, not coincidentally, aggrandizing the king whom Colbert served. Thus began a long, mutually beneficial relationship between Colbert and the artist.

In 1661, Le Brun received a commission that led to what he regarded as his greatest achievement. He began a series of four large canvases based on the life of Alexander the Great. The first of these paintings was the *Queens of Persia at the Feet of Alexander*. The picture displays Alexander in full battle dress in an open-handed gesture of magnanimity toward the cowering women. The gesture is, however, less about Alexander than it is about Louis XIV, who was preparing to subjugate Europe. Though Louis is not shown, he is nevertheless present, as is the notion of his generosity in victory. The painting is one of Le Brun's best: The sentiment is immediately understandable, and the allegory is uncluttered by the innumerable subtleties that beset Le Brun's later work.

The *Battle of Arbella* (1669) amply demonstrates the bombastic clutter that plagued Le Brun's large-scale historical paintings, which, over time, led to the steady decline of his reputation. The painting is a panorama of battle; the darkly painted leftmost third of the picture depicts a mass of struggling horsemen who, despite their existence in an improbably compressed space, manage to express the chaos of battle. Near the center right, a single Persian soldier looks over his shoulder while fleeing the maelstrom. Above, to the right, a doll-like Alexander surveys the battle. Higher still at the center left, a French imperial eagle hovers inexplicably above Alexander. As propaganda, Le Brun's painting succeeds admirably. Alexander sweeps toward Darius III, who shrinks visibly before him. Alexander typifies a model leader who remains heroic and calm while lesser men are undone by fear. Here again, while not physically present, the virtues of the French king may be inferred. Unfortunately, formal structure is subordinate to the message, so that approximately half of the painting is unnecessary; the story is completely understandable by viewing the center right portion of the picture. Le Brun tacitly admits as much by submerging the left and right sides of the painting in murky tones, reserving contrast and color as accents in the center.

Le Brun was the perfect artist-courtier. Unlike Poussin and *Claude Lorrain, who avoided the court of Louis XIV, Le Brun was willing to provide large scale decorations as needed since he lacked a strong personal aesthetic that might otherwise demand to be served; this virtue, of course, eventually led to the eclipse of Le Brun's reputation. Ever faithful to others, he never found himself. By the eighteenth century, changes in aesthetic and political sensibilities caused Le Brun to be reviled, even in France, for his subservience. Despite the efforts of a few French historians to resurrect his reputation based on the intellectual subtlety of his work, Le Brun remains a man out of time, whose ability was spent at the service of ideas that have little appeal or likelihood of resurgence. Le Brun's panel paintings are generally his most successful works, perhaps because the demands of the French court are less in evidence. In these paintings, we see glimpses of the Le Brun that might have been: *Daedalus and Icarus* (1645), the *Martyrdom of Saint Andrew* (1646), and *Chancellor Séguier* (1665–1657).

Bibliography: M. Gareau, *Charles Le Brun: First Painter to King Louis XIV*, 1992.

John Schmidt

LEEUWENHOEK, ANTONY VAN (1632–1723). This early maker of microscopes, whom kings and princes visited, was born on 24 October 1632, in Delft, the Netherlands. He was sent to grammar school in 1640, afterward living with an uncle, a town clerk, and an attorney, who may have taught him rudimentary mathematics and science. Leeuwenhoek never attended a university, nor did he learn any classical or modern European languages, deficiencies that hampered his later scientific pursuits. After an apprenticeship to a cloth merchant in Amsterdam, he returned to Delft, operated a shop, and married Barbara de

Mey in 1654. In 1660, he was appointed chamberlain of the Council Chamber of the sheriffs of Delft, a post he held for thirty-nine years. His wife died in 1666, and in 1669, he was appointed surveyor to the court of Holland. He married Cornelia Swalmius in 1671.

During the late 1660s, Leeuwenhoek learned to construct simple microscopes, which consisted of a tiny hand-ground lens fixed between small holes drilled in two thin plates of brass. Including its specimen holder, the device was four inches long. Several survive, the best one with a magnification of 270 power and a resolution of 1.4 micrometers (or 1.4 millionths of a meter). Leeuwenhoek had good eyesight; dexterous, steady hands; and patience, qualities that suited him well for fifty years of microscopy.

In 1673, he described microscopic observations of the bee stinger in a letter to Henry Oldenburg, secretary of the Royal Society of London, who published it in the *Philosophical Transactions*. Leeuwenhoek published all his findings in this manner. During the 1670s, he made many important discoveries, notably microscopic animals—infusoria—first observed in an infusion of pepper and water, reported in the *Philosophical Transactions* in 1676, and spermatozoa, which he first observed in 1677 in a specimen from a gonorrhea patient. He ultimately observed spermatozoa in a number of species, and by 1679, he had found them in the vas deferens of a hare. Concluding that they originated in the testicles, he accepted the doctrine of animaculism—the preformation of individuals in the spermatozoa—a view consistent with the mechanist philosophy that appealed to Leeuwenhoek.

Unaware of previous research by *Marcello Malpighi, Leeuwenhoek rediscovered the red blood corpuscles in 1674 and capillaries in 1683, and observed the corpuscles traveling through the capillaries in tadpole and eel tails. He thought, however, that the function of blood was only nutritive, not respiratory. Also in 1683, he inspected tooth plaque and discovered that it was teeming with life—bacteria. Observations on insect eggs and larva led him to reject spontaneous generation. He further speculated that microorganisms caused disease, an idea dismissed by the medical profession. His work also included microscopic studies of striated muscle and plant structures.

Leeuwenhoek's *Philosophical Transactions* letters were translated in French and Latin journals. He himself had some of his work translated and published in Latin. As his findings generated incredulity, he kept a number of microscopes with preparations and welcomed visitors to verify his observations for themselves. He was possessive of his microscopes, however, refusing to give some to an eager nobleman. Leeuwenhoek's tradesman origins, somewhat coarse manners, ungrammatical Dutch, and lack of formal education isolated him from scientific society, yet his skillful observations and discoveries earned him a place in that society. He was elected a fellow of the Royal Society in 1680 and a corresponding fellow of the Paris Académie des Sciences in 1699. In 1716, the Louvain College of Professors awarded him a silver medal. The city of Delft also gave him special awards in addition to a pension. Leeuwenhoek died on 26 August 1723.

Bibliography: E. Ruestow, *The Microscope in the Dutch Republic*, 1996.

Kristen L. Zacharias

LEIBNIZ, GOTTFRIED WILHELM VON (1646–1716). A German philosopher, mathematician, logician, scientist, lawyer, and diplomat, Leibniz was one of the towering figures of seventeenth-century thought. He was born in Leipzig on 1 July 1646 and received his education at the Universities of Leipzig, Jena, and Altdorf, taking a doctorate in law in 1666. During his studies, he also engaged the major philosophical and scientific ideas of the day, including those of *Bacon, *Galileo, *Kepler, *Gassendi, *Hobbes, and *Descartes, initiating his lifelong dream of reconciling modern philosophy and science with Aristotle and Scholasticism. During this period, he completed his first mathematical work, publishing in 1666 *The Art of Combination*, which lays the mathematical groundwork for computers.

Completing his legal studies, Leibniz entered the service of the Elector of Mainz (1667–1672) and later that of the Duke of Hanover (1676–1716). On a diplomatic mission to the court of *Louis XIV in 1672, Leibniz met *Nicholas Malebranche, the leading Cartesian philosopher of the day; *Antoine Arnauld, the Jansenist theologian and author of the *Port Royal Logic*; and later *Spinoza. Traveling to England in 1673, he visited thinkers, such as *Boyle and Oldenburg, and constructed a calculating machine that he presented to the Royal Society. Returning to Paris, he continued his mathematical studies, discovering the foundations for differential and integral calculus in 1676, publishing in 1684. This led to a controversy with *Sir Isaac Newton, who had already discovered differential calculus but not published. The issue came to a head years later in an exchange of letters between Leibniz and Samuel Clarke, representing Newton before the Royal Society. *The Leibniz-Clarke Correspondence* was published posthumously.

Leibniz's discovery of calculus emerged from his study of mechanics and the relationship between mind and extension of space. Cartesian physics supposed extension to be an independent substance. Galileo, on the other hand, had shown that two falling bodies move at a constant rate of acceleration no matter how large or small the unit of extension. Extension can therefore be thought of as an arbitrary construct of the mind. The underlying reality, Leibniz concluded, is a continuum composed of infintesimally small units of force.

In the service of the Duke of Hanover, Leibniz initiated educational reform and founded the Academy of Berlin. He also served as a mining engineer (1680–1685), laying the foundation for modern geology, among other things hypothesizing that the core of Earth was molten. He also published on binary systems of numeration and topology. Leibniz was also one of the first Western thinkers to look seriously at Asian philosophy, publishing his "Preface to the *Novissima*

Sinica" (1697–1699) and a number of subsequent papers. Amid his scientific and mathematical investigations and official court duties, Leibniz developed his mature philosophical system in a series of books and papers, especially *The Discourse on Metaphysics* (1685), *The New System* (1695), the *Theodicy* (1710), the *Monadology* (1713), and the *Nouveaux Essais* (1701–1709), a running commentary on *Locke's *Essay Concerning Human Understanding*. Leibniz's philosophy grew out of his desire to reconcile science and religion. As a rationalist, Leibniz concluded that the infinitesimal units of force comprised a basic substance, which, because substance by definition is independent of other substances, means that all reality is either one substance, as Spinoza had concluded, or an infinite number of substances, as he concludes. Thus, reality is composed of an infinite number of independent substances, or monads, each an alternative universe, a sort of spherical mirror reflecting the other monads from its own perspective. Because they are independent of each other, each acts according to its own intrinsic nature in a preestablished parallelism or harmony with the others. Each monad acts to realize its own potential. Collectively, reality is the unity of monads. It is like the music produced by a symphony orchestra. Each instrument makes music according to its own score and time signatures, rather than playing in a casual relation with the other instruments, though the final result is a single sound. Thus, from the myriad activities of infinite possible universes emerges a single optimum harmony, the best of all worlds satirized by Voltaire in *Candide*.

Leibniz fell into disfavor with the new Elector of Hanover, soon to ascend the English throne in 1714 as George I. Neglected, Leibniz died in relative obscurity on 14 November 1716 in Hannover, Hanover.

Bibliography: R. Adams, *Leibniz: Determinist, Theist, Idealist*, 1994; H. Carr, *Leibniz*, 1960; L. Loemker, *Struggle for Synthesis: The Seventeenth Century Background of Leibniz's Synthesis of Order and Freedom*, 1972.

Thomas L. Cooksey

LELY, SIR PETER (1618–1680). After the death of Van Dyck in 1641, Lely emerged as the most important portraitist in England, painting many of the more significant members of the royal court and high society. Born Pieter van der Faes in Westphalia, his father, Johan van der Faes, was a Dutch infantry officer. The paternal family was well-to-do, living in The Hague in a home whose name, *in de Lelye* (because of its decorative carved lily), the painter adopted by the time he was nineteen. By 1647, the family had moved to London, and three years later Lely had settled for good in the Covent Garden district. Lely steadily made a name for himself within the court, leaving behind his early landscapes and subject paintings for the more lucrative venue of society portrait work. His *Charles I with the Duke of York* (1648) and the *Children of Charles I* (1647) display the influence of Van Dyck, but his personal style clearly asserted itself under *Cromwell's protectorate and into the Restoration period, as seen in his paintings of the *Cotton Family* (1660), *Anne Hyde, Duchess of York* (1661–1662), and *Sir Thomas Isham* (late 1670s). Throughout his career, he developed a large and active studio whose apprentices completed portraits for which he had painted the heads. He became known as an impressive collector as well; when his holdings were auctioned after his death, the catalogue listed 87 original works and over 170 copies of his own paintings. He also amassed a large selection of Dutch, Flemish, and Italian paintings of the sixteenth and seventeenth centuries, together with an impressive assortment of prints and drawings. Well known in London cultural and social circles, he was mentioned in Alexander Pope's *Imitations of Horace*, was praised in a poem by *Lovelace, and was said by *Pepys to be "a mighty proud man and full of state." Lely was knighted in January 1680, but died at his easel of apopolexy and was buried in St. Paul's, Covent Garden, on 7 December.

Bibliography: R. Beckett, *Lely*, 1951; M. Talley, *Portrait Painting in England: Studies in the Technical Literature before 1700*, 1981.

Christopher Baker

LEMERY, NICOLAS (1645–1715). Born into a Protestant family in Rouen, Nicolas Lemery was the fifth of seven children. His father, who held a minor legal post, died when he was eleven, and at the age of fifteen Nicolas went to live and work with his uncle, an apothecary. After serving out his apprenticeship, Nicolas continued his studies in such places as Paris and, ultimately, Montpellier where he lived with a master apothecary and taught chemistry to his students. He continued teaching chemistry after his return to Paris, where he earned the praise of savants, amateurs, and, in particular, women. He ultimately transformed these lectures into his most famous work, the *Cours de chymie* (1675), a book that applied a Cartesian mechanical explanation to chemistry. This book also delineated the differences between chemistry and alchemy. Enormously popular, it was translated into several languages and went through multiple editions, causing Bernard de Fontenelle to comment that it was as popular as a romance or a satire.

In 1674, Lemery acquired the post of apothecary to the king, a position he had to give up after the Revocation of the Edict of Nantes in 1685. His conversion to Catholicism, along with the acquisition of a medical degree from the University of Caen, allowed him to reinvent himself, however, and he concentrated, in the 1690s, on pharmacy; this work culminated in his *Pharmacopée universelle* (1697) and *Traité des drogues simples* (1698). Lemery was admitted to the Académie Royal des Sciences after its reorganization in 1699 along with his son, Louis. He spent much of the last years of his life in a variety of studies, most important a treatise on antimony, the *Traité de l'antimoine* (1707).

Bibliography: M. Bougard, *La Chimie de Nicolas Lemery*, 1999; H. Metzger, *Les Doctrines Chimiques en France du début du XVIIe à la fin du XVIIIe siècle*, 1923.

Michael R. Lynn

LENCLOS, ANNE DE (1620–1705). One of France's most famous courtesans, Anne de Lenclos was born in Paris 10 November 1620, the daughter of Marie-Barbe de la Marche and Henri de Lanclos. The latter, a man with musical talent and a dissolute lifestyle, disappeared in 1632 after committing murder. The young Anne, who became famous under the name Ninon, had her father's gift for music and obtained charity for her family by entertaining ladies in the Marais district.

Ninon's first paramour was Jean Coulon, a counselor in the Parlement of Paris; subsequent conquests included the Marquis de Villarceaux, the Marquis de Sévigné, and the latter's son. Her lifestyle and vocation accorded her a reputation as a moral degenerate and, to a lesser extent, as a beauty. Beyond this, her intelligence, wit, and sincerity bestowed upon her equal, if not greater, renown as a hostess and a friend.

In 1656, the queen mandated Ninon's temporary imprisonment in a convent for women of disrepute. The courtesan's life changed afterward, partly as a result of her punishment, but even more because she achieved financial independence. Her amorous liaisons attracted less attention, and her position in society was solidified. A brilliant conversationalist and hostess, she was able to cultivate friendships in illustrious circles. Friends at various points in her life included Boisrobert, *Scarron, Françoise d'Aubigné, Saint-Évremond, Fontenelle, and Madame de la Sablière. In addition, as *Louis XIV aged and turned toward religion, many at his court came to prefer the charms of the home of Mademoiselle de Lenclos, as Ninon was later called. The courtesan's fame was further consecrated in the writings of Somaize, *Mademoiselle de Scudéry, Tallemant des Réaux, and later, Voltaire.

Although she was held to be a philosophical as well as moral libertine, Anne de Lenclos died a Christian death on 17 October 1705.

Bibliography: E. Colomby, ed., *Correspondance authentique de Ninon de Lenclos*, 1968; E. Magne, *Ninon de Lenclos*, 1948

Laura L. Dennis-Bay

L'ESTRANGE, SIR ROGER (1616–1704). L'Estrange was the most prolific royalist propagandist of Restoration Britain; L'Estrange's royalism began early on; in 1639, he accompanied *Charles I on his march toward Scotland. In 1644, L'Estrange hatched a plan to capture King's Lynn. The secret plot failed, and L'Estrange was caught, tried for treason, and sentenced to death. He received a fourteen-day reprieve and then waited in jail for three years. This experience caused L'Estrange to begin his lifelong career of using the press to influence public opinion: In 1664, he issued two broadsides in his defense. The broadsides had little effect, but the governor of the prison, considering L'Estrange harmless, allowed him to escape. The governor was mistaken. L'Estrange immediately began to rally royalists in Kent. This attempt failed, and L'Estrange fled to Holland where he issued a broadside defending his valor. In 1653, L'Estrange returned to England and, in 1659, he published broadsides advocating the return of monarchy.

After *Charles II's restoration, L'Estrange began to propagandize on behalf of Anglicans, blaming religious factions for the late troubles and claiming that the Presbyterians wanted to enslave king and country. He also derided liberty of conscience as a forerunner to liberty of practice. In 1663, L'Estrange became surveyor of the printing press, which gave him the duties of censor and publisher of the sole newsletter. The ardent royalist pursued the first task zealously, later claiming that he had suppressed 600 books. L'Estrange had mixed feelings about his second task. In the first issue of his *Intelligencer*, L'Estrange set forth the proposition that when the kingdom was calm there should be no news printed, for "it makes the multitude too familiar with the actions and councils of their superiors and [gives] license to the meddling with government." Only the marked civil disorders convinced L'Estrange of the need for a royalist newspaper.

L'Estrange reached full stride during the exclusion crisis. From 1679–1681, over 64,000 copies of his tracts were printed. The main themes in his publications were that those who urged exclusion truly aimed to reestablish the Cromwellian republic and that the Popish Plot was a sham. This last stand caused the London anti-Catholic crowd to despise L'Estrange. They burned him in effigy and forced him to flee to the continent. Once things calmed down, he returned to England and continued promoting ultra-Anglican views until his death.

Bibliography: G. Kitchin, *Sir Roger L'Estrange: A Contribution to the History of the Press in the Seventeenth Century*, 1913; D. Turner, "Roger L'Estrange's Deferential Politics in the Political Sphere," *Seventeenth Century*, 13, no. 1 (1998): 85–101.

Brian Weiser

LEYBOURN, WILLIAM (1626–1716). William Leybourn, a self-described "friend to all that are mathematically affected," was a teacher, writer, and publisher of some of the first printed textbooks of the scientific revolution, writing works of geometry, algebra, trigonometry, surveying, construction, astronomy, and navigation. At the beginning of his career, Leybourn was joint author with astrologer Vincent Wing (1619–1648) of the first book of Ptolemaic astronomy written in English, *Urania Practica* (1648). His later compendia, such as *Cursus Mathematicus* (1690), and *Pleasure with Profit* (1694), discussed both the Copernican and Ptolemaic systems, and *Kepler's discoveries, if nothing about *Newton's recently published *Principia* (1687). Leybourn was one of the first English mathematicians to write works of mathematical recreation; his *Pleasure with Profit* promised to "recreate ingenious spirits and to induce them to make farther scrutiny into these sublime sciences, and to divert them from following such vices, to which Youth (in this Age) are so much inclined." Leybourn also made contributions in surveying and the construction trade; his *Compleat Surveyor* (1653) was a major work of the period, and his *Platform Guide Mate* (1667) served as a guide for the brick construction of the Fortune Theater in London. Leybourn himself completed a comprehensive survey of London and Westminister, providing some of the maps for John Stow's *A Survey of the Cities of London and Westminster* (1720). He was the creator of the first English guide to thirty-year amortization (*Panarithmologia*, 1693). Leybourn's works were reprinted often in the seventeenth and eighteenth centuries due to their clarity and organization, his writing and publishing illustrating ingenuity and industry.

Bibliography: C. Kenney, *William Leybourn, 1626–1716*, 1950; H. Lowood and R. Rider, "Literary Technology and Typographic Culture: The Instrument of Print in Early Modern Science," *Perspectives on Science* 2, no.1 (1994):1–27.

Anna Marie Roos

LEYSTER, JUDITH JANS (1609–1660). Acknowledged as the most successful woman painter in early-modern Holland, Judith Leyster was the daughter, not of an artist but of a small-ware weaver, who later became the owner of a brewery. Baptized in Haarlem on 28 July 1609, she learned reading, writing, and arithmetic before studying painting with *Frans Hals and his brother, Dirck. Acquiring the status of a master painter without the benefit of "an artistic family" and establishing a workshop—with some paying pupils—as a young, single woman, she is considered to be unique among Baroque women artists. On 11 May 1636 she married the genre painter Jan Miense Molenaer. Molenaer had also been greatly influenced by the Hals brothers and had already had an independent workshop of his own. Perhaps because of an outbreak of the plague in Haarlem in 1636–1637, Leyster and Molenaer moved to Amsterdam. Leyster became a housewife and a mother of three children, taking charge of the administrative and financial aspects of her husband's workshop and, apparently, no longer pursued her own craft.

Leyster was eclipsed by Frans Hals and forgotten for some 250 years. She regained some renown in 1893, attracting the attention of feminist scholars since the mid-1970s. Until then, some of her autograph paintings were displayed and sold as though they had been created by the much more famous Hals.

Like the Hals brothers and Molenaer, but unlike other Dutch women painters such as Clara Peeters (1594–after 1657), known for flowers and still lifes, and Rachel Ruysch (1664–1750), who exclusively painted flowers, Leyster was a genre painter. Despite two extant autograph portraits (one an unsigned self-portrait at the National Gallery, Washington), a surviving signed still life, and an extant signed watercolor of a tulip, most of her paintings are of men and women making music, drinking, smoking, and playing games of chance in taverns, as well as of playful boys and girls. *Serenade*, presenting the half-length figure of a male lute player in mid-song, and *Concert*, portraying a group of two men making music and a woman singing, are subjects that originated in Italy with *Michelangelo Merisi Caravaggio and were adapted first by the Utrecht painters and then by Frans Hals. *Jolly Toper*, portraying the half-length figure of a man who has been excessively drinking and smoking, is easily confused as an autograph image by Frans Hals as well. The other two well-known paintings by Leyster—*Man Offering Money to a Woman*, also known as *The Proposition*, and the aforementioned *Self-Portrait*—are quite exceptional. They have been viewed as especially innovative and feminist. The former shows a young maiden who concentrates on her sewing in order to ignore the advances of an ardent suitor, and the latter depicts the exceptionally lively and relaxed image of Leyster who, stopping her painting of a flute player for a moment, boldly turns toward and smiles at the viewer.

Bibliography: F. Hofrichter, *Judith Leyster: A Woman Painter in Holland's Golden Age*, 1989; W. Slatkin, *Women Artists in History: From Antiquity to the Present*, 1997: 86–90; J. Welu and P. Biesboer, *Judith Leyster: A Dutch Master and Her World*, 1993.

Yael Even

LILBURNE, JOHN (1614–1657). One of the major pamphleteers of the English civil war period, this political theorist printed tracts that significantly anticipate the writings of *John Locke and Thomas Jefferson. John Lilburne, born in 1614 to Richard Lilburne of Durham, was educated in the grammar schools of Auckland and Newcastle before being apprenticed in 1630 to a London clothier. Along the way, he educated himself by reading the Bible and various Puritan writers, whose theological positions he embraced zealously. In 1636, he met the notorious Puritan pamphleteer Dr. John Bastwick and assisted in the publication of his *Litany*. For this and other other activities with the Puritan propagandists of the period, Lilburne found himself a fugitive, actually leaving England for a brief period. When he returned in 1637, he was arrested and sent before the Star Chamber, which found him guilty of seditious behavior and had him whipped and placed in the pillory. Throughout his punishment, Lilburne voiced his opposition to the practices of the Church of England and garnered sympathy from the observers. Even while in prison, Lilburne managed to continue his activities as a pamphleteer. When the Parliament of 1640 met, however, his petition for release was presented by no less a figure than *Oliver Cromwell, and Lilburne was quickly released. For a time after his release, Lilburne placed his political activities aside, long enough to establish himself as a brewer and to marry Elizabeth Dewell. By May 1641, however, Lilburne was again in trouble for speaking against the king and forced to explain himself to the House of Lords.

When civil war broke out, Lilburne was commissioned as a captain in the army of Parliament and fought at Edgehill, where he was captured by Royalist forces and taken to Oxford for trial. Although found guilty of treason, Lilburne was released in 1643 as part of a prisoner exchange between the Royalist and Parliamentarian forces. Lilburne served for a time in the army commanded by Manchester, but left military service in 1645. By this time, Lilburne had become involved with the Leveller movement, which was creating considerable dissension within the ranks of the New Model Army. The Levellers rejected essentially all traditional claims to position that England's hierarchical society demanded, but they found that even those in Parliament whom they had supported were unwilling to carry their rejection of the king's traditional position to this extreme. As a result, the Commons viewed Lilburne as a threat and had him imprisoned in the Tower in 1646. Again, despite imprisonment, Lilburne managed to continue his activities as a pamphleteer, issuing attacks now against both the monarchy and the House of Commons, even the leadership in the military, which made him an enemy to the increasingly powerful Oliver Cromwell.

In 1649, shortly after the execution of *Charles I, Lilburne was charged with treason and brought to trial, largely because of his participation in writing and submitting to Parliament a document entitled *An Agreement of the People*, which argued for the basic rights of all citizens and which in many ways anticipates some of the most profound components of the American Constitution. Ironically, his defense was the same employed by the king during his own trial. Both men rejected the legitimacy of the judicial power being exercised by the House of Commons. For Lilburne, the strategy worked. Although he was acquitted of the charge of treason, Cromwell had Lilburne banished from England in 1652. While traveling in the Low Countries, Lilburne actually became friends with Royalists who had taken refuge there after the execution of the king. In 1653, Lilburne felt that the political climate in England was changing and returned. He was, however, captured and imprisoned. Though again acquitted of all charges, Lilburne remained first in the Tower and later at Dover until almost the end of his life. After being released, Lilburne converted to the Quaker faith, then died in 1657.

Bibliography: P. Gregg, *Free-Born John*, 1986; H. Lubasz, ed., *The Levellers in the English Revolution*, 1975.

Gerald Morton

LILLY, WILLIAM (1602–1681). William Lilly was the most popular English astrologer in the seventeenth century, his career paralleling astrology's neo-Platonic revival and subsequent decline after the Restoration and the rise of science. Although his mother's ambitions led to an excellent early education, his family's poverty resulted in Lilly's apprenticing with a Master Salter in London. Lilly subsequently married his master's widow, and his wife's death in 1633 left him with a sufficient income to pursue astrology. Lilly was known for three types of publications: almanacs (1641–1681) of large circulation; pamphlets that fused astrological prediction with Parliamentarian propaganda, such as *Merlinus Anglicus Junior* (1644); and his *Christian Astrology* (1647), the first work in England in which astrologers' methods of calculation and interpretation of celestial signs were revealed in the vernacular. Lilly built a reputation for predictive accuracy via the *Starry Messenger* (1645), whose prophecy of a great Parliamentary victory appeared on the day that news reached London of *King Charles I's defeat at Naseby. Lilly's unique knack for astrology, as well as clashes in print with George Wharton, the Cavalier astrologer, gave him high name-recognition; he was referred to in praiseworthy and derogatory terms in *John Evelyn's and *Samuel Pepys' *Diaries*, as well as in works by *Samuel Butler, Daniel Defoe, *John Dryden and *Thomas Hobbes. However, Lilly's incorrect prognostication in the *Annus Tenebrosus* (1652) about the dire effects of a total solar eclipse called "Black Monday" damaged his reputation. Because of his previous prophecies about Charles I's death, in 1660 Lilly was examined by a Parliamentary committee enquiring into the execution of Charles I but subsequently discharged. His last years were spent writing his autobiography (1668), working with Elias Ashmole to recover the manuscripts of John Dee, and practicing medicine. Lilly died in Hersham in 1681.

Bibliography: W. Burns, " 'The Terriblest Eclipse That Hath Been Seen in Our Days': Black Monday and the Debate on Astrology during the Interregnum," in *Rethinking the Scientific Revolution*, ed., M. Osler, 2000:137–152; A. Geneva, *Astrology and the Seventeenth-Century Mind: William Lilly and the Language of the Stars*, 1995.

Anna Marie Roos

LIPSIUS, JUSTUS (LIPS, JOEST) (1547–1606).

A Flemish humanist scholar who became the foremost proponent of modern neo-Stoicism, Lipsius exerted a significant influence on Montaigne, *Francis Bacon, and *Joseph Hall. Born on 18 October 1547 to the burgomaster of Overyssche, near Brussells, he was reared as a Catholic and studied at the University of Louvain and at the Jesuit College at Cologne. But at age twenty-four he rejected the religion of his youth and began teaching history and rhetoric at the Lutheran University of Jena. Two years later, in 1574, he returned to Cologne and Catholicism, where he finished that year his first edition of Tacitus (*Taciti Opera*). By this time, his own prose had become so thoroughly imbued with Tacitean, Senecan, and anti-Ciceronean features that its curt, terse style became known as "Lipsian." He finished his training in law at Louvain in 1576, but political and religious conflict compelled him to move once again, this time to the new Calvinist University of Leiden, where he again abjured his Romanism.

Lipsius spent eleven years in Leiden (1579–1591), completing a variety of works that established his fame as an authority on the Stoics. In *De Constantia* (1584), he sought to resolve the differences between Roman Stoicism and Christianity by arguing that the stoic concept of fate was compatible with Christian free will. His *Politicorum* (1589) praised a monarchy with limited absolutist powers, but his harsh recommendations for curbing social discord aroused controversy, as did his support of the monarch's prerogative to determine the state religion. He left Leiden in 1591 and returned to Catholicism and a professorial post in Latin and history at Louvain the next year. In 1594, he became royal historiographer to Philip II, and in 1605, an honorary state councilor of Brabant. While at Louvain he also issued two essays on warfare, *De Militia Romana* (1595) and *Poliorceticon* (1596), as well as a pair of treatises on Stoicism, *Manuductionis ad Stoicam* (1604) and *Physiologiae Stoicorum* (1605), and an edition of Seneca's *Opera Omnia* (1605). Throughout his peripatetic scholarly career, Lipsius sought to employ neo-Stoicism as a bridge between quarrelling Reformation and Counter Reformation forces. This fact, together with his renewed emphasis upon natural philosophy, made him a contributor to the growing secularism of seventeenth-century Europe.

Bibliography: M. Morford, *Stoics and Neostoics: Rubens and the Circle of Lipsius*, 1991; J. Saunders, *Justus Lipsius: The Philosophy of Renaissance Stoicism*, 1955.

Christopher Baker

LLOYD, EDWARD (fl. 1688–1726). Edward Lloyd's name lives on in the London-based insurance company "Lloyd's." Lloyd opened a coffee house in Tower Street no later than 1688, moving the business to Lombard Street at Abchurch Lane in 1692. This location succeeded several Cornhill establishments as the center of maritime insurance, a commodity increasingly in demand due to the growth in English shipping and the war with France. Lloyd's direct involvement in insurance ventures was probably minimal: While the coffee house was the site of preliminary contacts, the actual underwriting took place at the Exchange. Above all, Lloyd's Coffee House provided an opportunity for individual underwriters to meet without the rules that governed companies, partnerships, and the Exchange. Lloyd can also be credited with employing runners to retrieve news from the wharves and compiling handwritten "ships' lists" for his customers. This practice developed into the publication of *Lloyd's News* of 1696–1697 (revived as *Lloyd's Lists* in 1726) and *Lloyd's List and Shipping Gazette* in 1734, which, under various names, has been published continually ever since. An Edward Lloyd who died in 1734 and was interred at St. Stephen's Walbrook is thought to be the former coffeehouse proprietor.

Bibliography: A. Ellis, *The Penny Universities: A History of the Coffee-Houses*, 1956.

Michael R. Hutcheson

LOBO, FRANCISCO RODRIGUES (1578-80?–1621). Francisco Lobo, one of the major literary figures of seventeenth-century Portugal, was known as the Portuguese Theocritus. Francisco Rodrigues Lobo was born in Leiria, Portugal, and died late in 1621. He accidentally drowned on a voyage on the Tagus River when the barge on which he was traveling overturned. Francisco Lobo received a degree in law at Coimbra, but never practiced. The pastoral poet spent most of his life in the rural retirement of Leira, although he was well known to the most prominent families of Philipine Portugal. After his graduation in 1602, Lobo joined the elegant circle of courtiers at the palace of the Duke of Bragança. His experiences in the royal court are described in his major work *Corte na aldeia e noites de inverno* (1619). While in his bucolic retirement at Leiria, he continued to receive the protection of the Bragança family, dedicating his epic poem *O Condestabre* (1610) to Duke Teodósio and *Corte na aldeía* to D. Duarte, the Duke's brother.

There are few bibliographical references of Francisco Rodrigues Lobo's life. It has been speculated that he experienced a supposed frustrated love affair or a family opposition to his matrimonial pretensions to a lady of the Marquis of Vila Real family. According to this theory, the grieving hero of his pastoral trilogy would be the alter ego of the poet. His personal political views in a time when Portugal was a Spanish province had also given rise to some controversy. Most critics considered Lobo's epic poem *O Condestabre* an expression of the author's opposition to the domain of the House of Austria through his portrayal of the fourteenth-century hero of national autonomy. However, Rodrigues also wrote an encomiastic welcome in verse to Philip III on the occasion of the monarch's visit to Lisbon in 1619: *La jornada que la Majestad Católica del Rey Felipe III hizo al reino de Portugal*, published posthumously in 1623.

What we know about the poet is that he was publicly accused of being a "new Christian," that is to say, having Jewish ancestry. The accusing finger in the time of post-Tridentine Portugal was Tomás Noronha, who composed a cruel, mock sonnet to the recently deceased poet in the name of Lobo's fictional shepherd Lereno. While Rodrigues was never denounced to the Inquisition, the Holy Office opened a case against his sister and his brother. The possible Jewish background may explain Lobo's relatively marginal place in the literary and social life of Portugal, as well as his early retirement. It has also been suggested that Lobo's secular tone and occasional misanthropy might have been the result of his awareness of his precarious situation.

Lobo's first book of poems, *Romances* (1596), a collection of ballads written in the Baroque style of the Spanish poet *Luis de Góngora, reveals a refined sensibility in describing the moods of nature. His best works are the eclogues interpolated in his trilogy of pastoral novels, *Primavera (Spring* 1601), *O Pastor Peregrino (The Wandering Shepherd,* 1608), and *O Desencantado (The Disenchanted,* 1614). These poems combine pleasing descriptions of the countryside with witty dialogues between shepherds and shepherdesses on the tricks of love, following the model established by Sannazaros' *Arcadia*, as well as Montemayor's *Diana*. Lobo's masterful work in prose is the lively and elegant dialogue in *Corte na Aldeia (Village Court)*, in which a young noble, a student, a wealthy gentleman, and a man of letters discuss questions of philosophy, letters, manners, and society. In *Corte na Aldeia*, the author demonstrated his reformist concerns and his disappointed vision of life.

Bibliography: R. Preto-Rodas, *Dialogue and Courtly Lore in Renaissance Portugal*, 1971.

Angela Morales

LOBO, JERÓNIMO (1593–1678). Lobo was a seventeenth-century Portu-
guese Jesuit missionary, renowned for his travels and literary accounts. Father
Jerónimo Lobo's *Itinerário*, the narration of his voyage to Abyssinia (modern
Ethiopia) during the years 1621 to 1635, was translated into English in Samuel
Johnson's first book, *A Voyage to Abyssinia* (1735). Johnson used a previous
French version of Lobo's manuscript published by Joachim Le Grand in 1728.
Jerónimo Lobo was born in Lisbon in early 1595, the third son of Francisco
Lobo de Gama, governor of Cape Verde. He began his studies of theology at
the College of Jesus in Coimbra in 1619. By 1621, his request to become a
missionary had been granted, and in that year he was ordained a deacon and a
priest. Lobo's first attempt to reach the Indies during the same year was a failure.
The ill voyagers had to return to Lisbon five months later, in October 1621.
The expedition was filled with dangers, misfortunes, and sickness that confined
the missionary to bed for a long recovery period.

Nevertheless, Lobo was ready the following year to try his second expedition
in 1622 with a new viceroy for Portuguese possessions in India, the Count of
Vidigueira. Lobo's experiences in this period are recorded in the *Itinerário*.
After rounding the Cape of Good Hope, the traveler arrived in Mozambique on
the east coast of Africa to continue for the settlement of Goa, command post
for the missions in India and eastern Africa. Lobo remained at Goa, on the west
coast of India, studying theology until 1624, when he sailed from India to the
coast of what is now Somalia in a frustrated attempt to penetrate Abyssinia.
Moving southward down the African coast, Lobo reached the island of Patta,
near the equator, where a Jesuit mission had long been established. From there
Lobo returned to Goa in 1625 to join the Patriarch Alfonso Mendes. Accompanying the Patriarch, Lobo crossed the Arabian Sea, landed at the Red Sea port
of Beilul, and started his dangerous expedition into the Abyssinian empire.

The Portuguese Jesuit spent the next years laboring for conversion and re-
baptism of the inhabitants of Abyssinia. In 1626, Lobo undertook successfully
the mission of recovering the bones of Cristovao da Gama who was murdered
one century before. Two years later, as he tells us in the account of his travels,
Lobo was bitten by a cobra and became seriously ill. The Jesuit also explored
and found the source of the Nile River, the stream known as the Blue Nile.
Lobo continued his missionary labors with the occasional protest from the natives. But when the Christian Abyssinian king died, all the Jesuits were finally
expelled from the Abyssinian territory. Lobo, along with the patriarch and other
missionaries, were imprisoned by the Turks and later captured by Dutch pirates.
Lobo returned to Lisbon in 1636 after a long, strenuous journey first eastward
to Goa and from there to Portugal, where he kept seeking help for the embattled
missions.

For centuries, Jerónimo Lobo's travel accounts have been the source of adaptations and translations. His reputation increased when the British ambassador
to Portugal, Sir Robert Southwell, put him in touch with the Royal Society.

Encouraged by Southwell, Lobo wrote short essays on topics such as the Nile River, the unicorn, the reason for the name "Red Sea," the palm tree, and the legendary priest and monarch Prester John. These works were translated into English and published anonymously in 1669 by the society.

Bibliography: M. da Costa, ed., *Lobo's Itinerário E Outros Escritos Inéditos*, 1971; J. Gold, ed., *Samuel Johnson's* A Voyage to Abyssinia, 1985; D. Lockhart, trans., *The Itinerario of Jeronimo Lobo*, 1984.

Angela Morales

LOCKE, JOHN (1632–1704). John Locke was born at Somerset in 1632 and educated at Christ Church, Oxford University, earning bachelor's and master's degrees in 1656 and 1658. He studied chemistry under *Robert Boyle and later participated in medical research with *Thomas Sydenham, but he did not take a doctor's degree. From 1661 to 1664, he held a lectureship in philosophy and classical languages at Oxford and, in 1665, served as secretary to Sir Walter Vane, British ambassador to Brandenburg. Following his return to England in 1666, Locke became a personal advisor, physician, and secretary to Lord Ashley, the first Earl of Shaftesbury. Through political connections, Locke was appointed to a variety of public offices, including secretary to the Council of Trade and Plantations. He was elected a fellow of the Royal Society in 1668. In 1683, after Shaftesbury had gone into political exile for opposition to the Stuart crown, Locke took political asylum in the Netherlands, where he wrote A *Letter Concerning Toleration*, which argues against religious absolutism, and completed his most influential work, *An Essay Concerning Human Understanding*, an empirical study of the human mind. After returning to England in 1689 under the safety of *William and Mary, he published *Two Treatises on Government*, which argues against political absolutism, and was appointed Commissioner of Appeals and then Commissioner of Trade in 1696. When his health deteriorated a few years later, he returned to a country estate with his friend Sir Francis Masham. He died in 1704, while still at work on *A Fourth Letter for Toleration*.

Locke wrote on politics, religion, education, ethics, and economics, but he is chiefly remembered for his pioneering investigation into the capabilities and limitations of human understanding. *An Essay Concerning Human Understanding* (1690), which Locke began drafting around 1670, holds a permanent place among the most influential works of cognitive philosophy in the Western canon. In it, Locke theorizes that in order to understand what we think, it is first necessary to understand how we think. Thus, he proposes to determine the origins and scope of human understanding by tracing the development of knowledge in the mind, from sensory perception of physical objects to abstract ideas. Locke believed that the mind acquires knowledge from four sources: inscription, tradition, revelation, and sensation. The first three are faulty because they constitute expressions of personal opinion that cannot be verified or falsified. All the human mind can verify as a true fact is that which can be tested through sensory experience, with sight and touch having the most accurate empirical abilities of the five senses. As a proponent of sensationalism as advanced by *Pierre Gassendi and *Thomas Hobbes, Locke concluded that the mind does not possess innate ideas. Instead, the mind works mechanically, in that ideas originate from sense impressions of the physical world that are imprinted upon human consciousness. Although ideas exist only in the mind, they do not occur independently from the sensory faculties. When the senses perceive a physical object, the mind proceeds to think, which gives rise to abstract ideas, which in turn intermingle to create higher forms of conceptualization. Seeing a tree, for example, can produce abstract ideas about mathematical shape and density.

Locke's empirical theory of the human mind underlies much of his religious and political thought. *A Letter Concerning Toleration* (1689), marking the religious phase of Locke's empirical philosophy, shows why religious differences should be tolerated unless they threaten the stability of society. The work combines empiricism and Cartesian rationalism to refute the intellectual grounds on which sectarian groups sought to establish and impose on society their own absolute articles of religious faith. According to Locke, articles of faith, such as the conviction that the Bible is historically infallible, are matters of opinion formed on untestable and, thus improbable, hypotheses. They cannot be proved or disproved; they can only be believed or disbelieved. From this premise, Locke argues that, while people have the right to form diverse religious doctrines, they do not have the right to force others to accept them. It follows that religious absolutism and persecution have no place in civil society because they are founded on faulty reasoning. Even so, Locke did not believe in tolerating atheists or Roman Catholics in England because of the social disruption they might cause.

While Locke's four treatises on religious toleration (1689–1704) are concerned primarily with placing limitations on religious absolutism, the need to limit absolute political power is the central thesis of his *Two Treatises of Government* (1690). In issues of political thought, Locke followed and expanded upon Hobbesian theories that governmental authority should be delegated by the subjects and that the state should be organized through a compact of obligation. Locke's objective in *Two Treatises* is to justify social resistance to the absolute power and authority of the monarchy. The main condition that justifies social resistance exists when a ruler abuses power and becomes a tyrant. For Locke, tyranny exists in two forms. First, tyranny is present when a ruler pursues selfish ambitions over the needs of the commonwealth. This pursuit constitutes tyranny because it violates the fundamental premise that all political power and authority must be selflessly exercised for the good of the people. Second, tyranny is present when a ruler exercises authority over what Locke calls Right. This occurs when a ruler, without consent of the law, deploys corporate force against citizens. Force of this nature violates the basic human rights of the individual. Locke's rational arguments for the rights of individuals to enjoy liberty and own property, his justification for revolution against tyranny, and his theories of a social contract between citizens and government influenced the colonial framers of the American Declaration of Independence and the Constitution.

Bibliography: G. Rogers, ed., *Locke's Philosophy: Content and Context*, 1994.

James Norton

LOCKE, MATTHEW (1621/22–1677). Matthew Locke received training as an Exeter Cathedral choirboy under Edward Gibbons (*Orlando's brother). At eighteen, Locke traveled in the Netherlands, possibly as a Royalist conscript; it was probably then that he converted to Catholicism. Locke's reputation as a composer of dramatic music began in 1653 with *James Shirley's masque *Cupid and Death*, whose music he and Christopher Gibbons (Orlando's son and fellow Exeter chorister) composed. In 1656, during Puritan rule, *William Davenant produced *The Siege of Rhodes*, called the first English opera, as a "moral representation," with music by Locke, Charles Coleman, Henry Cooke, George Hudson, and Henry Lawes. Davenant engaged Locke for music for further plays, including *The Cruelty of the Spaniards in Peru* (1658) and *The History of Francis Drake* (1659). During the interregnum, Locke was *Henry Purcell's mentor and was friends with other leading musicians. At the Restoration, Locke was evidently in favor with *Charles II, for the king granted him several positions— private composer in ordinary to the king, composer in the wind music, composer for the violins, and organist in the queen's Catholic chapel. Soon after Locke's death in 1677, Purcell assumed Locke's post as composer for the violins.

Bibliography: R. Harding, *A Thematic Catalogue of the Works of Matthew Locke with a Calendar of the Main Events in His Life*, 1971; L. Hulse, "Matthew Locke: Three Newly Discovered Songs for the Restoration Stage," *Music & Letters* 75 (1994): 200–213.

Susan Treacy

LORRAIN (LORRAINE), CLAUDE (1600–1682). Though Claude Lorrain is regarded as a French painter, he spent most of his working life in Rome. Lorrain was trained by Agostino Tassi; he was, however, more influenced by Paul Bril and Adam Elshiemer. Although Lorrain did not reside in France, he had numerous well-connected patrons there and throughout Europe. His painting, along with *Poussin's, became the basis for a distinctly French landscape painting style derived from Italian and Dutch sources.

Both Lorrain and Poussin created idealized settings at a distance from reality to ensure that the viewer would be inclined to first contemplate the narrative and then the beauty of the aesthetic presentation. Poussin stressed figure groupings and their consequential importance in the narrative over the beauties of the environment, whereas Lorrain, especially later in his career, allowed the ideal landscape itself to predominate so that the narrative revealed itself gradually.

During the 1630s, Lorrain painted a series of harbor views. *Morning in the Harbor* (1634) portrays a group of workmen in the foreground alongside a small group of travelers who appear to be waiting to board a ship. The rising sun illuminates the travelers and defines a deep space through the harbor to the open sea, alluding, perhaps, to distances as yet untraversed. By the 1640s, Lorrain's use of illumination had evolved considerably. *Landscape with Apollo and Marsayas* (1639–1640), while similar in composition to *Morning in the Harbor*, has a complex subtle lighting scheme. The sun's orb is no longer directly seen; rather, it is set outside the frame to the left, creating a raking light. Here, as in *Morning in the Harbor*, forms gradually lose detail in the distance, contributing to a sense of space, but in the latter painting, more levels of illumination and layers of space are evident. The volume of objects is preserved because light rakes over, or passes through them, preserving their surface characteristics. Apollo sits in judgment before Marsayas, who is about to be flayed for his presumptuous belief that his ability to play the flute was greater than Apollo's. Though softly illuminated, Marsayas's tightly bound body stands against a darkly wooded glade that serves as a reminder of his gloomy prospects.

Lorrain continued to expand the breadth of his painting to include all times of the day. *Night* (*Landscape with Jacob Wrestling the Angel*) (1672), one of Lorrain's more ambitious works, draws upon the story of Jacob in Genesis 32. As day breaks, Jacob is persuaded to release an angel, with whom he has struggled all night, in exchange for a blessing. The angel agrees, renaming Jacob "Israel." The newly renamed Jacob then pronounces, "I have seen God face to face; a soul has been saved!" From the east, at the extreme left of the picture, the breaking dawn illuminates Jacob and the angel who are standing in the left foreground. Weaker light falls on a group of spectators at the center of the picture. Finally, at the extreme right and right center, barely visible groups of shepherds lead their flocks to pasture. Israel, now transformed, goes on to father sons who found the twelve tribes of the nation of Israel. With this painting, Lorrain creates illumination that is descriptive and expository, signaling a new day and a renewed life.

Throughout his life, Lorrain remained a painter of ideal, narrative landscapes. Over time, he began to integrate the natural world more fully in his pictures. While he followed the usual custom of painting his finished pictures in the studio, Lorrain's close observation of nature is evident in the numerous sketches that he made out of doors. Lorrain strove to transcend the rhetorical models he inherited for landscape painting. He succeeded in painting in such a manner that ideal figures became a part of the environment and subject to its constraints. Certainly Poussin was a greater, more knowledgeable painter in the classical manner. It was, however, left to Lorrain to enlarge the scope of nature in painting. By the eighteenth century, classicism had faded as painters turned to the world of the eye. While Poussin is still admired, it is Lorrain who is emulated. Lorrain's engagement with nature inspired the direct confrontation of the natural world resulting in pictures done entirely out of doors by John Constable and

J.M.W. Turner in the late eighteenth and early nineteenth century. Their work, in turn, gave rise to the Barbizon School and to Impressionism. What began with Lorrain's movement away from purely ideal landscape toward a reverent, if somewhat parklike vision of nature, ironically culminates in the late nineteenth century with a detached view of nature from which anecdote is finally banished.

Other works by Lorrain of note for their advances in craft or content include *Coast View with Apollo and the Cumaean Sybil* (1648), *Christ on the Road to Emmaus* (1660), *Evening (Landscape with Tobias and the Angel)* (1663), and *Morning (Landscape with Jacob, Rachel and Leah at the well)* (1666).

Bibliography: S. Daniel and N. Serebriannaya, *Claude Lorrain: "Painter of Light,"* 1995; A. Mérot, *French Painting in the Seventeenth Century*, 1995.

John Schmidt

LOUIS XIV (1638–1715). Louis XIV became king of France at the age of five on 14 May 1643 upon the death of his father Louis XIII. The start of his reign was dominated by the Queen Mother, Anne of Austria, and Cardinal Mazarin, who had succeeded *Richelieu as prime minister after Richelieu's death in 1642. With the crushing of the five-year series of rebellions known as the Fronde in 1653, and the end of the Thirty Years' War in 1648, major internal and external disturbances to his reign were removed. He married his cousin, the Infanta Maria Theresa, daughter of *Philip IV of Spain, at the age of twenty-one, and in 1660—the same year in which *Charles II restored the English monarchy—he began the absolutist policies that led to France's preeminence as the leading European political and cultural center until his death. The Fronde had confirmed his distrust of the old nobility, and his assumption of greater personal power was defended by the divine-right principles espoused in the writings of *Jacques Bossuet. Mazarin's death in 1661 freed Louis to seek and gain unprecedented personal power, despite enjoying such effective ministers as Fouquet, *Colbert, and Lionne. He built upon the central authority of the monarchy, which Richelieu had crafted under his father's reign, extending it through such measures as the *lettre de cachet*, royal commands for imprisonment without trial. His court became the glittering center of both French and European culture, aware of its role as a continental arbiter of taste and led by a king who had famously declared in 1651 "L'état c'est moi" ("I am the state"). His wife had to contend with his various mistresses, of whom *Marie de Maintenon became his de facto wife after the death of Maria Theresa.

Seven years after his marriage, Louis became embroiled in a series of costly wars. The War of Devolution (1667–1668) sought to recapture Spanish territories on France's Belgian border. Louis was repulsed by an alliance of English, Dutch, and Spanish forces, but the Treaty of Aix-la-Chapelle (1668) did grant him several border towns. Four years later, in the Dutch War, Louis was foiled when William of Orange opened the dikes and flooded northern Holland. The treaty of Nijmegen gave France additional border territories. In 1688 in the War of the League of Augsburg, Louis attempted to annex lands on his eastern border. Despite his victories, the Treaty of Ryswick (1697) stripped Louis of most of his conquered land, though he retained the Alsace and the city of Strasbourg. The War of the Spanish Succession (1701–1714) was the longest of Louis's wars. Leopold, Hapsburg Holy Roman Emperor, declared that his son, rather than Louis' grandson Philip of Anjou, should succeed the childless Charles II of Spain. But the vast increase in French global influence that would have resulted was opposed by a coalition of England, Holland, and Leopold. England defeated the French at several key battles, notably at Blenheim under John Churchill, Duke of Marlborough. At the Peace of Utrecht in 1713, Philip of Anjou was decreed king of Spain, but France lost a number of its North American holdings to England, which kept Gibraltar and Minorca.

These wars of expansion left the French treasury sorely depleted, and his revocation of the Edict of Nantes also resulted in the emigration of French Huguenots, further aggravating France's relations with Protestant powers. The close of Louis's reign saw a court far less grand and a country much less confident than it had been two generations earlier. Having reigned longer than any other European monarch, Louis XIV died on 1 September 1715. The vast concentration of political power he had amassed within the monarchy was to be a key factor in provoking the anger of the French revolutionaries at the end of the century.

Bibliography: O. Bernier, *Louis XIV: A Royal Life*, 1987; F. Bluche, *Louis XIV*, 1990; P. Campbell, *Louis XIV, 1661–1715*, 1993; I. Dunlop, *Louis XIV*, 2000; A. Lossky, *Louis XIV and the French Monarchy*, 1994.

Christopher Baker

LOVELACE, RICHARD (1618–1657). Richard Lovelace was born to a venerable and affluent family in either Woolwich, Kent, or Holland. He was the consummate example of a Cavalier poet; he was a soldier and courtier first and a poet second. As a devout Royalist, Lovelace remained loyal to *Charles I until the throne was seized and the king beheaded in 1649. Lovelace's polished verse is infused with a courtly charm and grace, while his lyrics reflect the wit, charm, sensuality, and chivalry so indicative of the Cavalier genre.

Lovelace was educated at Charterhouse, Gloucester Hall, Oxford, and at Cambridge University. While at the university, Lovelace personified the quintessential Renaissance figure, with a dilettante's interest in painting, music, philosophy, and letters. While at Oxford, he wrote a comedic play entitled *The Scholars* (1635), of which only the prologue and epilogue survive. In 1639–1640 as civil unrest and insurrection against the court of Charles I escalated, Lovelace left England and accompanied General Goring's regiment on the first expeditions to Scotland. It was there that he wrote one of his most famous poems, "To Lucasta, Going to the Wars." In this same year, he wrote a play loosely based on personal experience entitled *The Soldier*. The fact that the play is a tragedy is indicative of Lovelace's mental state at the time of the play's composition. In 1641, Lovelace vehemently and publicly rejected a pro-Parliament, anti-Episcopacy petition at a civic gathering. Then, in 1642, with the nation on the verge of civil war, he presented an anti-Parliamentary, Royalist-supporting petition to an already angry House of Commons, for which he was imprisoned in the Gatehouse in London. There he wrote some of his most renowned and contemplative poetry, including "To Althea from Prison" in which he postulated "Stone walls do not a prison make / Nor iron bars a cage." In 1646, Lovelace was wounded while fighting for the French against the Spanish at Dunkirk. In June 1648, he was imprisoned once again, this time at Peterhouse in England. He was released ten months later in April 1649, poverty-stricken, unemployed, and thoroughly dejected. Ironically, also in 1649, his most celebrated series of poems entitled *Lucasta* was published. In 1657, Lovelace died of unknown circumstances, but it is speculated that he died of despondence due to his abject poverty. In 1659, *Lucasta, Posthume Poems* was published with his remaining writings added.

Bibliography: M. Weidhorn, *Richard Lovelace*, 1970.

David M. Rosen

LOWER, RICHARD (1631–1691). Richard Lower was born in Cornwall in 1631 and educated at Oxford, where he received a medical degree in 1665. He became a member of the Royal College of Physicians and was elected to the Royal Society in London in 1667. Though a very successful physician, Lower's physiological investigations are more notable. Before the achievements of Joseph Priestley and the revolutionary chemical work of Antoine Lavoisier, the English physician recognized the role of the lungs in discharging waste products and the absorption of oxygen, identified by him in vitalist terms as a "spirit of the air." His experimental research on heart muscles, published in *Tractatus de corde* (1669), secured a better understanding of that organ's contracting function, and he was probably the first to recognize the endocrine role of the pituitary gland.

Dogs were Lower's favored experimental subjects (he was a prominent vivisectionist), but after having transfused blood from one animal to another successfully, he repeated the experiment with humans. Though initially successful, Lower soon discovered the procedure's limitations and ended the practice. The physician's aggressive research expanded *William Harvey's work and helped lay the foundation for the close relationship that would emerge between the sciences of chemistry and biology. Among those with whom he collaborated and assisted were *Robert Hooke and *Robert Boyle, his contemporary at Oxford and in London.

Bibliography: R. Frank, Jr., *Harvey and the Oxford Physiologists: Scientific Ideas and Social Interaction*, 1980; E. Hoff and P. Hoff, "The Life and Times of Richard Lower, Physiologist and Physician," *Bulletin of the History of Medicine 4* (1936): 517–535.

Joan Klobe Pratt

LULLY, JEAN-BAPTISTE (1632–1687). The son of a poor Florentine miller, Jean-Baptiste Lully was taught the rudiments of music by a Franciscan monk. As a boy, he also learned to play the guitar and later the violin. Impressed by his musical talent, the Chevalier de Guise took Lully to Paris in 1646 as a page to Mademoiselle d'Orléans, a cousin of *Louis XIV. Lully rapidly adapted himself to the manners and ways of the French court and quickly mastered the language, although he could never rid himself of a marked Italian accent. When he set to music a sarcastic poem reflecting unfavorably on his patroness, he was promptly dismissed, but soon thereafter entered the service of the young Louis XIV. In his new position, Lully contrived to obtain instruction on the harpsichord and in composition. He regularly attended concerts and opera at the royal court, and developed a passion for the theater, which became the pivotal factor in his professional career.

Lully rose fast in royal favor, becoming the preferred composer of court ballets for numerous occasions. In several of these productions, Louis XIV himself participated next to Lully, who danced and acted as "M. Baptiste." In 1661, he received the lucrative post of composer to the king, and the next year was appointed as maître de musique of the royal family. A few weeks later, he married Madeleine Lambert, daughter of a prominent court musician. Between 1663 and 1671, Lully wrote music for several comic ballets by *Molière, which foreshadowed the development of opéra-comique. In 1672, he obtained letters patent for the establishment of an "Académie royale de musique," the precursor of what was to become the Grand Opera.

Lully was the first in a line of operatic reformers who attempted to elevate dramatic action above superficial musical effects. In the theater, he performed not only the composer's normal functions but also acted as director, stage manager, and mechanical technician. From 1672 onward, he worked in close cooperation with a congenial librettist, *Philippe Quinault, who proved to be a sympathetic collaborator over the next fourteen years. That an Italian-born musician should have become the virtual founder of French opera is one of the supreme ironies of music history.

In terms of his personal character, Lully was haughty, arrogant, and temperamental. With no tolerance for opposition, he was motivated primarily by selfish ambition, generally unrestrained by moral considerations. He knew how to be submissive to those in power, and he often achieved his goals by servility and flattery. Nevertheless, thanks to his irrepressible energy and his disregard of all obstacles, he effected revolutionary transformations in opera and changed the historical course of French music. His death resulted from what initially appeared to be a trivial accident. While conducting a *Te Deum* in celebration of the king's recovery from a severe illness, Lully inadvertently struck his foot with a sharp-pointed cane used as a baton. An abscess soon developed, and gangrene eventually set in. After refusing to allow amputation and after several months of suffering, he died in his own home of blood poisoning.

Bibliography: J. Fairleigh, "Lully as 'Secrétaire du Roi,' " *Bach* 15, no. 4 (1984): 16–22; J. Heyer, ed., *Jean-Baptiste Lully and the Music of the French Baroque: Essays in Honor of James R. Anthony*, 1989; J. Newman, *Jean-Baptiste Lully and His Tragédies lyrique*, 1979.

James P. Fairleigh

M

MABILLON, JEAN (1632–1707). Mabillon, Benedictine monk and diplomatic scholar, was born on 23 November 1632 at Saint-Pierremont in the Ardennes, educated at Reims, and ordained at Corbie in 1660. In 1664, Mabillon was placed at St. Germain des Pres to collaborate with Jean Luc d'Archery in editing the works of Saint Bernard, published in Paris as *Sancti Bernardi opera* (1667). Following d'Archery's plan, Mabillon collected and then organized documents pertaining to the lives and works of the Benedictine saints; these documents provided the basis for the chronologically organized *Acta Sanctorum ordini Sancti Benedicti* (Paris, 1668–1701). Mabillon's prefaces to the *Acta* placed these lives, for the first time, in the context of the ecclesiastical and civil history of the early middle ages. Beginning in 1672 with Champagne, Lorraine, and Flanders, Mabillon undertook a series of manuscript hunts. In Luxeuil, he came upon a Gallican Lectionary, which prompted the study set out in *Liturgia gallicana* (Paris, 1675). In *De re diplomatica* of 1681 and its 1704 supplement, Mabillon established the principles of authenticating and dating medieval documents, thus inaugurating a more organized, scientific methodology for historiography. His study of Latin paleography, also set out in this work, remains among the principal texts on the subject.

In 1682, Mabillon made his way through Burgundy; in 1683, through Germany and Switzerland, and in 1685 through Italy to purchase whatever books and documents he could for the king's library. These travels were recorded in *Itinerarium Burgundicum, Itinerarum Germanicum* (1685), and *Musaeum Italicum* (1687–1689). Mabillon twice defended the propriety of this sort of work in *Traité des études monastiques* and *Reflexions sur la réponse de M. l'abbé de la Trappe* (1691–1692). The publication of the *Annales ordinis sancti Benedicti* began under Mabillon's direction in 1703 and continued, after his death in 1707, under the guidance of R. Massuet and E. Martene. The work of this methodical monk set the patterns for manuscript study, established the principle of setting men and their deeds in the context of their time, and put in order the lengthy and complex story of the significant Benedictine contributions to Western European culture.

Bibliography: J. Bergkamp, *Dom Jean Mabillon and the Benedictine Historical School of Saint-Maur*, 1928; H. Lerclercq, *Dom Mabillon*, 1953–1957.

Martha Oberle

MACGREGOR, ROBERT or ROB ROY (1671–1734). Rob Roy MacGregor was a member of the Scottish Highland Clan Gregor and hero to Scottish nationalists, but he was regarded by the English as an outlaw and traitor.

Son of Donald MacGregor of Glengyle and his wife Margaret Campbell, Rob Roy received formal education, as well as grounding in oral culture, through evening ceilidhs. He was a cattle drover and maintained the watch against cattle raiders in return for black mail (rent), but also "lifted" (stole) cattle, an accepted Highland practice. Following the Glorious Revolution of 1688, his clan supported the ousted *James II against *William III and in 1689 fought with Viscount Dundee at Killiecrankie. After his father was imprisoned in 1690, Rob assumed a larger role. The notorious hership (raid) of Kippen in 1691 and other exploits brought him fame. However, in 1693, Parliament revived the lapsed penal statute against the clan for supporting the Jacobites; he then adopted his mother's surname. That year he married Helen Mary MacGregor of Comer, who bore him five sons (James, Coll, Duncan, Ranald, and Robin Og). For years he conducted cattle business with James Graham, marquis of Montrose, who in 1711 loaned him £1,000; however, when one of Rob's men stole the money, Montrose blamed Rob and had him outlawed, and he lost his lands. For a time he found refuge with John Campbell, earl of Breadalbane and took revenge by lifting Montrose's cattle. He supported James Edward the "Old Pretender" in the Jacobite rising of 1715, fought at Sheriffmuir, and was declared a traitor. Subsequently, he allied with John Campbell, earl of Argyll, feuded with Montrose and then John Murray, duke of Atholl, and eventually submitted to General George Wade and received a pardon in 1725, though A.H. Murray argues that he remained a Jacobite until his death.

Rob was a legend in his own time, thanks partly to Daniel Defoe's romantic portrayal in *The Highland Rogue* (1723), and his mythical status has grown, often at the expense of historical accuracy. William Wordsworth's poem "Rob Roy's Grave" (1803) compares him to Robin Hood. The main character in Sir Walter Scott's novel *Rob Roy* (1817) is not Rob but Francis Osbaldistone, an Englishman who has adventures in Scotland in which the "outlaw" is involved. Scott's long introduction describes Rob as motivated by plunder and reluctant to participate in battle, but recent scholarship refutes this. Rob is featured in a live-action Disney film, *Rob Roy—The Highland Rogue* (1954); the more serious *Rob Roy* (1995), with Liam Neeson in the title role; an animated film of the same name (1999); and even a CD-ROM, *Rob Roy, Legend of the Mist*. The 1995 epic, though critically acclaimed, contains numerous inaccuracies, for example, exaggerating the role (as villain) of Henry Cunninghame. There is no evidence that Rob was the subject of any portraits alleged to be his. A bronze statue by Benno Schotz (1975), commissioned by Elizabeth II, stands at Stirling.

Bibliography: W. Murray, *Rob Roy MacGregor: His Life and Times*, 1982.

William B. Robison

MAES, NICOLAES (1634–1693). Nicolaes Maes, the son of a wealthy Dordrecht merchant, was among *Rembrandt's most talented pupils. After a short but significant career as a genre painter in Dordrecht, Maes became a highly successful portraitist. Like his Dutch contemporaries *Jan Vermeer and *Pieter de Hooch, Maes painted domestic scenes, though only for a short period in the 1650s. Dutch scenes of domesticity corresponded to popular Reformed thought, which advocated tidy homes and vigilant housewives. Of this genre, Maes's most inventive are playful didacticisms of idle servants and female eavesdroppers. In *The Eavesdropper*, the lady of the house pauses in the stairwell while spying her maid in an amorous encounter with a male visitor. The preoccupied servant doesn't notice the cat stealing meat from the counter in the adjoining room. The eavesdropper, who engages the viewer and raises her hand in a gesture for quiet, exposes the negligence of her servant. Such works by Maes include complicated perspectives in which the witty satires are visually played out in a series of rooms. The soft chiaroscuro and intimate mood of these paintings show the influence of Rembrandt.

After his brief though productive period as a genre painter, Maes enjoyed a lucrative thirty-five year career as a portraitist. His patrons included Dordrecht and Amsterdam's elite, and his works were in high demand. His mature portrait style was influenced in part by his much acclaimed Flemish contemporary *Anthony van Dyck, and include the same formal settings and expressive posturing. By the time of his death in Amsterdam, Maes had a well-established following. Hundreds of his portraits survive as evidence of his popularity.

Bibliography: W. Robinson, *"The Eavesdropper"* and Related Paintings by Nicolaes Maes," *Jahrbuch Preussisher Kulturbesitz* 4 (1987): 283–313; P. Sutton, *Masters of Seventeenth-Century Dutch Genre Painting*, 1984.

Pamela Merrill Brekka

MAIMBOURG, LOUIS (1610–1686). A Jesuit from the age of sixteen, Maimbourg showed little talent for his first vocation, preaching, so later became a polemicist and historian in the service of his religious faith. He first battled against the Jansenists in France, attacking their translation of the New Testament, which appeared in 1667. He then turned his attention to combating Protestantism and convincing Protestants to return to the Catholic Church. To this end, Maimbourg wrote a series of books designed to show the superiority of Catholicism that include *Traité de la vraie Église* (1671), *Histoire des Croisades*

(1675), *Du grand schisme d'Occident* (1678), *Histoire du luthéranisme* (1680), and *Histoire du calvinisme* (1682).

Maimbourg was a practitioner of *histoire éloquente*, a manner of writing history that favors style over substance and frequently distorts historical events in order to present them according to the author's biases. Protestants of his day found his partisan view of their religion objectionable; Jurieu and *Bayle, for example, wrote works attacking Maimbourg's skewed account of the history of Calvinism. A strong supporter of *Louis XIV, Maimbourg compromised his position as a Jesuit by frequently espousing a Gallican point of view in his works and as a consequence was expelled from the Society of Jesus in 1682. Louis XIV rewarded him for his loyalty to the French cause by providing him with a pension until his death in 1686.

Bibliography: G. Declercq, "Un adepte de l'histoire éloquente, le père Maimbourg, S.J.," *Dix-septième Siècle* 143(1984): 119–132.

Patricia Armstrong

MAINTENON, FRANÇOISE D'AUBIGNÉ, MARQUISE DE (1635–1719).

Founder of Saint-Cyr in 1689, a school for poor girls of good family, and second wife of *Louis XIV, she influenced the court and the king in many ways. Her own biography alternates between constraining walls and the world outside. Born in prison to Jeanne de Cardilhac, who had married the turbulent and impoverished Constant d'Aubigné (son of the great Protestant poet and military officer Agrippa d'Aubigné), Françoise d'Aubigné was raised by relatives who tried to change her faith from Protestantism to Catholicism. As a result of a severe education, she remained all her life involved in theological issues and devoted to charities and education.

She first married the comic actor *Paul Scarron. The couple welcomed many intellectuals during their soirées; their wit and hospitality was renowned. When her husband died, she became governess of the children of nobility. Madame de Montespan, mistress of King Louis XIV, asked her to be the governess to her royal children in Versailles. With her excellent work, d'Aubigné acquired the domain of Maintenon in 1674 and became Marquise. Eventually, she supplanted Madame de Montespan in the king's favor. Louis XIV married her secretly in 1683, and it was largely under her influence that his court became devout. She played some part in public affairs behind the scenes, but devoted most attention to the school she founded, Saint-Cyr. Maintenon planned Saint-Cyr to provide a kind of subsidized education that she had not had, an education with a real intellectual basis that was oriented to convent and secular life. *Fénelon, one of Maintenon's advisers, was a clergyman with strong ideas on education in general and on the education of girls. He emphasized in his treatise *De l'éducation des filles* (1687) the desirability of preparing women for activities carried out entirely in the home and contrasted the good domestic woman with the fashionable and "public" woman of the salon. Saint-Cyr was at the same time the place where free observation of the world was encouraged through visits outside the school and plays to respond to the practical needs of modern life and to develop openness to social realities. In 1689, Maintenon commissioned *Racine to write "trois mille vers de piété" (three thousand verses of piety) and *Esther* to be played in the school. She expelled the superintendent Marie de Brinon after a month and a half of her representation when Maintenon started to see the dangers in the increased contact between the pupils and the court. In 1691, Racine's second biblical tragedy, *Athalie*, was given only three performances at Saint-Cyr in ordinary dress before a private audience. Maintenon's refusal to permit further open performances at the school marked an attempt to prevent the theatricality of politics from merging with the theatricality of the school. After the death of Louis XIV, she went to Saint-Cyr where she died. She kept writing all her life; her correspondence shows the evolution of her career and her goals for education. It presents vivid aspects of the daily life of the court and outside Versailles.

Bibliography: R. Chartier, D. Julia, and M. Compère, *L'éducation en France du XVIe au XVII*e *siècle*, 1976; J. Prévot, *La première institutrice de France: Madame de Maintenon*, 1981.

Martine Sauret

MALEBRANCHE, NICOLAS (1638–1715). Nicholas Malebranche was a major figure in French intellectual life during the period that gave rise to the methods of modern science and experimentalism. At age sixteen, Malebranche enrolled in the College de la March, University of Paris, and took the Master of Arts degree in 1656. Thereafter, he studied theology for three years at the Sorbonne, followed by four years at the Congregation of the Oratory, during which time he became expert in the texts of Plato and Augustine. Malebranche was ordained on 20 September 1664, beginning his priesthood with a year-long study of *Descartes' *Traite de L'Homme*. This study sealed his affinity for the Platonic tradition in philosophy and he became wedded to one of its main philosophical points: that knowledge is possible only by means of an intellectual grasp of necessary relations that obtain in the noncreated realm. This affinity makes Malebranche's intellectual career all the more interesting, for ultimately he was convinced that experimental inquiry into the relations among created things is a key element of knowledge, and he finished his career as a dedicated experimentalist.

In 1674, Malebranche was appointed professor of mathematics at the Oratory. In this same year, he published the first edition (there were seven in all, the last published in 1712) of his major work, *De la Recherche de la Vérité*. Here, in the tradition that goes back to Plato rather than originating with Descartes, Malebranche rejects sensory qualities as the foundation of knowledge in favor of the ideas that represent the sensory qualities as these ideas are grasped by the intellect. However, Malebranche argued also for decidedly non-Cartesian tenets, among which are his arguments for occasionalism with regard to created things, the pivot point in Malebranche's turn to experimentalism. The Doctrine of Occasionalism was not original to Malebranche, although the importance he gave to it in his systematic thought was groundbreaking and one may with reason read Malebranche's argumentation as a presage to David Hume's philosophical ideas. In brief, the notion of occasional cause stems from the following reasoning: If God is the cause of all relations among created things, then there is no causality in the created world, for God is not in this world. Given that every relation could be completely different, in the sense that there would be no logical contradiction in the case of such difference, relations among things have no metaphysical necessity but are sustained by God's will. Note that this reasoning merely presumes that in order for relations to be correctly deemed as causal, such relations must be metaphysically necessary. However, this presumption was typical of the times, and Malebranche accepted it without question. Whatever the ultimate propriety of the notion of occasional cause, conjoined with metaphysical necessity, Malebranche's occasionalism inspired his turn to experimental science, for him a natural turn. After all, if the only means of discovery available to the human mind are occasional causes, then experience of these causes is indispensable to an understanding of the relations of things in the created world. The understanding, that is the intellect, when supplemented with the fruits of experimental manipulation of created things, promises insight into the relations that God established in the created world, if not direct (which would be to say, true) knowledge of these relations. Malebranche's doctrine brought him into sharp controversy with the theologian *Antoine Arnauld.

Although Malebranche's academic appointment was in mathematics, he made no lasting, original contributions to that discipline. However, his influence across the field is notable. He influenced *John Wallis's work in algebra, and Malebranche knew *Gottfried Leibniz, whose far-reaching and innovative mathematical work Malebranche supported and promoted among French intellectuals. Malebranche's association with mathematical innovation and innovators in the late seventeenth century was the basis for his nomination in 1699 as an honorary member of the Académie Royale des Sciences.

Bibliography: N. Jolley, *The Light of the Soul: Theories of Ideas in Leibniz, Malebranche, and Descartes*, 1998; S. Nadler, *Cambridge Companion to Malebranche*, 2000; T. Schmaltz, *Malebranche's Theory of the Soul: A Cartesian Interpretation*, 1996.

Cassandra L. Pinnick

MALPIGHI, MARCELLO (1628–1694). Born near Bologna, Italy, Marcello Malpighi contributed to the fields of medicine and physiology. He was first interested in literature, but at the age of seventeen he began to study philosophy under Francesco Natali. Malpighi, dismayed at the death of both his parents, took up Natali's suggestion to study medicine, which he did under the guidance primarily of Barcolomeo Massari, eventually earning a doctorate in philosophy and medicine. He achieved a substantial reputation in medicine and was invited to become a professor of medicine at Pisa in 1656. Following the spirit of the new empiricism in philosophy, Malpighi rooted his approach to science firmly in experiments and observations as opposed to the philosophical or scholastic approach. Because of ill health, which he blamed on the hot and humid climate in Pisa, Malpighi returned to Bologna in 1660, where he taught theoretical medicine for two years.

During this stay at Bologna, Malpighi discovered and reported on his fundamental observations on the structures of the lungs and confirmed and completed *William Harvey's theory on the circulation of the blood. Harvey, through careful observations of the direction of blood flow and calculations of the blood pumped by the heat per unit time, hypothesized that blood was not continuously formed in the body. Rather, blood circulates within the body by traveling through the heart and lungs to various organs and then returns to the heart and lungs. However, Harvey did not identify the connecting system between the venous and arterial blood systems. In other words, Harvey's theory was basically a deduction and not an empirical demonstration. Malpighi made good use of the then novel microscope to study the structure of the lungs. He also used a variety of experimental techniques, using the frog as the experimental animal, such as inflating the lungs by way of the trachea and injecting ink into blood vessels. Malpighi learned that the lungs of the frog consist of small air sacs surrounded by a network of small blood vessels, the capillaries. Now called the "microcirculation," this network of capillaries was eventually discovered to be the pathway for oxygen transport within the body. Malpighi also studied the small vessels of the kidneys and of other organs and identified the red cell components of the blood. He discovered in the kidneys the structures now known as glomeruli (earlier called "Malpighian corpuscles"). Blood reaching the kidneys first enters the glomeruli and then leaves to be distributed over fiberlike structures called convulted tubules. Malpighi concluded that urine is formed in the glomeruli and that only the smaller components of blood pass through them in a filtration process.

Malpighi's stature as a physician and scientist grew considerably. In 1660, he was invited to become the first professor of medicine at the University of Messina, where he stayed until 1664. In 1666, Malpighi returned to Bologna to teach. In that same year, he published his first report on the kidney and its blood supply. Furthermore, Malpighi was able to outline the vascular supply and important structures in other organs, including the tongue, brain, heart, and various glands. In the following decade, Malpighi extended his studies to embryology, mostly in the chicken, and to congenital malformations. Furthermore, he formulated the hypothesis that development of large cysts in the kidneys could be related to the blockage of urine. Malpighi's work became widely recognized and appreciated, even though his work had earlier been often severely criticized because of the radical nature of many of his ideas and because of his intolerance of accepted dogma not solidly founded on observation or experimental data. In 1696, he was elected a fellow of the Royal Society of London. In 1691, at the height of his fame, he was appointed as the personal physician of Pope Innocent XII in Rome. Malpighi himself suffered from ill health, which included gout, heart palpitations, and probably kidney stones. He died of a stroke on 29 November 1694.

Bibliography: H. Adelmann, *Marcello Malpighi and the Evolution of Embryology*, 1966; L. Magner, *A History of the Life Sciences*, 1979.

Paul C.L. Tang

MANSART, FRANÇOIS (1598–1666). This French architect transformed classicism into a national idiom and founded a dynasty of architects that continued for over 150 years. The son of a master carpenter, Mansart received most of his training from his father and his brother-in-law, Germain Gaultier, a sculptor. He later apprenticed to Salomon de Brosse with whom he collaborated on the Chateau of Coulommiers in 1618. He was appointed architect to the king in 1636, but his most significant commissions came from lesser royalty.

Mansart's major contribution to architecture rests in combining traditional French forms (chateaux and hôtels) with Italian classicism, creating a distinct genre. His designs for chateaux clothed the pavilion and wing form that derived from castles and hunting lodges with an ingenious use of the classical orders, which gave these structures a sculptural quality. This highly decorative quality also enhanced the importance of the chateau as part of the surrounding park. His hôtels or townhouses employ a more restrained use of classical orders that was more in keeping with the streetscape. Both architectural forms incorporate the highly pitched roofline that became known as the Mansart roof. Mansart's work captured the attention of architectural theorists Claude Perrault and François Blondel, the latter of whom dubbed Mansart the "God of Architecture." His practice continued with his great nephew Jules Hardouin, who also assumed his surname, and through subsequent descendents until the turn of the nineteenth century.

Bibliography: A. Blunt, *Art and Architecture in France: 1500–1700*, 1954; A. Braham and P. Smith, *François Mansart*, 1973.

David D. McKinney

MARINO, GIAMBATTISTA (1569–1625). Marino was an Italian poet, born in Spanish-dominated Naples, the political and cultural capital of southern Italy. Marino's name has become synonymous with preciosity, far-fetched images, and verbal conceits, and his florid and grandiose style inspired a literary fashion called *marinismo*. The term designates the flamboyant technique of the poet and his seventeenth-century imitators, with its extravagant imagery and excessive ornamentation. When still very young, he knew and admired Torquato Tasso, the Italian Renaissance poet, and tried to emulate him. But the life of Giambattista Marino was very different from that of the melancholic Tasso. His father wanted him to pursue the legal career but eventually ordered his son out of the house, leaving him free to make his name among the literati of the academies and to seek patronage among the nobility. In 1598, he was suddenly thrown into prison for his affair with a young girl, the daughter of a rich Sicilian merchant. In 1660, Marino was imprisoned again for falsification of documents with the intention of getting a friend out of jail. He fled to Rome, finding a place in the service of Cardinal Pietro Aldobrandini who took Marino with him, first to Ravenna and then to Turin where he served as papal legate.

Carlo Emanuele, duke of Savoy, became the patron of the already famous poet and made him a knight of the Order of Saint Maurice, which thereafter Giambattista always paraded. From this arose the jealous anger of the duke's secretary, the poet Gaspar Murtola, who started a sonnet war against the Neapolitan. The literary results of the duel can be seen in two collections of sonnets: the *Murtoleide, fischiate (Exploits of Mutola: Boos!)*, Marino's answer to Murtola's libel the *Compendio della vita del cavelier Marino*; and Murtola's *Marineide (Chuckles)*. The nonliterary results of their duel were less diverting. The secretary decided to shoot his rival down in the street; the shot failed of its mark but hit a favorite of the duke. Murtola now found himself imprisoned and condemned to death, but Marino pleaded for his life, either out of generosity or as fine gesture, and Murtola's sentence was commuted to exile. But safe in Rome, the ex-secretary continued to annoy Marino.

Now more famous and prosperous than ever, Marino probably felt that he had seen the last of misfortune, but the ducal prison was awaiting him. It is believed that Marino offended the duke in some fashion, a woman has been suggested as the cause, but the early biographers point to a satiric poem written earlier in Naples called *La Cuccagna (Cockaigne)*. When this youthful poem of Marino's was suddenly brought to the duke's attention as a recently composed satire directed at him, the duke jailed the poet for a while. In 1615, Marino decided to move to France at the invitation of the wife of Henry IV, Queen Marguérite of France. His great patroness was Marguérite's successor, Queen Maria de Medici, and the Neapolitan poet intended to publish his *Adone* without interference, a poem which by that time had grown from a mythological tale to a long work, which he intended to equal Tasso's *Gerusalemme*. In France, Marino completed his *Adone*, which was presented by Maria de Medici to Louis III. In the French court, the poet received honors and became the favorite of the court salons, but Marino longed for Italy and wanted to enjoy fame in his own country, where his *Adone* had become the subject of acid polemics that made him still more renowned. His arrival was a triumph. A public statue was erected in his honor in his native Naples, but he died there shortly after his return, in 1625.

Giambattista Marino was the author of a large collection of lyric verse such as *La Lira* (1608–1614, *The Lyre*), *La Zampogna* (1620, *The Syrinx*), and his long mythological poem *Adone* (1623, *Adonis*). Marino derived inspiration from the poetry of the late sixteenth century, but his aim was to motivate wonder by novelty. His work is characterized by farfetched metaphors and conceits, or images of fantastic ingenuity. There is no doubt, judging from Marino's letters, that he devoted himself to his art with conscientiousness and ardor, and that poetry was the one and only worship of his life. His major work, the one that best reveals the literary trends of the century, is the *Adone*. It is an epic poem in twenty cantos, like the *Gerusalemme*, but the cantos are much longer than Tasso's. *Adone* brings the story of Venus and Adonis to a moral conclusion whereby the illicit love of Adonis is punished by the wronged husband Mars. But plot and moral alike are submerged beneath endless classical scenes and sensual descriptions. In the poetic expressiveness of the work, Marino demonstrated his mastery of the language and all the musical possibilities of verse.

Bibliography: P. Marzio, *Il Baroco, Marino e la poesia del seicento*, 1995; J. Mirollo, *The Poet of the Marvelous: Giambattista Marino*, 1963.

Angela Morales

MARQUETTE, PÈRE JACQUES (1637–1675).

French Jesuit missionary and North American explorer, Marquette was born on 10 June 1637 in Laon, France, and died near present-day Ludington, Michigan, on 18 May 1675. Educated at Pont-à-Mousson, Marquette then taught at Rheims, Charleville, and Langres. However, his desire was not for teaching but for the missions, and on 20 September 1666, Marquette arrived in Quebec. Ten days later, the priest journeyed to Three Rivers to learn the Indian languages, and in 1669, he moved much farther west to Chequamegon Bay. In 1671, the Sioux forced Marquette to return to the Straits of Mackinac where he founded the mission of St. Ignace. From this mission, in May of 1673, Marquette and *Joliet set out on their voyage down Lake Michigan, through Green Bay, over the Fox and Wisconsin Rivers to the Mississippi. These two were the first to confirm a river route from the St. Lawrence River to the Gulf of Mexico. Marquette recorded not only the details of the rivers and of the topography in general, but also the manners and customs of the peoples whose lands he and his companions passed through. His journal was published in 1681 in Thévenot's *Recueil de voyages*.

Marquette's interest was missionary; his journal speaks repeatedly of his wish to instruct the Native Americans in Christianity, but he also knew the importance of making his and Joliet's discovery known to the rest of the world, and so he and the band of explorers turned for home. The following year, Marquette retraced part of his journey; he set out in October 1674 for the tribes of Illinois, but illness forced him to give up his missionary effort. Sick, weak, and caught in severe weather, Marquette had to winter over with few provisions in what was little more than a hut. Despite the efforts of the Indians and of his companions, Marquette died. His death was of the sort he had wished for himself— virtually alone and in the service of his order's missions. Three years after his death, his body was returned to the mission at St. Ignace.

Bibliography: J. Donnelly, *Jacques Marquette, S.J. 1637–1675*, 1968; R. Hamilton, *Marquette's Explorations: The Narratives Reexamined*, 1970.

Martha Oberle

MARSTON, JOHN (?1575–1634). Satirist John Marston was born to Middle Temple lecturer John Marston, Sr., and Maria, daughter of Italian-born London surgeon Andrew Guarsi. After taking his B.A. from Brasenose College, Oxford, in 1594, Marston pursued legal studies at the Middle Temple, but seems to have really engaged in a writing vocation, since it is in these years, 1594 to 1607, that he composed all of his notable work. When he was ordained is uncertain, but in October 1616, he was given the living of Christchurch, Hampshire, which he held until about three years before his death on 25 June 1634. In addition to the connections he made at the university and the Middle Temple, Marston formed an essential one by marrying the daughter of *James I's chaplain, the Reverend William Wilkes.

Marston wrote satirical poems before moving on to his more currently read plays; all of his work is rough, obviously critical, often sarcastic, and so, to some, offensive. Indeed, because Marston's satire often was personal rather than social, and because the vices mocked were presented in graphic terms necessary for effective ridicule, John Whitgift, archbishop of Canterbury, ordered the verse satires, *The Metamorphosis of Pigmalions Image and Certaine Satyres* (1598) and *The Scourge of Villanie* (1599), burned. The royal court was so offended by the portrayal of Scottish men as unconscionably careerist in the play *Eastward Ho* (1605), written with *Ben Jonson and *George Chapman, that Marston and his collaborators were jailed. Such recrimination may account for Marston's early use of the pseudonym William Kinsayder, his cessation of writing in mid-life, and his insistence that his name be removed from a collected edition of his plays in 1633, about a year before his death.

Marston's best work is usually taken to be his plays, all written for children's acting troupes of the city, especially *The History of Antonio and Mellida* (1602), *The Malcontent* (1604), and *The Dutch Courtezan* (1605), though *The Insatiate Countesse: A Tragedie* (1613), the most carnal-minded of the works, probably begun by Marston and continued by the actor William Barksteed, seems to draw contemporary interest as well. *Antonio and Mellida*, a tragedy in two parts, with the first part ending comically and thereby accentuating the tragic catastrophe of the second part, effects an ingenious plot, even if it verges on the brink of the hackneyed in its subject matter, pace, and verse, and a typically unflattering theme of the essential corruptness of human beings, especially when involved with what is called romantic, selfless love. Equally, if not more, nodular are *The Dutch Courtezan* and *The Instatiate Countesse*, the former about how a privileged young man uses a prostitute to reform a friend's sexual interests, culminating in what is supposed to be the uproariously funny whipping and imprisonment of Franceschina the prostitute, and the latter, an unfinished play about Isabella, the countess who is governed by the salacious interests suggested by the title. Marston's best work, the tragicomedy *The Malcontent*, treats similar subject matter satirically, but it does so with less outsized actions and more psychologically interesting depictions of character, especially of the title character, the deposed Duke of Genoa, whose noble name is Altofronto but who assumes the malcontent's name of Malevole. Exploring the dramatic genres as fully as most of his noteworthy contemporaries, Marston is at his best when he circumscribes his expression with the basic formal demands of comedy, even though he seems to have held beliefs more in line with those expressed in his more melodramatic work. Marston was buried in Temple Church adjacent to his father on 26 June 1634, the day after his death. Marston's epitaph, no doubt suggested by himself, is "Oblivioni sacrum," a rather clear indication of Marston's love of this world and its people.

Bibliography: A. Davenport, ed., *The Poems of John Marston*, 1961; P. Finkelpearl, *John Marston of the Middle Temple*, 1969; H. Wood, ed., *The Plays of John Marston*, 1938.

Jesse G. Swan

MARVELL, ANDREW (1621–1678). An able administrator under both *Oliver Cromwell and *Charles II, Marvell won posthumous fame as a Metaphysical poet. Born into a Yorkshire rector's family on 31 March 1621, Marvell attended Hull Grammar School and then Trinity College, Cambridge, following up his 1639 B.A. with incomplete M.A. studies. After Cambridge—where, according to tradition, he toyed with Catholicism—he spent four years traveling in Holland, France, Italy, and Spain. Thus, he probably missed the outbreak of the civil war. It is unclear whether he ever fought among the Royalists, whose cause he later described ambiguously in a prose satire as "too good to have been fought for."

Records pinpoint Marvell's presence in England again in 1650–1651, when he became tutor to Mary, the daughter of Cromwell's lord-general, Sir Thomas Fairfax—who had resigned his post and retreated to his Yorkshire estate, Nun Appleton House, because of his Royalist sympathies. In 1650, Marvell wrote "An Horatian Ode upon Cromwell's Return from Ireland," his only poem to reflect the period's turmoil. In it, he combined admiration for *Charles I's courage at the scaffold with ambivalence toward Cromwell's triumph abroad. When the poem eventually appeared in print in 1681, Charles II's government censored it for its sympathy to the Protectorate.

In 1652, Marvell left his retired Yorkshire situation to pursue a political career in London, which meant entering Cromwell's service. On 21 February 1653, *John Milton, Latin secretary for the Commonwealth and an ardent Puritan, recommended Marvell for an assistant's post in the Council of State offices. But Marvell spent several years at Eton as tutor to Cromwell's ward, William Dutton, before being appointed Latin secretary in 1657, at £200 a year. Then, in 1659, he was elected a member of Parliament for his home town of Hull, which he continued to serve in the House of Commons for almost twenty years.

Despite a heartfelt poem of mourning written upon Cromwell's death in 1658, in Parliament, Marvell supported the Stuart restoration. With Charles II's arrival in 1660, Marvell optimistically worked on bills for religious tolerance and vigorously defended Milton against regicide charges for Charles I's execution. He also published various political satires anonymously, one notable exception being his signed plea on behalf of Protestant nonconformists, *The Rehearsal Transpros'd: The Second Part.*

Marvell's lyric poetry saw the light of day only after his death from a fever on 18 August 1678. In 1681, his housekeeper, Mary Palmer, came forward as his widow and published *Miscellaneous Poems*, a collection of thirty-seven English and fifteen Latin compositions. It is not as a politician or satirist but as a superlative lyric poet that Marvell stands out in his period. Master of the seventeenth-century's favored form, the closed couplet, he displayed the vigorous intellectual wit and urbane playfulness distinctive of the school of Metaphysical poets. Marvell's "To His Coy Mistress," a dramatic monologue rich in metaphysical irony, paradox, and elaborate conceits, is one of the most famous seduction poems in the English language.

Bibliography: J. Hunt, *Andrew Marvell: His Life and Writings*, 1978; T. Wheeler, *Andrew Marvell Revisited*, 1996.

Margaret Goscilo

MARY II (1662–1694). *See* WILLIAM III.

MASSINGER, PHILIP (1583–1640). Massinger, a prolific playwright of the Jacobean and Caroline eras, was born in 1583 and baptized on 24 November in St. Thomas's Church in Salisbury. His father, Arthur Massinger, had been a student at St. Alban's Hall, Oxford, later a fellow at Merton, and spent his life in the employ of Henry Herbert, second earl of Pembroke, and later William Herbert, the third earl. Massinger himself would later maintain this patronage connection with the Herberts. Massinger matriculated at St. Alban's in 1602, where both his father and his uncle had been educated. However, he left in 1606 without obtaining a degree. Some scholars have speculated that his departure may have been caused by his father's death and a consequent lack of funds.

There are no records of Massinger from his leaving Oxford until 1613, when, desperate for money and in debtor's prison, he wrote a letter with Nathaniel Field and Robert Daborne to Philip Henslowe, requesting that they be paid in advance for a play so that they could use the money for bail. It is widely accepted, however, that during this time he was working as a collaborator and reviser and that he became *John Fletcher's apprentice and chief collaborator after *Francis Beaumont married and retired from the theater in 1613. Hence, many of the "Beaumont and Fletcher" plays published in the 1647 folio are Massinger's collaborations with Fletcher or were later revised by him.

Scholars of Massinger and Fletcher are in disagreement as to which plays show signs of Massinger's work. However, he probably collaborated with Fletcher on *Sir John van Olden Barnavelt* (1619), *The Custom of the Country*

(1619), *The Double Marriage* (1620), *The Little French Lawyer* (1621), *The False One* (1621), *The Beggar's Bush* (1622), *The Prophetess* (1622), *The Sea Voyage* (1622), and *The Spanish Curate* (1622). He collaborated with Fletcher and others in *Thierry and Theodoret* (1613, possibly with Field), *The Queen of Corinth* (1617, with Field), *The Bloody Brother* (1617, with Field and Daborne, though some critics have argued for *George Chapman and *Ben Jonson), *The Jeweler of Amsterdam* (1616, with Field, now lost), and *The Knight of Malta* (1618, with Field). He also collaborated on *The Fatal Dowry* with Field, *The Virgin Martyr* with *Thomas Dekker, and *The Old Law* with *Thomas Middleton and William Rowley. *The Fair Maid of the Inn*, though it appears in the Beaumont and Fletcher folio, is probably mostly his, as is *Love's Cure*. He seems to have become the poet-in-ordinary to the King's Men upon Fletcher's death in 1625; however, he continued to write for Queen Henrietta's Men as well until 1630.

Massinger wrote twenty-nine plays on his own, a large number of which were for the King's Men. These plays include *The Duke of Milan* (1621), *The Bondman* (1623), *The Renegade* (1624), *The Parliament of Love* (1624), *The Unnatural Combat* (1625), *A New Way to Pay Old Debts* (1625), *The Roman Actor* (1626), *The Great Duke of Florence* (1627), *The Picture* (1629), *The Maid of Honour* (1630), *Believe as You List* (1631), *The Emperor of the East* (1631), *The City Madam* (1632), *The Guardian* (1633), *A Very Woman* (revision of a Fletcher play, 1634), *The Tragedy of Cleander* (1634), and *The Bashful Lover* (1636), all of which are extant and a number of plays which are now lost. His long apprenticeship with Fletcher enabled him to master and to extend the possibilities in tragicomedy, a genre especially suited to him.

The City Madam, *A New Way to Pay Old Debts*, and *The Roman Actor* are perhaps the most studied and best known of Massinger's plays. *A New Way to Pay Old Debts* is one of the last and best of the Jacobean city comedies, and his villain, Sir Giles Overreach, a classic of the kind. *The City Madam* is another excellent urban comedy, written in the somewhat different climate of the 1630s. *The Roman Actor* is a reflection on the benefits and pitfalls of dramatic patronage, which seems eerily prescient of the gradual appropriation of the King's Men by the Caroline court. Its long defense of acting, spoken by the actor hero, Paris, was a favorite of Shakespearean actor-managers and ensured the play a minor place in the classical repertory for two centuries.

Massinger died in 1640, quietly and in his sleep, and still active in the theater. He is known to have left a widow, but it is not certain whether or not he had any children. He was buried on 18 March 1640, in the parish church of St. Saviour's, now Southwark Cathedral, and at least one contemporary account suggests that he and Fletcher, his longtime coworker, are buried in the same grave. Massinger studies, for a long time focused almost exclusively on authorship, are enjoying a renaissance, as critics begin to examine the growing absolutism of the political climate in which they were written.

Bibliography: D. Adier, *Philip Massinger*, 1987; I. Clark, *The Moral Art of Philip Massinger*, 1993; D. Howard, *Philip Massinger: A Critical Reassessment*, 1985.

                                                           Melissa D. Aaron

MATHER, COTTON (1663–1728). An important Puritan minister, Mather came from a long ministerial tradition. He was born in Boston, Massachusetts, the son of *Increase Mather and the grandson of *John Cotton. He entered Harvard University in 1674, just shy of his twelfth birthday. While there he considered becoming a physician because he had a stutter, and, although he began preaching shortly after he earned his B.A. in 1678, science remained a passion for the rest of his life. In 1681, he received his M.A. from Harvard, where his father was serving as president. He was ordained at the Second or North Church in Boston in 1686 and worked there with his father until Increase died in 1723.

Mather married Abigail Phillips, the first of his three wives, in 1682. The couple had nine children. After Abigail died in 1702, he married Elizabeth Hubbard. They had six children. When she died in 1713, he remarried again, this time to Lydia Lee George. Unfortunately, she became mentally unstable. Mather also inherited financial obligations with his third marriage that left him on the verge of poverty and dependent on the charity of his friends. Of his fifteen children, only two survived him.

Mather saw it as his mission to preach the ideals of the original Puritan settlers of New England. His orthodoxy led him into conflict with another leading Puritan minister, Samuel Stoddard, over open communion. He was involved in politics, taking a role in the ouster of Governor Edmund Andros in 1689 following the Glorious Revolution. He also was instrumental in the Salem witch trials. While disturbed by the way the court proceedings were occurring, and like his father worried about the use of spectral evidence, Mather nonetheless justified the trials in his work *Wonders of the Invisible World* in 1692. In the wake of the trials, which saw hundreds of people arrested and twenty executions, Mather's standing in the community suffered.

Mather's interest in science led to a number of works. He published an account of native phenomena in his *Curiosa Americana* (1712–1724), which attracted the attention of the Royal Society of London. The organization made him a member in 1713. His scientific observations were influenced by his religious beliefs, as evidenced in his 1721 work, *The Christian Philosopher*. That same year he became embroiled in a controversy over using inoculation to prevent the spread of smallpox during an epidemic, an account of which was published by the Royal Society.

In all, Mather published more than 400 works. In his 1726 book, *Manuductio ad Ministerium*, Mather proffered advice to ministerial students. His largest work, *Magnalia Christi Americana*, which appeared in 1702, had to be published in England because of its length. In it, he attempted to record the early history of New England. At his death in 1728, he was involved in a twenty-year study on biblical learning.

Bibliography: K. Silverman, *The Life and Times of Cotton Mather*, 1984.

Christopher E. Hendricks

MATHER, INCREASE (1639–1723). Increase Mather, Puritan minister and author, was born in Dorchester, Massachusetts, the son of minister Richard Mather and Katherine Holt. He was the youngest of five brothers, four of whom became ministers. He received his B.A. from Harvard in 1656 and a M.A. at Trinity College, Dublin, in 1658. He returned to Massachusetts in 1661. The next year he married Maria Cotton, daughter of the influential Puritan minister *John Cotton. The couple had ten children, the oldest, *Cotton Mather, followed in his father's footsteps, becoming an important religious figure in the Massachusetts Bay colony.

During the 1660s, New England Puritans became embroiled in a debate over decreasing church membership. To increase numbers, a synod created the Half-Way Covenant, a system by which the children of church members could be baptized into the faith, even if they had not experienced conversion. Mather came out in opposition to the covenant against his father, Richard, who was an ardent supporter. In 1664, during the height of the controversy, Mather was ordained to lead the Second or North Church in Boston, a post he would hold for the rest of his life. He used the position to try to maintain a conservative position in church matters, trying, for example, to prevent Baptists from settling in the colony, and describing King Philip's War of 1675 in terms of God punishing the Puritans for straying from the true faith.

After combating a two-year bout of severe depression, Mather became a prolific writer, publishing forty-six books, beginning with an autobiography of his father, *The Life and Death of That Reverend Man of God, Mr. Richard Mather*, in 1670. He was a chief sponsor of John Foster, the first printer in Boston, who began operations in 1675. He also became involved in academics, being named temporary president of Harvard from 1685 to 1701.

From 1688 to 1692, Mather was in England lobbying for the return of the Massachusett's charter, which had been revoked in 1686. He met several times with *James II until his ouster in the Glorious Revolution, then, joined by Thomas Oakes, Elisha Cooke, and Sir Henry Ashurst, pressed *William III for the charter to be restored. Mather authored a new charter, and when it was rejected, worked to win concessions for Massachusetts in the Charter of 1691, such as including the Plymouth colony within its boundaries.

Mather returned to New England at the height of the Salem witch trials, the atmosphere for which he may have helped create with his 1684 work, *An Essay for the Recording of Illustrious Providences*. He was slow to criticize the proceedings, but eventually published a work decrying the use of spectral evidence, *Cases of Conscience Concerning Evil Spirits*. Ironically, his son *Cotton concurrently came out with a book supporting the trials. During the last years of his life, Mather's influence in Massachusetts waned, as people increasingly tired of his constant calls to religious orthodoxy and his role in the creation of the unpopular Charter of 1691.

Bibliography: M. Hall, *The Last American Puritan: The Life of Increase Mather*, 1988.

Christopher E. Hendricks

MERSENNE, MARIN (1588–1648). Spending the greater part of his life in the convent of the Order of Minims off Place Royale in Paris (now Place des Vosges) did not prevent this French monk from fostering a remarkable network of scientific relations. He did so by gathering figures such as *Gassendi (his close friend) and *Blaise Pascal. The voluminous correspondence with *Descartes, *Galileo, *Huygens and *Torricelli testifies to his energy in creating a community of thinkers united in their will to renew natural philosophy and its methodological foundations. Mersenne gathered and circulated information and responses, or proposed problems of his own initiative, since he conducted scientific investigation himself.

A distinctive aspect of his writings is the conflict between apologetics and science, which brought about the central issue of skepticism (*La Vérité des sciences contre les sceptiques et les pyrrhoniens*, 1625). Mersenne perceived the diffidence of Renaissance naturalism and libertine skepticism toward certainty of knowledge to be a grave threat to Catholic theology and Christian morality. He strategically embraced aspects of this skepticism, arguing, indeed, that the Aristotelian qualitative and essential understanding of phenomena was useless. Reason had to stop short at the surface of objects. However, this did not prevent a mathematical analysis of these appearances, which could then rely on the certainty of quantitative knowledge. With theology then an autonomous field, reason and religion could be reconciled. The approach is usually referred to as "mitigated skepticism." Dear's work suggests that, ironically, Mersenne's rejection of Aristotelian essential knowledge is grounded in the humanist probabilism of his masters, the Jesuits. Given the emphasis on motion and size to understand phenomena, Mersenne proceeded to carefully record the result of his experiments, sometimes correcting Galileo's own and establishing a standard of procedure. The range of interest in his work is wide, but because of his methodological stance he excelled mostly in the field of mixed mathematics, which included optics and music. With science distinct from metaphysics, Mersenne greeted Descartes' attempt to give method a radically new ontological foundation with suspicion. Nevertheless, his *Secondes Objections* to the *Méditations métaphysiques* (1641) remains further proof of his open-mindedness, together with his introduction in France of Galileo's new ideas on astronomy.

Bibliography: P. Dear, *Mersenne and the Learning of the Schools*, 1988; R. Lenoble, *Mersenne ou La Naissance du mécanisme*, 1971.

Jean-Vincent Blanchard

MIDDLETON, THOMAS (1580–1627). The prolific Middleton wrote prose satires, poems, masques, public pageants, almanacs, and journalistic tracts in addition to the comic, tragicomic, and tragic plays for which he is best known. Critical consensus ranks him as the best English Renaissance playwright after *Shakespeare and *Ben Jonson.

Middleton, a life-long Londoner, was born into a prosperous bricklayer's family in 1580. When his widowed mother married a man intent on appropriating her fortune, Middleton became embroiled in litigation whose combined gender and economic conflict eventually influenced his dramatic themes. He published his first book of poetry even before starting Queen's College, Oxford (1598), which he left without his degree, probably because of the family lawsuits. Since drama paid better than verse and needed no patronage, Middleton soon became a professional writer producing commissioned plays for theater-manager Philip Henslowe. His earliest works are lost. When the plague shut down the theaters in 1603–1604, Middleton composed prose satires bitterly critical of the privileged classes. Around 1602, he married Mary Marbeck, granddaughter of a musician almost burnt at the stake for his radical Protestantism. By 1603, Middleton had made his name writing for both the boys' and the adult acting companies (including Shakespeare's) in London and had collaborated with Thomas Dekker on the welcoming pageant for Elizabeth I's successor, *King James I. In 1620, his appointment as City Chronologer put him in charge of recording London's major events. Only seven years later, Middleton died and was buried in his parish church on 4 July.

In his city comedies—set in contemporary London—Middleton satirized fools and knaves from all classes, from country, town, and court alike. He often used intrigue plots to expose the intertwined sexual and financial corruption at the heart of the establishment. For instance, *The Roaring Girl* (1611), written in collaboration with Dekker, sympathizes with its criminal outcast heroine, Moll, while *A Chaste Maid in Cheapside* (1613) criticizes mercenary marriages and male control of women.

Middleton's three Italian tragedies likewise show economic, sexual, and power lusts causing universal corruption. *Women Beware Women* (1621) and *The Changeling* (1622, in collaboration with William Rowley) both expose court intrigues mired in betrayal, violence, and death. In *The Revenger's Tragedy* (1607), previously credited to *Cyril Tourneur but now generally accepted as Middleton's work, courtly vice contaminates even the self-appointed avenger who defends his family against aristocratic predators. Full of the incest, adultery, rape, and murder that characterize Jacobean revenge drama, the play also alludes to the Gunpowder Plot of 1605 in its depiction of treason and tyranny. Middleton's severe condemnation of immorality alongside his social radicalism has sometimes earned him the label of Puritan playwright or, in Julia Gasper's term, "militant Protestant" (1990).

Middleton's last play, *A Game At Chess* (produced in 1624), proved the most popular of the Jacobean period, setting a nine-day record at the Globe Theater. This political satire represented Catholic Spain as the black side on a chessboard, defeated by the White Knight symbolizing England's Prince Charles. A public sensation led to Spanish pressure on King James, who banned the subversive play despite Middleton's defense of it as "A harmlesse game raised meerely for delight."

Bibliography: W. Carroll, ed., *Women Beware Women*, 1994; S. Chakravorty, *Society and Politics in the Plays of Thomas Middleton*, 1996.

Margaret Goscilo

MILTON, JOHN (1608–1674). John Milton was born 9 December 1608 in Cheapside, London. There, it is said, his doting, Puritan father taught the boy prodigy to read by the age of three, using the Calvinistic Geneva Bible (1560) and the newly published Authorized Version (1611) as primers. Apocryphal or factual, the legend points not only to the young Milton's basal spirituality, but to his impressive gifts for eclectic assimilation of learning, languages, and literature as well. At St. Paul's School, he had command of Latin and Greek by age sixteen. At Cambridge, he mastered Hebrew and Italian, absorbed the classics, and was himself composing with facility in colloquial Latin (e.g., the seven *Prolusions* or oratorical exercises delivered annually before the student body) and formal Latin verse. The *Elegia Sexta*, addressed to his closest friend, Charles Diodati, coincides with the Christmas 1629 writing of *On the Morning of Christ's Nativity*. The latter poem underscores Milton's early commitment to English verse as well, begun with *At a Vacation Exercise* (1628) and continued in *L'Allegro* and *Il Penseroso* (1631).

Milton took the M.A. in 1632 and retired to his father's estate in Hammersmith to complete his formal education on his own. There, the legend continues, he read everything published in English since Caxton and virtually memorized Ovid, Virgil, and, of course, the Bible. In 1637, he wrote the strong English elegy *Lycidas*, celebrating the life of his late Cambridge classmate, Edward King, and denouncing the corruption of the English clergy. After his mother died that same year, Milton traveled for fifteen months on the continent, mainly in Italy, where his desire for friendship and cultural contact proved more compelling than religious differences. Among his Catholic hosts was the gracious and learned Giovanni Battista Manso (1560–1645), a former patron of the legendary poet Tasso. Milton paid tribute to the aged scholar in a one-hundred-line Latin poem (*Mansus*) written shortly before leaving Naples in 1638. Although Milton had hoped to visit Sicily and Greece, reports of civil strife in England cut short his travels. On his homeward journey, however, he took time to visit the blind and house-bound *Galileo in Florence. As he later pronounced in his *Areopagitica*, Milton considered the old, persecuted astronomer a voice of reason stifled by religious bigotry and political oppression.

While in Italy, Milton had learned of the early death of his friend Diodati, and when he returned home in 1639 he wrote and published privately the Latin poem *Epitaphium Damonis* (*Damon's Epitaph*). Therein the poet praises Diodati and his virtue, using the conventions of pastoral elegy he had employed two years earlier in *Lycidas*. After tutoring for less than two years, Milton then entered a twenty-year period of religious debate, political controversy, and personal stress that would engage his deepest emotions and invoke his most intense writing, mostly in passionate prose. In 1641, he became a central figure in the pamphlet warfare, in English, that followed the publication of *Bishop Joseph Hall's *An Humble Remonstrance to Parliament*, expressing the prelate's concern over ecclesiastical dissent that threatened to aggravate the growing civil hostility between Parliament and crown. Hall's tract was answered immediately by an anti-Episcopal pamphlet signed "Smectymnuus," an acronym for the five Presbyterian ministers who wrote it, chief of whom was Thomas Young, Milton's friend and boyhood tutor. After Hall countered with his "Defense," Milton entered the fray with a series of three anti-Episcopal tracts of his own, all in 1641: *Of Reformation Touching Church Discipline in England*, *Of Prelatical Episcopacy*, and *Animadversions Upon the Remonstrant's Defense Against Smectymnuus*. In 1642, he followed with *The Reason of Church Government Urged Against Prelaty*, his strongest and most meticulous argument yet against Anglican bureaucracy. The same year saw the publication of *An Apology for Smectymnuus*, Milton's answer to Hall's personal attack on him in the bishop's *Confutation* (1642) of the *Animadversions*.

Milton also married in 1642, an unlikely match at best. The bride was Mary Powell, the seventeen-year-old daughter of Royalist gentry, while the bridegroom was thirty-three and increasingly drawn to the Independent cause in the hostilities that were quickly escalating into civil war. Milton's ill-advised marriage lasted only two months. After his wife left him, he turned again to radical pamphleteering, issuing a series of controversial tracts on divorce. The most notorious, *The Doctrine and Discipline of Divorce* (1643), argued that divorce is acceptable in cases of spiritual incompatibility. In 1644, he published *Areopagitica*, a high-minded, oratorical discourse pleading against censorship and governmental control of printing. In 1645, Milton and his wife reunited and produced three daughters over the next seven years. A fourth child, John, died in infancy in 1652. In that same painful year, Milton's wife died as well, and his blindness became complete.

In the meantime, Milton was busy in the service of *Oliver Cromwell's new government. *The Tenure of Kings and Magistrates* appeared in January 1649, to coincide with the trial and execution of *Charles I. The *Tenure* argues the legitimacy of the regicide but declines to address the question of the legality of the "Rump Parliament," purged of Royalists before it condemned the king to death. The deed done, however, and the Commonwealth declared, Cromwell appointed Milton his Secretary for Foreign Tongues in March 1649. In this capacity he functioned not only as translator and correspondent, but also as apologist for the new regime. In 1651, he answered a widely published Latin tract entitled *Defensio Regia* written by the distinguished French scholar *Salmasius, who deplored the English revolution as a threat to monarchy everywhere. Milton's reply, *Pro Populo Anglicano Defensio*, was inflamed by a passion that prevailed over the sure knowledge that he was sacrificing his eyesight to the effort. In 1652, an anonymous Latin tract appeared (*The Cry of the Royal Blood to Heaven for Vengeance*) attacking Milton and the Puritan "parricides." Milton's 1654 response (*Pro Populo Anglicano Defensio Secunda*) was a highly personal, virulent counterattack against Alexander More, the editor and supposed author of *The Cry*. The author (later identified as Peter du Moulin) had savaged Milton and charged that his blindness was God's punishment for alleged licentiousness. Milton, in turn, blasted "Morus" for adultery with Salmasius's maid and defended his own moral integrity, arguing that his blindness was actually a source of enhanced spiritual insight unavailable to sighted fools.

Oliver Cromwell died in 1658, as did Milton's second wife, Katherine Woodcock, and their baby daughter Katherine. With the restoration of the monarchy in 1660, Milton's personal and political fortunes had descended to their nadir. He lost most of his property and was even imprisoned for a while, until *Marvell and other friends interceded on his behalf. The blind, aging poet married again, to Elizabeth Minshull in 1663, but essentially retreated into himself to rediscover the inner spirit that had always sustained him. Milton's writings had from the beginning been motivated, shaped, and directed by an extended value system that emphasized personal integrity and purity (e.g., *Comus*, 1634; *An Apology for Smectymnuus*); rigorous faith and strong, spiritual reserves (e.g., *When I Consider How My Light Is Spent*, ca. 1652; *Pro Populo . . . Secunda*); and a strong commitment to public concerns (e.g., *The Reason of Church Government*; *Areopagitica*). The latter passion, centered on personal dedication to the improvement of the culture, manifests itself in his lifelong ambition to become an epic poet in the tradition of Homer and Virgil. *At a Vacation Exercise, Elegia Sexta, Mansus, Damon's Epitaph, The Reason of Church Government* all make explicit reference to Milton's sense of destiny for himself and to his awareness of the preparation necessary to realize his aspirations. The young poet had always assumed that his epic achievement would center on English history and that King Arthur would be the likely hero. He even had a working title picked out, *The Arthuriad*. By the early 1660s, Milton's dream of a glorified England had been all but extinguished by disappointment and loss. The desire to contribute the great poem was still burning, however, and the result would be an epic not about patriotic or even temporal matters, but spiritual—an attempt to "justify the ways of God to men." *Paradise Lost*, in ten books, appeared in 1667, followed by *Paradise Regained* and *Samson Agonistes*, published together in 1671. A second edition of *Paradise Lost*, in twelve books, appeared in 1674, the year of the poet's death. Milton died on 8 November and was buried beside his father in St. Giles, Cripplegate.

Bibliography: S. Fish, *How Milton Works*, 2001; W. Parker, *The Life of John Milton*, 2 vols., 1968; J. Shawcross, ed., *The Complete Poetry of John Milton*, 1971; D. Wolfe, et al., *The Complete Prose Works of John Milton*, 8 vols., 1953–1982.

Dale G. Priest

MINSHEU, JOHN (fl. 1599–1623). John Minsheu, a teacher of languages associated for part of his life with Oxford, is recognized today for his massive polyglot work *Ductor in Linguas, The Guide into the Tongues*, published in 1617 in London. The work included some 20,000 lemmas, many of them in eleven languages: English, Welsh, Low Dutch, High Dutch, French, Italian, Spanish, Portuguese, Latin, Greek, and Hebrew. The *Ductor* has been described by D. Noland as marking "the beginning of general English etymological dictionaries" (1187), and it provided an important source for ensuing works. Minsheu has been considered a serious etymological scholar, but in fact his etymologies are not guided by any consistent theory of linguistic history, and he does not follow the usual procedure for scholarly works of citing sources in marginal notes. The work is an extensive compilation drawn from a variety of previous sources, including an English-Spanish dictionary by Richard Percyvall (*Bibliotheca Hispanica*, 1591); John Cowell's explication of law terms; *The Interpreter* (1607); John Gerard's *Herball* (1597), the concordance appended to the Geneva Bible after 1580; Thomas Speght's glossary of Chaucerian words in his 1602 edition of Chaucer; Randle Cotgrave's *Dictionary of the French and English Tongues* (1611); and various polyglot dictionaries published in Belgium. Minsheu had previously absorbed and much expanded Percyvall's 1591 Spanish-English dictionary into his *Dictionarie in Spanish and English*, published in 1599 and later appended (with a separate title page) to the 1617 *Ductor*. Minsheu's method involved concatenating his various sources to construct more thorough and explanatory glosses, sometimes with multiple (and conflicting) etymologies for a given word. His aim for Spanish words in particular seems to have been exhaustive documentation; he includes slight variants in structure and spelling conjugations of irregular verbs, and idiomatic expressions. His practices are only partially informed by the growing Renaissance awareness of language's historicity; for him, words and things do not have an incidental relationship but are inextricably and essentially linked.

Bibliography: D. Nolan, "John Minsheu's *Ductor in Linguas* and the Beginning of English Historical Lexicography," *Dissertation Abstracts International* 48 (1987): 1187A; J. Schafer, "John Minsheu: Scholar or Charlatan?" *Renaissance Quarterly* 26 (1973): 23–35; J. Schafer, "Introduction," *Ductor in Linguas, The Guide into the Tongues and Vocabularium Hispanicolatinum* (*A Most Copious Spanish Dictionary*) (1617), 1978.

Sara V. Fink

MINUIT, PETER (ca.1580–1638). Director general of New Netherlands and governor of New Sweden (modern New York and Delaware, respectively), Minuit was the son of Jean Minuit and Sarah Breil, French-speaking Walloons who moved to the Netherlands in the face of religious persecution. Minuit married Gertruudt Raets in 1613. In 1625, Minuit worked for the Dutch East India Company assisting Willem Verhulst, the provisional director New Netherlands, a colony established along the Hudson River the previous year. Verhulst was removed from office by the governing council, which made Minuit director general of the colony in 1626.

As its leader, Minuit sought to reorganize the colony and make it into a profitable enterprise. The company purchased Manhattan Island from a local Indian tribe for sixty guilders worth of cloth, tools, and other trade goods. In the nineteenth century, an antiquarian, using the exchange rate of the day, determined that the Dutch had cheated the Indians, giving them the equivalent of twenty-four dollars worth of merchandise, creating a widely held but wholly inaccurate myth. The leaders of the Delaware tribe, with whom the transaction occurred, were pleased with the trade, possibly because they may not have had the authority to sell the island. Minuit relocated most of the settlers into New Amsterdam, later New York City, and worked to encourage the fur trade and crop diversity. He became embroiled in a controversy as to the best way to encourage settlement and was removed from office in 1631.

Returning to Europe, Minuit grew interested in a Swedish attempt to establish a colony on the Delaware River. He was named governor of the colony by the New Sweden Company, a firm jointly funded by Swedes and Dutchmen. The colonists established Fort Christina (modern Wilmington) in 1638. In the fall of that same year, Minuit sailed to the Carribean to do some privateering in the Spanish shipping lanes. At the island of St. Christopher, a hurricane forced his ship out to sea where it was lost with all hands.

Bibliography: C. Weslager, *A Man and His Ship: Peter Minuet and the Kalmar Nyckel*, 1989.

Christopher E. Hendricks

MOLIÈRE [POQUELIN], JEAN-BAPTISTE (1622–1673).

French dramatist, comic genius, and court entertainer for *Louis XIV, Jean-Baptiste Poquelin was born in Paris to Marie Cressé and Jean Poquelin, a royal upholsterer. He received a classical education at the Jesuit Collège de Clermont and studied law at the Université d'Orléans. In 1643, Poquelin gave up his bourgeois life, cofounded the *Illustre Théâtre* in Paris with his companion Madeleine Béjart and her family, and subsequently took the name Molière. Unable to compete against the two permanent theaters in Paris, Molière and his troupe begin a twelve-year tour of the provinces in 1645. During this tour of apprenticeship, he began to write, stage, and perform in farces heavily influenced by the Italian *commedia dell'arte*. The wanderings of his theatrical company brought Molière in contact with a variety of people from different milieus. Noting their manner, their mores, and their language, Molière later used his observations to present comic but realistic portraits of them in his plays.

When Molière returned to Paris in 1658, he performed *Corneille's *Nicomède* and his own farce, *Le Docteur amoureux*, before the king. Pleased with the farce, Louis XIV authorized the troupe to perform regularly in the *Théâtre du Petit Bourbon* at the Louvre. Molière experienced his first real success in 1659 with *Les Précieuses Ridicules*, his earliest comedy of manners in which he satirized the pretentious behavior of Parisian and provincial society. By 1661, Molière and his troupe had moved into the best theater in Paris, the prestigious Palais Royal. Although their inaugural play, *Dom Garcie de Navarre*, was a disaster, it was quickly followed by the triumph of *L'École des Maris*, the first of Molière's comedies to include a moral thesis, in this instance that trust and tenderness can compensate for a great age difference between husband and wife. In that same year, Molière also wrote *Les Fâcheux*, his first *comédie-ballet*, which he presented as part of the festivities given by Fouquet at Vaux in honor of the king.

With the enormous popularity of *L'École des Femmes* in 1662, Molière's artistic reputation was firmly established, but his life was not without difficulty. The king's patronage aroused jealousy on the part of the actors at the *Hôtel de Bourgogne*, who criticized the play, instigated a *Querelle* (quarrel) as important as the *Querelle du Cid*, and accused Molière of incest because of his marriage earlier that year to nineteen-year-old Armande Béjart, who they claimed was Madeleine's daughter and not her sister. Moreover, Armande's infidelities caused Molière much unhappiness, a situation which many scholars contend is reflected in his plays, particularly in *L'École des Femmes*.

In 1664 Molière staged his first version of *Tartuffe*, a play which would prove to be even more controversial than *École des Femmes*. Heavily criticized by the devout because it attacked religious hypocrisy, *Tartuffe* was subsequently banned. After several revisions, Louis XIV finally authorized its performance in 1669. But Molière had learned his lesson. Now in failing health, he returned to the light comedies and farces preferred by the king. On 17 February 1673, during the fourth performance of *Le Malade imaginaire*, Molière suffered a convulsion while playing the role of the imaginary invalid. He was taken home, where he died an hour later from a lung hemorrhage. A Christian burial was refused to him, but Armande asked the king to intervene and Molière was quietly buried in a church cemetery the evening of 21 February. In 1680, Louis XIV merged Molière's troupe with another to form the *Comédie Française*.

A frustrated tragedian whose physiognomy was more suitable for playing comic roles, Molière elevated comedy to a new level. Using simple plots, complex and nuanced characters, realistic settings, serious moral subjects, and a mixture of comic registers and devices, he attacked all hypocrisy and imitations of virtue. Molière took universal types, such as the hypocrite, and made them look ridiculous by exaggerating their behavior and by pairing them with other characters who exhibited common sense and moderation. According to Molière, comedy should correct people while entertaining them. Indeed, some of the audience may have recognized themselves in Molière's characters, but they laughed anyway and they continue to laugh 300 years after his death.

Bibliography: R. Fernandez, *Molière: The Man Seen Through the Plays*, 1958; J. Palmer, *Molière: His Life and Works*, 1930, rpt. 1970; H. Walker, *Molière*, 1971.

Leslie A. Sconduto

MOLINOS, MIGUEL DE (1628–1696). Spanish priest, resident of Rome, author of the popular *Spiritual Guide* (1675), Molinos and his teachings were condemned as heretical by the Roman Catholic Church. Molinos was the last of the Spanish mystics, and he formulated the doctrine of Quietism. Molinos was born in Muniesa, in the diocese of Zaragoza, Spain. He studied in Valencia in the Jesuit school of San Pablo. Ordained in 1652, he was sent to Rome as mediator in a cause of beatification during Clement X's papacy. In Rome, he became an esteemed spiritual director. In 1675, his *Spiritual Guide* was published by Fra Juan de Santa María. The *Guide*, a small handbook teaching that Christian perfection is achieved by a mixture of contemplation and divine assistance, caused a sensation. In the next few years, more than twenty editions in different languages were published. As proof of his prestige, such notable personalities as Queen Christina of Sweden and *Pope Innocent XI became attracted to his ideas and considered him a wise theologian.

Miguel de Molinos advocated an extreme form of Quietism, a doctrine of Christian spirituality that holds that perfection consists in the passivity (quiet) of the soul. Molinos's spiritual principle focuses on *One Act*, that is, the teaching that the will, once fixed on God in contemplative prayer, cannot lose its union with the divine. For Molinos, the way of Christian perfection was the interior way of contemplation, which anyone with divine assistance can attain. This contemplation entails a vague view of God that inhibits man's interior powers. The soul remains in *dark faith*, a state of passive purification that excludes all thought and all interior action. Inactivity brings the soul to its principle, the divine being, into which it is transformed. Molinos believed that men must banish their individual wills so that God's will can work unrestricted. According to Quietism, the devil can make himself master of the contemplative's body and force one to perform acts that seem sinful, but because the contemplative does not consent, they are not sins. Quietistic elements have been discerned in several religions, both Christian and non-Christian, through the centuries. But Quietism is generally associated with Miguel Molinos and the renewed conflict over mysticism with the rise of the Quietist controversy during the seventeenth century.

In 1685, at the height of Molinos's influence, and when his friend Innocent XI was pope, the Spanish priest was arrested and subjected to a two-year trial. The Holy Office also decreed the arrest of around 200 Quietists. Among the suspects was the pope's nephew, who was not prosecuted, but as a result was forced to abandon public life. Molinos was accused by the Jesuits and by one of his friends, the powerful French Ambassador Cardinal D'Estrées. It is believed that Louis *XIV was persuaded that the Quietists were acting in Rome against France's interest in its dispute over the future of the Spanish crown with the House of Austria. Louis ordered his ambassador to persecute the Quietists' leader. During the long legal process, and since no heresy was found in his *Guia* and no sound argument was presented against his writings, the judges started to search his private letters and personal life. Some 20,000 letters were presented and numerous witnesses were interrogated with the intention of bringing discredit to Molinos. He was accused of personal immorality and sexual misconduct, charges that have been questioned by modern scholars. Molinos was sentenced to life imprisonment for heresy and died in prison in 1696.

Bibliography: P. Dudon's, *Le Quiétiste espagnol Michel Molinos*, 1921; H. Lea, "Molinos and the Italian Mystics," *The American Historical Review* 11 (1906): 243–260; J. Valente, *Ensayo sobre Miguel de Molinos*, 1974.

Angela Morales

MONTEVERDI, CLAUDIO (1567–1643). Born in Cremona on 15 May 1567, Monteverdi and his brother were musically educated by Marc'Antonio Ingegneri, maestro di cappella of Cremona's cathedral. By 1582, Monteverdi already had an anthology of three-voice motets issued by the Venetian publisher Gardane. From then until 1592, several publishers released collections of Monteverdi's music, culminating with his third book of madrigals. By this time, he had a position as a string player with the duke of Mantua, Vincenzo I Gonzaga, whose music director was the Flemish composer Giaches de Wert. Upon de Wert's death in 1596, Benedetto Pallavicino, who had seniority, succeeded him, and Monteverdi would wait until 1601 for promotion to this post. Nevertheless, this did not impede Monteverdi's career, for so popular was his third book of madrigals that it was reprinted in 1594, but not until 1603 did he have any more music published. During this time, Monteverdi's reputation grew in Mantua and elsewhere, and in 1599, he married court singer Claudia de Cattaneis.

In 1600, the theorist Giovanni Maria Artusi published an attack on what he considered unwarranted harmonic innovations in Monteverdi's music, citing madrigals that were as yet unpublished. Such notoriety was beneficial to Monteverdi's career and marked him as a modern composer who sought to heighten the expressiveness of the lyrics through daring harmonies. The works cited by Artusi then appeared in the fourth and fifth books of madrigals, published in 1603 and 1605. Book 5 also contained his draft of a riposte to Artusi, which was later expanded by Monteverdi's brother, Giulio Cesare, in his *Dichiaratione*, an appendix to the madrigal anthology *Scherzi musicali* (1607). The Monteverdis argued that conventional counterpoint and harmony, the prima prattica, were to be revered, but that Claudio's innovations represented something new, the seconda prattica, in which music serves the words.

Another landmark was the premiere, in February 1607, of Monteverdi's *L'Orfeo*, the first truly great opera, at Mantua. In August, Monteverdi was inducted into the Accademia degli Animosi of Cremona, but on 10 September his wife died. The year 1610 saw the publication of Monteverdi's celebrated *Vespers of the Blessed Virgin*. The volume actually contains, along with a musical setting of Vespers, a complete six-voice Mass, and some sacred concertos. These three different components illustrate Monteverdi's Janus-like position in music history, because musically they represent both conservative and progressive traits. The Mass, based upon a motet by Nicolas Gombert (ca.1495–ca.1560), is the most conservative and features a dense texture of imitative polyphony, the prima prattica; however, the Gregorian psalm tones of the *Vespers* are elaborated with polyphony and instrumental accompaniment, a forward-looking trait. The sacred concertos, works for solo voices and small instrumental ensemble, are the most progressive, with characteristics most like the new genre of opera, and secular vocal chamber music, the seconda prattica. During this time, Monteverdi was becoming increasingly disenchanted with the Gonzagas. His release finally came in July 1612, after the death of Vincenzo Gonzaga, when his son and heir Francesco fired Monteverdi.

In August 1613, Monteverdi was summoned to Venice, where he successfully auditioned for maestro di cappella of St. Mark's, vacant since the death of Giulio Cesare Martinengo on 10 July. Monteverdi possessed not only great musical gifts but also the administrative skills that Martinengo evidently had lacked, and he revived the basilica's moribund musical establishment. While at San Marco, Monteverdi had the choir sing not only his own music, but also music by sixteenth-century masters like Palestrina and Lassus, and music by his own chorister-composers, like Francesco Cavalli, Alessandro Grandi, and Giovanni Rovetta. These last two served successively as deputy maestro di cappella to Monteverdi, thus providing the composer with more time to answer requests for dramatic works and madrigals, like those in his forward-looking Book 7. Monteverdi also accepted commissions for music from religious confraternities like the Scuola di San Rocco. From about 1630 to 1638 Monteverdi composed little, and some time around 1632 he was ordained a priest. The opening, in 1637, of Venice's Teatro San Cassiano, the first public opera house, sparked the final phase of Monteverdi's creative life. After the 1640 revival of *Arianna*, Monteverdi composed three new operas, including his final masterpiece, *L'incoronazione di Poppea* (1642). He died on 29 November 1643.

Bibliography: D. Arnold, *Monteverdi*, 1990; D. Arnold and N. Fortune, eds., *The New Monteverdi Companion*, 1985; G. Tomlinson, *Monteverdi and the End of the Renaissance*, 1987.

Susan Treacy

MORE, HENRY (1614–1687). Born into a family of well-to-do Calvinists, More earned an education at the same school in Grantham, Lincolnshire, that later would train *Sir Isaac Newton. After this initial training, and some further education, More was admitted to Christ College, Cambridge, where much to the distress of his family, his Calvinist upbringing continued to lose its hold. Upon graduation, More became a fellow of Christ College and remained there in residence for nearly fifty years until his death. A scientist, theologian, and poet, More remains best known for his prominent role as the most prolific of a group of scholars known as the "Cambridge Platonists." These men—Cambridge University theologians, philosophers, and scientists all—relied on Platonic philosophy to bridge the gap between religion and science and between faith and reason. They championed Plato as a pre-Christian epitome of the successful philosopher: a thinker who used reason to illuminate faith in things beyond reason and faith to illuminate reason applied to material reality, all then applied in active, communal endeavors and displays. Specifically, Henry More and the Cambridge Platonists added the influence of Christianity and of new scientific enterprise to Plato's concepts. They neither denied God a primary role in natural processes, as *Hobbes and *Descartes had, nor embraced a Puritan "enthusiasm," the experience of God as a revelation apart from reason or action.

It was to just those relationships that More—by then, a committed supporter of the Church—devoted his writing. Oddly enough, More's earlier years found him holding great appreciation for Descartes, with whom (before ultimately coming to reject Cartesian philosophy, seeing it as promoting atheism) More had initiated correspondence. The letters between the men circulated and helped to secure a following for Descartes in England. More also corresponded with and influenced the younger *Joseph Glanvill, himself an important member of the Royal Society, whose work on spirits changed the way Protestants understood spiritual mysteries. More worked though copious letters, prose, and poetry to uncover and argue against both the atheism he perceived in an emerging empiricism and the enthusiasm he saw in the Puritan opposition. His intentions are clear from just a sample of his many works: *Philosophical Poems* (1647), *Antidote against Atheism* (1652), *Enthusiasmus Triumphatus* (1656), *The Immortality of the Soul* (1659), and *Enchiridion metaphysicum* (1671). More's volumes testify to his commitment to his vocation, but his willingness to turn down more lucrative and powerful positions does as well: He was offered the post of dean at Christ's Church, Oxford, and at least two opportunities to serve as bishop. In the view of historians of science and literature, More's success derives not from beauty of written expression—which some rightly claim is lacking in his work—but rather in his rigorous wrestling with ideas, his attempt to provide a sound guide to navigating the opposing systems of belief and practice in his time: progress and tradition, science and faith, mechanistic philosophy and spiritual practice, and action and contemplation.

Bibliography: S. Hutton, ed., *Henry More (1614–1687) Tercentenary Studies*, 1990; A. Lichtenstein, *Henry More: The Rational Theology of a Cambridge Platonist*, 1962.

Rebecca Totaro

MORLEY, THOMAS (ca.1557–1602). Morley inherited the musical legacy of his teacher *William Byrd, both in style and range of output. Born in Nor-

wich, Morley became master of the choristers and organist of the cathedral there
in 1583. He earned the Bachelor of Music degree at Oxford in 1588 and was
listed as the organist at St. Paul's Church in London in 1589. Like Byrd twenty
years earlier, Morley served as a Gentleman of the Chapel Royal from 1592
until his death ten years later. In a 1591 correspondence between Charles Pages
and government agent Thomas Phelippes, Morley was implicated in espionage
activity against the English Catholics. It has been suggested that although Morley may have in fact been a Catholic himself, he favored the Church of England
as a better alternative to the possibility of Spanish Catholic rule.

Morley's vocal music is notable for its extremes in both the gravity of the
sacred and bouyancy of the secular works. The early motets were probably
modeled after Byrd but later ones after Italian madrigals, which style Morley
adopted and developed in the English language, examples being the *Canzonets*
(1573) and *Madrigalls* (1594). His instrumental compositions, not as numerous
as the vocal, include keyboard dances and variations, and at least one work for
English consort. He also produced several Anglican services. Although Morley
resided near *Shakespeare and set "It was a lover and his lass" (from *Twelfth
Night*), there is no evidence that they ever collaborated.

Successor to Byrd also in the arena of early English music printing, Morley
expanded this effort to a greater extent in both number and variety of
publications, but he was the last to hold such a monopoly in England. Although
Morley's espousal of Italian madrigal style did not take permanent hold in England, he did spread continental musical ideas there by means of his influential
manual *A Plaine and Easie Introduction to Practicall Musicke* (1597, dedicated
to Byrd) as well as through three volumes of Italian music he published between
1595 and 1598.

Bibliography: P. Brett, "Thomas Morley," *The New Grove Dictionary of Music and
Musicians*, ed. Stanley Sadie, 1980; T. Morley, *A Plain and Easy Introduction to Practical Music*, ed. R. Harman, 1973.

Joyce Lindorff

MUGGLETON, LODOWICKE (1609–1698). Lodowicke Muggleton was
born in London in 1609. As a journeyman tailor, Muggleton became involved
in the Puritan circle of his cousins William and John Reeve. Like many radicals
during the civil war and interregnum, around 1650 Muggleton entered a period
of religious despair and "seeking." With John Reeve, he associated with Thomas
Tany and John Robins, prophets who claimed to possess miraculous powers and
direct communication with God (indeed, Robins claimed to be God!). From 3
to 5 February 1652, Reeve received a revelation proclaiming Muggleton and
himself as the two last Witnesses of the Spirit from Revelation. Engaging in
debates with sectarian leaders like Robins and *George Fox, proclaiming the
coming apocalypse and damning their enemies, Reeve and Muggleton gathered
a small number of followers (approximately 250 at Muggleton's death) around

them. Their rejection of all of the formal elements of religion led to a community based on irregular, informal meetings in public houses. The two Witnesses were imprisoned for blasphemy for six months in 1653, but thereafter, with the exception of another sentence for blasphemy in 1677, Muggleton managed to avoid imprisonment. After Reeve died in 1658, Muggleton consolidated his leadership of the sect, fending off a challenge from the one-time Ranter, Laurence Clarkson, around 1660.

Until his death in 1698, Muggleton devoted himself to publishing his own works, including an autobiography, as well as revisions of Reeve's writings, and to counseling his flock through his extensive correspondence. Unlike many of the other sectarian leaders of the time, he was not an evangelist, but he provided his followers with practical encouragement and advice. It is to his organizational ability and quiet, realistic temperament that we must attribute the tiny sect's longevity; the last Muggletonian died in 1979, leaving behind an astonishing record of the survival of seventeenth-century radical religion.

Bibliography: C. Hill, B. Reay, and W.M. Lamont, *The World of the Muggletonians*, 1983; E. Pearlman, "Typological Autobiography in Seventeenth-Century England," *Biography* 8 (1985): 95–118.

Janet Bertsch

MURILLO, BARTOLOME ESTEBAN (1617–1682).

A painter and draftsman of the Spanish Baroque, famous for religious paintings, portraits, and sentimental genre paintings of children and poor people, Murillo studied in the studio of his relative Juan del Castillo. His early paintings show knowledge of the work of Francisco de Herrera the Elder, *Diego Velazquez's Sevillian paintings, and the tenebrism of *Francisco Zurbaran's canvases. Paintings from 1649–1650 have greater naturalism inspired by *Jusepe Ribera and Flemish and Dutch prints. Using black chalk or pen, Murillo made quick drawings, while he combined red and black chalk and used pen with wash for more finished sketches. Over 100 drawings and some oil sketches survive.

Murillo depicted the Virgin's Immaculate Conception many times throughout his career. Two examples dating from 1650 for the convent and church of San Francisco in Seville, portraying a beautiful young Virgin surrounded by cherubim, were to elicit the religious emotion of devout viewers by combining naturalism with the spiritual world. In 1650, Murillo became president of the Real Academia de Belles Artes de Santa Isabel de Hungria, but by 1663 he was less active in the academy. He also became a lay member of the Venerable Orden Tercera de San Francisco in 1662 and joined the Confraternity of the Santa Caridad in 1665.

His images of children, popular from mid-century, include *Boy Killing Fleas* and *Boys Eating Grapes and Melon*. The decade before he died, Murillo worked primarily on large religious compositions. It was while painting the high retable altar of the *Mystic Marriage of Saint Catherine* for Santa Catalina in Cadiz that Murillo fell from the scaffolding and died a few months later.

Bibliography: N. Mallory, *Bartolome Esteban Murillo*, 1983; A. Palomino, *Lives of the Eminent Spanish Painters and Sculptors*, 1987.

Deborah H. Cibelli

N

NAPIER, JOHN (1550–1617). John Napier was a landed Scottish laird who lived most of his life on the family estate near Edinburgh, Scotland. Beyond managing his family's extensive holdings, Napier's passions were religion and politics. In service of these passions, he wrote extensively and provided counsel to both *King James VI and the general assembly of the Scottish Church. He spent his leisure time studying mathematics, and out of these studies came the invention of logarithms, a computational system that greatly facilitated the progress of science for the next three centuries.

Napier was born in Merchiston Castle on his family estate in 1550 and died in that same dwelling in 1617. Little is known of his childhood and youth other than a brief attendance at St. Andrews University and probable travels abroad. He was a violent anti-Catholic and an avid supporter of John Knox. In both pamphlet and in person, he petitioned James VI of Scotland to support the Scottish church and avoid entanglements with Papists. In 1593, he published *A Plaine Discovery of the Whole Revelation of Saint John*, a commentary on the biblical book of Revelation in which he attacked the church at Rome, vilified the pope as the Antichrist, and counseled James to purge court and family of connections to either. As befits his troubled view of his time, he also designed instruments of war intended for the defense of his country, inventions that included burning mirrors, artillery pieces, and metal chariots akin to tanks.

In contrast to his broad and passionate public interests, Napier's leisure investigations of mathematics were very focused and private. A major obstacle to the accelerating progress of science was the difficulty of completing the arithmetic computations associated with advances in astronomy, physics, and engineering. Astronomers in Denmark had noted that trigonometric identities allowed one to convert certain multiplications to additions. Since additions of large numbers are much easier than multiplications, Napier spent twenty years devising a system to allow such a conversion in all circumstances. In 1614, he published *Mirifici Logarithmorum Canonis Descriptio (Description of the Marvelous Canon of Logarithms)*. The publication was greeted enthusiastically. Professor Henry Briggs, who was to become the first Savilian professor of geometry at Oxford, traveled to Scotland to consult with Napier. Together they made modifications to Napier's system that resulted in the Common Logarithms or base-ten logarithms that revolutionized scientific computation in the decades that followed. Napier also popularized the use of the decimal point in representing real numbers, made contributions in spherical trigonometry, and devised a mechanical calculating device, Napier's rods, that foreshadowed the invention of the slide rule. Napier believed that his reputation had been established by the publication of his religious tract *A Plain Discovery*, in which he prophesied the impending end of the world. It is ironic that his creation of logarithms that served the computational needs of scientists until the invention of the computer was his real contribution to posterity.

Bibliography: C. Boyer, *A History of Mathematics*, 1968; C. Knott, *Napier Tercentenary Memorial Volume*, 1915.

Ed R. Wheeler

NAUDÉ, GABRIEL (1600–1653). Although primarily remembered today as a freethinker and skeptic, Naudé in his own lifetime was better known as the personal physician to Louis XIII and an accomplished book collector and librarian. He gained his reputation as the latter with the publication of *Avis pour dresser une bibliothèque* (1627), which he wrote while serving as librarian to the Président de Mesmes. Naudé made a career of collecting books for notable people and served as librarian to *Cardinal Richelieu, Cardinal Mazarin, and Queen Christina of Sweden.

As librarian to Mazarin, Naudé purchased some 40,000 volumes for the cardinal's collection, which, as a result of political turmoil during the Fronde, was ordered sold by the *parlement* (law court) of Paris. Upon his return to favor, Mazarin again called upon Naudé to rebuild his collection of books. The librarian did so and was able to convince the cardinal to open his collection to scholars on Thursdays, thus introducing the first public library in France. This collection forms the basis of the present-day Bibliothèque Mazarine, located at the Institut de France.

Naudé wrote books in his own right, including an *Apologie pour les grands personnages faussement soupçonnés de magie* (1625), a response to a virulent attack on suspected *libertins* by Garasse, and *Considérations politiques sur les coups d'état* (1639), written while serving the Cardinal di Bagno in Italy and which shows the influence of Machiavelli. A friend of Guy Patin and a member of the *"tétrade"* of freethinkers along with Diodati, *Gassendi and *La Mothe le Vayer, Naudé is considered an important *libertin érudit* of the seventeenth century.

Bibliography: R. Pintard, *Le libertinage érudit dans la première moitié du XVIIe siècle*, 1943; C.-A. Sainte-Beuve, *Portraits littéraires*, 1843.

Patricia Armstrong

NEWCOMEN, THOMAS (1663–1729). The man often credited with the invention of the steam engine, Newcomen was born in Dartmouth, England, into an aristocratic family that had lost its wealth. The son and grandson of religious dissenters, Newcomen was active in the sect of English Baptists and would later serve as a lay preacher from time to time. Little is known about Newcomen's education or training, but he entered into business in Dartmouth in 1685 as an ironmonger (blacksmith) with a partner, John Calley, a fellow Baptist. He did not marry until twenty years later. In Dartmouth, Newcomen had ample opportunity to learn about the technology of mining, frequently visiting the mines at Cornwall and Devon.

In 1698, Thomas Savery, another native of Devonshire, had invented a primitive steam pump for the rapidly growing coal mining industry. The mineshafts, some of which extended under the sea, were likely to fill by seepage with water, which had to be removed. Savery's pump involved filling a pipeline with steam, then condensing the steam to form liquid water, creating a vacuum that would pull the water upward. With more steam then applied from below, the water would then be forced to the surface.

Newcomen's design included a cylinder and piston, which could be filled with steam pushing the piston one way and then cooled so that the steam condensed, allowing air pressure to push the piston back. The first successful engine was demonstrated in 1712. Because Savery had obtained a very broad patent covering any sort of heat engine to lift water, Newcomen and his partners were not able to gain a patent of their own.

Nonetheless, the owners of the Savery patent did not prevent the building of Newcomen engines. Newcomen died in London on 5 April 1729 at the age of sixty-nine. In 1765, the Scottish engineer James Watt introduced an engine with a separate condenser, which was in turn far more efficient than Newcomen's and became the standard design for the next century.

Bibliography: L. Rolt, *Thomas Newcomen*, 1963.

Donald R. Franceschetti

NEWTON, ISAAC (1643–1727). Isaac Newton, the preeminent figure in the scientific revolution, displayed in his work both poles of scientific inquiry. Newton was a brilliant theoretician who built on the preparatory work of several talented mathematicians and seized the universal principle that defines the calculus, massaged the experimental data of his contemporaries, and announced universal principles that described the physical world. But Newton was also the skillful experimentalist who built instruments when they could not be bought and who entered the laboratory with vague intuitions, designed experiments to direct those intuitions, and emerged with a cogent physical model.

Isaac Newton was born in Woolsthorpe, England, in 1643, the son of a farmer who had died before Isaac's birth. Reared first by a grandmother, then returned to the home of his mother with whom relations were strained, Newton's childhood and youth were distinguished primarily by his facility at building mechanical toys. Nonetheless, he survived difficult family circumstances and attempts to make him a farmer and, at age eighteen, entered Trinity College, Cambridge. Despite the fact that the curriculum at Cambridge was still largely Aristotelian in focus, Newton quickly immersed himself in the new natural philosophy that was providing the foundation of the scientific revolution. His readings of *Descartes, *Kepler, and *Boyle launched his scientific inquiries; his reading of Descartes' *La Geométrie* and *Wallis's *Arithmetica infinitorum* began his investigations in mathematics.

Shortly after Newton's completion of the B.A., the plague struck Cambridge, and faculty and students left the university and the city. In the ensuing months of enforced retreat at his rural home, Newton intensified his investigations in mathematics, mechanics, and optics. When he returned to Cambridge in 1666, he was in possession of the calculus and had made major first steps in his understanding of motion and the nature of light. However, he was also unknown to the scholars across England and the continent who would have been intensely interested in his investigations. His anonymity began to fade when in 1669 he circulated to a limited circle a tract entitled *De Analysi per Aequationes Numeri Terminorum Infinitias* followed by a revision in 1671 entitled *De methodis serierum et fluxionum*. In these papers, Newton recasts the mathematical work of the first half of the century and "discovers" the calculus by making the crucial observation about the inverse relationship between the problems of finding the slope of a curve and the area bounded by a curve.

In 1671, his studies of optics led to his invention of a reflecting telescope different from *James Gregory's design. This invention drew the admiring attention of the Royal Society of London and, as a consequence, Newton presented to them a paper on light and color. This paper was generally well received, but also received some criticism. Newton's angry reaction to this criticism foreshadowed many other angry exchanges in response to the public discussion of his work. Further, the distracting correspondence that accompanied his several public presentations of his work reinforced his natural tendencies to work alone and to withhold publication. This unwillingness to publish led to more controversy later in his career when questions of priority arose, because Newton's unpublished work predated the published discoveries of other mathematicians and scientists. Perhaps the most notable of these controversies was his disagreement with *Leibniz as to who should take primary credit for the calculus; mathematicians today acknowledge both men.

By 1676, Newton had withdrawn almost completely from public discussion of the sciences and retreated to the privacy of the academic life. Moreover, his intense studies were focused not on the natural sciences but on alchemy and theology. However, he could not isolate himself completely from the domains in which he had excelled. Correspondence with contemporaries, excitement about observations of comets, and a return to studies of geometry all conspired to draw him back to his scientific inquiries. In 1684, *Edmund Halley asked Newton about the connections between Kepler's laws for planetary motion and the principles of dynamics. Shortly thereafter, Halley received a nine-page treatise showing that the inverse square law for gravitational force would result in elliptical orbits for the planets. More importantly, the question touched off two years of intense intellectual activity that resulted in publication in 1687 of Newton's *Philosophiae naturalis principia mathematica*, the masterwork of the new science. In its pages, Newton proposed the theory of universal gravitation, worked out its implications for the motion of planetary bodies, and stated his three laws of motion. These were (a) a body at rest tends to stay at rest, while a body in motion tends to stay in motion unless acted upon by an outside force; (b) the acceleration of a given mass by a force is directly proportional to the force and inversely proportional to the mass; and (c) for every action there is an equal and opposite reaction. In 1704, his *Opticks* was published, in which he asserted that light was composed of particles or "corpuscles." This corpuscular theory dominated the science of optics until the nineteenth century, when it was superseded by the wave theory, and then again in the twentieth century when the wave and particle theories were combined by quantum mechanics.

Subsequent to publication of the *Principia*, Newton's interests shifted. Perhaps because of his new prominence, he was selected to represent Cambridge University in several political assemblies. Contacts made in these political duties led to his appointment as Warden of the Mint, an administrative post he took quite seriously. As his involvement with political and administrative issues increased, his creative work in science progressively diminished. However, he became the patriarch of English science. In 1703, he was elected president of the Royal Society; in 1705, he was knighted for his scientific achievement; and during the next two decades he supervised the publication of many of his prior works, some for the first time.

Bibliography: C. Boyer, *A History of Mathematics*, 1968; R. Westfall, *Never at Rest: A Biography of Isaac Newton*, 1980.

Ed R. Wheeler

NICOLE, PIERRE (1625–1695). Nicole was an important polemicist for Jansenist Catholicism. Born in Chartres and educated in Paris, he taught at Port-Royal, where from 1655 to 1668, in conjunction with his colleague *Antoine Arnauld, he edited and disseminated the doctrines of *Cornelius Jansen. His *Logique de Port-Royal* (1662) reflected the influence of *Pascal and *Descartes. *Préjugés légitimes contre les calvinistes* (1671) is typical of his anti-Calvinist stance; in other works, he asserted the Catholic understanding of the nature of eucharistic transubstantiation. Skeptical of mystics, he criticized *Molinos and Guyon in his last work (composed at the urging of *Bossuet) *Réfutation des principales erreurs des quiétists* (1695). He is most noted for his *Essais de* *morale* (four volumes from 1671–1678, ten more posthumously), which explored a variety of religious and ethical issues. In later years, Nicole moderated his Jansenist views somewhat, embracing a more universalist view of divine redemptive grace.

Bibliography: J. Carrevre, "Pierre Nicole," in A. Vacant et al., eds., *Dictionnaire de Théologie Catholique*, vol. 11, 1931.

Christopher Baker

NICOLET [NICOLLET], JEAN DE BELLESBORNE (ca.1598–1642). Jean Nicolet, son of a French royal courier, was born in Cherbourg, probably in 1598. In 1618, at age twenty, he sailed to Quebec as an employee of the Company of Merchants of Rouen and Saint Malo. He worked as intermediary between the French and various tribes of Indians living in the Hudson Bay area, learning their languages and customs and keeping them loyal to French interests. Following two years with the Algonquins and Hurons, and nine with the Nipissings, during which he kept a valuable journal (now lost), Nicolet returned to Quebec in 1633. Though newly employed as clerk of the Company of One Hundred Associates there, he agreed to go west when *Samuel de Champlain asked him to locate the East China Sea and to establish friendly relations with the Winnebago Indians. French understandings about Great Lakes geography were imprecise, so Nicolet took along a colorful robe of Chinese damask in order to impress any Chinese officials he might encounter. Wearing this robe, Nicolet signed a treaty with the Winnebagos near what is now Green Bay, Wisconsin. He became the first European to visit much of the Upper Midwest. From the Winnebagos, he heard of and visited the Sioux and the Potawatomi. Unhappy because he had not reached China, he returned to Quebec in 1635. With two other men, he acquired several grants of land near Trois Rivières and married Marguerite Couillard in 1637; they had two children. He had earlier (in 1628) fathered a child by a Nipissing woman. Respected for his knowledge of Native American life, he served as interpreter for missionaries bringing religion into the region. Working temporarily in Quebec in 1642, he set out for Trois Rivières to save an Iroquois prisoner of the Hurons from torture. His little boat overturned, and he drowned.

Bibliography: C. Butterfield, *History of the Discovery of the North-west by Jean Nicolet in 1634, with a Sketch of his Life*, 1881, rpt. 1969; L. Kellogg, *Early Narratives of the Northwest, 1634–1699*, 1917; L. Kellogg, *The French Regime in Wisconsin and the Northwest*, 1925.

Keir B. Sterling

NIKON (1605–1681). Russian churchman, patriarch, and reformer of the Russian Orthodox Church from 1652 to 1666, Nikita Minin, afterward known as Nikon, was born in Veldemanovo, Russia. He married prior to entering the clergy, and he and his wife soon settled in Moscow. Tragedy struck when the couple lost all three of their children. Nikon sought penitence through solitude, going into a self-imposed exile, living first as a monk, then a hermit, and ultimately becoming an abbot and moving among various northern Russian provinces from 1634 to 1646. While in Moscow on an independent religious business matter, Nikon so impressed Tsar Alexis that in 1652 he accepted the tsar's appointment as patriarchate of Russia on the condition that he have absolute authority in all church matters. However, Nikon's aggressive efforts to reform church disciplines, particularly his introduction of a new prayer book in 1654, combined with his hostile intolerance of those who opposed his new ideologies, soon created opposing factions within the church. The result was the formation of a major opposing sect, the Raskolniki (the Old Ritualists or Old Believers), who retained the traditional religious fundamentals. By 1658, Nikon's usurpation of the tsar's power during his absence, his insistence upon church despotic patriarchy over state authority, and his brutal treatment of those who challenged him caused sufficient dissension within the church hierarchy to bring about his discharge and ultimate banishment. In 1666, he was disposed and disgraced. His reforms however, remained intact long after his displacement. Nikon died on 27 August 1681 while en route to Moscow.

Bibliography: N. Lupinin, *Religious Revolt in the XVII Century: The Schism of the Russian Church*, 1984.

David M. Rosen

NORDEN, JOHN (1548–1626). Chorographer and estate surveyor, John Norden played a crucial role in early modern Britain's cartographic revolution. He was born in Somerset, the son of a yeoman. In 1564, he entered Hart Hall, Oxford, receiving his M.A. in 1573. Early in his career, Norden sought to produce a cartographic and chorographic redescription of England, one that would surpass Christopher Saxton's *Atlas of England and Wales* (1579). Urged on by Lord Burghley, Norden completed his first map and description of an English county, Northamptonshire, in 1591. Unlike Saxton, however, Norden failed to secure patronage. Thus, his ambitious desire to assemble an atlas that would include, among other things, road maps and town plans (which Saxton's maps lacked) was thwarted. In fact, only two of the county maps to be included in the proposed *Speculum Britanniae* (*A Mirror of Britain*) were published during his lifetime: Middlesex (1593) and Hertfordshire (1598). Norden did, however, go on to produce surveys of five other counties; his maps of Essex, Hampshire, and Cornwall were published well after his death.

Norden's map of Middlesex has drawn the attention of map historians because it was the first map engraved in England to include an explanation of signs. This innovative table of signs included, for example, signs for market towns, houses, and palaces of Queen Elizabeth, houses of nobility, castles and forts, and hospitals. Many of these innovations are explained in Norden's *Preparative to his Speculum Britanniae* (1596), a justification of his, by then, foundering project. Among Norden's other publications are *The Surveyors' Dialogue* (1607) and his road book *England, An Intended Guyde for English Travailers* (1625), in which he discusses one of his most important innovations: the triangular tables used to measure distances between towns.

That Norden's maps were incorporated into both the sixth edition of *William Camden's *Britannia* (1607) and *John Speed's *Theatre of the Empire of Great Britaine* (1611–1612) is evidence of his cartographic expertise.

Bibliography: S. Mendyk, *"Speculum Britanniae": Regional Study, Antiquarianism, and Science in Britain to 1700*, 1989.

Christopher Ivic

NORTON, JOHN (1606–1663). Born in Hertfordshire on 6 May 1606, Norton earned his B.A. from Cambridge's Peterhouse College in 1624 and his M.A. from the same college three years later. Attracted both to Puritanism and to the New World, Norton landed in Plymouth in October 1635. The following spring, he moved to Massachusetts Bay and to the church at Ipswich. Antinomianism, the belief that the Gospel freed the believer from adherence to earthly or spiritual laws of any kind and that faith alone was necessary to salvation, was the theological controversy of the day; Norton held for the orthodox Puritan position that God's grace, and God's grace alone, determined Election. *Anne Hutchinson, in particular, disagreed, and Norton found himself among the members of the Massachusetts Synod of 1637 that dealt with the controversy raised by her resistance. In 1648, Norton and others wrote the Cambridge Platform to guide the Massachusetts congregations, and in 1656 he succeeded *John Cotton as minister of the First Church in Boston according to Cotton's wish.

After the Restoration, Norton and Governor Simon Bradstreet renegotiated the Massachusetts charter with *Charles II. The new terms included ownership of property, not Church membership, as a qualification for voting; court proceedings to be held in the name of the King; toleration of Dissenters; and the enforcement of the Navigation Acts. The changes were not popular, for they limited the independence of the colony; Norton and Bradstreet were regarded as having betrayed the trust of their fellow colonists. The first textbook written in the colonies was Norton's; entitled *Responsio ad totam questionum syllogen* (1648), it responded to questions concerning the nature of church government. Two of his other works, *The Orthodox Evangelist* (1654) and *The Heart of New England Rent at the Blasphemies of the Present Generation* (1659), set out the beliefs of the founders of Massachusetts. Both works remain necessary for a full understanding of the period. In 1658 appeared *Abel Being Dead yet Speaketh: or the Life and Death of . . . John Cotton*, the first separately published biography of an American. In the New World, Norton accumulated a considerable fortune (some £2000), to say nothing of a library of 729 volumes. He died without issue on 5 April 1663. Norton was an ancestor of the famed nineteenth-century Harvard scholar Charles Eliot Norton.

Bibliography: D. Hall, ed., *The Antinomian Controversy, 1636–38: A Documentary History*, 1968; P. Miller, *The New England Mind: The Seventeenth Century*, 1954.

Martha Oberle

O–P

OLDFIELD, ANNE (1683–1730). Born in Pall Mall in 1683, Anne Oldfield was the daughter of a tavern owner. Though his ambition bought him a position in the Horse Guards, his early death left Anne and her mother relying on relatives for their survival. In 1699, at sixteen, Oldfield began her theatrical apprenticeship at Drury Lane; by 1703, she gained notoriety in John Crowne's *Sir Courtly Nice*, and soon her comic talent and her flair for portraying fashionable upper-class women earned her great favor with playwrights like Colley Cibber and *George Farquhar. She took up over 100 roles during her career, over sixty of them original; Cibber claims to have written the notable roles of Lady Betty Modish in *The Careless Husband* (1704) and Lady Townly in *The Provoked Husband* (1728) specifically with Oldfield in mind. Much of Oldfield's notoriety, however, stems from life beyond the stage. It was widely rumored that Oldfield had borne a daughter out of wedlock during her apprenticeship. In 1703, moreover, she began living openly and out of wedlock with Arthur Maynwaring, whose active involvement in promoting both Oldfield's career and Whig politics gained public attention. Though Oldfield and Maynwaring raised their son Arthur together and remained devoted to one another until Maynwaring's death in 1712, they never married. Soon after, Oldfield began living just as openly with another Whig lover, Charles Churchill, in 1714, and bore him a son as well. She died in 1730, and her autopsy revealed that she had long suffered from ovarian cancer.

Bibliography: R. Gore-Brown, *Gay Was the Pit: The Life and Times of Anne Oldfield*, 1957; J. Lafler, *The Celebrated Mrs. Oldfield: The Life and Art of an Augustan Actress*, 1989.

Hillary Nunn

OPITZ, MARTIN (1597–1639). Opitz was celebrated by his contemporaries as the "father of German poetry." He was an important representative of late humanism which, in Germany, is characterized by the tension between confessionalism and the emerging absolutism. His biography is marked by the turmoil of the Thirty Years' War (1618–1648). Opitz was born on 23 December 1597 in Bunzlau (Silesia) as the son of a butcher. Because of his unusual talents, young Opitz was given the opportunity to attend the distinguished Maria-Magdalena Latin school in Breslau. In 1619, he continued his studies at Heidelberg, where he developed an affinity to Calvinism. During this time, he also made the acquaintance of the Strassburg professor Matthias Bernegger, whose polyhistoric eclecticism influenced an entire generation of young German intellectuals. After the Battle of the White Mountain in 1620, which marked the end of Calvinism in Germany, Opitz fled to Leiden, where he met the influential Dutch writer *Daniel Heinsius.

In 1621, Opitz returned to war-torn Silesia, but accepted a teaching position at the Calvinist Latin school in Weissenburg (Transylvania) in 1622. From 1623 onward, he held several administrative and diplomatic positions with Protestant and Catholic noblemen and princes in Silesia—although he never abandoned his own humanist and Calvinist leanings. He was crowned poet laureate by Emperor Ferdinand II in Vienna in 1625. In the years between 1626 and 1632, he served as private secretary to Count Karl Hannibal von Dohna who, as imperial representative, oversaw the bloody Catholic Counter Reform in Silesia. In 1627, Opitz was knighted as "Opitz von Boberfeld," and in 1629, he was initiated into the most significant German language society, the *Fruchtbringende Gesellschaft* (Fruitbearing Society). Opitz undertook many diplomatic missions, most notably to Paris in 1630, where he befriended the Dutch philosopher *Hugo Grotius. When Dohna was forced to leave Silesia in 1632, Opitz reentered the service of the Protestant dukes of Liegnitz and Brieg. After the Emperor regained control, a politically compromised Opitz had to leave Silesia in 1635. He entered the service of the Polish King Vladislav IV. Opitz died of the plague on 20 August 1639 in Danzig while on a diplomatic mission for the King.

The most significant and influential of Opitz's works is the 1624 *Buch von der Deutschen Poeterey* (Book of German Poetry), a poetological treatise in the late humanist tradition of Julius Caesar Scaliger and Pierre de Ronsard. He legitimized poetry as a respectable occupation and justified it through its social and political function. He pioneered the use of word stress in German metrics and thus reconciled the conflict between classical metric systems and the natural stress pattern of the German language. His linguistic and poetic norms, which remained authoritative throughout the German seventeenth century, contributed decisively to the creation of a German national literature and of a standard literary language in the German vernacular and anticipated the aesthetic needs of the absolutistic state.

Between 1625 and 1638, Opitz published several collections of poems, which served to illustrate his own poetic norms. Many panegyrical poems are dedicated to princes whose services he sought to enter. A similar agenda underlies his *Schäffery von der Nimfen Hercinie* (*Pastoral of the Nymph Hercinie*, 1630), the first original bucolic eclogue in prose in the German language. With *Dafne* (1627) and *Judith* (1635), both collaborations with the composer *Heinrich Schütz, Opitz created the first German opera librettos. Opitz translated a wide range of texts, from John Barclay's novel *Argenis* (1626) to Sophocles' *Antigone* (1636) to the militant Jesuit pamphlet *Becanus Redivivus* (*The Revised Becanus*, 1631). Opitz was instrumental in raising German literature to European standards. He stood at the beginning of a bourgeois, urban, learned culture in the German vernacular, a culture that became increasingly political and courtly in orientation.

Bibliography: B. Becker-Cantarino, "Martin Opitz," *Dictionary of Literary Biography*, vol. 164 (1996): 256–268; B. Ulmer, *Martin Opitz*, 1971.

Peter Hess

OTWAY, THOMAS (1652–1685). Thomas Otway, Restoration dramatist, was born 3 March 1652 at Trotton, the only son of a Sussex curate. He matriculated at Winchester College in 1618 as a commoner, but the next year entered Christ Church, Oxford. While enrolled at Oxford, he went to London and was drawn to the public theater upon meeting *Mrs. Aphra Behn, who invited him to play the part of the king in her *Forc'd Marriage* (1671) at Dorset Gardens. The experience was a disaster. After leaving Oxford without a degree in 1672, the failed actor moved to London to try his skill at writing plays.

As a dramatist, Otway enjoyed modest success with his first verse tragedy, *Alcibiades* (1675). His second heroic play, *Don Carlos* (1676), was a Dorset Gardens hit, although *Dryden judged it inferior to *Alcibiades*. A third tragedy in rhymed verse, *Titus and Berenice*, adapted from *Racine, followed in early 1677. The latter play, with its popular mixture of love and heroism, is said to have influenced Dryden to sentimentalize his version of Antony in *All For Love* (1677). Otway also made an impression in comedy, especially *The Soldier's Fortune* (1681), a racy, popular play based partially on the author's own brief and unspectacular stint in the military, 1678 to 1679. But it is with tragedy that Otway makes his mark, most notably in *Venice Preserv'd* (1681), written in blank verse—perhaps reflecting Dryden's own turning away, in *All For Love*, from heroic rhyme. Fourteen years after the appearance of *Venice Preserv'd*, Dryden still admired it and wrote, "nature is there . . . the passions are truly tricked in it."

Otway died destitute in a public house, reportedly choking on a roll purchased with a borrowed shilling. His fondness for strong drink, especially punch, was legendary, and one biographer has it that "the last thing (Otway) made before his death (was) an excellent song on that liquor." Otway was buried 16 April 1685 in the churchyard of St. Clement Danes.

Bibliography: J.C. Ghosh, ed., *The Works of Thomas Otway*, 1932, 1968; R.G. Ham, *Otway and Lee: Biography from a Baroque Age*, 1931.

Dale G. Priest

OUGHTRED, WILLIAM (1574–1660). William Oughtred was born in Eton, Buckinghamshire, England. He taught mathematics at King's College, Cambridge, before assuming the vicarage of Shalford, Surrey, in 1604. Later, he became rector of Albury. An Anglican priest, Oughtred weathered the Commonwealth, keeping his position in Surrey until his death. Mathematics prevailed as a major preoccupation, and Oughtred's contributions were substantial and practical. Best known for his invention of an early slide rule, the mathematician also provided in *Clavis mathematicae* (1631) the "times" symbol for multiplication and the abbreviations *sin*, *cos*, and *tan* for those trigonometric functions.

Until digital computing superseded the use of logarithms and slide rules, these developments represented the most important elements in practical computation. A contemporary of Oughtred's, the fellow English mathematician *Edmund Gunter, first suggested a device to simplify navigational calculations, using applicable logarithms. Gunter's design was a straight scale that employed removeable dividers to multiply and divide. The first really workable version of a slide rule was Oughtred's circular scale of numerical logarithms with the proper trigonometric functions. Two arms attached at the center "slid" into place, providing mechanical calculations. The more familiar linear instrument arranged logarithmic scales on rulers mounted against one another. *John Napier's tables of logarithms, published in their first edition in 1614, were thus adapted to practical and portable usage. Though the originality of Oughtred's invention of the circular slide rule was challenged, credit is usually assigned to him for both it and the linear rule. Oughtred published a description of the latter instrument in 1633.

Bibliography: F. Cajori, *A History of the Logarithmic Slide Rule and Allied Instruments*, rpt. 1992; F. Cajori, "On the History of Gunter's Scale and the Slide Rule During the Seventeenth Century," *University of California Publications in Mathematics* 1 (1920); 187–209, rpt. 1992.

Joan Klobe Pratt

OVERBURY, SIR THOMAS (1581–1613). A writer who associated with *Ben Jonson and *John Webster and whose poem "The Wife" enjoyed great popularity in Stuart England, Overbury died as the result of a plot that scandalized the English royal court and made him one of the seventeenth century's most famous murder victims.

Overbury was born in 1581 in Warwickshire to Sir Nicholas Overbury and his wife Mary. He attended Queen's College, Oxford, where he received the B.A. in 1598, and studied law at the Middle Temple. In 1601, he befriended Robert Carr, who between 1607 and 1614 was the favorite of *James I and thus became Viscount of Rochester in 1611 and Earl of Somerset in 1613. Overbury benefited from his friend's good fortune, becoming sewer (waiter) to the king and obtaining knighthood in 1608. When Rochester became romantically involved with Frances Howard, Overbury wrote letters and poems for him to give her. However, when she sought to divorce Robert Devereux, Earl of Essex, and marry Rochester, Overbury objected, fearing loss of influence (one historian has suggested that Overbury and Rochester were lovers). This angered Rochester, the Howards, the king, and Queen Anne. Frances's great uncle, Henry Howard, Earl of Northampton, urged the king to offer Overbury a diplomatic post overseas, and James agreed because of jealousy of Overbury's influence over Rochester. The traditional account states that Overbury's refusal landed him in the Tower of London on 26 April 1613; however, it has been argued that Overbury was incarcerated because his Protestant views on foreign policy fell afoul of the pro-Spanish faction at court. Tradition also has it (probably correctly) that Frances instigated a scheme whereby Overbury was gradually poisoned, dying on 15 September, though some accounts exonerate her or even implicate Overbury. Two years later, with George Villiers (the future Duke of Buckingham) ascendant at court, Somerset and his new bride were tried for Overbury's murder (with *Sir Francis Bacon as prosecutor), but they were merely imprisoned and in 1621 exiled from court, though several lesser plotters were executed. It was perhaps poetic justice that their marriage was unhappy.

The circumstances of Overbury's death probably enhanced his literary reputation, though Jonson credited him with fostering greater appreciation of the arts at the Jacobean court. Jonson also claimed that Overbury wrote "The Wife" because he was smitten with the Countess of Rutland (Sir Philip Sidney's daughter), a matter that ended their friendship; others claim he wrote it to persuade Rochester not to marry Frances. First published in 1613, it went through many editions and inspired numerous imitations, including Jonson's "The Husband" (1614). Overbury authored other poems; dozens of character sketches ranging from "A Wise Man" and "A Good Woman" to "A Golden Asse" and "A Very Whore"; and "Observations Upon the XVII Provinces [the Netherlands] As They Stood A.D. 1609." His death inspired works by Jonson, Webster, and others and influenced Nathaniel Hawthorne's *The Scarlet Letter*.

Bibliography: A. Somerset, *Unnatural Murder: Poison at the Court of James I*, 1997.

William B. Robison

PACCHIONI, ANTONIO (1665–1726). Pacchioni studied medicine in his native Reggio nell' Emilia, Italy, and received a medical degree in 1688. In 1689, he moved to Rome and attended the Santo Spirito hospital. Between 1693 and 1699, he was the town doctor in Tivoli. He returned to Rome in 1699 and established a successful medical practice; eventually, he became head doctor at the Hospital of San Giovanni in Laterano and then at the Ospedale della Consolazione. Interested in anatomy, he was guided by *Malpighi (who lived in Rome betwen 1691 and 1694) and collaborated with Lancisi. Among Pacchioni's dissertations, from 1701 on, dealing with the structure and functions of the dura mater, the *Dissertatio epistolaris de glandulis conglobatis durae meningis humanae* (1705) is particularly well known and contains his description of the arachnoidal, or so-called Pacchioni, granulations. Pacchioni attributed to these bodies the faculty of secreting lymph for lubricating the sliding movement between the meninges and the brain. He believed that the contraction of the dura mater, then considered to be muscular in nature, served to compress the glands, which, according to Malpighi's doctrine, constituted the cerebral cortex. Pacchioni also collaborated with Lancisi on the explanatory text to Eustachi's *Tabulae anatomicae* (1714).

Bibliography: J. Chiappelli, "Notizie intorno alla vita di Antonio Pacchioni da Reggio," in *Raccolta d'opuscoli scientifici e filologici*, vol. 3 (1730); Gillespie, ed., *Dictionary of Scientific Biography*, 1981.

Rosi Prieto Gilday

PACHELBEL, JOHANN (1653–1706). Johann Pachelbel was born and died in Nuremberg. From 1674 onward, he was a pupil of Johann Caspar Kerll in Vienna and served as Kerll's assistant at the organ of the Imperial Chapel. In 1677, Pachelbel was appointed court organist at Eisenach and in the following year moved to Erfurt as organist of the Predigerkirche. Between 1690 and 1695, he held successive positions as court organist at Stuttgart and Gotha, after which he was appointed to St. Sebald's Church in Nuremberg. Pachelbel composed most of the music for which he is today remembered while he was organist at St. Sebald's. His importance in music history derives mainly from those of his compositions that display elaborate settings of Lutheran chorale melodies. Also of historical significance are the almost 100 organ fugues on the Magnificat that he wrote for use in church services. Well before Pachelbel's birth, the Magnificat had been taken from the Roman Catholic liturgy and inserted into the Lutheran order of worship. As Kerll's student in Vienna and later as an organist in various centers to the north, Pachelbel served as a bridge between the southern Catholic tradition and the newer Protestant tradition of central Germany. Taken as a whole, his compositions furthered the establishment of procedures which yielded rich results in the hands of subsequent composers such as Bach and Handel.

Bibliography: J. Fairleigh, "Pachelbel's Magnificat Fugues: Models for J.S. Bach," *The American Organist* 14, no. 6 (1980): 34–37; J. Perrault, *Johann Pachelbel: A Thematic Catalogue of His Musical Works*, rev. ed., 2000.

James Fairleigh

PARNELL, THOMAS (1679–1718). Parnell, Irish-English poet, was the son of Thomas Parnell of Congleton, Cheshire, who left England after the Restoration and settled in Dublin. There he married and had sons John (ancestor of Charles Stewart Parnell) and Thomas. The latter was born in 1679 and in 1693 was admitted to Trinity College, Dublin, where he took the B.A. in 1697 and M.A. in 1700. He was ordained a priest in 1703, rose in preferment, and was made archdeacon of Clogher in 1706. Soon thereafter, Parnell married Anne Minchin of Tipperary, with whom he had three children. In 1709, he was chosen to head the committee appointed by the House of Convocation to make recommendations concerning the conversion of the Irish Catholics to Anglicanism. In August 1711, he reported the committee's resolutions to the House and headed a deputation to *Queen Anne later that year.

His wife having died in late 1711, Parnell chose to remain in England for an extended visit that lasted over four years. Settled in London, he contributed in 1712 two earnest, moral allegories to the *Spectator*, published by the Whig writers *Addison and *Steele, and several of his poems would appear in Steele's *Poetical Miscellanies* (1713). By late 1712, however, Parnell had begun to discard the low-church political views of his forebears and followed Jonathan Swift in the shift from Whig to Tory loyalties. He became friends with the circle of Tory wits surrounding Oxford and Bolingbroke, and in 1713, Swift was instrumental in paving the way for the publication of Parnell's *Essay on the Different Styles of Poetry*, with effusive compliments to Bolingbroke attached thereto. Swift was said to have taken pride in "making the Ministry desire to be acquainted with Parnell, and not Parnell with the Ministry."

In 1713, Parnell became an active member of the "Scriblerus Club" comprised of Berkeley, Swift, Pope, *Gay, *Congreve, Oxford, and *Dr. Arbuthnot, whose purpose was to ridicule pedantry and the pretensions of modern learning. His association with the club evoked satirical resources in Parnell to complement his innate seriousness. Pope says that the Scriblerian *Essay Concerning the Origin of Sciences*—which purports that all knowledge is derived from Ethiopian monkeys—was largely Parnell's work. Parnell's actual learning was considerable and enabled him to assist Pope in the translation of Homer. After Oxford's fall and Queen Anne's death in 1714, Parnell stayed on several months with Pope. In 1715, the first volume of Pope's *Iliad* appeared, with an introductory essay on Homer by Parnell, who returned to Ireland that same year.

In 1717, Parnell succeeded to the vicarage of Finglas. In that year, he finished and published an earlier Scriblerian project entitled *Homer's "Battle of the Frogs and Mice," with the Remarks of Zoilus*. To the translation of the Homeric poem Parnell appended mock-learned notes by a modern critic, and the prose *Life of Zoilus* included in the volume was probably designed to ridicule Lewis Theobald and *John Dennis. The satire delighted Pope. The two friends met again in London in the summer of 1718 to discuss Parnell's unpublished poetry, which the latter had left with Pope for examination and correction. In the fall, Parnell left for Ireland by way of Chester, where he was suddenly taken ill and died on 23 October 1718. Pope was unable to locate Parnell's grave, but he erected a permanent monument to his friend with the publication in 1721 of *Poems on Several Occasions, by the late Dr. Parnell*. The volume contains twenty of Parnell's best poems and concludes with his much-admired *Night Piece on Death*. Parnell's verse in general, and the *Night Piece* in particular, reveals an earnest sense of Christian solemnity. The sweetness and sentiment often ascribed to Parnell's poetry were compelling to writers of the next generation, and his work exerted considerable influence on Young, Blair, Goldsmith, and Collins.

Bibliography: L. Robinson, ed., *Poems of Thomas Parnell*, 1927; T. Woodman, *Thomas Parnell*, 1985.

Dale G. Priest

PASCAL, BLAISE (1623–1662). Mathematician, philosopher, and religious controversialist, Pascal was a mathematical genius who began contributing to the study of solid geometry and number theory during his teens; he was also an experimental physicist whose researches into the vacuum mark a turning point in the history of science. At age twenty-five, he had overturned Aristotle's rejection of the possibility of a vacuum and reported the results of his work in *Experiences nouvelles* (1647). Controversies that followed prompted further experimentation and defenses of his methods, work that earned Pascal a place at the forefront of a scientific renaissance. In mathematics as well, where his principal contributions are in geometrical method, Pascal's reputation as initiator of a new era rivals those of *Descartes and *Leibniz.

Outside of mathematical and scientific circles, Pascal is best known as a religious writer. Although his thought outside of these scientific realms has been considered unphilosophical, in reality Pascal is as original and philosophical as many other thinkers who did not leave comprehensive systems. The questions, in fact, that drove him to speculation were recognizably epistemological ones; he is as much a skeptic as *Gassendi and Descartes. As a mathematician, Pascal found the greatest certainty in geometry; as an observer of the human condition, he found uncertainty and despair dominant among his contemporaries. It was the need for existential certainty that Pascal addresses in his philosophical writings, particularly the *Pensées* (begun in 1657).

Pascal's *Pensées*, even though unfinished and composed in a combination of aphorisms and essays in a style reminiscent of Montaigne (whom he repeatedly attacks), represent a notable contribution to philosophy and Christian apologetics. Identifying the heart, rather than reason, as the locus of knowledge of truth, Pascal isolates religious propositions from rational accountability, at the same time subordinating the "spirit of geometry," or reason, to the "principles" derived from the heart. These principles, moreover, are the means for intuiting both the infinite and the infinitesimal; it is the geometrical spirit that mediates these and gives order to finite spaces and objects.

The "wager" for which Pascal is probably best known is found in the *Pensées* and asserts that since God is infinitely incomprehensible, it is beyond human reason to know divine existence or nonexistence, or perforce any of God's attributes. If, however, a choice must be made (and Pascal insists on this), then one makes it according to what there is to gain or lose. The possibility that God, eternal life, and blessedness exist is for Pascal a compelling reason to choose to believe in God's existence. The alternative, in Pascal's view, is the simple absence of life and eternal blessedness, hence nothing. The choice between nothing to lose and everything to gain appears obvious to Pascal; this "wager" would prove an enduring legacy to Christian apologetics. Pascal's religious sensibility originates in experience rather than reflection. The turning point in his awareness of God came during the night of 23 November 1654, when for about two hours he experienced what he called "fire" and became aware of the God of the biblical patriarchs (not the deity of the "savants") and of Jesus Christ. The joy he experienced at that time provided solace and certainty for the rest of his days.

Theologically, his religious loyalties were with the doctrines of *Cornelius Jansen, on whose behalf he was an active propagandist. In the *Provincial Letters* (1656–1667), Pascal offers a scathing attack on Jesuit doctrine and methods. Cast in the form of correspondence to an imaginary Jesuit provincial, with a few equally fictitious replies, the *Letters* seek to expose the Society of Jesus for its lax casuistry, that is, the propensity to adjust the norms of pious and penitential life to accommodate the faithful. In Pascal's depiction, this tendency becomes a craven capitulation to the laity's baser interests. At the heart of seventeenth-century Jesuit thought was the assumption, associated with *Luis de Molina (1535–1600), that God grants grace only to those foreknown as able to persevere and merit their own salvation. Jansenist teaching, on the other hand, stressed the incapacity of the human will and the necessity and sufficiency of grace for salvation. Its rigorous spirituality, rooted in the Augustinian doctrine of grace, enjoyed the patronage of many in the French hierarchy, but papal condemnations began in 1653, and by the end of the century, Jansenism would be a minor radical sect within the church.

Bibliography: E. Caillet, *Pascal: The Emergence of Genius*, 1961; E. Mortimer, *Blaise Pascal: The Life and Work of a Realist*, 1959; B. Pascal, *The Pensées*, trans. J. Cohen, 1961.

Ralph Keen

PAUL V, POPE (1552–1621). Born Camillo Borghese, Pope Paul V (pontificate, 16 May 1605–23 January 1621), related to St. Catherine of Siena, became a cardinal in 1596 and was noted for his expertise in canon law. A native Roman, he took special interest in expansion of St. Peter's Basilica and the Vatican palace, renovation of aqueducts and fountains, and enlargement of the Vatican library. He approved the creation of *Francis de Sales' visitation convent and *Pierre de Berulle's French oratory, commissioned *Vincent de Paul to function as papal emissary to Henry IV of France, and beatified Teresa of Avila. Aided by *Robert Bellarmine, he contended against Venice's challenges to papal authority by excommunicating for one year the Venetian senate (whom *Paolo Sarpi defended) for defying church prerogatives. His opposition to *James I's Oath of Allegiance, which rejected the pope's right to unseat monarchs, led to a debate between James and Bellarmine. It was also under his pontificate that *Galileo was censured on 5 March 1616 for advocating the heliocentric cosmology of Copernicus. Paul's papacy also saw the start of the Thirty Years' War in 1618. His indulgence in nepotism was his greatest flaw; otherwise, he was an able administrator, one who took a special interest in missionary work, granting permission for the use of Chinese in the liturgies performed in the Jesuit missions in China.

Bibliography: L. Pastor, *The History of the Popes from the Close of the Middle Ages*, vols. 25–26, 40 vols., 1891–1954.

Christopher Baker

PECQUET, JEAN (1622–1674). Known primarily as an anatomist, Jean Pecquet was born in Paris on 9 May 1622. His youth was spent in Normandy, where he lived in Dieppe and then Rouen. In 1642, Pecquet moved to Paris where his interests in the sciences were stimulated by other members of the early scientific circles. In 1646, he enrolled in the Paris Faculty of Medicine, but unhappy with the conditions there he left and matriculated at Montpellier, receiving the M.D. in July 1651. In February 1652, he was officially licensed, defending his doctoral thesis in March of that year. Returning to Paris, he became the physician to Nicolas Fouquet and to the family of the *Marquise de Sévigné. Pecquet was a proponent of *anatomia animata*, which advocated an empirical approach to investigations and involved numerous dissections. His most significant discovery was the thoracic duct, the primary trunk in the lymphatic system that allows chyle from the lacteals to bypass the liver and flow directly to the venous system. This had immediate effects on discussions of the liver and helped solidify *William Harvey's theory of circulation. He published an account of the discovery in his *Experimenta nova anatomica* of 1647. In 1666, Pecquet was among the first seven individuals chosen to be part of the new Académie Royale des Sciences. There he participated in the transfusion experiments of 1666–1667, worked on parasites of the liver, and investigated the physiology of the eye. An accident on a horse resulted in his death in February 1674.

Bibliography: C. Gillispie, ed., *Dictionary of Scientific Biography*, vol. 10 (1974), 476–478.

Nicole Howard

PELL, JOHN (1611–1685). John Pell was born in Southwick, Sussex, England, on 1 March 1611 to John Pell, the vicar of Southwick, and Mary Holland. After attending Steyning School in Sussex, he entered Trinity College, Cambridge, in 1624. He received a B.A. degree in 1629 and an M.A. in 1630, whereupon he became a schoolmaster in Horsham and subsequently in Samuel Hartlib's academy Chichester. He married Ithamaria Reginalds in 1632. Upon closing the academy, Hartlib arranged a position for Pell in London as a mathematics instructor to the Comenian group, a position Pell held until 1642. *Idea* *of Mathematics* was published in English in 1638, earning Pell recognition as a mathematician from *Marin Mersenne and *René Descartes, among others. In 1643, Pell relocated to Amsterdam and in 1646 moved to Breda, teaching mathematics in both places. With an established reputation as a mathematician, Pell published his *Controversiae de vera circuli mensura* in 1647, a critique of C.S. Longomontanus's work on the value of pi. Included in the work were corroborating proofs by *Hobbes, as well as endorsements by *Cavalieri, *Roberval and Descartes. In 1652, Pell returned to London to teach mathematics, but two years later he was sent by *Cromwell to Zurich where he worked as a Commonwealth agent until 1658. Upon returning to England, he worked as a clergyman, and in 1663, he was elected to the Royal Society. He died in poverty on 12 December 1685.

"Pell's equation," which was incorrectly attributed to him by Leonard Euler, is the Diophantine quadratic $x^2 = 1 + py^2$. It has been thought that Euler attributed the work of Lord Brouncker to Pell; however, it seems that it was a 1668 translation by Thomas Brancker, found in a volume Pell edited, which caused the confusion. Nevertheless, his reputation as a premier algebraist was deserved and his contributions were significant.

Bibliography: O. Dick, ed., J. Aubrey, *Brief Lives*, 1949; L. Wickham Legg, ed., *Dictionary of National Biography*, vol. 15 (1949), 706–708.

Nicole Howard

PENN, WILLIAM (1644–1718). Penn, son of a wealthy admiral, was educated at Oxford and Lincoln's Inn before serving in the English army in Ireland (1665). There, he associated closely with the Quaker Thomas Loe. After getting arrested at a Quaker meetinghouse, Penn was recalled to London by his father, where he immediately took up preaching and writing. In 1668, *Sandy Foundation Shaken* was published without license, and Penn was committed to the Tower of London, where he wrote *No Cross No Crown* (1669). In these and other works, Penn advocated liberty of conscience and religious toleration, the principles that would earmark his life. He was released through the intervention of his father, but found himself arrested several times in 1670 for infringing the Conventicle Act. After the death of his father in 1670, Penn established himself as a country gentleman, married, and, under the Declaration of Indulgence (1671), undertook preaching tours. Over the next decade, Penn served as principal spokesperson and propagandist for Quakers.

Penn toured America in 1677 with *George Fox. In 1681, he was awarded the land west of the Delaware (to which the region that became the state of Delaware was eventually added) by *Charles II in repayment of an old loan made by Penn's father to the crown. The land was named Pennsylvania ("Penn's Woods") after the admiral. Penn arrived there in 1682, took formal possession, entered into a land treaty (the first of its kind) with the Delaware Indians, drew up a set of architectural plans for Philadelphia, and framed laws that provided for large, liberal political bodies and allowed all modes of Christian worship. By 1685, Penn was back in England and close to *James II, but during the Revolution of 1688 he was styled a Jacobite (an adherent of James II) and found no favor under *William III. He returned to Pennsylvania in 1699 to find that his liberal constitution had faltered and that the colony was governed by a committee of five persons. In 1701, a bill to convert proprietaries into crown colonies was before Parliament, and Penn returned to fight for his colony. The bill lapsed on the death of the king (1702), and *Queen Anne's liberal religious views encouraged Penn to remain in England. Despite his best efforts, his contributions to religious toleration, and his importance to American religious and political identity, Penn's colony lost him a fortune, and the remainder of his life was spent in pecuniary discomfort.

Bibliography: R. Dunn and M. Dunn, *The World of William Penn*, 1986.

Ken MacMillan

PEPYS, SAMUEL (1633–1703). Renowned diarist and bureaucrat of Restoration England, Pepys provided an intimate and detailed portrait of life and events in the first decade of the reign of *Charles II and contributed greatly to the administrative reform of the British navy.

From a modest background, Pepys was a graduate of Cambridge and a devotee of the arts. Through his service to Sir Edward Montagu, Pepys participated in the restoration of the king to the throne and became acquainted with James, Duke of York. Pepys was able to secure the naval post of Clerk of the Acts and quickly became a clerk of the Privy Seal and a justice of the peace. Pepys attended the king's coronation in 1661 and, as the war with the Dutch began, became the commissioner for Tangier and Surveyor of Naval Victuals. He dedicated himself to the professionalization of the navy and received a number of commendations during the Dutch war from the Duke of York, making regular reports to him and to the king.

Pepys continued to work in London as Surveyor-General of Victuals for the navy during the height of the plague in 1665. By 1666, Pepys was growing very concerned about the laxity in naval administration and only a subsequent peace removed the possibility of a Dutch invasion. The dramatic event of 1666 was the great fire of London, and Pepys wrote extensively about it in his diary. Pepys himself was in charge of creating firebreaks to keep the flames from spreading and is credited with saving the navy office from destruction. In the fire's aftermath, Pepys faithfully recorded the damage and the loss of many of the city's structures.

Pepys faced parliamentary inquiries over his handling of naval administration in 1668 and 1669, but emerged from them unscathed. Trouble with his eyes led him to abandon the diary in 1669, but when his sight improved, he began to write history and to amass a collection of books and engravings. Pepys focused all his energies on business after his wife's death late in 1669, and this dedication led to his appointment as secretary of the Admiralty in 1672. The following year, Pepys was elected to Parliament and used his seat as a platform from which to achieve further naval administrative reform, such as a code of promotion. Denouncement as a secret Catholic in 1679 led to a brief committal to the Tower, but in the aftermath of the Exclusion Crisis and the king's proroguing of Parliament, the inquiry died a natural death, though Pepys remained unemployed until 1683. He was restored to his post as secretary of the Admiralty in 1684 and pushed forward with a complete administrative reform of the navy; the accession of *James II to the throne in 1685 only enhanced Pepys' standing.

Pepys also returned to Parliament but played little role in the political machinations that eventually unseated James from his throne in 1688. With *William III's accession, Pepys was deprived of his offices and lost his seat in Parliament as well. Though he was jailed twice more for his suspect ties to the deposed James II, Pepys enjoyed a rather normal retirement, pursuing almost exclusively intellectual activities. Suffering various ailments through much of the 1690s, Pepys fell seriously ill and died on 26 May 1703. The esteem and respect in which he was held was evidenced by the great crowd that attended his funeral and validated his enormous contributions to literature, social history, and naval administration—contributions that are still vastly admired today.

Bibliography: R. Ollard, *Pepys*, 1974; I. Taylor, *Samuel Pepys*, 1989.

Connie S. Evans

PERGOLESI, GIOVANNI BATTISTA (1710–1736). A composer of the Neapolitan school whose originality led to a new style of composition in the eighteenth century, Giovanni Battista Pergolesi was born in Iesi on 4 January 1710 and died in Pozzuoli near Naples on 16 March 1736. Following the study of composition in Naples with Gaetano Greco, Leonardo Vinci, and Francesco Durante, Pergolesi made his career in the service of Neapolitan noblemen and religious institutions. Although he composed opere serie, cantatas, and instrumental music, he is known primarily for his comic operas, especially the intermezzo *La Serva Padrona* (1733) and his sacred work *Stabat Mater* (1736). In addition, twentieth-century audiences are familiar with several tunes from Pergolesi's *Lo Frate 'nnamorato* (1732), a comic opera in Neapolitan dialect, which were adopted by Igor Stravinsky for his ballet *Pulcinella*. Pergolesi's opera serie are notable for the subtlety and flexibility of their recitatives and the variety of their arias and ensembles. They include *La Salustia* (1732), *Il Prigionier Superbo* (1733), *Adriano in Siria* (1734), and *Olimipiade* (1735). Besides *La Serva Padrona* and *Lo Frate 'nnomaorato*, Pergolesi's comic works for the stage include *Livietta e Tracollo* (1734) and *Il Flaminio* (1735). His intermezzi, originally written to be performed between the acts of his opere serie, were crucial to the development of comic opera as an important genre in the eighteenth century. They were frequently performed in Italy, Austria, and Germany from 1733 until 1743. In 1752, the presentation of *La Serva Padrona* in Paris precipitated the *Querelle des Bouffons* between supporters of traditional French opera and the new Italian comic opera.

In his *Lettre sur la musique française*, J.J. Rousseau championed the style of Pergolesi, hearing in its melodies a new simplicity and directness of expression. Pergolesi's most famous sacred work, the *Stabat Mater*, is a setting of the Latin hymn for Good Friday, which includes twelve arias and duets with string orchestra accompaniment. It attracted the attention of J.S. Bach, who made an arrangement of it.

Pergolesi's music became so popular after his death that many compositions, including many instrumental works, were forgeries incorrectly attributed to him. Recent scholarship, especially that done under the auspices of the Pergolesi Research Center at the City University of New York, has focused on the publication of complete and accurate editions of his compositions.

Bibliography: B. Brook et. al., *The Complete Works of Giovanni Battista Pergolesi*, 1985– ; F. Degrada, ed., *Pergolesi Studies/Studi Pergolesiani: Proceedings of the International Symposium on Pergolesi and His Times*, 1986; M. Paymer, *Giovanni Battista Pergolesi, 1710–1736: A Thematic Catalogue of the Opera Omnia*, 1977.

Maria Archetto

PERI, JACOPO (1561–1633). Singer, composer, and a pioneer in the formation of early Baroque vocal style, Peri was born in Rome but spent most of his life in Florence, where he studied with Cristoforo Malvezzi, maestro di cappella at the Florentine Cathedral. He was appointed organist at the Badia in 1579 and as singer (tenor) at San Giovanni Battista in approximately 1586. Peri was associated with the Medici court beginning in 1588, serving Grand Duke Ferdinand. His most important contributions were to the first operas *Dafne* (performed in 1598) and *Euridice* (1600), the latter for the wedding of Marie de' Medici to King Henri IV of France. Peri belonged to Jacopo Corsi's circle of humanists interested in creating a new dramatic style expressive of the emotional content of the poetry. In collaboration with the poet Ottavio Rinuccini, Peri devised a flexible recitative style ranging from simple narrative to deeply expressive lines. His goal was to "imitate in song a person speaking" (preface to the music for *Euridice*, 1601) to thus achieve an effect similar to that of the "Ancients" in their tragedies. Along with other innovators, especially his rival Giulio Caccini, Peri influenced composers of the early seventeenth century, notably *Claudio Monteverdi in his renowned score of *Orfeo* (1607).

Bibliography: T. Carter, *Jacopo Peri (1561–1633); his Life and Works*, 1989.

Olga Termini

PERRAULT, CHARLES (1628–1703). Charles Perrault was born a lawyer-magistrate's youngest son on 12 January 1628 in Paris. At fifteen, Perrault left school and started writing poetry—including a satiric *Aeneid* translation. After receiving a law degree from the University of Orleans in 1651, he barely practiced law; instead, he served as secretary-clerk to his brother, tax collector for Paris, and produced verses celebrating *Louis XIV's reign. By March 1660, the financial minister *Jean-Baptiste Colbert was aware of Perrault's talents, praising them in a letter to the king's prime minister, Cardinal Mazarin. Three years later, Colbert became Superintendent-Director General of Buildings for a monarch eager to use architetural grandeur for restoring the royal power and prestige that had been threatened during the Fronde. In 1663, Colbert made Perrault his secretary and appointed him to the Little Academy, a new department responsible for designing inscriptions for royal buildings and monuments. Although Perrault's public service often kept him from composition, in 1671, he was elected a member of the French Academy on his third nomination. Ten years later, he became its director. With Colbert's death in 1683, however, Perrault lost his government job and turned all his energies to writing. Over the next two decades, he attended the academy meetings regularly. Perrault started his memoirs a few years before his death in May 1703.

It was Perrault's poem "The Age of Louis the Great" that provoked 1687's Quarrel of the Ancients and the Moderns. By praising Louis's civilization as a peak of artistic creativity, Perrault suggested that contemporary culture could equal and surpass the classical heritage—an unprecedented notion. The French Academy split into those supporting his belief in progress (the Moderns) and those defending classical achievement as unsurpassable (the Ancients). One of Perrault's foremost opponents in this debate was fellow academician and satire-writer *Nicolas Boileau-Despréaux. Between 1688 and 1697, Perrault produced the *Parallel of the Ancients and Moderns*, four volumes of dialogues justifying his challenge to tradition. His greater contribution to world literature, however, was a dozen fairy tales preceding the Grimm Brothers' work by over a century. In a 1695 preface to several tales, Perrault insisted on their superiority to ancient fables because of their moral purpose. By 1697, he had collected eight stories for *Histories or Tales of Past Times, with Morals*—a book fondly known as *Tales of Mother Goose*. His transformation of folk tales like "Puss in Boots" and "The Sleeping Beauty" into Christian models of behavior for upper-class children shaped both the literary fairy tale and his era's growing conception of childhood as a separate developmental stage.

Bibliography: J. Barchilon and P. Flinders, *Charles Perrault*, 1981; J. Zarucchi, ed. and trans., *Charles Perrault: Memoirs of My Life*, 1989.

Margaret Goscilo

PETER I (1672–1725). Born on 30 May 1672, Peter I ("the Great") reigned as emperor of Russia from 1682 until 1725. Enamored of western European mores and industrial progress, he sought to modernize Russia by reorganizing its government, improving economic growth, and exerting greater control over the Russian Orthodox church. Shipbuilding and engineering were two childhood interests that lasted a lifetime, and at twenty-two, he launched a ship he had built himself. His interest in naval affairs led to his war against Turkey in 1695 to gain a foothold on the Black Sea, which he succeeded in doing with the capture of Azov in 1696. The next year, he visited Prussia, Holland, and England, fascinated with the social and technological innovations he witnessed. However, in his absence the *streltsy* (the garrison of musketeers guarding Moscow) revolted against him. Peter returned and suppressed the rebellion, executing 1,200 men as a bloody reminder of how threats to his authority would be treated.

Still seeking a wider maritime influence, he campaigned against Sweden in the Great Northern War (1700–1721) in an effort to break its control of the Baltic. The Swedish Charles XII defeated Peter at the Battle of Narva (1700) but failed to exploit this victory to his advantage. To anchor his expansion on the Baltic, Peter also founded the city of St. Petersburg on the Gulf of Finland. At the Battle of Poltava in 1709, Peter conquered the Swedes, but when he invaded Turkish-controlled Moldavia (now Romania), he was defeated and forced to relinquish Azov in 1711. Charles XII died in 1718, and the war ended with the Treaty of Nystadt (1721) by which Russia gained several more former Swedish territories, now Peter's "windows on the West."

Internally, Peter embarked on a vast reorganization of both government and church. Imitating a Swedish model, he created nine "colleges" to govern varying governmental bureaus, each college directed by an eleven-member committee; he also set up a nine-member Senate to govern during his trips away from court. Though prone to cumbersome deliberations, the new system improved efficiency, and Peter's authority over his ministers was affirmed when, on 22 October 1721, he was declared "Father of the Fatherland, Peter the Great and Emperor of all Russia." He employed a committee as well in his refashioning of the church, by setting up a Holy Synod that replaced the older office of Patriarch. His wars and reforms were costly, leading to pervasive taxation on everything from caviar to coffins, even a head tax on every male child called the "soul tax." But his innovations were almost as numerous as his taxes. He forced the *boyars* (nobles) to cut their beards and wear European-style clothing, demanding that they also serve the state in either the armed forces or civil service (a system similar to that adopted in Prussia by *Frederick William). He set up schools, hospitals, an Academy of Science, and several military service academies, and he continued the practice of encouraging the immigration of foreign technicians to bring fresh expertise to Russian industry.

The revolutionary renown he gained through these reforms was strengthened by the notoriety of his personal behavior. Nearly seven feet tall before he was twenty, Peter was known for his youthful nightly carousing in Moscow and was responsible for a drunken, destructive spree at *John Evelyn's London home in 1698. He personally cut off the beards and trimmed the jacket sleeves of his courtiers. As the *streltsy* had learned, he also had a bloodthirstly streak: He attended the torture death of his alchoholic son Alexis to quash potential opposition to his rule. Peter's brand of absolutism did result in broad, if uncomfortable, improvements for the Russian ruling class, which he reformed on the basis of skill, not ancestry. But the gap between them and the growing population of serfs grew wider, and Russian class distinctions became stronger.

Bibliography: M. Anderson, *Peter the Great*, 1978; R. Massie, *Peter the Great: His Life and His World*, 1980.

Christopher Baker

PETTY, WILLIAM (1623–1687). William Petty was an economist, inventor, and physician. During the first twelve years of his life, Petty learned a variety of crafts but signed up as a cabin boy at the age of thirteen. Aboard ship, the satirical bent later remarked upon by *John Evelyn probably landed the teenager in trouble: After he broke his leg, Petty's unsympathetic shipmates abandoned him in the port of Caen. Luckily, he gained admission to the Jesuit College there. From 1636 until 1646, Petty stayed on the continent, except for a three-year stint in the royal navy, studied at the universities of Leiden and Paris, and became an amanuensis of *Thomas Hobbes. In 1646, Petty returned to England where he entered Brasenose College of Oxford to study medicine, earning his doctorate in 1650 and becoming a full professor in 1651. Petty joined the London Philosophical Society, a forerunner of the Royal Society of which he was a founding member. University life did not suit Petty, however, and in 1652 he took up the position of physician to the army in Ireland. While in Ireland, Petty began to work on his famous land survey of Ireland. The very name of this endeavor, "the Down survey," came about because of Petty's insistence on writing everything down. The survey was rewarding for Petty in two ways: He acquired 9,000 acres of land and the empirical data necessary to apply his famous principles of political arithmetic.

When Petty returned to England, three tasks consumed him: defending his property from numerous lawsuits; inventing, with limited success, a double-keeled ship and a copying machine; and economic theory. In 1662, his most important work, *A Treatise of Taxes and Contributions*, appeared. In this work, Petty established his principle of surplus labor, or the part of a country's labor, in terms of labor-hours, which is not necessary to produce the basic needs of life. In his *Treatise*, Petty used the concept of surplus labor to determine the proper concerns for government expense, to contend for the superiority of sales taxes over direct taxes, and to decry "unproductive" workers such as clerics and lawyers. Later in life, Petty was reluctant to publish, but some of his posthumously published work was influential to economists, particularly the eighteenth-century French physiocrats. The most famous of his later works were *The Political Anatomy of Ireland* (1690) and *Political Arithmetick* (1691), which used quantitative data to evaluate all sorts of social phenomena. Petty drew upon the earlier work of *John Graunt in his research, and he also wrote on monetary theory and formulated the concept of velocity, the speed at which money circulates. It is for all of these contributions that some call Petty the father of economics.

Bibliography: T. Aspromourgos, *On the Origins of Classical Economics: Distribution and Value from William Petty to Adam Smith*, 1996; A. Roncaglia, *Petty, the Origins of Political Economy*, 1985.

Brian Weiser

PHILIP IV (1605–1665). The history of Spanish influence in western Europe after the expansive reign of Philip II (1527–1598) is one of gradual decline, as its role as a cultural and colonial leader was eclipsed by France and England. His fourth son, Philip III, inherited the largest geographical empire in history, but Philip III's minister, the Duke of Lerma, involved Spain in the Thirty Years' War (1618–1648), thus beginning its long military and economic slide. Philip IV (who reigned from 1621 to 1665) proved unable to stem this decline. His minister, Gaspar de Guzman, Count of Olivares, sought to play *Richelieu to Philip's *Louis XIV and embarked on an absolutist program of centralized administration with Castile as its hub. But the peninsular states of Catalonia, Aragon, and Valencia proved too independent to be easily coerced into Guzman's plan. A key point of contention was the cost of Spain's involvement in the war, despite its victories against the Dutch at Gibraltar and in South America, and against North African pirates in the Mediterranean. When French forces invaded the border region of Rousillon in 1639, the Catalans resisted more bravely than did their Castilian reinforcements, thus confirming their opposition to the wisdom of submitting to Castilian rule. They rejected Spain's authority, becoming a French province for sixteen years. In 1640, Lisbon insurgents accomplished the independence of Portugal; Spain's thirty-year effort to reclaim Portugal drained its treasury further. Despite the Treaty of Westphalia (1648), France and Spain remained in conflict after the end of the Thirty Years' War; this Franco-Spanish fighting finally ended with the Treaty of the Pyrenees in 1659, but the result had been an additional erosion of the already weakened Spanish economy.

These military and political failures of Philip were, however, at least balanced if not outshone by the cultural achievements of his era, accomplishments he did much to patronize. This was the era of *Velazquez, *Cervantes, *Lope de Vega, and *Calderon, among others, who flourished during the waning years of the Spanish "Golden Age," which had begun in the latter half of the sixteenth century. Philip died on 17 September 1665 and was succeeded by Charles II, a weakly ineffectual figurehead whose reign was strongly influenced by his Austrian mother and his bastard brother, Don John of Austria. Charles proved unable to contend effectively against Louis XIV's military efforts to acquire Spanish territories. With Charles's death on 1 November 1700, the Hapsburg dynasty in Spain ended. He willed the throne to Louis's grandson, Philip of Anjou, but the prospect of Spanish and French possessions united under a single monarch was too threatening to England, Holland, and the Holy Roman Empire, a threat they reacted against in the War of the Spanish Succession (1701–1713).

Bibliography: R. Stradling, *Philip IV and the Government of Spain, 1621–1665*, 1988.

Christopher Baker

PHILIPS, KATHERINE (1632–1664). Katherine Fowler was born in London to middle-class parents and, while in school, began writing poetry and attracting female friends who adopted literary names in an effort to emphasize the Platonic flavor of their group. Katherine became "the Matchless Orinda," writing poems to "Rosania" (Mary Aubrey); men too were included, such as her poetic mentor Charles Cartwright ("The Prince of Phansie") and James Philips ("Antenor"), a Welsh member of parliament and—like her—a Puritan, whom she married in 1648. Others in her circle were Anne Owen ("Lucasia") and Sir Charles Cottrell ("Poliarchus"), but she was also known to *John Dryden, *Abraham Cowley, *Henry Vaughan and *Jeremy Taylor, who addressed a work to her the title of which captures the spirit of her group: *A Discourse on the Nature, Offices, and Measures of Friendship* (1657). Her group lacked the influence of Lady Mary Sidney's Wilton circle a century earlier, but it combined a more eclectic set of personalities, drawn not simply from the aristocracy. Her own work (*The Letters of Orinda to Poliarchus* [1705]; and a translation of *Corneille's *Pompey* [1663]) was less noticed in the eighteenth century, but Keats admired her poem *To My Excellent Lucasia, on our Friendship*. She has been called "the first real English poetess," and she blended the Renaissance ideals of Platonic friendship with a neoclassical poetic style. *John Aubrey included her in his *Brief Lives*. She died of smallpox on 22 June 1664 in London.

Bibiography: P. Thomas, ed., *The Collected Works of Katherine Philips, "The Matchless Orinda,"* 1990.

Christopher Baker

PICARD, JEAN (1620–1682). Astronomer Jean Picard was born at La Flèche and entered the priesthood, becoming the prior of Rillé in Anjou. But Picard's major interest was the science of astronomy and the new emphasis on accurate instrumentation and data recording, which had been inaugurated by the great Danish astronomer Tycho Brahe (1546–1601). His schooling included becoming a pupil of renowned astronomer *Pierre Gassendi, whom he assisted in the observation of the solar eclipse of 1645. Picard succeeded him at the Collège de France as professor of astronomy in 1655 and would proceed to become the essential founder of modern astronomy in France. His major work centered on contributions to the accuracy of astronomical measurement with new instruments and techniques.

The graduated naked-eye astronomical and measuring instruments of the pre-telescope sixteenth century were still very useful through the seventeenth century for observational measurement. Picard constructed special versions of these and developed many adaptations of telescopes and micrometers to such instruments beginning in 1667. Picard introduced other instruments with astronomical application, including the pendulum clock. His most precise quadrant (angle measuring instrument with 90° arc) of thirty-eight inches in radius was graduated such that angles to one quarter of a minute could be measured. In 1659 he constructed and used a six-foot radius sextant (measuring instrument with less than 90° arc) to measure accurate positions of stars along the meridian (circle from the zenith or vertical overhead passing through the earth's North and South Poles) integrated with a telescope. With this same sextant, Picard determined a precise measurement of the degree of an arc of a meridian, his most important contribution to astronomy, between Sourdon (near Amiens) to Malvoisine (south of Paris) in 1669–1670. His measurement was 57,060 toises (a toise equals 6.4 feet)—only 14 toises shy of the modern value. He retained the constancy of this standard of measurement with a method that enabled it to be reproduced at any time. English physicist *Isaac Newton used the results in calculation revisions and validation of his universal law of gravitation. Determining the latitude and longitude of Brahe's observatory at Ven (1671, now Landskrona, Sweden), Picard used Brahe's positions of stars and planets with his collection of observations as an important legacy to astronomy. Picard also founded the Paris Observatory.

Bibliography: V. Abalakin and J. Lieske, eds., *Inertial Coordinate System on the Sky*, 1990; E. Grant, *Planets, Stars, and Orbs: The Medieval Cosmos, 1200–1687*, 1993.

William J. McPeak

POCAHONTAS (ca.1595–1617). Pocahontas (translated "playful one") was the nickname of Matoaka, an Algonquin Indian important in the survival of the Jamestown colony. The daughter of Powhatan, the principal chief of the coastal Indians, Pocahontas was about eleven in 1607 when the English arrived in Virginia. She had frequent contact with the colonists and participated in what was probably a ritual adoption ceremony of *Captain John Smith, where the Indians were demonstrating Powhatan's authority. As Smith was about to be beheaded, Pocahontas dramatically threw herself over his body to "save" him. Often serving as an intermediary between the Algonquins and the English, Pocahontas carried food and trade goods to the colonists, and worked to settle disputes. Smith returned to England in 1609 after being injured in a gunpowder explosion, and without his mediation, Anglo-Indian relations deteriorated.

In 1613, in the midst of a conflict with the Indians, Captain Samuel Argall took Pocahontas hostage and held her captive in Jamestown. During her imprisonment, she became a member of the Anglican Church and was baptized with the name Lady Rebecca. As a way to end the crisis, Pocahontas married tobacco pioneer John Rolfe the next spring, despite a purported previous marriage to an Indian warrior named Kocoum. Governor Thomas Dale and other Virginia leaders, excited by Pocahontas's conversion to Christianity and anxious to take advantage of her potential to promote interest in the Virginia colony, sent the Rolfes and their infant son Thomas to England in 1616. In London and at the court of *King James I, Pocahontas was feted as a royal princess. As the couple was about to return to Virginia the next year, Pocahontas became sick and died. She was buried at Gravesend, Kent. John Rolfe returned to Virginia, leaving Thomas in England to be raised as a gentleman. Thomas later returned to Virginia in 1640 and became a tobacco planter. After Pocahontas's death and the death of her father in 1618, relations between the Indians and the colonists broke down again. On 22 March 1622, the Algonquins, now under the leadership of Pocahontas's uncle, attacked the colony and killed a third of its population. The attack led to the bankruptcy of the London Company, and Virginia was made a royal colony.

Bibliography: P. Barbour, *Pocahontas and Her World*, 1972; F. Mossiker, *Pocahontas: The Life and the Legend*, 1976.

Christopher E. Hendricks

POUSSIN, NICOLAS (1594–1665). One of the leading exponents of classicism in painting, Poussin was born in Les Andelys, Normandy, and baptized in June 1594. The details of his early life and career are obscure, but he probably received some training from Quentin Varin. To continue his development as an artist he moved to Paris in 1612, where he was influenced by the mannerism of the Second School of Fontainebleau. While in Paris, he met the Italian poet *Giambattista Marino, who commissioned a series of drawings based on Ovid.

Encouraged by Marino, Poussin moved to Rome in 1624. There, except for a brief interlude in Paris (1640–1642), he lived and worked until his death. His early patrons included Cardinal Francesco Barberini, nephew of *Pope Urban VIII, and Barberini's secretary, Cassimo dal Pozzo, a noted classicist and antiquarian. Influenced by the *Bacchanals* of Giovanni Bellini and Titian, Poussin worked to combine the warm colors of the Venetian painters with classical form, developing a poetic representation of nature. This resulted in works such as *Venus and Adonis* and the *Arcadian Shepherds* (*"Et in Arcadia Ego"*). In 1628, Poussin received a commission to paint *The Martyrdom of St. Erasmus* for a Vatican altarpiece.

Poussin turned to smaller works based on classical sculpture to create allegorical figures who have an independent significance, separate from the larger narrative context of the painting. This is exemplified in works such as *Dance to the Music of Time* (1634–1636), arguably his masterpiece, and the two *Seven Sacraments* series. Embodying Christian Stoicism and a meticulously executed philosophy, his late work was more contemplative and mystical in mood, expressing a near pantheistic vision of nature as in *Funeral of Phocion* and the incomplete *Apollo and Daphne*.

Bibliography: A. Blunt, *Nicolas Poussin*, 1967, rpt. 1995; J. Thuillier, *Nicolas Poussin*, 1994.

Thomas L. Cooksey

PRAETORIUS, MICHAEL (ca.1571–1621). Praetorius was an organist, composer, scholar, and theorist known today chiefly for his great compendium of musical knowledge, *Syntagma Musicum* (1614–1620). A native of Creuzburg, near Eisenach, he was interested greatly in Lutheran chorales (hymns), perhaps due to paternal influence. He was organist at St. Marien in Frankfurt (1587–1590); later, he entered the service of the Duke of Brunswick-Wolfenbuttel, after whose death he went to Dresden to serve Johann Georg of Saxony. After 1616, he worked in various places such as Halle, Magdeburg, Kassel, and Sondershausen, often involved in reorganizing the music of the respective courts. His compositional output is staggering, including the *Musae Sionae* (over 1,200 chorale arrangements in sixteen volumes), *Polyhymnia* (fifteen volumes of choral music), and *Terpsichore* (two volumes of secular instrumental music). Praetorius's works show extraordinary diversity of compositional styles including Renaissance motet style and traditional Lutheran musical practices. In approximately 1613, he incorporated Venetian practices such as polychoral and concertato style, basso continuo, ornamentation, and instrumental ritornelli. The three parts of *Syntagma Musicum* contain invaluable information on early vocal music for the church (psalmody, mass, offices) on instruments (terminology, construction, illustrations) and on compositional practices and theoretical topics. Praetorius combined the musical tradition of the Reformation with modern trends in seventeenth-century music.

Bibliography: F. Blume et al., eds., "Afterword," in *Michael Praetorius: Gesamtausgabe der MusikalischenWerke*, vol. 20, 1960.

Olga Termini

PRIOR, MATTHEW (1664–1721). Son of London joiner George Prior and Elizabeth Pennefather Prior, Matthew Prior overcame ignoble birth to achieve diplomatic and literary success. After his father's death in 1675, Prior worked at his uncle Arthur's Rhenish Tavern in Whitehall. There he met his lifelong friends Charles and James Montague, and Charles Sackville, sixth earl of Dorset, through whose patronage he attended Westminster School as a King's Scholar (1681–1683). Prior attended St. John's College, Cambridge (1683–1687), where his gifts for languages and versification culminated in his *Satyr on the Poets: In Imitation of the Seventh Satyr of Juvenal* (May 1687). Recognition as coauthor with Charles Montague of *The Hind and the Panther Transvers'd* (July 1687), a send-up of Dryden's *Hind and Panther*, led to desirable opportunities, including tutelage of Lord Exeter's sons and election as a Keyton fellow of St. John's College, Cambridge. Eventually Prior secured secretarial appointments to embassies at The Hague (1690–1697) and Paris (1697–1699) and in absentia to the lords justices of Ireland (1698–1699), Royal Society membership (1698), and service as undersecretary of state (1699–1700), board of trade and plantations commissioner (1700–1707), commissioner of customs, and plenipotentiary to Paris (1712–1715). As Whig Parliamentarian turned Tory (1701), he discredited former political associates and incurred their relentless animosity. Enemies exulted in his humiliating failure to negotiate a secret treaty with France in 1711 but resented his successful mediation of the Treaty of Utrecht in 1713. In 1715, the secret committee of the Whig-dominated House of Commons detained Prior on vague charges of treachery, then released him eighteen months later without vindication or hope of employment. A 1718 subscription volume of poems provided financial security that had eluded him throughout his life. He died on 18 September 1721 and was buried in Poets' Corner, Westminster Abbey, London.

Throughout his public career, Prior wrote and associated with literary notables of his day. His early Whig politics provided access to Kit-Cat Club members, including *William Congreve, *Nicholas Rowe, *Joseph Addison, and *Richard Steele. After his shift to Toryism soured these relationships, he befriended *John Gay, Alexander Pope, *John Arbuthnot, and most significantly, Jonathan Swift. Prior's "To the Honourable Charles Montague, Esq." and "An Ode in Imitation of the Second Ode of the Third Book of Horace" were published in 1692, and eighteen poems appeared in Jacob Tonson's *Miscellanies* (Winter 1703–1704). *Poems on Several Occasions* (1707), Edmund Curll's pirated collection of Prior's poems, prompted Tonson's authorized editions in 1709, with subsequent ones in 1711, 1713, and 1717, and the 1718 subscription edition.

Twentieth-century scholars established Prior's influence on Swift, Pope, Jonson, and others. Generally regarded as a prolific but minor poet, his reputation attributes most immediately to his mastery of Augustan *vers de société*, or light verse—a witty, often delightful if unevenly executed variety of forms including satires, epigrams and epitaphs, odes, songs, and metrical tales. In addition to British occasions, royalty, and nobility, Prior celebrated three mistresses—Jane Ansley, Anne Durham, and Elizabeth Cox—as "Flanders Jane" or "Jinny," "Cloe," and "Lissette" in these verses. His contemporaries also favored his longer poems *Henry and Emma*, an updating of the medieval *Nut-brown Maid*; the contemplative *Solomon On the Vanity of the World*, Prior's preference; and the Hudibrastic *Alma: or, The Progress of the Mind*, which Jonson proclaimed tedious but Pope considered a masterpiece. Eighteenth-century appreciation of these poems has not endured, and his prose works "Essay upon Learning," "Essay upon Opinion," and "Dialogues of the Dead" remained unpublished until 1907.

Bibliography: C. Eves, *Matthew Prior, Poet and Diplomatist*, 1939; F. Rippy, *Matthew Prior*, 1986; H. Wright and M. Spears, eds., *The Literary Works of Matthew Prior*, 2 vols., 1971.

Sallye Sheppeard

PRYNNE, WILLIAM (1600–1669). As one of *Charles I's major adversaries, the career of this pamphleteer demonstrates the degree to which the written word was used as a political weapon during the English Civil War. Born in Swanswick, Somerset, in 1600 to Thomas Prynne and his wife Marie Sherston, William Prynne lived through and became a significant figure in one of England's most tumultuous periods of social and political change. Prynne's activities as a pamphleteer contributed to the controversies that occurred as England moved toward, engaged in, and recovered from civil war. Like many idealists of the period, however, Prynne found himself persecuted not just by Charles I before the war, but, after Charles's execution, by former allies *Oliver Cromwell and the Rump Parliament.

William Prynne was educated at the Bath Grammar School before attending Oriel College, Oxford, in 1618. He took his B.A. in 1622 then entered Lincoln's Inn. While pursuing his legal studies, Prynne embraced his first social cause, attacking the Saturday revels held at Lincoln's Inn while condemning all forms of theatrical performance. Probably Prynne's most famous publication, of the over 200 that he produced, was the 1632 *Histriomastix*, over 1,000 pages of venomous attacks on theatrical events, including dancing. This work contains assaults on the virtues of women who perform in masques. This specific allegation, and its implications about Queen Henrietta, whose love of performing in masques was well known, initiated Prynne's first conflict with Charles I's government. Archbishop of Canterbury *William Laud banned Prynne's work and in 1633 brought to the Star Chamber charges against its author. Prynne was convicted of libel and sedition then sent to the pillory where he had his ears cut off before being imprisoned in the Tower.

In 1640, when Charles called his first Parliament in eleven years, the House of Commons declared Prynne's sentence to have been illegal and released him. Over the next several years, particularly after the beginning of the civil war, Prynne energetically pursued the cause of Parliament through pamphlets attacking the king, his royalist sympathizers, and the episcopacy. Prynne avenged himself on Laud by leading the case against the then-imprisoned clergyman who was executed in 1645. His theological positions, however, would alter Prynne's relationship with Parliament as the Commons' efforts to forge an alliance with Scottish Presbyterians violated his own beliefs.

Prynne's concerns, however, did not prevent him from issuing pamphlets against the politics of the New Model army and its leaders when they broke with Parliament in 1647. In 1648, he was elected MP for Newport in Cornwall and urged the Commons to accept the terms negotiated with the then-imprisoned Charles I and to declare those in the army opposing that position to be rebels. Pride's Purge, however, removed Prynne from Parliament and placed him once again in prison. For the next several years, Prynne found himself in and out of custody, constantly advocating his various positions, including a denunciation of the trial of the king. Prynne came to terms with his enemies in 1653 and was released unconditionally. Over the next several years, Prynne advocated positions that seem to contradict his earlier stances. He consistently attacked the Rump Parliament, defended the cause of *Charles II, even contributing to the efforts to secure the Restoration. In 1662, he was appointed Keeper of Records at the Tower. Though active during this period and never ceasing to write, Prynne faded from the scene until his death in 1669. Little that Prynne offered to the debates of the period was unique. What his career points up, however, is the impact of propaganda during the civil war, Commonwealth, and Protectorate years. At no previous time in history had the written word contributed so much to shaping events, nor had a writer as prolific as Prynne used the power of printing to advance social, religious, and political causes.

Bibliography: E. Kirby, *William Prynne: A Study in Puritanism*, 1931; W. Lamont, *Marginal Prynne*, 1963.

Gerald Morton

PUFENDORF, BARON SAMUEL (1632–1694). Born in Saxony, Samuel Pufendorf studied Lutheran theology at Leipzig University, intending to follow his father into the ministry. However, he developed an interest in other subjects, moving to Jena in 1656 to study natural law and moral philosophy under Erhard Weigel. During his life, Pufendorf pursued three separate, though related, careers and lived in four different countries. He began, in 1658, as tutor to the family of an influential Swedish politician in Copenhagen. In 1661, he took up the chair of international law and philology at Heidelberg University. He then moved to Lund in 1670 as professor of natural and international law. Six years later, he went to Stockholm to work for King Charles XI as privy counsellor, secretary of state, and royal historian. In 1688, he took up a similar position in Berlin under Frederick William I and his son Frederick III. It was on his return to Germany from a trip to Sweden, during which time he had received a barony from Charles XI, that Pufendorf died on 26 October 1694.

Pufendorf's writings encompass moral and political philosophy, comparative politics, international relations, and history. While some of his later works remain well known (particularly his *History of Frederick III*, which included a commentary on the Glorious Revolution in England), it is primarily as a natural law theorist that he is remembered today. Modern natural law theory developed, against the background of the religious divisions of post-Reformation Europe, as an ethical code that would be acceptable to all, regardless of nationality or religious beliefs. Its early proponents included *Hugo Grotius and *Thomas Hobbes. Pufendorf first set down his ideas on natural law in his *Elements of Universal Jurisprudence*, which he wrote in 1658 and published in 1660. However, his main exposition appeared in his magnum opus of 1672 *On the Law of Nature and Nations*. A year later Pufendorf produced *On the Duty of Man and Citizen*, an epitome of the larger work. It was perhaps his most influential work, being used as a university textbook throughout Protestant Europe during the eighteenth century. Grotius had built his theory upon two fundamental features of human nature: the instinct for self-preservation and the innate propensity for society. Hobbes had accepted the former but rejected the latter, arguing that sociable behavior could only be imposed through civil laws. For Pufendorf, sociability played a crucial role, but was itself built upon the desire for self-preservation and the inability of human beings to survive alone. His theory influenced many subsequent writers, notably *Locke, members of the Scottish Enlightenment, Rousseau, and Kant.

Bibliography: L. Krieger, *The Politics of Discretion: Pufendorf and the Acceptance of Natural Law*, 1965; M. Seidler, "Introduction" to S. Pufendorf, *On the Natural State of Men*, 1990; J. Tully, "Introduction" to S. Pufendorf, *On the Duty of Man and Citizen*, 1991.

Rachel Hammersley

PURCELL, HENRY (1659–1695). Henry Purcell is considered one of England's finest composers, especially for his wide array and imaginative treatment of vocal genres. He was a noted organist and also composed instrumental music. Purcell's versatile musical talents showed themselves early, and he held a number of positions in London in his youth. These included chorister of the Chapel Royal, organ tuner and copyist at Westminster Abbey (1674–1678), and composer-in-ordinary for the violins (1677). He became organist of Westminster Abbey in 1679, and in 1682 became an organist of the Chapel Royal. The following year, he was named organ maker and placed in charge of the king's instruments, serving during the reigns of *Charles I, *James II, and *William III. He built organs for the coronations of James and William. Purcell was buried next to the organ in Westminster Abbey.

Purcell synthesized a wide array of influences. He greatly admired Italian music, but also made use of French models, both in dances and in overtures. He knew and absorbed the works of earlier English composers such as Thomas Tallis, *William Byrd, and *Orlando Gibbons, all predecessors at the Chapel Royal. Direct influences include Pelham Humfrey, *Matthew Locke, and Purcell's teacher, *John Blow. Orlando Gibbons's son Christopher was a family friend and may have also taught Purcell. His style was unique to him and remarkable for its blend of lively creativity and technical genius. He made particularly skillful and natural use of canon, ground bass, and written-out vocal ornamentation, of which there are numerous examples throughout his work. Henry Playford, anthologist of his songs, referred to Purcell's reputation for vocal mastery in 1698: "he was especially admir'd for the Vocal, having a peculiar Genius to express the energy of English words, whereby he mov'd the Passions of all his Auditors." Purcell composed a large quantity of incidental music for theater, which was again popular in London after the Restoration. The most significant musical contributions were composed from 1690 to 1695 for *Dioclesian, The Fairy Queen, The Tempest, The Indian Queen,* and *King Arthur.* The latter was a collaboration with *John Dryden. Purcell's compact masterpiece *Dido and Aeneas* is completely set to music, the closest to an opera of all his works, although opera as such was not established in London at that time.

Odes and welcome songs were customary during the reigns of *Charles II and James II. These cantatas of royal praise were scored for solo voice, chorus, and orchestra. Purcell had begun composing anthems for the Chapel Royal by

1678. Some of these employed the earlier polyphonic anthem style while others made new use of solo voice and instruments in verses, especially after his Chapel Royal appointment, since during the reigns of Charles II and James II strings were customarily employed in the presence of royalty. These anthems were characterized by typical French overtures and frequently demonstrate an animated character. Other sacred works include two Anglican services, one in B-flat and the Evening Service in G-minor.

Purcell's secular songs were striking for their unity of music and text. More than 100, in addition to the theater works already mentioned, were published in his day. The songs display a fantastic range of forms, including sectional, rondeau, ground bass, and da capo. Exceptional ornamentation is often incorporated into the vocal lines. Purcell presents a lighthearted and sometimes bawdy side in his many catches, which were rounds to be sung by three voices.

Much of Purcell's instrumental writing is incorporated into works for theater, odes and anthems, in the form of overtures, dances, obbligato parts, and accompaniments. His expert handling of string orchestra is understandable in view of the Chapel Royal environment. The fantasias were written in 1680. They hark back to the earlier dominion of the viol consort and make use of Purcell's elegant polyphonic gifts. The *Ten Sonatas in Four Parts* and the *Sonatas of III Parts*, written soon after, employ the more modern trio sonata scoring of two violins with bass viol and keyboard accompaniment. Purcell's keyboard music includes eight suites, many small pieces such as airs and dances, and transcriptions of his other works. Curiously, he wrote very little for organ.

Bibliography: J. Westrup, *Purcell*, 1980; F. Zimmerman, *Henry Purcell 1659–1695: His Life and Times*, 1983.

Joyce Lindorff

PURCHAS, SAMUEL (?1575–1626). Samuel Purchas, a clergyman and compiler of travel literature, was born in Essex, England, circa 1575 and educated at Cambridge. He married his wife Jane in 1601. In addition to holding other benefices, Purchas was chaplain to George Abbot, Archbishop of Canterbury, and later rector of St. Martin's Church, London. He was also assistant to the geographer, cosmographer, and clergyman Richard Hakluyt, whose three-volume prose epic *Principal Navigations* (1598–1600) was the authoritative collection of travel literature produced in Elizabethan England. In 1613, Purchas published *Purchas His Pilgrimage*, a survey of peoples and religions of the world. Its success led to the publication of *Microcosmus, Or the Historie of Man* (1619) and the four-volume *Hakluytus Posthumus, or Purchas his Pilgrimes* (1625). The latter work was composed chiefly from Hakluyt's papers, which came into Purchas's hands upon Hakluyt's death in 1616, and its renown results partly from the fact that many of these manuscripts are no longer extant. The four volumes (the work of 1613 is sometimes considered a fifth volume) deal with ancient explorers; the circumnavigations of the world; and the voyages of

Englishmen to Africa, Asia, the East and West Indies, and North and South America. The tales in the epic are often unfaithful to the originals when these are available for consultation and show Purchas to have been more interested in demonstrating that European society was vastly superior to others than arriving at accurate knowledge.

Bibliography: L. Pennington, ed., *The Purchas Handbook*, 2 vols., 1997.

Ken MacMillan

Q–R

QUARLES, FRANCIS (1592–1644). Francis Quarles was baptized on 8 May 1592 in Romford, Essex, England. The son of prosperous parents, Quarles was educated at Christ's College University of Cambridge and at Lincoln's Inn, London. In 1620, he published his first book of sober religious poetry entitled *A Feast for Worms set forth in a Poem of the History of Jonah*. In 1626, Quarles became secretary to *James Ussher, archbishop of Armagh in Ireland. He returned to England in 1633, and in 1635 he published *Emblems*. Traditional sixteenth-century emblem poems were didactic poetic composites: A short motto (*inscriptio*) introduced the theme, an illustration (*pictura*) embodied the topic, and an epigram (*subscriptio*) reiterated the idea once again. Quarles embellished the emblem's format by placing an exegetical illustration first, followed by its relevant scriptural passage. Then a poem elucidated the scripture and finally an epigram would recapitulate the religious virtue. In 1638, Quarles wrote a second book of emblematic poetry entitled *Hieroglyphics of the Life of Man*. These two works were combined in 1639, producing *Emblems divine and moral: together with Hieroglyphics of the Life of Man*, quite possibly the most popular book of verse of the seventeenth century. The book had a huge following of Puritan readers, which is ironic considering Quarles maintained devout Royalist ideologies and affiliations. In 1640, Quarles became the chronologer to London, where he virtually abandoned writing poetry for more lucrative prose ventures, such as his highly successful book of aphorisms, *Enchiridion* (1640). In 1644, Quarles wrote a pamphlet defending *Charles I, entitled *The Loyal Convert*. This allegiance is reputed to have cost Quarles his reputation. He died on 8 September 1644 in London.

Bibliography: A. Cousins, *Emblems (1635) by Francis Quarles: A Facsimile Reproduction*, 1991; A. Witherspoon, ed., *Seventeenth-Century Prose and Poetry*, 1982.

David M. Rosen

QUEVEDO Y VILLEGAS, FRANCISCO GOMEZ DE (1580–1645). Francisco de Quevedo y Villegas was born in Madrid in 1580. He began his studies with the Jesuits and later attended the universities of Alcalá de Henares and Valladolid. Quevedo's favor in court rose and fell with the fortune of his friend, the Duke of Osuna. When the duke was designated as viceroy to Naples, Quevedo accompanied his noble friend to Naples in an administrative capacity, which sometimes proved quite dangerous. After the Duke of Osuna lost favor in the Spanish court, Quevedo was imprisoned for four years in Leon for having written a scathing epistle to the king against his minister, the Duke of Olivares. Quevedo was released from prison once Olivares fell from favor. However, Quevedo died shortly thereafter in 1645. Quevedo was married at age fifty-four, but the marriage did not last very long.

Quevedo's vision of the world includes a high, moral sensibility and a merciless critique and ridicule of human baseness. The Baroque themes of scorn toward vanity, the fugacity of worldly pleasure, and the inevitable change of fortune are present in his works. His poetry exemplifies doctrinal intention, amorous themes, and the preoccupation of the spiritual and material decadence of Spain. As a prose writer, Quevedo is satiric and cynical. *El Buscón* (1603) is a picaresque novel that presents a grotesque and comic version of a *pícaro* and his life experiences. *Los Sueños* (1627) provides a critical review of the social classes and customs. Quevedo is considered Spain's foremost satirist and wit of his time. He is noted for openly dealing with the decline of Spain in his satiric and more serious literary works.

Bibliography: D. Bleznick, *Quevedo*, 1972; J. Olivares, *The Love Poetry of Francisco de Quevedo*, 1983.

M. Ellen Blossman

QUINAULT, PHILIPPE (1635–1688). Quinault, a dramatist whose highly emotional plays marked the generation between those of *Corneille and *Racine, was born in 1635 to the son of a Parisian baker. Educated by the poet Tristan l'Hermite, Quinault later became a lawyer. He soon associated himself with the court and gained influential protectors. In 1660, he married a young and rich widow, Louise Goujon, and subsequently purchased a post as valet to the king. During the 1660s, Quinault wrote multiple comedies, tragedies and tragicomedies, which were all well received. In 1671, he collaborated with *Molière and *Corneille to write the comédie-ballet *Psyche*. That same year, Quinault was admitted into the French Academy, presumably to mark the end of his literary career. In 1672, *Louis XIV provided a pension of 2,000 livres to Quinault on condition that he write the librettos for *Lully's operas. Amid frequent bickering, Quinault and Lully composed multiple operas—*Cadmus et Hermione* (1673) being their most successful—until 1686. Lully, not wishing to jeopardize his continued operatic successes, retained Quinault despite urgings that he be replaced by *La Fontaine. After his retirement in 1686, Quinault, perhaps in deference to the by then austere court, publicly announced his regret for having portrayed sensual and unbridled emotions on stage.

Quinault's first plays were derivative comedies based on the Italian and Spanish traditions. After Thomas Corneille's successful tragicomedy *Timocrate* (1656), Quinault wrote his own, including *La Mort de Cyrus* (1656) and *Amalasonte* (1657). The latter offers a character who dies but then revives, a typical occurrence in these extravagant plays, whose insistence on pathos attempts to compensate for the lack of dramatic interest. Quinault's plays are indicative of the mid-century preference for literature that underscores the violence and gallantry of love. With convoluted and fantastic plots, these plays offer characters who sacrifice all in the name of love. Their passion is far from the heroic and idealist love portrayed by Pierre Corneille but one that overpowers, becoming the characters' obsession. Principal characters, oftentimes women, rarely face dilemmas but rather wax poetic on their blind love, which betrays any principles they may have once had. This vision of love adapts better to Quinault's librettos in which the music sustains these often continuous elegies to the supremacy of passionate love. Quinault's comedies today are considered his best works, among which *La Mère Coquette* (1665) is the most memorable. *Dryden would adapt *L'Amant indiscret* (1654) for his version of *Molière's *L'Etourdi*.

Bibliography: W. Brooks, *Bibliographie critique du théâtre de Quinault*, 1988; E. Gros, *Philippe Quinault, sa vie et son oeuvre*, 1926.

Margaret Harp

RACINE, JEAN-BAPTISTE (1639–1699). French playwright and historiographer for *Louis XIV, Jean-Baptiste Racine was born in La Ferté-Milon, a small town in the duchy of Valois, seventy-six kilometers from Paris, to Jean Racine, a local solicitor, and Jeanne Sconin. Orphaned when he was barely two years old, Racine lived with his paternal grandparents until the age of ten. After the death of his grandfather in 1649, Racine accompanied his grandmother, Marie des Moulins, to the Jansenist convent-monastery at Port-Royal des Champs near Paris, where he received a solid education in the classics of Latin and Greek literature. In 1653, Racine left Port-Royal to study at the Collège de Beauvais. He returned to Port-Royal two years later to study rhetoric. Racine was profoundly influenced by the doctrines of *Cornelius Jansen at Port-Royal, which were considered heretical because they proclaimed that God did not give man free will. According to Jansenist doctrine, man's fate was predetermined and could not be influenced by his actions.

In 1658, Racine's world started to drift away from that of the Jansenists when he was sent to study philosophy at the Collège d'Harcourt in Paris. There he associated with intellectuals and began a friendship with *Jean de La Fontaine. In 1660, Racine published his first poem, *La Nymphe de la Seine*, an ode written in honor of the king's marriage. The following year, Racine left Paris in an unsuccessful attempt to obtain an ecclesiastical benefice in Uzès. He returned to Paris and his literary career in 1662 or 1663. Racine's poem *Renommée aux*

Muses, which he wrote for the king, attracted the attention of *Nicolas Boileau, who would become his lifelong friend and mentor.

Racine's first tragedy, *La Thébaïde ou les Frères Ennemis*, which was based on a Greek legend, was produced by *Molière at the Palais Royal in 1664. His second, *Alexandre*, which established Racine in popular opinion as *Corneille's equal, was produced first by Molière in December 1665 and then two weeks later by Molière's rivals at the Hôtel de Bourgogne, possibly because Racine did not think Molière's troupe did his play justice. Racine's disloyalty, as well as the fact that he persuaded Mademoiselle du Parc, Molière's leading actress, to leave Molière's troupe, join the company at the Hôtel de Bourgogne, and become his mistress, marked the end of the friendship between Racine and Molière. In 1666, the Jansenists urged Racine to leave the sinful world of the theater and threatened him with excommunication. Instead, Racine broke his ties with them by engaging in a virulent pamphlet war.

Racine's first masterpiece, his tragedy *Andromaque*, was produced in 1667. In the next ten years, it was followed by one comedy, *Les Plaideurs* (1668), and six other tragedies: *Britannicus* (1669), *Bérénice* (1670), *Bajazet* (1672), *Mithridate* (1673), *Iphigénie* (1674), and *Phèdre* (1677). In 1673, Racine was admitted to the *Académie française* as its youngest member. Although Racine's first six tragedies were all enormous hits with the public, his last, *Phèdre*, which is considered by many scholars to be his greatest, encountered difficulties when it had to compete against another version of the Phèdre legend that opened only two days after Racine's. That same year, Racine married and was appointed Louis XIV's court historiographer, along with Boileau. Racine reconciled with the Jansenists, and, with the exception of two biblical dramas written at the request of *Madame de Maintenon, the king's mistress, never wrote another play. The devoted father of seven children, Racine died in 1699 after a long illness and was buried at Port-Royal.

Racine elevated French classical theater to a new level of perfection. Reduced to the essentials, his plays adhere faithfully to the unities and achieve a psychological intensity not seen before. Using realistic situations, Racine depicts his heroes and heroines as sympathetic characters engaged in a hopeless struggle with a force that is more powerful than they. Profoundly human in their suffering, they are crushed by fate as they fall victim to their own passion. Corneille's plays exalt the triumph of human will and inspire us; Racine's plays, however, dramatize the weakness of that same will and arouse our pity. According to *Jean de la Bruyère, Corneille painted men as they should be, but Racine portrayed them as they are.

Bibliography: B. Knapp, *Jean Racine: Mythos and Renewal in Modern Theater*, 1971.

Leslie A. Sconduto

RALEGH, SIR WALTER (1554–1618).

Contemporary spelling of Sir Walter Ralegh's name as "Raleigh" will lead to confusion, since that spelling is the only one of roughly forty alternatives that Ralegh himself did not use. He signed his name numerous times and with great range, including "Rawleigh," "Rawley," and "Raley." For consistency, libraries and scholars use "Ralegh" in cataloguing and publishing. The pronunciation—"raw lee"—is not disputed. Born in Hayes Barton in Devonshire, England, Sir Walter Ralegh stands as an example of a Renaissance man, securing fame, although not always fortune, in literature, at court, in war, and in adventure. Ralegh was a poet, although most of his work remained unpublished during his life, and it is difficult to this day to know how much of the writing that was attributed to him was actually his own. The only poems that are known to be his own are *The Eleventh and Twelfth Books of the Ocean to Cynthia*—a little known work of courtly love dedicated to the queen—and *The Nymph's Reply to the Shepherd*—a poem responding to Marlowe's sonnet of Petrarchan love, *The Passionate Shepherd to His Love*. Although he attended Oriel College, Oxford, he turned from the university without completing his degree.

Ralegh earned greater fame from his many successful battles, playing in all a leading part in England's expansion of her territories, particularly in America. His desire, whether fighting against the Irish or the Spanish, was to spread Protestant doctrine and English power through colonization. As early as 1569 and at the age of fifteen, Ralegh was a soldier, fighting on the side of the French Huguenots. In 1580, he helped to subdue the Irish in Ireland to increase England's power. Ralegh sponsored an expedition to North Carolina in 1584, which secured two Native American Indians for the queen's perusal. This success, among other achievements, earned Ralegh knighthood on 6 January 1585. Later that year, Ralegh sponsored another American colonial venture. Although the expedition successfully gathered information about regions as far north as the Chesapeake Bay, the team returned from the trip with little in goods or land to claim. In the same year, Ralegh completed one of his most famous, though least successful, expeditions. Recorded in prose by Ralegh himself and proclaimed a glorious triumph in the poem *De Guiana* written by *George Chapman, the venture to the Orinoco River in Guyana in search for gold proved a failure. It did, nevertheless, increase his fame, as the writings about the voyage enticed Englishmen to take to the sea to find El Madre del Oro, the Mother of Gold. Ralegh and Chapman made it very clear in their writings that England could rule the world if she secured the gold from those distant lands. In 1587, Ralegh tried again to colonize America, but this time the colony at Roanoke led by Governor John White disappeared. The fate of the colonists remains a mystery to this day. Ralegh's battles and adventures continued, but although historical legend suggests Ralegh helped to defeat the Spanish Armada in 1588, it was more likely that he helped to orchestrate the naval operations but did not actually carry them out. By 1596, Ralegh was back on the sea, but instead of securing England's wealth in gold, he was joining Lord Essex and others in triumph at Cadiz. In this endeavor, England and Ralegh were successful again. His career at court only added to his fame. By 1582, Queen Elizabeth established

Sir Walter Ralegh as her "favorite"; in exchange for companionship, she supplied him with titles and land—including 40,000 acres in Ireland forfeited to England.

By 1589, Ralegh had left the court and taken up residence in Ireland, where he became patron for Edmund Spenser as Spenser completed *The Faerie Queene*. Some conclude that Ralegh left the court because he did not get along with another of the queen's favorites, the Second Earl of Essex, Robert Devereux. Yet, in 1592, he was back at court after the queen's argument with Essex. Soon again, Ralegh and the queen were at odds; he secretly wed Elizabeth Throckmorton, one of the queen's ladies in waiting, and the queen sentenced Ralegh to a prison term in the Tower. Shortly thereafter he was released, but his trouble with the monarchy would not end there; it increased with Elizabeth's death when in 1603 *James I came from Scotland to rule. Ralegh found himself in the Tower on charges—on all accounts well-founded—of conspiracy with Spain against England. His sentence included death, but the charges were changed to life imprisonment. During the more than a dozen years Ralegh spent in the Tower, he grew especially fond of alchemical experimentation, hoping he might turn base metal into gold. He also began his last work of literature, *The History of the World* (1614). This work he had dedicated to Prince Henry in hope of release. In 1616, he was free again, but only so that he could return with an expedition for El Madre del Oro in Guyana. Unsuccessful in this endeavor, the king had him arrested and Ralegh found himself again in prison with time to practice alchemy. In 1618, James changed his mind again about Ralegh's worth; he ordered Ralegh's execution, presenting the severed head to Ralegh's wife, who had stayed with him in the Tower all of those years. Ralegh never finished his *History*, but his life as adventurer, warrior, poet, and courtier tells its own interesting history of a Renaissance man.

Bibliography: S. Greenblatt, *Sir Walter Ralegh: The Man and His Roles*, 1973; S. May, *Sir Walter Ralegh*, 1989.

Rebecca Totaro

RANDOLPH, THOMAS (1605–1635). Thomas Randolph was born in Newnham-cum-Badby in Northamptonshire in 1605. His father was a steward to Edward, Lord Zouch. He became a king's scholar at Westminster School and, later, attended Trinity College, Cambridge, where he received his B.A. in 1628 and his M.A. in 1631. Much of Randolph's short literary career centered around Cambridge. Verses on the marriage of *Charles I (1625), a Latin poem on the death of *Francis Bacon (1626), his "Salting" (written for the Cambridge initiation rites), and his satirical monologue "The Conceited Pedlar" (1627) were written during this time. Randolph's plays *Aristippus, or The Jovial Philosopher* and *The Jealous Lovers* were written for performance at Cambridge. The latter play was performed for Charles I and Henrietta Maria upon their visit to Cambridge in 1632. During the late 1620s, Randolph spent much time in London.

Around 1630, he wrote two comedies for the King's Revels company, the pastoral *Amyntas* and a satire against Puritan objections to the theater, which contains an allegorical play-within-a-play, *The Muses' Looking Glass.*

As an undergraduate, Randolph was adopted into the "Tribe of Ben" *Jonson, and a strong Jonsonian influence is visible in his plays and poetry. Until his early death in 1635, rumored to be the result of drunken dissipation, Randolph was regarded as Jonson's heir. His interest in classical literature appears in his Latin verse and in his adaptation of Aristophanes' "Plutus" in the play *Hey for Honesty, Down with Knavery* (published in 1651). Randolph's literary accomplishments, indeed, lie primarily in innovative adaptation rather than originality. He uses the dramatic models of Aristophanes, Plautus, Terence, and the Italian pastoral and betrays the clear influence of both the English playwrights and poets who preceded him and those who were his contemporaries and friends. Sadly, the popular impression of his work remains one of unfulfilled promise and potential.

Bibliography: Charles M. Kovich, "Thomas Randolph," *Dictionary of Literary Biography Vol. 126: Seventeenth-Century British Nondramatic Poets (Second Series),* 1993, 239–245; Jill L. Levenson, "Thomas Randolph," *Dictionary of Literary Bibliography Vol. 58: Jacobean and Caroline Dramatists,* 1987, 231–240.

Janet Bertsch

RAVENSCROFT, THOMAS (ca.1590–ca.1623). A chorister at St. Paul's in London and music master at Christ's Hospital from 1618 to 1622, English composer Thomas Ravenscroft published four collections of compositions during his lifetime. His *Pammelia* (1609) is the first English printed collection of "rounds" and "catches," songs in which the melody is repeated at specific points by several singers, thus passing the tune "round and round." *Pammelia* contains 100 such works in both English and Latin. *Deuteromelia* (1609) collects thirty-one songs, including what is probably the first publication of "Three Blind Mice." *Melismata* (1611) contained twenty-three compositions. In 1614, Ravenscroft published *A Briefe Discourse of the true (but neglected) use of Charact'ring the Degrees, by their Perfection, Imperfection and Diminution in Measurable Music, against the Common Practise and Custome of these Times.* This treatise condemns the breaking of musical rules by popular musicians and minstrels, and it was dedicated to the directors of Gresham College in London. Unlike the music in his other compilations, the tunes used as examples in this work are by Ravenscroft himself. He is best known for his last work, *The Whole Book of Psalms* (1621), also known as "Ravenscroft's Psalter." Forty-eight of the 100 psalm settings are his own, with the remainder by a variety of contemporaries, such as *Dowland, *Morley, and Tallis. Ravenscroft combined a scholarly and traditional musical background with a wide-ranging interest in the popular forms and melodies of his day.

Bibliography: D. Mateer and I. Payne, "Thomas Ravenscroft," *The New Grove Dictionary of Music and Musicians*, ed., S. Sadie, 2001.

Christopher Baker

RAY, JOHN (1627–1705). A preeminent botanist who espoused natural theology, Ray was born on 29 November 1627. He acquired an excellent grounding in Latin at the grammar school in Braintree. In 1648, he received a B.A. from Trinity College, Cambridge. In 1651, he earned an M.A. and was appointed Greek lecturer. His earliest botanical studies, begun in 1650, produced *The Cambridge Catalog* (1660), which critically assessed older botanical works and clarified the names, descriptions, and practical uses of plants. Ordained an Anglican priest in 1660, Ray remained at Cambridge until 1661, when he refused to take the Oath of Conformity instituted with the Restoration. Now jobless, Ray became associated with Francis Willughby, a former student who supported him financially. Together they traveled through England and Wales and spent three years on the continent collecting materials with the intent that Willughby would write about animals and Ray about plants. When Willughby died in 1672, Ray completed his works *Ornithology* (1676) and *History of Fishes* (1683). He returned to animal studies at the end of his life.

In *A New Method of Plants* (1679), he revolutionized the classification of plants, separating those with imperfect flowers from those with perfect flowers, which were then subdivided into dicotyledons and monocotyledons. Ray's major work, the three-volume *The History of Plants* (1686–1704), aimed at describing all known plants and provided a new criterion to define species: descent from a common ancestor. While its cost and paucity of plates prevented it from selling well, it paved the way for Carolus Linnaeus, the famed eighteenth-century Swedish biologist and taxonomist. Updating his work on English botany, Ray published *Synopsis of British Flora* (1690), which stopped listing plants alphabetically and arranged them according to his new classification. Unlike his contemporaries, Ray preferred a more natural classification based on a combination of characteristics rather than a single feature. His most popular work, influential until the nineteenth century and read by Charles Darwin, was *The Wisdom of God Manifested in the Works of Creation* (1691), enlarged three times. Ray started with *Henry More's arguments for the existence of God based on natural phenomena and accepted *Ralph Cudworth's concept of "spiritual plastic powers." The result was a survey of the adaptations of structure to function in natural phenomena ranging from physics to human anatomy, which demonstrated God's goodness and wisdom. Ray continued to publish works until shortly before his death on 17 January 1705.

Bibliography: C. Raven, *John Ray, Naturalist: His Life and Works*, 1950.

Kristen L. Zacharias

REGNARD, JEAN-FRANÇOIS (1655–1709). A highly regarded dramatist of the late seventeenth century whose witty verse was judged to approach that of *Molière, Regnard was born 8 February 1655 in Paris. The son of meat merchants, Regnard received an excellent education and then was engaged as a jeweler's apprentice. He consequently traveled to Italy and Constantinople and after his second trip to Italy was kidnapped by Algerian pirates October 1678. Released eight months later, Regnard culled his experiences as a captive for his romance *La Provençale*. In 1681, he undertook an extensive trip throughout Sweden, Poland, and Austria. The great uncle of the Romantic poet Alfred de Vigny, Regnard in his travel writings himself displays a tendency for romantic meditation, expounding upon the natural wonders seen during his trip. Regnard also wrote an entertaining and evidently embellished account of his travels, *Voyage de Laponie*, as well as various epistles. Regnard's fifteen-year public literary career began in 1688 with his composition of the *Divorce joué* for the Italian Theater. For the next six years, Regnard wrote eight plays for this troupe; after 1697, he wrote for its imposing rival, the more prestigious Comédie Française. His plays *Attendez-moi sous l'orme* (1693), *Le Joueur* (1696), and *Le Légataire universel* (1708) were particularly popular. The clever verbal play, strong female characterization, and ingenious use of the five-act comedy mark these last two, along with *Le Distrait* (1697), *Démocrite* (1700), *Les Ménechmes* (1705), and *Folies amoureuses* (1704) as his best plays. The latter is a marked imitation of Molière's *L'Ecole des Femmes* and *L'Amour médecin*. Regnard died suddenly 4 September 1709, leaving vast wealth and numerous titles.

Aware of but undaunted by the imposing dramatic tradition to which Parisian audiences were accustomed, Regnard did not hesitate to weave transpositions of famous lines from *Racine into his comedies. Such examples of wit and gaiety characterize his plays. With most depicting a debauched society, Regnard's works nonetheless refrain from moralizing and are more faithful to the Italian Theater tradition of elaborate intrigue and finely honed comic scenes than to Molière's preoccupation with the portrayal of character types.

Bibliography: A. Calame, *Regnard, sa vie et son oeuvre*, 1960; D. Medlin, *The Verbal Art of Jean-François Regnard*, 1966.

Margaret Harp

REMBRANDT HARMENSZOON VAN RIJN (1606–1669). Known to be the greatest artist of the Dutch school, Rembrandt Harmenszoon van Rijn was born in Leiden on 15 July 1606 and died in Amsterdam in October 1669. Rembrandt's painting had an enormous impact on contemporary and later artists. His understanding of the natural world and the effects of light was equaled by his empathy for the human subject, and his chiaroscuro, or use of shadow, rivaled his use of *impasto*, or the building up of layers of paint. Technical excellence is apparent in Rembrandt's work, from drawings of contemporary life in Amsterdam to etchings of the crucifixion, and from self-portraits to the large-scale paintings of his mature period.

In 1620, at the age of fourteen, Rembrandt, the son of a miller, entered the

Leiden University, only to leave soon afterward to study painting first in Leiden with Jacob van Swanenburch and then with Pieter Lastman in Amsterdam. His paintings of the 1620s show the influence of Lastman in the choice of dramatic subjects and contrasts of light and shadow. By 1628, he had established a studio in Leiden, then the second most populous city in Holland, and began taking pupils, among them Gerrit Dou. The success of his studio prompted a move in 1631 to Amsterdam, where his painting was in high demand. In 1634, he married Saskia van Uylenburgh, who figured prominently in his work as a model, until her untimely death in 1642. Hendrickje Stoffels, his housekeeper, became his common-law wife in 1649 and modeled for later paintings.

For some years Rembrandt was the most successful portrait painter in Amsterdam; his group portrait of the members of the Amsterdam Surgeon's Guild, the famous *Anatomy Lesson of Dr. Tulp* dates from 1632. In the painting, the professor of anatomy instructs a group of seven of the guild's members while holding with forceps the muscles and tendons of the forearm of the cadaver, lit from above. The dramatic concentration on the expressive left hand of Tulp recalls ancient gestures of oratory, and the variety of the representation of the listeners' attention to the lecture not only enhances their characterization, but also increases the illusionistic quality of the painting.

Another famous group portrait, the so-called *Night Watch*, was finished in 1642 and depicts a military company preparing for an exercise. The correct title of the painting is *Officers and Men of the Company of Captain Franz Banning Cocq and Lieutenant Willem van Ruijtenburgh* and owes its misnomer to nineteenth-century critics who, deceived by Rembrandt's strong chiaroscuro, believed it to be a nocturnal scene. Rembrandt used a large scale to portray the members of the company, and by avoiding static grouping of the figures achieved a dramatic naturalism. The painting signals a development in Rembrandt's art by the increase in figures, more complex spatial and pictorial design, and a maximum amount of action. The many individual commissioned portraits within the painting not only reveal the artist's physiognomic skill, but also his financial success. His mythological and religious works were also highly in demand during this period, and he painted such masterpieces as *The Blinding of Samson* in 1636. Over one-third of Rembrandt's work depicts scenes from the Bible, mythology, and ancient history. The strength of his history painting lies in the perceptive characterizations of the protagonists, whose emotional responses to their circumstances have validity and truthfulness. They also show Rembrandt's love of exotic costume, as in *Belshazzar's Feast* (1637).

The artist's middle period is marked by a gradual shift from the dramatic Baroque style to a more classical treatment, although he never fully embraced the classical vogue. Instead of scenes portraying strong dramatic elements, the paintings of the 1640s display more domestic settings, where human relationships are subtly expressed. *The Holy Family with a Curtain* (1646) depicts a scene from the everyday life of the holy family and glows with domestic warmth. During the 1640s, Rembrandt also took a great interest in landscape, as in *The Rest on the Flight into Egypt* (1647), in which the members of the holy family are shown as tiny figures in a vast and mysterious nocturnal setting.

Rembrandt's greatest paintings were created during the last two decades of his life. His works of this period reveal a concern with mood and the spiritual qualities of contemplation. The portraits and historical paintings of these years concentrate on psychological drama rather than on the excitement of the narrative element, and they probe the surface of external appearance to reveal the soul. In 1654, he painted his common-law wife Hendricjke as *Bathsheba*, displaying an unprecedented concern for revealing the innermost thoughts of his subject as she awaits her inevitable personal tragedy. Rembrandt's final painting, the 1669 *Return of the Prodigal Son*, expressed the idea of Christian mercy in a work that combined the artist's power of realism with an evocation of religious mood, human sympathy, psychological insight, and spiritual awareness. Rembrandt died in 1669 (six years after the death of Hendrickje and one year after the death of his only surviving child Titus) having created nearly 2,000 drawings, some 300 etchings, and over 600 paintings.

Bibliography: A. Blankert, *Gods, Saints and Heroes: Dutch Painting in the Age of Rembrandt*, 1980; S. Schama, *Rembrandt's Eyes*, 1999; S. Slive, *Dutch Painting 1600–1800*, 1995.

Mark Williamson

RENAUDOT, THÉOPHRASTE (1586–1653). A man of great curiosity and imagination, Renaudot was a physician, a philanthropist, and Europe's first journalist. After completing his medical studies in Montpellier, he went to Paris in 1612 and became both secretary and physician to Louis XIII. In 1618, Renaudot was named the "Commissioner General of the Poor of the Kingdom" and would, in 1637, create the first *mont-de-piété*, or pawnshop, whose profits were used to help the poor. He also directed France's first free health clinic and, in 1630, opened a *bureau d'adresses*, where people of all backgrounds could go to look for a job, ask for advice, or simply catch up on the latest news.

His desire to disseminate information led him to found, with the patronage of Louis XIII and *Cardinal Richelieu, the *Gazette de France*, France's and Europe's first regularly appearing newspaper. Published on Saturdays, the *Gazette* reported news from Paris, the provinces, and throughout Europe, and was almost immediately a huge success. It is reported that Richelieu and the king himself contributed articles to the *Gazette*. Renaudot later founded a publication entitled *Conférences du Bureau d'adresses* in which he provided more specialized news for businesspeople and intellectuals of his day. This would later become the *Journal des Savants*, a precursor to the modern scholarly journal. The patronage of Richelieu and Louis XIII did not prevent Guy Patin, a noted physician, from bringing charges of misconduct against Renaudot and having him barred from practicing medicine for several years. Renaudot also wrote several historical accounts of famous men, including an *Abrégé de la vie et de la mort* *de Henri de Bourbon, prince de Condé* (1647) and a *Vie de Michel de Mazarin, cardinal* (1648).

Bibliography: C. Bailly, *Théophraste Renaudot: Un homme d'influence au temps de Louis XIII et de la Fronde*, 1987; C. Laurioux, *Théophraste Renaudot: 1586–1653*, 1988.

Patricia Armstrong

RENI, GUIDO (1575–1642). Recognized in his own time as one of the greatest painters of seventeenth-century Italy, Reni was born in Bologna on 4 November 1575 to a prominent musician, Daniele Reni, and his wife Ginevra Pozzi. By the age of nine, he was accepted into the local workshop of Denys Calvaert, a Flemish Mannerist painter. Approximately ten years later, Reni transferred to the Carracci Academy, which emphasized a more naturalistic representational approach, evidenced in Reni's first public altarpiece, the *Coronation of the Virgin* (1595).

In search of further patronage and artistic inspiration, Reni left for Rome in 1601. His *David with the Head of Goliath* (1605) reveals the adaptation of *Caravaggio's dramatic tenebrism while maintaining a Mannerist elegance. The artist created two of his best-known works for the powerful Borghese family: *St. Andrew Led to Martyrdom* (1608–1609) and the glorious *Aurora* for the ceiling of the Borghese villa (1613–1614). The *Aurora* demonstrates another stylistic shift to a more graceful classicizing manner, influenced by Reni's study of ancient Greek sculpture.

In 1614, the artist returned to Bologna, where he established himself as a leading painter of religious subjects, emulating the upturned, yearning faces of Raphael's saints. Reni received commissions from an international clientele, including the politically charged *Abduction of Helen* (1627–1629) as well as an *Immaculate Conception* (1627) for the Spanish court, with other paintings sent to Henrietta Maria of England and Marie de' Medici of France. Such works greatly influenced other seventeenth-century artists, including *Murillo, Maratti, and Le Sueur, while the numerous students trained in Reni's workshop were another significant means by which his style was disseminated.

Reni's style underwent a profound transformation in the 1630s, characterized by desaturated colors and loose brushwork. Often referred to as his "silvery style," this late manner has been the subject of much art historical speculation. Was the change due to economic hardship, exacerbated by the artist's compulsive gambling habit, or the result of an aesthetic progression toward a more idealized approach? Whatever the impetus, such works validate the praise offered by a fellow artist, the Cavaliere d'Arpino, that while most artists painted like mortals, Guido Reni painted "like an angel."

Bibliography: S. Pepper, *Guido Reni*, 1984.

Julia K. Dabbs

RIBERA, JUSEPE DE (1591–1652). "Lo Spagnoletto," as he is sometimes known, was Spanish by origin and is said to have been born in the town of

Jativa, Valencia, to the shoemaker Simon Ribera. Scholars have disputed his apprenticeship to the Valencian painter Francisco Ribalta from 1601 to 1607. From 1613 until 1616, he was a member of the Academia de San Luca and an associate of the Caravaggesti, those younger artists in Rome from the Netherlands and Italy influenced by the realist style of *Caravaggio. From 1616 until his death at the age of sixty, he lived and worked in the capital city of the Kingdom of Naples, patronized by the Spanish viceroys, the Neapolitan aristocracy, and the church, and never returned to Spain.

The affective quality of religious paintings, and their ability to instruct, move, and edify, had been decreed to be of prime importance by the 1563 Council of Trent, and seventeenth-century painters of religious subjects worked under restrictive iconographic and visual guidelines. Despite these constraints, Ribera was preoccupied with obtaining a completely accessible and comprehensible interpretation of religious experience. To accomplish this, he used contemporary Neapolitans as models, depicting and ennobling their natural qualities with an exacting verisimilitude. In the Caravaggesque tradition, he brought the subjects forward in the picture plane, involving the spectator in the imagined world of the artist's creation. He also employed a strong Tenebrism, a style of painting characterized by the use of chiaroscuro, the use of deep variations and subtle gradations of light and shade to enhance the dramatic effect. Best known for his paintings of elderly martyrs or penitents (e.g., *Martyrdom of St. Andrew* [1628]), he also has a reputation for the representation of violent martyrdom. However, the bloody aspects of his art do not justify his infamy as a gruesome artist. Instead, his paintings reveal a high degree of scientific observation and bear witness to the intensity of emotive content Ribera achieved through imitation.

Bibliography: J. Brown, *Jusepe de Ribera: Prints and Drawings*, 1973; C. Felton, *Jusepe de Ribera: A Catalog Raisonne*, 1971; A. Perez-Sanchez and N. Spinosa, eds., *Ribera 1591–1652*, 1992.

Mark Williamson

RICHELIEU, ARMAND JEAN DU PLESSIS DE (1585–1642). Richelieu, Roman Catholic cardinal and cunning statesman, was the architect of French absolutism under Louis XIII, laying the basis for *Louis XIV's later assumption of unmatched personal authority in the French monarchy. Born on 9 September 1585, he was consecrated Bishop of Luçon in 1607 when only twenty-two. He attached himself to the powerful Concini, whose murder in 1617 blunted his rise in government. He became an advisor to the Queen Mother Marie de' Medici, then in exile; when she emerged from exile in 1619, he had greater latitude for advancement. He was designated a cardinal in 1622 and two years later became chief minister of the king.

Richelieu frankly declared his intent to destroy the Huguenots, crush the power of the nobility, and augment the king's influence both in and out of

France. These ambitions aroused plots against him by the aristocracy, which he ruthlessly put down, executing Montmorency (the governor of Languedoc), Françoise Auguste de Thou, and the Marquise de Cinq-Mars.

He moved against French Protestants with equal vigor, defeating the Huguenots at the siege of La Rochelle (1627–1628), allowing them to hold to their religious practices but severely curtailing their political prerogatives. The king disliked Richelieu but respected his power and made an effort to benefit from his strategy, thus alienating his mother who left France for Holland, leaving the cardinal with one less opponent in his quest for authority. In 1631, Richelieu became a duke. His primary goal in foreign policy was to nullify the power of the Catholic Hapsburgs. To this end, he persuaded Gustavus Adolphus, king of Sweden, to battle the Germans, who were later defeated in 1630 at Pomerania. In 1635, he declared war on the Spanish Hapsburgs and on the Holy Roman Emperor in Germany. France conquered the Alsace and other territories shortly before the cardinal's death on 4 December 1642.

Imperious, ambitious, proud, and feared, Richelieu was in many ways a French Machiavelli, leaving the monarchy of France in a position of power that was broadened under his successor Mazarin and honed by Louis XIV. He had largely bypassed the French Parliament, extended the king's bureaucracy deep into the French heartland, and taken control of the emerging French press. Even his creation of the French Academy was as much a political as a cultural decision. Patron of *Corneille, he authored plays himself and wrote a *Defense of the Main Principles of the Christian Faith*, but his *Memoires* and letters are his most notable literary legacy.

Bibliography: J. Bergin, *Cardinal Richelieu: Power and the Pursuit of Wealth*, 1985.

Christopher Baker

ROBERVAL, GILLES PERSONNE DE (1602–1675). Little is known about the early life of Roberval. Apparently the son of a farmer, Roberval began the study of mathematics at age fourteen, traveling throughout France and giving lessons. By 1628, he had moved to Paris and joined the circle of mathematicians that had formed around the French mathematician and Minimite friar *Marin Mersenne. While not gaining the acclaim of his contemporaries *Blaise Pascale or *René Descartes, Roberval nonetheless was an important contributor to the body of mathematics that would become formalized as the calculus by *Newton and *Liebniz. He would become a charter member of the Académie des Sciences in 1666.

In 1628, Roberval was named Professor of Philosophy at the Collège de Maître Gervais. In 1634, he won a competition that entitled him to occupy the Ramus Chair in mathematics at the College Royal in Paris. As holder of the chair, Roberval would determine the problems for the subsequent competitions, which he was able to win to remain in the chair for the rest of his life. This unusual procedure may have been the reason Roberval kept much of his work secret. He could thus preserve an advantage in the competition, but at some cost to his eventual reputation.

Roberval succeeded in evaluating what modern mathematicians would call the integral of a number of simple functions. He also was able to find the area under the "cycloid" curve and the volume of a cycloid of revolution. In his investigations of kinematics, or the description of motion, he introduced a method for finding the tangent to a curve, anticipating the modern notion of the derivative of a function. His most important practical contribution was the principle of the Roberval balance, which allows the accurate weighing of goods using pans set above the pivot of the balance beam, a design that would be used well into the twentieth century. Roberval died in Paris on 27 October 1765 at the age of seventy-three.

Bibliography: L. Auger, *Un savant méconnu: Gilles Personne de Roberval (1602–1675): Son activité intéllectuelle dans les domaines mathématique, physique, mécanique et philosophique*, 1962; C. Boyer, *A History of Mathematics*, 1968.

Donald R. Franceschetti

ROEMER, OLAUS (1644–1710). Olaus Roemer was born on 25 September 1644 in Jutland. He studied astronomy and mathematics at the University of Copenhagen, where he later taught mathematics and astronomy and directed the Danish Royal Observatory. Roemer also held public office in Copenhagen, serving as mayor.

Drawn into the heady scientific circle of *Louis XIV's court, Olaus Roemer joined *Jean Picard at the Paris observatory in 1672. It was there that he made his best known achievements in astronomy and physics. At work on the practical task of arriving at a universal time standard, Roemer discovered discrepancies within *G.D. Cassini's predictive schedules for the eclipses of Jupiter's moons. Cassini had established the schedules to confirm the latitude and longitude of Uraniborg, Denmark (the site of Tycho Brahe's observatory), and Paris. The accurate prediction of eclipses and occultations of moons allowed scientists to establish a time reference for measuring terrestrial longitude, a stubborn navigational problem that occupied many at that time. Roemer found that the schedules varied systematically by several minutes from his own observations of Jovian lunar eclipses. The variance was clearly related to the distance of the earth from Jupiter at the time of the eclipse. The closer the space, the shorter the intervals between eclipses and vice versa. Light, therefore, had measurable speed, an observation that effectively challenged the long-held theory of instantaneous light transmission. In 1676, the Danish astronomer calculated the speed of light within two-thirds of the modern value.

Roemer also contributed to instrument making. The first really useful transit telescope was erected at his home in 1690. These small refractor telescopes narrowed the range of deep space observation, allowing very accurate determination of stellar and planetary positions. The advantage of their stability outweighed the disadvantage of waiting for the celestial object to move into view, that is, to transit.

Bibliography: I. Cohen, "Roemer and the First Determination of the Velocity of Light," *Isis* 31 (1940): 327–379; A. Van Helden, "Roemer's Speed of Light," *Journal for the History of Astronomy* 14 (1983): 137–141.

Joan Klobe Pratt

ROTROU, JEAN DE (1609–1650). More or less a contemporary of France's great seventeenth-century playwrights *Corneille, *Molière, and *Racine, Jean Rotrou was born 21 August 1609 into a family of magistrates in Dreux. Little is known of him beyond his evident passion for the theater and his heroic, premature death on 28 June 1650, the consequence of an epidemic to which he fell victim while executing his duties as a local functionary.

Although he was born and died in Dreux, Rotrou spent much of his life in Paris, where his first play, a tragicomedy entitled *L'Hypocondriaque*, was performed in 1628. From 1629 to 1636, he worked as a salaried playwright for the Hôtel de Bourgogne. This period was the most prolific of his career; it is thought that he wrote more than forty plays in seven years, of which approximately twenty have survived. Critics have suggested that while these plays are unequal in quality, they are excellent evidence as to which types of theater were popular at the time. During Rotrou's tenure at the hôtel, he composed one of the first "regular" French tragedies, *Hercule mourant*. A comedy, *La Bague de l'oubli* (1629), attracted the attention of both the royal family and *Richelieu, who in 1635 selected Rotrou to participate in the "Five Authors" theatrical experiment.

In 1639, Rotrou purchased the aforementioned post and was appointed *lieutenant de baillage*. He married in 1640 and had six children; three survived. He returned to his native town to attend to his civil and familial obligations, factors that may have slowed his literary production. However, as the quantity decreased, the quality improved, and he created his best works, including *La Soeur*, an Italian-style comedy (1645); *Le Véritable Saint-Genest*, a tragedy (1645); *Venceslas*, a tragicomedy bordering on tragedy (1647); and *Cosroès*, a tragedy (1648).

Many of Rotrou's plays exhibit traits of more than one type of theater, including pastoral, tragicomedy, comedy, and tragedy, and are consequently difficult to classify by genre. Moreover, Rotrou seems to have oscillated between Baroque tendencies, such as exaggerated language and a taste for surprise, and the classical aesthetics of "regular" theater, which were promoted by Richelieu, as well as by another of Rotrou's mentors, Jean Chapelain. The varying blends of genre, style, and aesthetics could be read as marks of inconsistency, but they should also be viewed as a meaningful barometer of the dominant trends and evolution of French seventeenth-century theater.

Bibliography: J. Morel, *Jean Rotrou, dramaturge de l'ambiguïté*, 1968; J. Morello, *Jean Rotrou*, 1980.

Laura L. Dennis-Bay

ROWE, NICHOLAS (1674–1718). Son of prominent barrister John Rowe and Elizabeth Edwards, Nicholas Rowe was born on 25 September 1674 at Little Barfield, Bedfordshire. He attended Highgate School; entered Westminster School (1688), where he excelled in classical languages, literature, and poetics; and studied law at the Middle Temple (1690–1695). Whether he practiced law remains unsubstantiated, but that he also pursued his aesthetic interests is undeniable. The success of his first play, *The Ambitious Step-Mother* (1700), established him as a rising young dramatist. He married Antonia Parsons (1693), daughter of revenue auditor Anthony Parsons. Of their seven children, only their second son John (1699) survived, and Antonia died in 1712. Having failed to gain desirable employment with George Savile, Marquis of Halifax, in 1705, Rowe sold part of his Devonshire estate to clear his father's debts and became secretary to the Duke of Queensberry (1709–1711). Continuing to write and produce plays, he held positions as Land Surveyor of Customs, Clerk of the Council to the Prince of Wales, and Secretary for Presentations to the Lord Chancellor. In 1715, Rowe succeeded *Nahum Tate as Poet Laureate to King George I and married Anne Devenish, daughter of Joseph Devenish of Buckham, Dorset. Several months after their daughter Charlotte's birth (May 1718), Rowe became ill and died on 6 December 1718. He was buried near Chaucer in Poets' Corner, Westminster Abbey, London. At Anne's request, Pope provided the epitaph for his monument.

Long celebrated as the first modern editor of Shakespeare's plays (1709), a more accessible than textually authoritative collection, and for his translation of Lucan's *Pharsalia*, published posthumously in 1719, Rowe also produced a large body of nondramatic work, inclusive of occasional verse—witty, often topical poems, translations (*Golden Verses of Pythagorus* and *Nicolas Boileau's Lutrin*), and *vers de société*—as well as dramatic epilogues and prologues, much of them included in *Poems on Several Occasions* (1714). His Lucan and Shakespeare notwithstanding, Rowe's enduring reputation obtains from the seven tragedies that made him the prominent writer for the early eighteenth-century stage: *The Ambitious Step-Mother* (1700), *Tamerlane* (1701), *The Fair Penitent* (1703), *Ulysses* (1705), *The Royal Convert* (1707), *The Tragedy of Jane Shore* (1714), and *The Tragedy of Lady Jane Gray* (1715). Of these plays, Rowe's tragedies of Calista (his fair penitent), Jane Shore, and Lady Jane Gray have proved the most durable. Known as his "she-tragedies," these plays reflect Rowe's fine-tuning of pathos and other conventions of the period's heroic and sentimental drama and his recasting of familiar stories to portray courageous heroines affected by the double standards recognizable in, if not as, Rowe's own England. That Rowe's strong convictions and grasp of contemporary politics, religion, and social issues inform all his tragedies and render some too topical for the appreciation of later generations has been demonstrated. So, too, has his importance to the development of the English drama despite the fluctuations of popular and critical taste in subsequent centuries. His only comedy, *The Biter*

(1704), has rarely delighted anyone and over time has remained virtually untouched by human thought.

Bibliography: A. Hesse, "Some Neglected Life-Records of Nicholas Rowe," *Notes & Queries* 220 (1975): 348–353, 484–488; A. Jenkins, *Nicholas Rowe*, 1977.

Sallye Sheppeard

RUBENS, PETER PAUL (1577–1640). Rubens, born in the German province of Westphalia, is considered one of the most important painters of the seventeenth century. The son of a lawyer, Rubens settled in Antwerp with his mother and surviving brother and sister in 1587. After studying the classics in a Latin school and serving apprenticeships to three minor Flemish painters, Rubens was accorded the rank of master painter of the Antwerp Guild of St. Luke in 1598. Following the example of many northern artists of the time, Rubens traveled to Italy in 1600, and in Venice was inspired by the paintings of Titian, Veronese, and Tintoretto. Later, while in Rome, he was influenced by the works of Raphael and *Michelangelo. During his eight years in Italy, Rubens was employed by the Duke of Mantua, Vincenzo Gonzaga, and, besides copying Renaissance paintings for the ducal collection, painted many original portraits of the Italian aristocracy. Upon his return to Antwerp in 1609, Rubens had become a painter of international reputation, and after marrying his first wife, Isabella Brant, became the court painter to the Archduke Albert and his wife Isabella, who ruled the Low Countries as viceroys for the king of Spain. He established a workshop that ultimately produced over 2,000 works under his supervision.

Between 1622 and 1630, Rubens's value as a diplomat equaled his importance as a painter. As the envoy of the Archduchess Isabella, in 1622 he visited Paris, where Queen Marie de Medicis commissioned him to paint a series of allegorical paintings depicting her life for the Luxembourg Palace, finished in 1625. In 1628, the Flemish viceroys sent him to Spain, and while in Madrid he received several commissions from *King Philip IV. In 1629, *King Charles I knighted Rubens, after he successfully handled peace negotiations between Spain and England. From 1630 until 1640, Rubens remained in Antwerp, married to his second wife Helene Fourment, living primarily at his country residence, Castle Steen. During the final decade of his life, he continued executing commissions for the Hapsburg monarchs of Austria and Spain and also painted many pictures of personal interest, especially of his wife and child, and the Flemish countryside.

Rubens's style of painting became the international definition of the Baroque, with exuberant, animated figures in active motion in an excited emotional atmosphere. Bold brushwork, luminous color, and shimmering light infuse his paintings with energy. He created a vibrant art depicting biblical scenes, animal hunts, clashing battle scenes, and powerful evocations of religious spirit, all with equal dedication to high drama. Many of Rubens's finest works were painted for the

Jesuits, who, as the dominant religious force of the Counter Reformation, sought in their commissions to propagate the Catholic faith through passionate, energetic, and often theatrical expressions. One of Rubens's most important religious commissions for the Jesuits was to provide three altarpieces and thirty-nine ceiling paintings for the church of St. Charles Borromeo in Antwerp, which was struck by lightning and burned in 1718, leaving only the three altarpieces to survive. Another of Rubens's greatest religious masterpieces is the 1620 *Fall of the Damned*, commissioned for a Jesuit church in Neuberg, Germany. It is a panoramic scene in which masses of naked figures are strung across the painting, a river of tortured bodies tumbling toward a mass of devils that bite and claw at them, pulling them into the fires of Hell. Rubens's joyous spirit emerges nowhere so clearly as in his portrayals of the nude female. They are equally erotic, sensual, and wholesome, without being lascivious or banal. In his view, the human body was as much a work of God as the life of the holiest saint, and although he often placed his nudes amid settings of the pagan past, he painted them with a forthrightness that reflected a strong religious conviction. From these nudes we come to have the term Rubenesque, which defines the sumptuous flesh of his well-endowed women, a type not only preferred aesthetically at the time, but also one whose curves and folds of flesh captured the subtle nuances of light and color. His feminine ideal is captured in the intimate portrait of his wife Helene from 1638, in the delicacy of her expression and the softness and warmth of her body, wrapped in fur.

   Although Rubens suffered in the last few years of his life with arthritis in his right hand, he continued to work. He executed with assistants an assignment for King Philip IV of Spain of a comprehensive series of pictures illustrating all the legends in the *Metamorphoses* of Ovid for the king's new hunting lodge and summer palace near Madrid. He died from heart failure on 30 May 1640 and was buried in the parish church of St. Jacques in Antwerp.

Bibliography: K. Belkin, *Rubens*, 1998; P. Sutton, *The Age of Rubens*, 1993.

Mark Williamson

RUDBECK, OLOF (1630–1702). The son of Bishop Johannes Rudbeckius, the chief churchman of his time, Olof Rudbeck was destined for a career in medicine rather than the church. He entered the University of Uppsala in 1648 and, influenced by the writings of *William Harvey, began to study animal anatomy; at the age of twenty, he discovered the lymphatic system. He demonstrated his discovery before Queen Christina but did not publish it until a few years later (in *Nova exercitatio*, 1653), a delay that led to a bitter priority dispute between Rudbeck and Thomas Bartholin, who simultaneously found the same system. Rudbeck completed his education at the University of Leiden before returning to Uppsala, where he accepted a chair on the medical faculty. A tireless supporter of the university, Rudbeck donated some of his own money to found a botanical garden and helped construct an anatomical theater. In addition, he worked as an architect, printed his own books, and composed music while academically being an early supporter of Cartesian natural philosophy.

He worked on several lengthy projects during his career. Beginning in the 1670s, he composed a four-volume patriotic historical work, the *Atlantica*, published in Swedish and Latin, in which he developed the idea that Sweden was the birthplace of modern civilization. He also worked extensively on a multivolume botanical study with the goal of creating an illustrated book that would describe all plants. He hired craftsmen to create thousands of woodcuts and illustrations for his book, which would have run to thousands of pages and included around 7,000 species. Unfortunately, a fire destroyed many of the blocks and much of Rudbeck's botanical collections, manuscripts, and books. Thus, only two volumes of his work, the *Campus Elysii* (1701 and 1702) ever appeared. Although Rudbeck died in 1702, his son continued his work and was himself a patron of a rising botanist named Carl von Linnaeus.

Bibliography: G. Eriksson, *The Atlantic Vision: Olaus Rudbeck and Baroque Science*, 1994; J. Hoyrup, "Reflections on the Baroque in the History of Science," *Physis* 34 (1997): 675–694.

Michael R. Lynn

RUISDAEL, JACOB ISAACKSZOON VAN (ca.1628–1682). Van Ruisdael, a Dutch landscape painter active in Haarlem and Amsterdam in the second half of the seventeenth century, is known for his dramatic forest scenes and panoramic landscapes. He entered the painter's guild in Haarlem in 1648, when that city dominated landscape painting. Although his early work still resembled the monochromatic "tonal" landscapes of his uncle Salomon van Ruisdael, in the 1650s, he moved decisively toward a richer palette and strong, monumental forms. This shift is reflected in the *Castle of Bentheim* (1653), painted during travels in Holland and Germany. Although his travel companion was Nicolaes Berchem, he never adopted that painter's Italianate light effects.

He settled in Amsterdam in 1656. In addition to the local topography, he adopted elements from the Scandinavian landscapes of Allaert van Everdingen, who returned to Haarlem in 1645. Van Ruisdael tended to manipulate and combine elements for greater dynamic effect, mood, or monumentality. His proto-romantic *Jewish Cemetery* (1668–1672), for example, combined the local graveyard, dead trees, and a ruined castle with a waterfall and rainbow to suggest the cycle of death and rejuvenation. To add grandeur to the *Mill at Wijk near Duurstede* (ca.1670), he greatly enlarged the windmill and widened the river.

His later panoramic views of Haarlem are the culmination of his progress toward dramatic effects of atmosphere rather than topography. The vertical format of the *View of Haarlem from the Dunes at Overveen* (ca.1675), for instance, makes it a portrait of billowing clouds and sunlight mottled over bleaching fields, its low horizon interrupted only by the tower of the Grote Kerk in the distance.

The biographer Arnold Houbraken noted van Ruisdael's education in Latin and his practice of medicine. Although he was registered in Amsterdam as having received his medical degree at Caen in 1676, this entry is the only one crossed out. Van Ruisdael never married and died in Amsterdam in 1682. He was buried in the Grote Kerk in Haarlem that he had portrayed so often.

Bibliography: A. Houbraken, *De Groote Schouburgh III*, 1753; J. Rosenburg, *Jacob van Ruidsael*, 1928; S. Slive and H. Hoetink, *Jacob van Ruisdael* (Exhibition Catalogue), 1981–1982; E. Walford, *Jacob van Ruisdael and the Perception of Landscape*, 1991; W. Wiegand, *Ruisdael-Studien: Ein Versuch zur Ikonologie der Landschaftsmalerie* (Dissertation), 1971.

Lloyd DeWitt

RYMER, THOMAS (1641–1713). Born in Yorkshire, England, Thomas Rymer left Cambridge without taking a degree and then studied law at Gray's Inn, London. Although called to the bar in 1673, he almost immediately turned his attention to history and literature. He was a rigidly dogmatic critic who dismissed prose history as amoral, but found poetry, when properly written and infused with truthful historical events, both ethical and philosophical. Rymer adhered to the principles of French neoclassicism, criticizing playwrights who failed to apply the unities of time, place, and action, and demanding that dramatic action be probable and reasonable so that it could instruct by moral precept and example. He expounded these views in a preface to René Rapin's *Reflections on Aristotle's Treatise of Poesie* (1674) and applied them in a play that was never performed, *Edgar, or the English Monarch* (1678). In *The Tragedies of the Last Age* (1678), Rymer criticized plays by the Jacobean dramatist *John Fletcher, and in his amusing though vituperative *A Short View of Tragedy* (1693), he termed Shakespeare's *Othello* "a bloody farce without salt or savor," criticizing the trivial and unrealistic nature of the plot by wryly questioning "Why was this not called the *Tragedy of the Handkerchief?*" In addition to writing a few poems, Rymer also translated Plutarch's *Lives* (1683–1686) and might have been the author of *An Essay Concerning Critical and Curious Learning* (1698). Rymer's criticism was praised by *John Dryden, although Samuel Johnson thought him a tyrant and T.B. Macaulay called him the worst critic who ever lived.

In 1692, Rymer was appointed historiographer royal. When *William III's government decided to publish for the first time copies of all past treaties entered into by England, Rymer was made editor of the project. By the end of Rymer's life, fifteen of the twenty volumes in this series, known as the *Foedera* ("Treaties"), were published, with the remainder being published posthumously. Although bland in comparison to Rymer's efforts as a literary critic, the *Foedera*

(1704–1735) is a considerable and valuable achievement that remains useful to historians, and it is for this work that Rymer is best known today.

Bibliography: G. Reedy, "Rymer and History," *Clio* 7 (1978): 409–422.

Ken MacMillan

S

SALMASIUS, CLAUDIUS (1588–1653). Born in Burgundy on 15 April 1588 and known among seventeenth-century scholars as the author of the work that prompted *John Milton's *Pro populo anglicano defensio (Defense of the English People* [1651]) Salmasius (known originally in France as Claude de Saumaise) is far more than a disputant in a matter of regicide. Salmasius began his study of Latin and Greek early on, and by the age of sixteen he had come to the attention of *Isaac Casaubon in Paris and had also turned Protestant. In 1604, Salmasius was given the chance to study in the library of the count Palatine in Heidelberg; his work there included the discovery of a tenth-century manuscript of Greek poetry and the transcription of epigrams from that manuscript for Scaliger, whom Salmasius succeeded at Leiden in 1631. Salmasius's publication of theology, commentary, and classics began in 1608 with an edition of Nilus Cabasilas's fourteenth-century tract against the primacy of the pope together with a similar work by the Calabrian monk, Barlaam. His 1609 edition of *Rerum Romanorum* by Lucius Annaeus Florus was followed by Salmasius's return to Burgundy and study for the magistracy; although academically qualified, Salmasius's Protestantism precluded his succeeding his father as magistrate.

In 1620, he published, with his own annotations, Casaubon's notes on Augustan history. In 1629, having learned Arabic, Salmasius brought out his commentary on Solinus's *Polyhistor*; in actuality, Solinus had drawn heavily on Pliny, and Salmasius indicated this indebtedness. In 1631, Salmasius accepted Scaliger's old post at Leiden and remained a member of that faculty until his death at Spa, Belgium, on 3 September 1653. His publications reflect an extraordinary breadth of knowledge. For Prince Frederick of Nassau, Salmasius wrote *De re militarii Romanorum* (1657). His justification for the charging of interest in *De Usuris Liber* (1638) and *De modo usurarum* (1639) persuaded the Dutch Church to permit moneylenders, both clerical and lay, to return to the Sacrament, and his 1645 edition of Cabasilas' *De Primatu Papae*, went surprisingly unsuppressed in France. His *Defensio regia pro Carlo I* (1649), which supported the theory of divine right, drew the attention of the English and Milton's well-known response. Salmasius, in turn, responded to Milton, but before what might have been a celebrated feud between two men of enormous learning and literary skill could really get under way, Salmasius died. Nevertheless, Salmasius's abilities in languages, law, and history made him a provocative controversialist of the era.

Bibliography: D. Roberts, "Claudius Salmasius," *A Milton Encyclopedia*, ed. W. Hunter, 1979.

Martha Oberle

SANCHES, FRANCISCO (1551–1623). Physician and philosopher, Sanches was born probably of converso (New Christian) parents on 16 July 1551 in the small Spanish town of Tui on the left bank of the river Miño. Despite his mixed Spanish, Portuguese, and Jewish origins, Sanches should be considered a French author because of his schooling and later career in France. From 1562 to 1571, he attended one of France's most prestigious Latin grammar schools, the Collège de Guyenne in Bordeaux, where Michel de Montaigne had studied between 1539 and 1540. From 1571 to 1573, Sanches traveled through Italy, spending most of his time in Rome University ("La Sapienza"), where he was initiated into the Galenic tradition of medical philosophy and into Vesalius's and Fallopio's critiques of Aristotelian science. Back in France in 1573, he attended the University of Montpellier, where in the course of a single academic year he took the exams and was awarded a doctorate by the faculty of medicine in 1574.

Due to the growing influence of the Huguenot community in Montpellier and the meager prospects for a chair at its university, in 1574 Sanches moved to Toulouse, where he was to spend the rest of his life, alternating, and often combining, his work as a practicing physician with his teaching and writing projects. Although he also finally earned a chair in philosophy in 1584 after a disputed public competition, he was to wait until 1612 to fulfill his lifelong ambition of a chair in medicine. Sanches advocated a new type of experimental medicine, which was taught in his lifetime only at select Italian schools. It consisted in the integration of botany and anatomy in a single course of study, one that favored the research into new therapies derived from experimentation with pharmacological substances obtained from plants. This new emphasis on the physician's contact with nature and the body parallels the importance that the new scientists of the late Renaissance granted the mechanical workers and artisans, who had preceded them in advocating the empirical, hands-on investigation of natural phenomena.

Sanches's philosophical and medical works were published in Toulouse in a single collection in 1636, *Opera medica*. The medical writings proper include several lengthy commentaries on select key treatises by Galen, two on drugs, an additional treatise on anatomy drawn mainly from Fallopio and Vesalius, and a fragment of a projected treatise on bones. His extant philosophical writings include *De longitude* and *Brevitate vitae*, two commentaries on Aristotle, and his only two works published in his lifetime, the *Carmen de cometa anni M.D.LXXVII* (1578) and *Quod nihil scitur* (1581).

Sanches's modern reputation rests exclusively on *Quod nihil scitur* (*That Nothing Is Known*), which bears comparison with *Francis Bacon's *The Advancement of Learning* (1605) and *René Descartes's *Discours de la méthode* (1637). Partly because of his reiterated use of the locution "Quod nihil scitur" throughout his treatise of the same title, and partly because he devoted much of his writing to casting received ideas into doubt, Sanches has been called a "Pyrrhonian" and a "skeptic." Sixteenth-century French Pyrrhonism is best represented by Montaigne's recalcitrant suspension of judgment and his willingness to carry on with the free investigation of particulars unencumbered by a priori judgments. Yet from the moment Sanches moves beyond the Pyrrhonian cognitive impasse, stating his optimistic belief in the future discovery of a universal method of research into the causes of natural phenomena, he makes himself deserving of the label of rational empiricist, an immediate forerunner of Bacon and Descartes. Although he was not a medical researcher in the modern sense, in *Quod nihil scitur* Sanches contends that knowledge can be acquired exclusively through the use of reason and only when reason's modus operandi is to organize the factual evidence presented by the senses and to make a judgment based on that experience. Concomitantly, he made no bones about contesting the ascendancy of ancient authorities such as Aristotle, as when the latter highlights the centrality of syllogistic reasoning to the detriment of sensory experience.

Bibliography: E. Limbrick, ed., Introduction to *That Nothing Is Known* [*Quod nihil scitur*], trans. D.F.S. Thomson, 1988; R. Popkin, "The Revival of Greek Scepticism in the Sixteenth Century," in *The History of Scepticism from Erasmus to Spinoza*, 1979.

José María Rodríguez-García

SARPI, PAOLO PIETRO (1552–1623). Sarpi's work spanned theology, politics, and science. Upon the death of Sarpi's father at a young age, the responsibility for his education shifted to his uncle, a priest named Ambrogio Morelli. Morelli placed him under the tutelage of a Servite friar, who instructed him in philosophy, theology, and logic, and facilitated his entry into the Servite Order at the age of fourteen. Sarpi received a baccalaureate in theology in 1574 and later taught philosophy in the Servite monastery in Venice. In 1578, he received a doctorate in theology from the University of Padua and was appointed state theologian by the Venetian Senate in 1606. The following year, he became adviser to the Venetian State, a position he held until his death.

Some of Sarpi's writings are politically motivated. Perhaps most important is his *History of the Council of Trent* of 1619, which is a highly biased and at times satirical account of the council's proceedings. Although a cleric of high rank, he often found himself at odds with the Roman Curia and the papacy: He was once excommunicated and later was the object of an attempted assassination that he accused the Curia of engineering. This can be understood in the context of his advisorial position to the Republic of Venice at a time when the papacy tried to control the city and its dominions.

Sarpi was also interested and involved in the scientific thinking of his times. He was highly praised for his mathematical and speculative abilities by contemporaries such as *Galileo, but all that survives by which one can judge the originality of his scientific thought are some letters and notebooks containing his philosophical, physical, and mathematical speculations. The notebooks do not form a consistent philosophical system; instead, these are a hodgepodge of miscellaneous thoughts about the nature of the physical world and chronological documentation of the scientific experimentations of the period.

Bibliography: A. Campbell, *The Life of Fra Paolo Sarpi*, 1869; C. Gillespie, ed., *Dictionary of Scientific Biography*, 1981; P. Sarpi, *Pensieri naturali, metafisici e matematici*, 1996.

Rosi Prieto Gilday

SAUVEUR, JOSEPH (1653–1716). Joseph Sauveur, the son of a notary, grew up and studied in La Flèche, France, before moving to Paris in 1670, where he studied theology and attended the Cartesian lectures of Jacques Rohault. Although he was born with a hearing and speech problem, Sauveur decided upon a teaching career in mathematics and geometry. His enormous aptitude in this field, along with his serendipitous mathematical work on games of chance at a time when nobles and courtiers spent much of their time at Versailles gambling, led to a rapid accumulation of rich and powerful patrons, students, and positions. His first marriage was witnessed by, among others, the famous *Bishop Bossuet, Philippe d'Orléans, and Prince de Condé. In 1686, he received an appointment to the chair in mathematics at the prestigious Collège Royal. The Académie Royal des Sciences admitted him as a member in 1696 and his enduring popularity can be seen in the fact that he survived the reform of the academy in 1699 and remained a member until his death in 1716. His most important scientific contribution centers on his development of a new branch of physics called acoustics. In particular, Sauveur determined how to identify the pitch of a note by assessing the frequency of its vibrations. He also asserted that harmonics are the component parts of all musical sound. Sauveur's work provided the basis for the science of acoustics and the development of musical theory.

Bibliography: L. Auger, "Les apports de J. Sauveur (1653–1716) à la création de l'Acoustique," *Revue d'histoire des sciences et de leurs applications* 1 (1948): 323–336; D. Sturdy, *Science and Social Status: The Members of the Académie des Sciences, 1666–1750*, 1995.

Michael R. Lynn

SAVAGE, RICHARD (ca.1697–1743). Neither the exact date nor the parentage of Richard Savage, poet and dramatist, is known. He was born around

1697, most likely in London. In 1718, his comedy *Love in a Veil* was produced in Drury Lane. Savage also wrote a tragedy based on the life of Sir Thomas Overbury; it was produced in 1723 with the author in the lead role. In 1729, he published a prose satire on Grub Street entitled *An Author to be Let*. Throughout his life, Savage claimed to be the illegitimate son of the Earl of Rivers and the countess of Macclesfield, and he attempted without success to force the countess to acknowledge him as her son. Subsequent research has largely discredited this claim, but Savage's attack on his supposed mother in the poem *The Bastard* features the often-quoted line, "no tenth transmitter of a foolish face." *The Bastard*, along with his discursive poem of five cantos entitled *The Wanderer*, is considered Savage's best work.

But Savage's literary reputation has little to do with his own work and rests almost entirely with a work of which he is the subject: Dr. Samuel Johnson's *Life of Richard Savage* (1744). Johnson met Savage between 1737 and 1738, when Johnson first came to London. After Savage's death, Johnson wrote his famous *Life*, a groundbreaking work in the field of literary biography. Johnson is sympathetic but also demonstrates his usual balance in response to the character of his subject; he praises Savage as a moral poet but expresses frustration with Savage's erratic and self-destructive behavior. An example of such behavior occurred in 1727, when Savage and three friends were arraigned for the murder of John Sinclair in a barroom brawl. Savage was found guilty, but he received a pardon through the influence of the Countess of Hereford. In 1730, he nearly received the laureateship but was beaten out by Colly Cibber. He was, however, granted a pension by Queen Caroline that continued until her death in 1737. Thereafter, Savage proved himself unable to secure regular income. The nephew of Lady Macclesfield offered him a pension, as did Pope and other friends, but Savage could not abide by the stipulations of either offer and lost both. He was arrested for debt and died in prison in 1743.

Bibliography: R. Holmes, *Dr. Johnson and Richard Savage*, 1999; S. Johnson, *The Life of Richard Savage*, 1748, rpt. 1971.

Charles Pastoor

SAVILE, GEORGE, MARQUIS OF HALIFAX (1633–1695).

George Savile was born in Thornhill, Yorkshire, in 1633 and was educated at home at Shrewsbury School and by private tutors in Europe. Savile was elected a member of Parliament for Pontefract in 1660, but instead of standing for reelection, he became a deputy lieutenant for the militia in Prince Rupert's regiment. In 1667, he was appointed to captain, which led to his being named Baron of Savile and Viscount of Halifax in 1668. Halifax was named to the Privy Council in 1672, an event that marked the beginning of his rather unique political career. Halifax opposed *Charles II's pro-Catholic policies but also opposed the Earl of Clarendon's 1673 Test Act that sought to force Catholics into the Church of England. In 1681, Halifax successfully defeated the Whig attempt to pass the Exclusion Act that would prevent Charles's Catholic brother *James from succeeding the throne. For his service, Halifax was elevated to the rank of marquis in 1682. Following this victory, however, Halifax irritated his Tory supporters by calling for limitations on any Catholic monarch. When James took the throne in 1685, Halifax was removed from the Privy Council for opposing the repealing of the Test Act and the Act of Habeas Corpus. Through the publication of largely anonymous pamphlets, Halifax opposed James's absolutist politics and fostered a congenial correspondence with *William of Orange. Halifax proves to be a contradictory figure in seventeenth century politics and philosophy since he continually lobbied for one side, then promptly turned around and supported his opposition. In 1688, Halifax opposed the revolution that would soon place William of Orange on the throne, and instead, he acted as an advisor to James II. When James fled to France and Spain, Halifax sided again with William and Mary and was reappointed to the Privy Council and became William's chief adviser. After 1690, the growing number of enemies gained through his shifting political career essentially forced Halifax into retirement.

Halifax's political writings are largely ephemeral and stand as a clear representation of his political wavering. In 1667, he worked out his ideas for naval reform in *A Rough Draft of a New Model at Sea*—a work that would not be published in its final form until 1694. Halifax anonymously attacked James II's plea for indulgence in the 1687 pamphlet *A Letter to a Dissenter*, then followed up the next year with another pamphlet leveled at James's religious policy titled *The Anatomy of an Equivalent*. Perhaps his most famous and controversial work, *The Character of a Trimmer*, was circulated anonymously in manuscript in 1685 and published in 1688. The text provides both a rebuttal to the negative definition of a trimmer (anyone who "trims" a political position to suit the prevailing situation) used frequently by *Roger L'Estrange in the *Observator*, but more importantly, the work provides a defence of Halifax's political philosophy of moderation. L'Estrange presented trimmers as a third party that attempted to gain power by taking in the liberal members of both the Whig and Tory parties. The danger to L'Estrange, a firm supporter of high Tory politics, was not in a third party itself as much as in the lure of political moderation. While Halifax's political career appears to be controlled by flights of fancy, he is consistently moderate in his views, all of which oppose absolutism on either party side. Beyond this, the *The Character of a Trimmer* offers a series of recommendations to the king that focus on toleration and maintaining a political balance both in England and Europe. On a more general level, the text calls for a pragmatic style of government that seeks compromise over an inflexible system of abstract ideals.

Bibliography: M. Brown, "Trimmers and Moderates in the Reign of Charles II," *Huntington Library Quarterly* 37, no. 4 (1974): 311–336.

Michael W. Jackson

SCARLATTI, (PIETRO) ALESSANDRO (GASPARE) (1660–1725).

Known especially as an operatic composer, Alessandro Scarlatti is credited with originating the "Italian overture" of three contrasting sections, which began as an introduction to his operas but over time developed into the symphony. Born in Palermo, Scarlatti came from a musical family; his son Domenico (1685–1757) is also a major composer. When he was twelve, his parents sent him to Rome for schooling. Scarlatti's first operatic success was *Gli equivoci nel sembiante*, which premiered at Rome in 1679. It was noticed by Queen Christina of Sweden, a Roman resident since her 1655 conversion to Catholicism; she made Scarlatti a protégé and retained him as her maestro di cappella until he moved to Naples in 1683. Scarlatti was also favored by Cardinals Pietro Ottoboni and Benedetto Pamphili, both patrons of the arts.

Scarlatti's move to Naples, significant for the history of opera, was prompted by an invitation from Spain's ambassador to the Vatican. Scarlatti would be a major player in transforming Naples from a cultural hinterland to a musical capital of Europe. Scarlatti was named director at Teatro San Bartolomeo—a company of nine singers, five instrumentalists, and a music copyist. He was also required to compose, rehearse, and conduct two or three operas annually. Not only were patrons clamoring for cantatas, serenatas, and oratorios, but Scarlatti frequently traveled to Rome to oversee premieres of his works, such as the opera *La Rosmene*, at Palazzo Doria Pamphili in 1686.

In 1702, after nineteen years in Naples, Scarlatti took a leave of absence and with his son Domenico traveled to Florence to try for a post with Prince Ferdinando de' Medici. Although the prince had been a promoter of Scarlatti's music for over twenty years, the composer departed without a contract. In Rome by late 1703, Scarlatti became assistant maestro di cappella at Santa Maria Maggiore and was also apparently employed by Cardinal Ottoboni. Here Scarlatti composed little opera, because, since 1700, the public opera houses had been closed and there were scant opportunities for private performances. Instead, he sent new operas to Prince Ferdinando and to Venice, where in 1707, he oversaw the premieres of two new works, including *Il Mitridate Eupatore*, one of his best operas. Scarlatti also composed many cantatas, serenatas, and oratorios for Cardinals Ottoboni and Pamphili, Prince Ruspoli, and other patrons. Meanwhile, in 1704, Scarlatti's Naples position was announced as vacant. In 1706, Scarlatti was elected to the Arcadian Academy, and in 1708 was promoted to maestro di cappella at Santa Maria Maggiore. In 1709, Scarlatti accepted an offer from Cardinal Grimani, now Austrian viceroy, to return to his old job at Naples. He stayed for ten years, but kept in touch with his Roman friends. This Neapolitan sojourn resulted in eleven known operas by Scarlatti, the most famous being *Tigrane*, which premiered in 1715. Once again Scarlatti went on leave in Rome early in 1718. New operas during these Roman years were *Telemaco* (1718 carnival season), *Marco Attilio Regolo* (1719), and *La Griselda* (1721). He returned to Naples in 1722, where he lived until his death on 22 October 1725.

Bibliography: E. Dent, *Alessandro Scarlatti*, 2nd ed., 1960; D. Grout, *Alessandro Scarlatti: An Introduction to His Operas*, 1979.

Susan Treacy

SCARRON, PAUL (1610–1660). Author Paul Scarron was born 4 July 1610, shortly after Henri IV's death, and baptized with his father's name. In 1617, four years after the death of Scarron's mother, his father remarried, giving his children a domineering, uncaring stepmother and an unpleasant home situation to which the young Scarron reacted with the satirical laughter that was to become his trademark. In 1624, he was sent to a Parisian academy, where he stayed until 1629, when he became an abbé and sojourned in Le Mans and Rome.

In 1638, Scarron was forced to abandon a predilection for sensual pleasures because of a painful, paralyzing illness that rendered his body, by his own description, "not unlike the letter Z." This does not seem to have diminished his good humor, however, for he enjoyed a reputation as a witty conversationalist and entertaining host. This was only strengthened by his 1652 marriage to Françoise d'Aubigné, as their home became a popular social and intellectual center where writers, artists, and notable court figures gathered.

The literary vocation of the self-dubbed "malade de la Reine" extended to many genres, including the novel, short story or *nouvelle*, theater, and poetry. He is especially noted for being the father of the burlesque, a comic style that presents its subject in an atypical, even contrary, manner to that traditionally employed. His most famous works in this vein are the *Typhon* (1644) and the *Virgile travestii* (1648–1649), the latter of which was never completed, apparently because Scarron wished to move beyond the burlesque. His most widely read work, the *Roman comique* (1651, 1654), displays not only burlesque elements, but also evidence of his fascination with the romanesque and the theater. The romanesque finds fuller expression in works such as the *Nouvelles tragic-comiques* (1655) and a tragicomedy, the *Écolier de Salamanque* (1654). His other theatrical works are primarily comedies; the most successful include *Jodelet ou le maître valet* and *Dom Japhet d'Arménie*, performed in 1643 and 1647, respectively.

Although Scarron's fame is derived primarily from his comic and burlesque works, his corpus also reveals a more serious aspect seen in his adaptations of the romanesque, his increasing adherence to classical aesthetics, and the publication of texts like the *Épîtres chagrines* (1658–1659). Perhaps these directions would have become more significant in Scarron's work had he not died at age fifty, on 5 or 6 October 1660.

Bibliography: F. de Armas, *Paul Scarron*, 1972; E. Magne, *Bibliographie générale des œuvres de Scarron*, 1924.

Laura L. Dennis-Bay

SCHEFFLER, JOHANN (ANGELUS SILESIUS) (1624–1677). Johann Scheffler was the most significant mystical poet of the German seventeenth century. He is known today for his collection of mystical epigrams. Scheffler was born in Breslau (Silesia) in December 1624 into the chaos of the Thirty Years' War (1618–1648). The war, which was especially violent in his native Silesia, and the imperial efforts to re-Catholicize Silesia profoundly affected Scheffler's development and influenced his writings. During the years 1639–1643, he attended the distinguished Elisabeth Academy in Breslau and studied with the *Opitz disciple Christoph Köler. He studied in Strassburg (1643–1644) and in Leiden (1644–1646) where he was exposed to mystical and pietistic thinking. He continued his studies in Padua in 1647 where he received his M.D. the following year.

Back in Silesia, he befriended Abraham von Franckenberg, who was a devotee of the visionary mystic *Jakob Boehme and the influential mentor of the Silesian mystical circle, who left Scheffler his library of mystical and occult writings upon his death in 1652. On 12 June 1653, Scheffler acted upon his long-standing misgivings about Lutheran orthodoxy and converted to Catholicism; he assumed the name Angelus Silesius (Silesian Angel). It is during this period of inner turmoil that Scheffler wrote most of his significant poetic works. In May 1661, he was ordained a priest so he could, as learned and spiritual doctor, heal both body and soul. By this time, Scheffler had become an active participant in the renewal of Catholicism in Silesia, and his writings turned aggressively polemical. In 1666, Scheffler entered the St. Mathias convent in Breslau where he died on 9 July 1677.

Both his major collections of poetry were published in 1657. The *Geistreiche Sinn- und Schluβreime* (*Ingenious Epigrams and Apophthegms*, 1657), a collection of 1,676 distichs influenced by Daniel Czepko's *Monodisticha* (manuscript 1647), were republished in 1675, expanded by a polemical sixth book, as *Cherubinischer Wandersmann* (*Cherubinic Wanderer*). The wanderer's goal is the union of the human soul with the pure divine essence that is reached by overcoming physical needs and human will and by achieving a state of eternal stillness and calm. The epigram with its tendency toward brevity and pointedness is a suitable genre to cope with the aesthetic problem of the ineffability of the mystical experience. The epigram's extreme economy of expression allows for a wealth of allusions and reverberations and leads to an obscure and arcane quality of the mystical text. The cherubic path leads to insight through cognitive means and through intellectual rhetorical strategies such as antithesis and paradox. The *Heilige Seelen-Lust Oder Geistliche Hirten-Lieder* (*Holy Joy of the Soul, or Spiritual Pastorals*, 1657) chooses the seraphic, affective path to the mystical experience. It redirects erotic feeling and its poetic expression toward the love of God. The contradiction between the cherubic and the seraphic paths is in appearance only; in Scheffler's view, they represent complementary aspects of a single path.

Bibliography: J. Sammons, "Johann Scheffler," *Dictionary of Literary Biography*, vol. 164 (1996): 279–288. J. Sammons, *Angelus Silesius*, 1968.

Peter Hess

SCHEINER, CHRISTOPH (1573–1650). Christoph Scheiner was born in 1573 in the Swabian region of Germany. Educated through the Jesuit system, he joined the order in 1595. In 1600, he completed his education at Ingolstadt, studying mathematics, philosophy, and theology. In 1610, he became professor of mathematics and Hebrew at Ingolstadt. He had already demonstrated his scientific talents by devising the pantograph, an instrument for copying and enlarging drawings (later presented in *Pantographice, seu ars delineandi* [1631]). In 1610, after confirming *Galileo's astronomical discoveries, Scheiner undertook telescopic observations of the sun, discovering the sunspots in March 1611. As the sunspots challenged Aristotelian natural philosophy—specifically, the sun's perfection and the impossibility of celestial generation and corruption—he delayed publication of his discovery. During the same period, Galileo in Italy, Fabricius in Germany, and *Harriot in England independently observed the spots. In early January 1612, Scheiner published his discovery in *Tres epistolae de maculibus solaribus scriptae ad Marcum Velserum*. In these letters and in the subsequent *De maculis solaribus et stellis circa Iovem errantibus accuratior disquisitio ad Marcum Velserum* (1612), Scheiner interpreted the sunspots as planets orbiting the sun. When Galileo received Scheiner's earlier work in April 1612, a bitter dispute over the priority of discovery ensued. Afterward, Scheiner continued his astronomical investigations, publishing *Sol ellipticus* (1615) and *Refractiones caelestes* (1617); he also studied the eye's physiology, detailing the retina's function in human vision and published as *Oculus, hoc est: fundamentum opticum* (1619). From 1624 through 1633, he was in Rome completing his most important work, the *Rosa Ursina* (1626–1630). Building upon fifteen years of observations, Scheiner demonstrated that the sunspots are in the sun's atmosphere and are there generated and destroyed, and determined also the period of solar rotation and the solar axis's inclination. Above all, the *Rosa Ursina* effectively silenced the defense of Aristotelian cosmology by other Jesuit philosophers through an authoritative use of scriptural and observational evidence. After 1633, Scheiner returned to Germany and died in 1650; his *Prodromus pro Sole mobili et Terra stabili contra Galileum a Galileis* appeared in 1651.

Bibliography: W. Shea, "Scheiner and the Interpretation of Sunspots," *Isis* 61 (1970): 498–519.

Renzo Baldasso

SCHÜTZ, HEINRICH (1585–1672). Heinrich Schütz, the most influential musician of seventeenth-century Germany, was baptized near what is now Bad Köstritz on 9 October 1585. Despite his gifted singing voice as a young man, his comfortably middle-class parents initially opposed his interest in a career in music, and he entered the University of Marburg in 1608. However, in 1609, he left to study with *Giovanni Gabrieli in Venice, for whom he composed a group of madrigals. Returning to Germany in 1613, he traveled to Dresden in 1615 at the invitation of the elector of Saxony, eventually becoming his kapellmeister (bandmaster), succeeding *Michael Praetorius as the primary composer for court occasions.

In 1617, Schütz composed a ballet for the Emperor Matthias's visit, as well as musical settings for the Psalms (*Psalmen Davids*). His fame growing, he also was in demand as an influential musical advisor throughout Germany. He composed two wedding concertos in 1618 and, in June 1619, married the daughter of a court official. For the next ten years he composed a variety of works for weddings, birthdays, and funerals, the *Syncharma musicum* (1620) commemorating the end of a battle in Silesia, a book of motets (*Cantiones sacrae*, 1625), and in 1627 the music for *Dafne*, the first opera in Germany (with libretto by *Martin Opitz). His wife died in 1625, and Schütz remained unmarried for the rest of his life.

In about 1628, he journeyed to Florence where, in close association with *Claudio Monteverdi, he absorbed the newest innovations in Italian music. He returned to Germany in 1629, also publishing *Symphoniae sacrae*. His musical activities at the Dresden court were greatly hampered by Saxony's involvement in the Thirty Years' War in 1631. In 1633, he was invited to oversee the music for the marriage of the Crown Prince of Denmark to the daughter of the elector of Saxony, composing music for two comic stage productions there. Upon returning to Dresden, he published a range of works including *Musicalische Exequien*, a significant funeral composition. After publishing no major collections for eight years, in 1647 he issued *Symphoniarum sacarum secunda pars*, dedicated to the Danish prince.

Advancing age and lack of support for his court programs caused Schütz to request a retirement from his duties as kapellmeister as early as 1645. Though his several requests were repeatedly ignored, he continued to be musically active, though on a limited scale, his musicians often going without pay. When the elector died in 1657, his son granted Schütz reduced duties and a pension. From this point until his death on 17 November 1672, he composed largely religious compositions, including a *St. John Passion*.

Bibliography: H. Moser, *Heinrich Schütz: His Life and Work*, trans. 1959.

Christopher Baker

SCUDÉRY, MADELEINE DE (1607–1701).

Madeleine de Scudéry was born into an ancient, noble family 15 November 1607. Upon the death of her parents in 1613, an uncle conscientiously raised and educated Madeleine and her older brother, Georges. Madeleine was known for her prominent role in society and for her vocation as a writer. However, her prolific literary output was never published in her name, but rather in her brother's or, in later years, anonymously. While the siblings did indeed coauthor books, particularly multivolume novels such as *Le Grand Cyrus* and *Clélie*, most commentators, then and now, believe that Madeleine did much of the writing.

Judgments pertaining to Madeleine's societal role have fluctuated over the centuries following her death. This is primarily because of her significant contributions to the social and literary movement that has come to be known as Preciosity, concerning which there are divergent conclusions. Dismissed by many as frivolous, insipid, pedantic, or dull, Preciosity, a pastime for cultured members of the upper classes, is now being redeemed by cultural and literary historians. The movement featured a passion for diversions such as witty conversation and in many ways reached its epitome with Madeleine's celebrated *Samedi*, or Saturday salons.

Her critical fortune as a writer has been equally inconsistent. Even as many scholars have rehabilitated Preciosity and Madeleine's contributions to it, her books, particularly the novels, which are massive in scope and intricate in their attention to detail, remain for these same reasons easy targets for the critic's pen. Fortunately, scholars are now recognizing that these works, which were immensely popular, are significant for the insight they provide into early modern society and for their complex, psychologically accurate depictions of love and other relationships.

Madeleine died 2 June 1701, having, she claimed, preserved her freedom by never marrying, a remarkable decision for the time.

Bibliography: N. Aronson, *Mademoiselle de Scudéry*, 1978; R. Lathuillère, *La Préciosité,étude historique et linguistique*, 1966.

Laura L. Dennis-Bay

SELDEN, JOHN (1584–1654). Educated as a lawyer, John Selden was a distinguished historian, legal scholar, and politician. He wrote several works that examined the structure of "British" government before the Norman Conquest (1066) and dabbled in Orientalism, publishing works on Persia, Turkey, and China. His legal writings include *Titles of Honor* (1614) an edition (1616) of Sir John Fortesque's (1394–1476) writings and *History of Tithes* (1618), in which he concluded that tithing was a matter of common law rather than divine law. In 1618, *James I commissioned Selden to respond to the Dutch legal scholar *Hugo Grotius's *Mare Liberum*, which argued in favor of freedom of the seas. In *Mare Clausum* (finally published in 1636), Selden argued that Britain exercised dominion in all waters surrounding the British Isles and in the Atlantic Ocean north and west of Britain. In 1647, he published an important law treatise in which he showed the influence of Roman law on English jurisprudence. This relationship between common and civil law would occupy much of Selden's political life as well.

Selden sided with the House of Commons' increasing desire to legislate according to common law precedents, rather than at the discretion of the king's prerogative, advocating a system of mixed monarchy. As a parliamentarian from

1626 to 1629, Selden worked to impeach George Villiers, duke of Buckingham, the chief minister believed to have brought about the malfunctioning political system. He defended John Hampden during the Five Knight's Case, when the latter was imprisoned for refusing *Charles I's forced loan from taxpayers, and he took part in the parliamentary debate on habeas corpus. These events led to the Commons' *Petition of Right* being delivered to the king in 1628, which argued against arbitrary taxation and imprisonment. The crown took reprisals, and Selden spent most of 1629–1631 in prison for his contempt. Selden also sat in the Long Parliament (1640–1660). He is best known for *Table Talk*, published in 1689. In a series of conversations taken from the last two decades of his life, Selden made reference to the relationship between sovereignty and the state, suggesting that a legal contract existed between monarchs and subjects, a theory similar to that later espoused by *John Locke.

Bibliography: P. Christianson, *Discourse on History, Law, and Governance in the Public Career of John Selden, 1610–1635*, 1996.

Ken MacMillan

SÉVIGNÉ, MARIE (DE RABUTIN-CHANTAL), MARQUISE DE (1626–1696).

French letter writer and chronicler of her times, Marie de Rabutin-Chantal, normally referred to as Madame de Sévigné, was born into a noble, intellectual family. Her parents were Celse-Bénigne de Rabutin, Baron de Chantal and Marie de Coulanges. Orphaned at the age of seven, Madame de Sévigné lived for three years with her maternal grandparents. After they both died, she was raised by her maternal uncle, Christophe de Coulanges, the Abbé of Livry. In his care she received a solid Christian education and learned Latin, Spanish, and Italian. When she was eighteen, she married Henri Marquis de Sévigné. A gambler and an unfaithful husband who flaunted his infidelities, Sévigné was killed in a duel in 1651 defending the honor of his mistress. Madame de Sévigné enjoyed the freedom that widowhood gave her and never remarried but devoted herself to her two children. After returning to Paris from her husband's estate in Bretagne in 1654, Madame de Sévigné participated in the intellectual life of the capital and became friends with *François de La Rochefoucauld, *Madame de Lafayette, and Nicolas Fouquet, *Louis XIV's finance minister. In 1669, her daughter Françoise-Marguerite married the Comte de Grignan and moved with him to Provence, where he had been appointed governor. Desperate to maintain ties with her daughter, Madame de Sévigné began to write letters to her in which she vividly described life and major events in Paris. During the last thirty years of her life, Madame de Sévigné's correspondence numbered in the thousands; of those written to her daughter, 1,500 remain today. After Madame de Sévigné's death from smallpox while visiting her daughter in 1696, her letters were published by her granddaughter. They set the standard for the epistolary genre and serve as a window to seventeenth-century France.

Bibliography: J. Ojala and W. Ojala, *Madame de Sévigné: A Seventeenth-Century Life*, 1990.

Leslie A. Sconduto

SHABBETAI ZEVI (1626–1676). Shabbetai Zevi, one of the most colorful figures in the history of Jewish mysticism, was born in Smyrna and studied Kabbalah and Talmud and, on the basis of his learning and asceticism, acquired a number of followers, to whom he imparted supposedly secret teachings. While still in his twenties, he began to think of himself as the Messiah, a conviction that brought him numerous followers but guaranteed him a peripatetic life, since Judaism had tended to be skeptical of Messianic claims and hostile toward claimants. But after revealing his secret identity at first only to his followers, in 1665 he declared openly that he was the Messiah, and in a wave of apocalyptic enthusiasm, his followers began abandoning their worldly goods in the expectation of immediate relocation to the Holy Land and the return of the biblical deity to Zion. Instead, Shabbetai Zevi was arrested in Constantinople in 1666 and denounced by leaders of the Jewish community. He thereupon converted to Islam, losing many but not all of his followers. He finally went into exile in Albania, his only followers being a sect of crypto-Jewish Muslims. Shabbetai's Messianism was connected to expectations of God's return to Zion that originated with the expulsion of Jews from the Iberian peninsula in the 1490s and that stimulated a mystical movement in Safed in the latter sixteenth century. Certain of Shabbetai's charismatic traits would in turn contribute to the central role of the "Zaddik," or righteous person, in Eastern European Hasidism in the eighteenth century.

Bibliography: G. Scholem, *Sabbatai Sevi: the Mystical Messiah, 1626–1676*, 1973.

Ralph Keen

SHAFTESBURY, ANTHONY ASHLEY COOPER, THIRD EARL OF (1671–1713). An important writer of the early Enlightenment whose work was of seminal significance in the realms of moral philosophy, aesthetics, and political thought, Shaftesbury was the scion of a very prominent and rather recently ennobled family. His grandfather became the first earl of Shaftesbury in 1672, but had won fame and public esteem as a supporter of the Parliamentarian cause during the civil wars and was to be a founding father of the Whig opposition to the policies of the royal court of *Charles II. The second earl suffered from such poor health that he consigned his sons to the care of their grandfather when the future third earl was but four years of age, who consequently soon found himself under the tutelage of *John Locke. The third earl venerated his grandfather's memory and was to remain a stalwart defender of the Whig cause throughout his career, though he was unable to avoid conflict with the Court Whigs during his brief but intense tenure as a member of the House of Commons. He became earl on his father's death in 1699 and was active in the last Parliaments of *William III's reign, but greatly reduced his direct involvement with Westminster politics from 1702, partly because of his failing health and partly because the political climate under *Queen Anne was favorable to his Tory enemies. He now devoted much more attention to his writing, and in 1711 published *Characteristicks of Men, Manners, Opinions, Times*, easily his most important work.

Characteristicks is a compendium of revised versions of five essays, which had appeared in print separately from 1699 to 1710, to which Shaftesbury adds a lengthy commentary. The third earl takes issue with *Thomas Hobbes's vision of human nature as essentially selfish and argues for a natural inclination toward social intercourse; he also repudiates Locke's ridicule of innate ideas, insisting that everyone is born with a "moral sense" as well as a love of beauty. Shaftesbury contends that our natural inclination toward sociability has important consequences for the conduct of inquiry and endeavors aimed at stimulating moral improvement, which he maintains are undertaken most effectively in the context of refined but morally serious conversation. He is unconvinced that sermons and abstract philosophical treatises are apt to disseminate genuine devoutness more widely or foster the profitable consideration of important questions. The refined and classically educated gentleman is much better equipped to engage in socially useful conversation than the authoritarian cleric or the learned pedant.

Shaftesbury argued that philosophy and the arts in England had been hobbled by the tyranny of crown and church; the "Glorious Revolution" of 1688 had for Shaftesbury inaugurated an era in which virtue and consequently the arts could flourish. Even such recent achievements as *Christopher Wren's St. Paul's Cathedral had in Shaftesbury's estimation been sullied by the tyrannical circumstances of their gestation; the Baroque was but a variation on the execrable "Gothicism" of the Middle Ages. The arts for Shaftesbury were properly a profoundly public concern, not a matter for private connoisseurship. Artist and patron were obliged to promote the dissemination of virtue, and this goal was to be attained by adopting a severely classicist and stoic aesthetic that Shaftesbury associated with ancient Athenian democracy and ancient Roman republicanism.

Shaftesbury was an eloquent advocate of latitudinarianism. Though harshly critical of the "tyranny" of "priestcraft," he was also troubled by the potential "enthusiasm" of dissenters, whose claims to privileged insight into God's will were a threat to that social harmony Shaftesbury saw as an essential prerequisite to building a virtuous society. He was nonetheless convinced that a divine predilection for beauty and harmony is reflected in nature.

Bibliography: P. Ayres, ed., *Characteristicks of Men, Manners, Opinions, Times*, 2 vols., 1999; L. Klein, *Shaftesbury and the Culture of Politeness*, 1994; R. Voitle, *The Third Earl of Shaftesbury 1671–1713*, 1984.

Matthew Koch

SHAKESPEARE, WILLIAM (1564–1616).

William Shakespeare was baptized on 26 April 1564 in Holy Trinity Church, Stratford-Upon-Avon, the son of Mary Arden, daughter of a local landed family, and John Shakespeare, a glover who also held various positions of civic importance. Shakespeare's birthday is often given as 23 April, partly because he also died on that day and partly because it is the feast of St. George, the patron saint of England, and hence a fitting day for the birth of a national poet. Shakespeare may have been educated at Stratford Grammar School, where he would have received a basic grounding in classics, but as the records from that era were burned, there is no way to corroborate this. During Shakespeare's childhood, the family appears to have fallen into debt and difficulty. A number of scholars have suggested that these difficulties may have been caused by John Shakespeare's putative Roman Catholic sympathies.

Shakespeare married Anne Hathaway, the daughter of a local farmer, in November 1582, when he was eighteen and his bride some six or seven years older. Anne must have been pregnant already, because the first of their three children, Susanna, was baptized on 26 May 1583, about six months later. The twins Hamnet and Judith were baptized on 2 February 1585. Only Shakespeare's two daughters survived to maturity; Hamnet was buried on 11 August 1596, at the age of eleven.

The first professional reference to Shakespeare occurs in Robert Greene's *Greene's Groatsworth of Wit* (1592), in which he describes Shakespeare as "an upstart crow, beautified with our feathers, . . . his *Tiger's heart wrapped in a Player's hide*." This passage echoes one from Shakespeare's own early play *3 Henry VI*, "Oh tiger's heart wrapp'd in a woman's hide!" (I. iv. 137). In 1593, he wrote the epyllion or narrative poem *Venus and Adonis* and dedicated it to Henry, earl of Southampton; this was followed by *The Rape of Lucrece* in 1594. No other narrative poems have been definitely attributed to Shakespeare, although recently the *Elegy by W.S.* has been analyzed by Donald Foster and included in some of the most recent editions of the complete works. His sonnets were not published until 1609, but since Francis Meres refers to his "sugar'd sonnets" in his essay *Pallidis Tamia* of 1598, one can safely assume that at least some of them had been written by then, particularly as the fashion for sonnets was at its height in the mid-1590s.

Shakespeare was a working actor and a sharer in the Chamberlain's Men by 1594; he may have had a contract something like Richard Brome's, in which he agreed to write a certain number of plays, probably two, every year. In 1598, the Burbages lost the ground-lease on which the theater stood, and since they were not permitted to play in the Blackfriars and needed capital, Shakespeare and some of his fellow sharers bought a share in the new theater, the Globe, built in 1599. In 1608, he bought a portion of the Blackfriars as well. While the profits Shakespeare made from his work in theater have sometimes been exaggerated, he was able to buy a coat of arms for his now-deceased father and a considerable amount of real estate in Stratford. The company became the King's Men on the accession of *James I in 1603. By 1613, Shakespeare seems to have spent more time in Stratford and wrote his last plays in collaboration with the new house playwright *John Fletcher. He died on 23 April 1616, leaving rings to his old colleagues John Heminges and Henry Condell, the second-best bed to his wife Anne, and the bulk of his estate to his daughter Susanna and her husband Dr. John Hall.

Some of Shakespeare's plays were published in quarto form during his lifetime. In 1623, after his death, Heminges and Condell authorized a folio of his theatrical works to be printed, "according to the True Originall Copies." It is from the First Folio's example that the plays have subsequently divided into Tragedies, Histories, and Comedies; the subcategory "Romances" was coined later to refer to *Pericles, The Winter's Tale, Cymbeline*, and *The Tempest*. Other categories include the "problem plays" (*All's Well That Ends Well, Measure for Measure, Troilus and Cressida*), which combine comic and tragic elements in unconventional ways; the "Roman plays" (*Titus Andronicus, Julius Caesar, Antony and Cleopatra, Coriolanus*); and "chronicle histories" (to distinguish the earlier history plays from the nostalgic Tudor play *Henry VIII*).

Unfortunately, much of what passes for Shakespearean biography is based on legend, much later accounts, and speculation based on his works as to what sort of man he was. Samuel Schoenbaum's study *Shakespeare's Lives* alludes to this multiplicity of portraits, dividing the historical record, contemporary gossip, later stories by writers like *John Aubrey, and information on early Shakespeare tourism into separate sections. There are traditions that Shakespeare played kingly parts and roles like old Adam in *As You Like It*, traditions that he left the Stratford area because he poached a deer, and Aubrey's assertion that he made the startling sum of £1,000 a year.

With scholarly editions stretching back to *Nicolas Rowe's in 1709 and critical commentaries being published at the rate of one every eleven minutes, it is almost impossible to do justice to the wide range of Shakespearean scholarship. Work on textual criticism continues, including debates as to whether or not to include Shakespearean apocrypha such as the above-mentioned funeral elegy or whether *King Lear* should be considered as two separate plays because of its two differing texts, which have survived. Shakespeare's stature as perhaps the preeminent author of the European Renaissance is attested by numerous studies that have examined his work in relation to such topics as race in the Early Modern era, queer theory, Marxism, gender, economics, and many other critical approaches.

Bibliography: A. Kernan, *Shakespeare, the King's Playwright: Theater in the Stuart Court, 1603–1613*, 1995; S. Schoenbaum, *Shakespeare's Lives*, 2nd ed., 1991; B. Vickers, *Appropriating Shakespeare: Contemporary Critical Quarrels*, 1993.

Melissa D. Aaron

SHIRLEY, JAMES (1596–1666). James Shirley's career as a professional playwright and poet reflected the rapidly changing atmosphere of seventeenth century England. Born in London to a merchant father, he was educated at the

Merchant Taylor's School at St. John's College, Oxford, and at St. Catherine's Hall, Cambridge. His first literary work, *Echo and Narcissus*, appeared in 1617–1618; no copies survive, but the work is thought to be an early version of *Narcissus, or the Self-Lover* that appeared in Shirley's *Poems* (1646). By 1618, Shirley was actively pursuing a career in the Anglican church, when the designation "clarke" followed his name in the records of his marriage to Elizabeth Gilmet, daughter of a prominent St. Albans politician (when this marriage came to an end is unclear). He settled for a time in St. Albans, working as a schoolmaster at the local grammar school.

In 1624, however, Shirley returned to urban life, moving to London with his family, perhaps with a job already secured in the city. On 11 February 1625, less than ten months after his arrival, Shirley's first known drama, *Love Tricke, or the Schoole of Complement*, was licensed to appear in the Cockpit Theatre in Drury Lane. Shirley's association with the Cockpit lasted more than a decade, resulting in the staging of some of his best known works such as *The Traitor* (1631), *Love's Cruelty* (1631), *Hyde Park* (1632), and *The Lady of Pleasure* (1635), among others. His dramas routinely commented on the tastes of the Cockpit's socially ambitious audiences, both appealing to and satirizing their concerns and aspirations. The closing of the theaters due to plague in 1636 brought Shirley's relationship with the Cockpit to its end; Shirley, however, did not abandon theatrical writing, moving instead to Ireland, where he helped establish the Werburgh Street theater in Dublin. He returned to London in 1640, succeeding *Philip Massinger as playwright for the King's Men at the Blackfriars, where his play *The Cardinal* (1641) debuted in 1641. When Parliament closed the London theaters in 1642, Shirley's Royalist politics found expression in *The Triumph of Peace* (1633–1634), a masque created in collaboration with *Inigo Jones and William Lawes and sponsored by the Inns of Court to endorse *King Charles I and Queen Henrietta Maria.

Shirley fought for the Royalist cause under William Cavendish, earl of Newcastle, with whom he had earlier collaborated to write *The Country Captain* (1639–1640); after the war, Shirley confessed to his actions and was granted clemency. It has been suggested that Shirley had in fact converted to Roman Catholicism, but such arguments remain inconclusive, though he did dedicate a number of poems to devoted Catholics. After the war, Shirley appears to have returned to a quiet life as a schoolmaster, this time at Whitefriars. His earlier writings, however, continued to appear in print, among them a variety of masques and his *Poems* (1646). According to *Anthony à Wood, Shirley and his second wife died within a day of witnessing the Great Fire of London in 1666, which left them both "overcome with affrightments, disconsolations, and other miseries."

Bibliography: S. Burner, *James Shirley: A Study of Literary Coteries and Patronage in Seventeenth-Century England*, 1988; B. Lucow, *James Shirley*, 1981.

Hillary Nunn

SIRANI, ELISABETTA (1638–1665). Like most Italian women artists of
the Early Modern period, including the more famous *Artemisia Gentileschi
(1593–1652/3), Elisabetta Sirani was the daughter of a painter, Giovanni Andrea
Sirani, from whom she probably learned the rudiments of her art. Following the
example of her celebrated predecessor Lavinia Fontana (1552–1614), she practiced her craft in Bologna from the age of seventeen, producing many commissioned portraits and small-scale devotional images. However, unlike many
female counterparts, Sirani never married and died at the age of twenty-seven.
She supported her family by taking over her father's workshop, teaching female
pupils, and completing more that 190 recorded paintings and drawings. Her
sisters, Anna Maria (1645–1715) and Barbara (recorded in 1678), were also
painters.

Among Sirani's most well-known representations of women are *Judith with
the Head of Holofernes*, *Portia Wounding her Thigh* (1664), and *Penitent Magdalene* (1660). Both Judith's triumph and Mary Magdalene's repentance were
popular subjects in early-modern Italy, but they were of special interest to
women artists. The former was represented several times by Fontana and Gentileschi. Yet Sirani's iconography and style have been primarily evaluated with
regard to her portrayals of Roman female heroes as the aforementioned Portia
and Timoclea (1659). The former was Brutus's wife; she appears in Sirani's
aforementioned painting wounding her thigh in order to prove that she is as
strong as a man is considered to be and able to keep her husband's secret.
Attesting to Sirani's popularity in Bologna were such major gender-neutral commissions as those for *St. Jerome* (1660) and *St. Anthony of Padua* (1662).
Among her most distinguished patrons was Grand Duke Cosimo II de' Medici.
Despite a distinguished career, no monographs have yet been written about
Sirani.

Bibliography: *The Age of Correggio and the Caracci: Emilian Painting of the 16th and
17th Centuries*, 1986–1987, 534–537; W. Chadwick, *Women, Art, And Society*, 1996,
96–105; E. Fine, *Women & Art*, 1978.

Yael Even

SMITH, CAPTAIN JOHN (1580–1631). The first of the great English explorers of North America who did not come from an aristocratic home, Smith
spent the inauspicious first sixteen years of his life in Lincolnshire, first working on the family farm and later as an indentured apprentice to a shopkeeper.
Only after his father's death in 1596 did he manage to prepare himself for a career as mercenary soldier in Eastern Europe and the Orient. Between 1596 and
1600, he traveled widely through Great Britain and the Continent, teaching
himself in Lincolnshire the arts of war and horsemanship and joining the English army in the Netherlands; between 1600 and 1602, he became a mercenary
soldier, fighting with a Hungarian battalion against the Turks, with a Transyl-
vanian regiment against the Hungarians and the Turks, and with the Habsburg

Imperial army against the Ottoman Empire and its Tartar and Walachian allies, until he was finally taken prisoner by the Turks in November 1602. Between this date and 1604, he experienced further adventures, escaping from his enslavement in Constantinople and making his way back to England through the Mediterranean, Germany, and France. In 1601, he had reached the rank of cavalry captain in the Hungarian army; he would later sign all his published works as Captain John Smith, often claiming that he was first and foremost a professional soldier.

Throughout his writing career, Smith showed the same opportunistic habits of composition and publication. On the one hand, several of his shorter writings are either first drafts of eyewitness accounts interspersed with geographical and scientific observations obtained in extant sources or else hurried rewritings of administrative documents conceived in bureaucratic contexts, often specifically conceived as promotional tracts in which the author sought the advancement of his country, the advancement of colonial projects, and the advancement of his own career. On the other hand, his longer works contain only a small portion of his own original writing sandwiched between loosely connected chunks of narrative authored by several hands, including the scholars and travelers from whom he borrowed freely elsewhere.

Smith's experiences and writings are indispensable to any investigation of the insufficiently documented presence of English settlers in America between the founding of the short-lived Jamestown in 1607 (he was president of the governing body between 1608 and 1609) and the founding of Plymouth Colony in 1620. Smith made only two successful voyages to America: the first one, in 1607, took him to the West Indies and Virginia, where he remained until October 1609; and the second one, in 1614, centered on the exploration of the Maine and Massachusetts seaboard. He attempted three other expeditions that proved failures: In 1614, the Plymouth Company appointed him Admiral of New England (a geographical name he coined), but his ship was in a calamitous condition and he had to turn back; in the same year, he set sail again in a smaller vessel, which was captured by pirates; and in 1617, an expedition for New England was called off because of persistent bad weather. He could still have become the first colonist of New England had Lord Chancellor *Francis Bacon complied with his petition in 1618 for a grant to establish a settlement north of the Chesapeake Bay, or had the Pilgrims accepted his offer of guidance in 1620, when he also published *New England's Trials*. His frustration at the Puritans' indifference helps explain his animadversion on *Bradford and the Separatist Church in his later writings.

Smith's first substantial work was written in America at the beginning of 1608 and published in London later that year under the title *A True Relation of Such Occurrences . . . as Hath Happened in Virginia*. In the same year, he was elected president by the Governing Council of the Jamestown Colony, but in October 1609, he had to resign his command and go back to England after he sustained severe injuries in a gunpowder explosion. In 1612, he published *A Map of Virginia*, whose first part, often referred to separately as the "Description," is considered by many to be his masterpiece, mainly because Smith makes up for his lack of scholarly training with his earlier, firsthand acquaintance with Near-Eastern realities (used as terms of comparison to describe the novelty of America) and his natural talent for observation. These qualities allowed Smith to come across not just as a writer possessed of a sympathetic imagination, but also as an improvised natural historian and a comparative ethnographer of sorts. The concluding pages of the Description and the whole of the second part, known as the Proceedings, advance almost exclusively one of Smith's recurrent arguments: the need for a resourceful leader who (like himself) combines the qualities of a courageous soldier, a diligent administrator, and a tenacious entrepreneur. Here we also find such related topics as the nobility of hard work and administrative transparency, and his firm belief that America should provide an opportunity for the honest advancement of untitled yet talented gentlemen such as himself, a topic also prominent in Book IV of the *Generall Historie*. In recent years, Smith has been acknowledged as a precursor of a new type of cognitive authority for his direct experiencing of, and active participation in, new political and scientific realities.

The Generall Historie of Virginia, New England, and the Summer Isles (1624) is a compilation of firsthand accounts of previously published sources on Virginia and other places, among them large portions of Smith's revised versions of his own True Relation, Proceedings, Description, and *A Description of New England* (1616). In its conception and use of extant travel accounts, this work is also indebted to Richard Hakluyt and *Samuel Purchas. Smith's original contribution as history writer appears in Book III, which covers events that took place in Virginia from 1607 to 1609, while Book IV centers on the fated decline of the Jamestown settlement immediately following Smith's forced return to England. Even the work that the reader would assume to present the personal Smith, *The True Travels, Adventures, and Observations of Captain John Smith, in Europe, Asia, Affrica, and America, from Anno Domini 1593 to 1629* (1630), written in the Caesarean third person, turns out to be another recasting of previously published texts: Smith's earlier self-narration in Book VIII of his friend Samuel Purchas's *Purchas His Pilgrimes* (1625), other minor travel accounts first collected in Purchas and in Hakluyt's *Principall Navigations* (1599), and other contemporary histories and travelogues.

Smith's last work is entitled *Advertisements for the Unexperienced Planters of New-England, or Any Where* (1631). This unfinished short tract was the third and best of his books on New England, written in awareness of *John Winthrop's success in attracting backers and settlers to New England. This might be the reason why he advocates for the first time the spread of religious communities in America as a vehicle for social cohesion. The final Smith attenuates somewhat his penchant for narcissistic and hyperbolic self-narration, focusing the second part of the *Advertisements* mainly on giving practical advice on building defensive structures, farming, and warfare to the would-be colonists. It is as if he finally came to accept, in the last stretch of his life, that he could serve better the cause of colonization as a writer of promotional works than as an explorer and administrator. He died, unmarried and childless, in June 1631.

Bibliography: P. Barbour, *The Three Worlds of Captain John Smith*, 1964; E. Emerson, *Captain John Smith*, rev. ed., 1993; M. Jehlen, "The Literature of Colonization," in *The Cambridge History of American Literature, vol. 1, 1590–1820*, ed. S. Bercovitch, 1994.

José María Rodríguez-García

SNELL (or SNELLIUS), WILLEBRORD VAN ROIJEN (1580–1626).

Snell was born in Leiden, where he was a law student at the university at which his father Rudolf was professor of mathematics and where he too developed an interest in that subject. Between 1600 and 1602, he traveled through Europe, pursuing an interest in astronomy, especially at Paris toward the end of his tour. In 1604, he traveled to Switzerland with his father and succeeded him in the Leiden professorship. Based on research begun in 1615, he published in 1617 his *Eratosthenes Batavus*, a work advancing his method for measuring the size of the earth through the use of triangulation, a concept first proposed by the sixteenth-century scientist Tycho Brahe. Snell used thirty-three triangles with carefully measured baselines to measure an arc of meridian of the globe and thus calculate the earth's dimensions, erring by only 3.4 percent. He also calculated the value of pi to seven places by using ninety-six-sided polygons. In 1621, he discovered Snell's law, the principle of refraction. It expresses the relationship between the path of a ray of light passing through the boundary of two adjacent substances and their respective refractive indices. The law is expressed by the formula: $n \sin(theta) = n' \sin(theta')$, where n and n' are the refractive indices of the adjoining materials and *theta* and *theta'* are the angles of incidence and reflection of the light ray. The manuscript of Snell's unpublished equation was read by *Christiaan Huygens and perhaps by *Descartes as well, who in 1637 published it and also attempted to physically verify it. Huygens also published it in 1703 in his *Doptrica*. Snell also described the path that makes a constant angle with the meridians of a sphere, the loxodrome, in his study of navigation *Tiphys batavus* (1624), and he discovered the sine law. Snell died on 30 October 1626 in Leiden.

Bibliography: L. Beek, *Dutch Pioneers of Science*, 1985; N. Haasbroek, *Gemma Frisius, Tycho Brahe and Snellius, and their Triangulations*, 1968.

Christopher Baker

SPEED, JOHN (1552–1629).

Born in Cheshire in 1552, the son of a tailor, John Speed became a Freeman of the Merchant Taylors' Company in 1580. It was through the generous patronage of Fulke Greville that Speed began seriously to pursue his antiquarian interests. When, on 15 June 1598, Queen Elizabeth granted Speed the use of a room in the Custom House, he fully devoted himself to historical and cartographic research, research that led to the publication of his *Theatre of the Empire of Great Britaine* and his *Historie of Great Britaine* (1611–1612).

Dedicated to *King James VI and I, the *Theatre* is a compilation of beautifully engraved and elaborately ornamented maps of English and Welsh counties and Irish provinces. Supplementing the *Theatre*'s cartographic images are chorographic descriptions. Engraved in Amsterdam by Jodocus Hondius, printed in London by William Hall and John Beale, and published by John Sudbury and George Humble, the *Theatre* bears the traces of an international collaborative project. It was to his compatriots, however, that Speed was most indebted. His county maps of England and Wales are close copies of the work of other Englishmen, in particular, Christopher Saxton and *John Norden. As a member of the Society of Antiquaries, Speed was well acquainted with *William Camden and Robert Cotton, and he certainly benefited from these acquaintances: Both his *Theatre* and his *Historie* (1612) owe much to the various editions of Camden's *Britannia*; moreover, he made frequent use of Cotton's magnificent library. With their inset town plans, costumed figures, and coats of arms, Speed's maps became the most coveted cartographic possessions in seventeenth-century England, and they remain invaluable cultural artifacts.

Although Speed is best remembered as the compiler of the earliest comprehensive atlas of the British Isles, he saw himself as a historian; indeed, he viewed *Historie* as his magnum opus. In fact, the *Theatre* was originally conceived as a topographical section of the *Historie*—the signatures and pagination of the first edition of the *Theatre* and *Historie* are continuous. While the *Theatre* remained popular throughout the seventeenth century and even into the mid-eighteenth century (a Latin edition first appeared in 1616), the same cannot be said of his *Historie*, which was last printed in 1650. Perhaps Speed's greatest achievement was to invite early modern subjects to visualize, as opposed to merely read about, the British Isles. Just before his death in 1627, Speed published *A Prospect of the Most Famous Parts of the World*, the first world atlas produced by an Englishman.

Bibliography: *The Counties of Britain: A Tudor Atlas by John Speed*, 1995; T. Kendrick, *British Antiquity*, 1950.

Christopher Ivic

SPINOZA, BARUCH DE (1632–1677).

A Dutch Jewish philosopher, Spinoza is one of the major figures in Continental rationalism. Spinoza was born in Amsterdam on 24 November 1632 of Portuguese Marrano parents. He received a Talmudic education, but was also encouraged to study Latin and other secular subjects. Contact with dissident Christian groups and Cartesian philosophy led him to abandon Jewish orthodoxy, culminating in his excommunication in 1656.

Latinizing his name to Benedict, Spinoza supported himself by lens grinding and teaching. Moving to the village of Rijnsburg, he composed his *Short Treatise on God, Man, and his Well Being* and the *Treatise on the Improvement of the Intellect*. In 1663, he published *The Principles of Descartes' Philosophy*, the only book to appear under his own name. While admiring *Descartes' mathematical method, he found Descartes' use of radical doubt problematic, arguing that certain metaphysical propositions were axiomatic. Taking this as his starting point, Spinoza began to cast his own philosophical position in geometrical terms. The result was his most important and controversial work, the *Ethics*.

Moving to Voorburg, he became acquainted with Jan de Witt, the liberal opponent to the House of Orange. This association led to the appearance of the *Tractatus Theologico-Politicus* in 1670. Influenced by *Thomas Hobbes, and anticipating biblical criticism, he advocated religious toleration and freedom of thought, based on a historical reading of the Bible. The work was condemned by both Catholics and Protestants.

In 1673, Spinoza refused the chair of philosophy at the University of Heidelberg. Nevertheless, he continued to develop his philosophy in the company of a small circle of friends and correspondents. Moving to The Hague, he began work on a Hebrew grammar, which he soon abandoned to continue work on the *Ethics*, probably completing it around 1675. The *Ethics* is divided into five parts. In the first part, Spinoza offers an ontological proof for the existence of God, whom he understands as an infinite and self-dependent substance. This leads him to conclude that there can only be one such substance, one fundamental reality, God. God and existent nature are thus the same—*deus sive natura*—a conclusion that led the Enlightenment to condemn Spinoza as a materialist and the Romantics to praise him as a pantheist. In the second part, Spinoza argues that individual things are just modes or modifications of the basic substance. Individual mind and matter are not distinct substances as Descartes had concluded, but simply two characteristics of the same reality. For every mode of thought there is a parallel object under the mode of extension, and for every mode of extension, there is a parallel idea under the mode of thought. Because the individual human mind is a finite mode of God under the attribute of thought, its comprehension is necessarily fragmentary, often confused. For Spinoza this is the cause of error and immorality.

In the third part, Spinoza develops a psychology to explain the origin of the basic passions—desire, joy, and sadness. Central to this psychology is the *conatus* doctrine, the principle that things strive for self-preservation and power. Parts four and five prescribe an ethics in which the passions are brought under the rule of reason. Fundamental to this is the recognition that the love of one's neighbor is the love of God. In turn, anticipating modern psychotherapy, Spinoza believed that the rational understanding of the passions offered a means of ameliorating their harmful effects and healing the soul.

Spinoza died of tuberculosis on 21 February 1677. Thereafter, his friends arranged the publication of the *Ethics*, the *Treatise on the Improvement of the*

Intellect, the *Tractatus Politicus*, the incomplete Hebrew grammar, and his correspondence, which offers a valuable commentary on his work. Most contemporary thinkers dismissed him as an atheist and materialist, following the lead of *Pierre Bayle. In turn, the Enlightenment's focus on empiricism and Hume's and Kant's attacks on the ontological argument eclipsed Spinoza's rationalism until the rise of Romanticism.

Bibliography: E. Curley, *Behind the Geometrical Method: A Reading of Spinoza's Ethics*, 1988; S. Hampshire, *Spinoza*, 1962; G. Lloyd, *Spinoza and the Ethics*, 1996.

Thomas L. Cooksey

STAHL, GEORG ERNST (1660–1734). Stahl was born at Ansbach and completed medical studies at the University of Jena. From 1687 to 1694, he served as court physician for the duke of Weimar before taking a chair in medicine at the new University of Halle. Stahl left the university in 1715 and became an advisor and later royal physician to the Prussian king. Beginning in 1725, he also served as president of the medical board for the whole of Prussia. Stahl is best known for his contributions to chemistry, particularly as a staunch defender of the phlogiston theory that dominated chemical thinking in the early eighteenth century. Stahl embraced, republished, and expanded upon the works of Johann Joachim Becher (1635–1682), who originated the idea that a specific principle was responsible for the chemistry of combustion. Stahl renamed this principle phlogiston, which he said left metallic ores and entered the atmosphere during combustion. Phlogiston also explained the reverse reaction, for it could cause calxes, or oxides, to return to metal when it reabsorbed phlogiston from the atmosphere. Because pure metals weigh less than oxides, the theory implied that the addition of phlogiston reduced chemical weight. Positivist historians therefore discount Stahl's theories as a hindrance to real progress and highlight Lavoisier's triumphant rejection of the phlogiston theory in the 1770s. Yet Stahl's theories served for decades as systemic and plausible explanations for a number of chemical questions, such as the fundamental division between organic and inorganic substances, the differences between alkalinity and acidity, and the emerging distinction between notions of the natural chemical "reaction" and the unnatural alchemical "transmutation."

Stahl helped legitimize chemistry as a valid and rational science independent from alchemy and iatrochemisty. He attacked alchemy as gravely flawed, virtually immoral, and of no value to the state. Stahl's connections in the political arena led to increased support for chemistry in medical schools and greater study of chemistry's practical implications for mining, brewing, dyeing, and other industries. Stahl also practiced medicine throughout his career and developed sophisticated and influential theories of medicine and physiology. He was very interested in the notion of the *anima*, an immaterial vital principle that was responsible for the motion, behavior, and teleological purpose of all living bodies. Stahl assumed that medical intervention required a holistic approach, using diet, hygiene, and psychotherapy in order to allow the free flow of animistic forces. These ideas placed him squarely opposed to those who saw materialist and mechanist explanations for medical ailments, and Stahl denied that anatomical, pharmacological, "heroic" interventions could cure disease. Like his chemical theories, Stahl's ideas on medicine and physiology were influential throughout Europe in the eighteenth century.

Bibliography: J.R. Partington, *History of Chemistry*, vol. II, 1961; I. Strube, *Georg Ernst Stahl*, 1984.

Mark Finlay

STEELE, SIR RICHARD (1672–1729). Steele was born in Dublin in March 1672, the son of an attorney (also Richard) and his wife Elinor. We are told in *The Tatler* No. 181 that the elder Steele died shortly before his son reached the age of five, at which time the young Richard Steele entered into the care of his uncle Henry Gascoigne, private secretary to James Butler, the Duke of Ormonde. Through Gascoigne's influence, Steele obtained a nomination to the Charterhouse in November 1684, where he first met *Joseph Addison in 1686. Steele matriculated at Merton College on 13 March 1690, but, caught up in the excitement surrounding *William III's campaigns against the French, left Oxford in 1694 without taking a degree in order to enter the army as a gentleman-volunteer in the second troop of lifeguards under the second Duke of Ormonde. Steele's aspirations for a military career were enhanced when he dedicated his celebration of Queen Mary's funeral, "The Procession," to John, Lord Cutts, colonel of the Coldstream guards, who subsequently employed Steele from 1696 to 1697 as his secretary and, ultimately, gave him a standard in his regiment.

Later in 1697, Steele was involved in a duel with a Captain Kelly, whom he wounded seriously, but not fatally. This incident was not only the instigation of Steele's famous dislike of dueling, but was also a motivating factor behind his first major dramatic endeavor, *The Christian Hero*, an effort that so altered his reputation among his military colleagues for the worse that he authored his first comedy, *The Funeral*, in order to mend his reputation among them. In December 1703, Steele produced his second comedy, *The Lying Lover*, in which he attempted to actuate the precepts of Jeremy Collier's *Short view of the Profaneness and Immorality of the English Stage*. The attempt was unsuccessful and, according to Steele, was "damned for its piety." Steele's third comedy, *The Tender Husband*, an imitation of *Moliere's *Sicilian*, was also unsuccessful, though greatly enhanced by contributions from Addison.

The failure of *The Tender Husband* marked a temporary departure from Steele's attempts to write for the stage, though he seemed intent from that point onward on a career in literature and politics. In March 1705, Steele married Margaret Stretch, who died little more than a year afterward. During the brief period of his married life with Stretch, Steele became gentleman waiter to Prince George of Denmark and in April or May 1707 was appointed gazetteer by

Harley. The death of Prince George in October 1708 deprived Steele of his position of gentleman waiter, and after several unsuccessful bids for other posts, he began his career as an essayist with the inauguration of *The Tatler* in April 1709. *The Tatler*, which appeared three times weekly until 2 January 1711, was written anonymously under the persona of Isaac Bickerstaff (a name borrowed from Jonathan Swift's satires of the popular astrologer John Partridge) and addressed all questions of manners and taste. When Steele halted *The Tatler* project because his identity as the voice behind Bickerstaff had been perceived, he was not out of the periodical business for long. On 1 March 1711, Steele commenced production of *The Spectator* with Addison. Like *The Tatler*, *The Spectator* was concerned with issues of taste and morality, especially among the middle class. Unlike *The Tatler*, however, *The Spectator* typically avoided charged political issues and disputes and appeared daily until 6 December 1712.

Despite the apolitical nature of *The Spectator*, Steele was becoming increasingly involved in Whig politics, a trend typified by his first projects after the termination of *The Spectator*, the political periodicals *The Guardian*, *The Englishman*, and his 1714 Hanoverian apology *The Crisis*, which resulted in his expulsion from the House of Commons in March, 1714. Steele's personal fortunes improved with the death of *Queen Anne and the Whigs' rise to power, when his party loyalty was rewarded with an appointment to several posts (including supervisor of Drury Lane Theater) and a knighthood in 1715. Steele did not enjoy his good fortune for long, however, losing both his position at Drury Lane and his friendship with Addison because of his 1718 denouncement of Lord Sunderland's Peerage Bill in *The Plebian*. Steele continued an active involvement in both the literary and political worlds until the last few years of his life, publishing *The Theater*, a biweekly periodical from January until April of 1720 and a series of pamphlets directed against the South Sea mania also in 1720. His last play, the comedy *The Conscious Lovers*, was produced in 1722. Plagued by mounting debts, Steele retired to Carmarthen in 1724, where he died quietly in 1729. Despite the anonymity of his death, however, Steele is still considered to be an important Augustan literary figure because of his introduction of the sentimental comedy (exemplified in such dramatic efforts as *The Tender Husband*), which stood as Steele's moralistic response to bawdy Restoration comedy, as well as his development of the "brief essay" prose style in *The Tatler* and its immediate predecessors, which today represents Steele's primary claim to literary fame.

Bibliography: G. Aitken, *The Life of Richard Steele*, 1889; rpt. 1968; W. Calhoun, *Captain Steele; The Early Career of Richard Steele*, 1964; W. Calhoun, *Sir Richard Steele, M.P.: The Late Career of Richard Steele*, 1970.

Walter H. Keithley

STEEN, JAN HAVICKSZOON (1626–1679). A key Dutch genre painter of the seventeenth century, Jan Steen is known primarily for interior scenes of feasts, brothels, and dissolute households. Steen attended Leiden University, and J.C. Weyerman, an early biographer, was probably correct in noting his teacher as the Haarlem painter Adrian van Ostade, a follower of Adrian Brouwer.

By including himself and his family members in his works, he humanized his paintings' strong moralizations. These inclusions, however, prompted the biographer Houbraken to label him a drunkard and his family dissipated (*Easy Come, Easy Go*, 1661; *As the Old Sing, So Pipe the Young*, 1663–1665). He often based his genre scenes on folk expressions or themes derived from the works of the popular chambers of rhetoric, or *Rederijkers*. He consistently sought to improve and update his style, imitating such fashionable peers as the young van Mieris, Metsu, and ter Borch in an effort to increase his prices. This met with little success. After the brewery his father leased for him in 1654 foundered, he moved to Warmond (near Leiden) and lived there from 1656 until 1660; he then moved to Haarlem for his most artistically productive years (1661–1670).

Although he imitated the precisionist effects of such Leiden school "Fijnschilders" as Gerrit Dou, his brushwork always remained loose and dashing. Eschewing preparatory drawings, he seems to have composed his unconventionally elaborate and theatrical genre scenes directly on the canvas. The didactic clarity of such works as *The Feast of St. Nicholas* (1665–1668) has secured Steen's reputation as a master storyteller in paint.

Bibliography: H. Chapman et. al., *Jan Steen, Painter and Storyteller* (Exhibition Catalogue), 1996; A. Houbraken, *De Groote Schouburg*, I, 1718–1721; W. Martin, *Jan Steen*, 1954; J. Weyerman, *De levens-beschryvingen der Nederlandsche kunst-schilderen, II*, 1729.

Lloyd DeWitt

STENO, NICHOLAS (NIELS STENSEN) (1638–1686).

Danish anatomist and geologist Nicholas Steno studied in his native Copenhagen and then at age twenty-two (1660) traveled to the Netherlands (University of Leiden) to continue anatomical studies and medicine. He gained further experience in Paris and Montpelier and then moved to Florence in 1665, staying there for the remainder of his short life. In Florence he was patronized by Grand Duke Ferdinand II and the scientific community of the Accademia del Cimento (Experimental Academy). Steno's anatomical studies led to the examination of a shark. Reminding Steno of stony objects known as glossopetrae ("tongue stones") found in rock, the shark's teeth led him to conclude that tongue stones were the teeth of shark remains buried in mud and sand, and later solidified by time. Traditional notions of such objects were that they fell from the sky or just grew in the rocks. Others before had noted teeth and animal-like fossils (defined as anything dug up) as being just that, but Steno wondered how these and other objects, such as mineral deposits and layers of rock, came to be embedded in other solid layering.

His conclusions were collected in a book known as the *Prodromus* (1669).

Steno concluded that animal fossils were buried in soft matter that hardened, whereas mineral veins and crystals in rocks formed from chemical fluids percolating after rock solidification, because their irregularities showed conformity to stresses in the rock. Steno decided that layers or strata form from an initial liquid flow with progressive fallout of matter that builds into horizontal layers, his Principle of Original Horizonality. But layers can change by later geological forces. Any rock layer conforms to the shape of lower layers and from this follows Steno's most important formulation, his Law of Superposition, layers of rock are arranged in a time sequence from youngest on top to oldest on the bottom, unless geologically disturbed. In this regard, Steno realized that the collapse of a subterranean cave or broken layers forced up or down or molten rock undermining layers could mix up the normal sequence of layering. Noticing two rock types in the Apennines near Florence, Steno made the first geological guess at time periods by noting the upper fossil layers formed after the lower fossil-less bedrock. Steno is known as the Father of Stratigraphy. Giving up scientific study, Steno became a priest (1675) and then a titular bishop (1677).

Bibliography: S. Gould, "The Titular Bishop of Titioplis," *Hen's Teeth and Horse's Toes: Further Reflections on Natural History*, 1983; R. Laudan, *From Mineralogy to Geology: The Foundation of Science, 1650–1830*, 1987.

William J. McPeak

STILLINGFLEET, EDWARD (1635–1699). Born on17 April 1635 at Cranborne and educated at Cambridge, Stillingfleet met with early and continued academic and ecclesiastical success. He earned his B.A. in 1652 at St. John's college, Cambridge, and a fellowship there in 1653; he later took an M.A. in 1656, a B.D. in 1663, and a D.D. in 1668. In the mid 1650s, he served as tutor first in the house of Sir Roger Burgoyne in Warwickshire and then in the home of the Honorable Francis Pierrepont in Nottingham. In 1657, Stillingfleet became rector of Sutton in Bedfordshire, the gift of Burgoyne. In 1661, Stillingfleet published, or possibly republished, *Irenicum*, which proposes some common ground, and thereby compromise, between Episcopacy and Presbyterianism. In 1663 came *Origines Sacrae* and in 1664, *A Rational Account of the Christian Faith as to the Truth and Divine Authority of the Scriptures*, a work that, in discussing Old Testament figures, clearly demonstrated Stillingfleet's learning and logic. In 1665, Humphrey Henchley, the bishop of London, retained Stillingfleet to respond to the Jesuit version of the Laud-Fisher controversy. The same year, thanks to the earl of Southampton, Stillingfleet became rector of St. Andrew's, Holborn, and preacher at the Rolls Chapel. In 1667, he became prebendary at St. Paul's and in 1668, chaplain to *Charles II. The year 1669 found Stillingfleet as canon of Canterbury; by 1677, he had become archdeacon of London and a year later, dean of St. Paul's.

The Unreasonableness of Separation (1681–1682) seemed to mark a shift from the tolerance of the early *Irenicum*, but responsible opposition brought

Stillingfleet's return to his earlier, tolerant position. Stillingfleet's 1685 *Origines Brittanicae, or the Antiquities of the British Church* reexamined the sources of British ecclesiastical history. In 1689, he became bishop of Worcester and stood with six other bishops in opposition to *James II's Declaration of Indulgence. When revision of the *Book of Common Prayer* was proposed, Bishop Stillingfleet was a member of the commission. In 1695, Stillingfleet was called on to mediate a dispute regarding Antinomianism among the nonconformists. In 1697, he published *A Discourse on the Vindication of the Doctrine of the Trinity* and, in a related tract, *The Bishop of Worcester's Answer to Mr. Locke's Letter*, rebutted *John Locke's criticism of the doctrine of the Trinity. In sum, Stillingfleet spent his life as scholar and teacher, counselor and conciliator. He died at Westminster on 27 March 1699. Robert Harley, later earl of Oxford, purchased his library and Narcissus Marsh, archbishop of Armagh, his manuscripts.

Bibliography: R. Carroll, *The Common-Sense Philosophy of Religion of Bishop Edward Stillingfleet, 1635–1699*, 1975.

Martha Oberle

STRADIVARI, ANTONIO (1648–1737). Antonio Stradivari is considered the greatest maker of violins. His instruments have never been surpassed in quality of tone and beauty of design and workmanship. Stradivari was probably born in Cremona, where he spent his life and developed his reputation, and where he died on 18 December 1737. Scholarly consensus is that he studied his craft under *Nicolo Amati (1596–1684), an outstanding member of the Cremonese family, which had been known for making excellent stringed instruments since the sixteenth century. Sometime before Amati's death, probably between 1660 and 1680, Stradivari is believed to have become the manager of the Amati workshop. He soon began to make design modifications, which made his instruments more powerful and brilliant in tone than the instruments made by the Amati, and by the eighteenth century, Stradivari's instruments had surpassed those of the Amati in popularity. In 1667, Stradivari married Francesca Feraboschi. Two of their six children, Francesco (1671–1743) and Omobono (1679–1742), became violin makers in the shop of their father.

Although Stradivari probably made more than 1,000 instruments, only about 650 of them are extant. These include violins, violas, cellos, and guitars. Although scholars do not agree on exact dates, they usually divide Stradivari's career into several periods. In the early period extending from about 1660 through about 1690, Stradivari's instruments show the influence of the Amati in their measurements, arching, corners, sound holes, and scrolls. After 1690, Stradivari's instruments show changes in style developed by him, including a longer length and a new, deeper-colored varnish. The violins of this period are known as "Long Strads." The period from 1700 through 1720 is generally known as the "golden period" of Stradivari's designs. Several outstanding instruments from this period survive, including the violins known as the "Betts," the "Alard," the "Medici," and the "Messiah," as well as the cello known as the "Duport." All of these instruments exhibit great richness and brilliance of tone, as well as responsiveness. Many instruments made by Stradivari in his later career also display an orange-brown varnish, which is believed to be an important factor in their tone production. From 1720 through 1727, Stradivari made some minor design changes, which further improved the tone of his instruments.

During the last ten years of his life, some instruments with the labels of Antonio Stradivari may have been made by his sons Francesco and Omobono as well as by his pupils. Recent scholarly work on Stradivari has focused on the search for the "secrets" of his designs and varnish. This has resulted in the use of music-historical and scientific methods to investigate the composition and workmanship of instruments in order to authenticate and date them, and to provide data from which modern instrument makers may work.

Bibliography: C. Bonetti et al., *Antonio Stradivari: Reports and Documents*, 1999; W. Henley, *Antonio Stradivari*, 1961; S. Sacconi, *The Secrets of Stradivari*, 1979.

Maria Archetto

SUÁREZ, FRANCISCO (1548–1617). Suárez, a leader among the second generation of Jesuits, was an architect of Counter Reformation Scholasticism and a catalyst in the acceptance of Thomas Aquinas as a normative thinker in the Roman Catholic tradition. Born in Granada, Suárez was trained in law at Salamanca before entering the Society of Jesus in 1564, and thereafter studied philosophy and theology. He became professor of theology at Valladolid in 1576 and wrote a commentary on the first part of the *Summa Theologiae* of Thomas Aquinas. Retiring from that position in 1580, Suárez moved to Rome, where he completed his *Summa* commentary. Leaving Rome for Alcalá for health reasons, Suárez composed a number of theological works, many Christological in theme (among them *De verbo incarnato* [1590] and *De mysteriis vitae Christi* [1592]). Some of these works brought him under attack and scrutiny, sometimes from members of his own order, sometimes from the Inquisition. Prohibited by Pope Clement VIII (1592–1605) from writing about confession, Suárez's authority was restored by Pope Paul V (1605–1621). Treatises on the Trinity, grace, and faith followed. In addition to theological works, Suárez wrote a number of philosophical treatises; substantially Thomistic in character, they nevertheless go beyond Aquinas on several topics and take on a number of Thomas's later critics. Suárez's legal thought is the most innovative and "modern" strain in his philosophical work.

Bibliography: J. Mora, "Suárez and Modern Philosophy," *Journal of the History of Ideas* 14 (1953): 528–547; F. Suárez, *Selections from Three Works of Francisco Suárez*, trans. G.L. Williams et al., 1944.

Ralph Keen

SWAMMERDAM, JAN (1637–1680). Jan Swammerdam was a Dutch biologist who is often considered a founder of both comparative anatomy and entomology. He was also a brilliant microscopist. Swammerdam was born in Amsterdam, the Netherlands. He graduated with a medical degree from Leiden University in 1667 but never practiced medicine. Instead, he pursued his interests in natural history. His father, an apothecary, wanted the younger Swammerdam to become a priest and withdrew his financial support when his son continued to study biology. Jan Swammerdam suffered severe privations and eventually became both physically and mentally ill. In 1763, he became deeply embroiled in religious controversy, which lasted until he died.

Swammerdam contributed to many areas of biology, particularly to entomology. He accurately described and illustrated the life cycles of bees, mayflies, and dragonflies. From observations of metamorphic development, Swammerdam also classified insects into four major groups, three of which, in modified form, are still used in insect taxonomy. Futhermore, he refuted many false beliefs about insects, including the belief that insect bodies are structureless, fluid-filled cavities without fully formed internal organs.

Swammerdam also investigated vertebrates and provided much new knowledge, most of which was subsequently shown to be correct. He showed that the lungs of newly born mammals sink in water if the lungs are taken before breathing has started, although the lungs taken from young mammals after respiration has been established will float. He also demonstrated that muscle removed from a frog could be stimulated to contract but that when muscle (including heart muscle) contracts, it does not increase in volume. In addition, Swammerdam was probably one of the first scientists to anticipate the role of oxygen in respiration by postulating that a volatile element could pass from the lungs to the heart and then to the muscles. Swammerdam was also probably the first to discover red blood cells when he observed oval particles in the blood of frogs.

Swammerdam also worked on human and mammalian anatomy. He discovered the valves in the lymphatic system that still bear his name: "Swammerdam valves." He was one of the first biologists to examine the human reproductive system and to argue that female mammals produce eggs analogous to birds' eggs. Swammerdam's work was not published in full until 1737, when Herman Boerhaave published *Biblia Naturae* (*Bible of Nature*), a two-volume Latin translation of Swammerdam's Dutch text that included illustrations engraved from Swammerdam's own drawings, many of which remain unsurpassed.

Bibliography: R. Porter, *The Biographical Dictionary of Scientists*, 2nd ed., 1994; E. Ruestow, *The Microscope in the Dutch Republic: The Shaping of Discovery*, 1996.

                                                                              Paul C.L. Tang

SWEELINCK, JAN PIETERSZOON (1562–1621). The Netherlands musician Jan Pieterszoon Sweelinck was a prominent composer, organist, and teacher. Born into a musical family in Deventer, he received his early training from his father and later from a music master in Haarlem. Sweelinck became organist (possibly as early as 1577) at the Oude Kerk in Amsterdam, a post held by his father and, later, by Sweelinck's own son Dirck (1591–1652). There Sweelinck's musical duties were limited because of the Calvinist prohibition against use of the organ during worship. However, his playing was prized by the city of Amsterdam, his actual employer, and he became famous for his improvisatory skill. He held this position until his death. Sweelinck was teacher to the finest organists of his day. He is considered the father of the north German school of organ playing, which led directly to J.S. Bach. Sweelinck's keyboard style was influenced both by Italian toccatas and English virginal fantasias. His own contrapuntally rich and imitative keyboard Fantasias and Toccatas predict the later keyboard fugue. Vocal works are most numerous among his output and also include secular songs (chansons and madrigals). Psalm settings (to French texts) occupied him throughout his life. His *Cantiones sacrae* (1619) were Catholic liturgical motets; possibly these demonstrate that Sweelinck held to Catholicism even under the Calvinists.

Bibliography: F. Noske, *Sweelinck*, 1988.

Joyce Lindorff

SYDENHAM, THOMAS (1624–1693). Thomas Sydenham, often called "The English Hippocrates" because of his careful clinical observations, was the foremost clinical physician of his time. He was born into a family of devout Puritans and baptized in Wynford Eagle, Dorset, England. At the age of seventeen, he entered Magdalen College, Oxford, but left after only two months in order to join his father and his brothers who were fighting in the parliamentary army against *King Charles I. During this period of the English Civil War (1642–1649), his mother and two of his brothers were killed.

In 1647, Sydenham returned to Oxford, receiving his Bachelor of Medicine degree in 1648. In 1651, he left Oxford to join the army and, as a Captain of Horse, fought with *Oliver Cromwell's army in Scotland. He subsequently returned to Oxford where he and *Robert Boyle became friends. Boyle did much to further Sydenham's career as a physician. Sydenham married Mary Gee in 1655 and settled in London, where he established a private practice. Shortly after *Charles II restored the monarchy in 1660, and with support from Boyle, he began his clinical study of the London epidemics. In order to establish his practice as a physician in London, Sydenham took the examinations to become a licentiate of the Royal College of Physicians there.

When the plague broke out in London in 1665, he fled the city with his wife and three sons into the countryside. He was able to finish his five-year study of the London epidemics, published in 1666 as *Methodus curandi febres (Methods of curing fevers)*, which he dedicated to Boyle. In 1676, Sydenham received an M.D. degree from Pembroke College, Cambridge, and in the same year published *Observationes medicae (Medical observations)*. Among his medical students were *John Locke, the noted empiricist philosopher and political theorist, and Hans Sloane, whose collections later formed the basis of the British Museum.

Sydenham was well known for his sometimes nasty arguments with other leading physicians of the day, such as Thomas Willis, concerning the various treatments of diseases. Some of Sydenham's irritability—even bad temper—may have been due to the painful gout that he had suffered from since the age of thirty. In 1683, he published *Tractatus de podagra et hydrope* (*Treatise on gout and dropsy*) in which he vividly described the many remedies with which he had experimented. Sydenham died in London in 1689; his last book, *Processus integri in morbius fere omnibus curandis* (*The complete method of curing almost all diseases*) was published posthumously in 1694.

Sydenham's general approach to knowledge was heavily influenced by his Puritan faith, which embodied the principle that increasing useful knowledge was a religious duty and that experience was the best teacher. He followed Hippocrates and Francis Bacon by making careful and systematic general histories of the diseases that he encountered while caring for the sick in London hospitals; he wanted to classify these illnesses as one would classify plants. Contrary to so many of his contemporaries, Sydenham believed that diseases were identifiable as collections of symptoms that could be sorted into categories using close clinical observations. He emphasized the value of bedside observation and integrated his personal observations of patients with seasonal and atmospheric changes in order to draw up a detailed classification of fevers. Previously, for example, fevers associated with smallpox, measles, and typhus had only been described by such general and vague terms as "continued," "intermittent," and "eruptive." However, over a period of fourteen years, Sydenham used case histories to build more precise disease histories and taxonomies, carefully noting the prevailing maladies of each season in London. From his data, Sydenham concluded that it was the differing epidemic constitutions of the atmosphere that caused certain outbreaks of fever. This "ontogenic" concept of disease became very influential. Because of his publications, later physicians became more precise in distinguishing between different diseases with similar symptoms.

Bibliography: K. Drewhurst, *Dr. Thomas Sydenham (1624–1689): His Life and Original Writings*, 1966; D. Porter and R. Porter, *Patient's Progress; Sickness, Health and Medical Care in England, 1650–1850*, 1988.

Paul C.L. Tang

T

TASMAN, ABEL JANSZOON (ca.1603–ca.1659). Abel Janszoon Tasman was born in Lutjegast, Groningen Province, the Netherlands, probably in 1603. Little is known of his life before 1632. In either that year or the next, he arrived in Batavia (modern Jakarta) in Java, now part of Indonesia, and entered the service of the Dutch East India Company. Soon, he was leading voyages of exploration to various places in the Southwest Pacific and Indian Ocean. His first trip was made in 1634 as captain of the *Mocha*. He went east to Ceram (now Seram), an island lying just to the west of Irian Jaya (Western New Guinea) in the Molluccas. Over the next decade, Tasman made many trading voyages to Japan, Formosa (now Taiwan), Cambodia, and around the Dutch East Indies (now Indonesia). He briefly returned to the Netherlands in 1637. In 1639, under the command of Mathijs Quast, he went to the Northern Pacific, seeking gold and silver on islands east of Japan. During this same period, other Dutch explorers explored sections of the western coast of Australia, but the interrelationships of the land masses in the Southwest Pacific were still imperfectly understood. Tasman was directed to explore the Indian Ocean, then sail east to determine whether a sea route to Chile was feasible. He was also directed to visit the Solomon Islands, originally discovered by the Spanish, and New Guinea. The company also wanted him to find, if possible, suitable new trading partners in the region.

On 14 August 1642, in command of two ships, the *Heemskerk* and the *Zeehaen*, Tasman departed Batavia and sailed to Mauritius, off the southeast coast of Madagascar, thence south and east along latitude 44° South. At the beginning of December 1642, Tasman anchored off a large island, now Tasmania, which he named Van Diemen's Land after the Dutch governor-general of the Dutch East Indies, Anthony van Diemen. Continuing east, Tasman next encountered the northern coast of South Island, New Zealand. He continued to the northeast, entering what is now Cook Strait, then continued up the western coast of North Island to North Cape. Tasman concluded that this territory might somehow be connected to Staten Island (Staten Island), which is located east of the island of

Tierra del Fuego at the tip of Chile and Argentina. Believing that he had found the desired route to Chile, Tasman continued to the north, discovering Tonga on 21 January 1643 and Fiji on 6 February 1643; he then returned to Batavia via the Solomons and New Guinea.

Having traveled all the way around Australia without actually making landfall there, Tasman demonstrated that Australia was not part of some much more massive southern continent, as some authorities had suggested. But his dissatisfied superiors directed him to try again and find out how New Guinea related to Western Australia, Van Diemen's Land, and the alleged "South Land." On 29 February 1644, Tasman sailed east from Batavia, crossed to Cape York, entered the Gulf of Carpentaria, then turned west, hugging the often dangerous Australian coast until he reached Dirk Hartog Island, whereupon he returned to Batavia. Although much of what he had seen was new to the Dutch, company officials were still displeased with Tasman. Clearly, the natives he had encountered were not suitable trading partners, and he had not penetrated into the Australian interior. Hence, disappointed company officials published no account of his expeditions. Nevertheless, he was made a commander and member of the colony's Council of Justice. Tasman made two more voyages for the company, one to Thailand and the other an attack on the Spanish-held Philippines. He retired in 1653 and died in Batavia, probably in 1659.

Bibliography: G. Henderson, *The Discoverers of the Fiji Islands: Tasman, Cook, Bligh, Wilson, Bellinghausen*, 1933; A. Sharp, *The Voyages of Abel Janszoon Tasman*, 1968.

Keir B. Sterling

TATE, NAHUM (1652–1715). Tate, born in Dublin, was the son of a dissenting minister who was the pastor of Ballyhaise and known as "Faithful Teate." After Tate received his B.A. in 1672, he traveled to London and dropped the internal "e" from his name. He began his career as a playwright with *Brutus of Alba or The Enchanted Lovers* (1678). Then followed *Loyal General* (1680), *The Sicilian Usurper* (1681), *History of King Lear* (1681), *Duke and No Duke* (1685), and a libretto for *Purcell's *Dido and Aeneas* (1689). Tate was known for his adaptations of Elizabethan dramatists, especially Shakespeare. Although *The Sicilian Usurper or the History of Richard II*—which depicted Richard sympathetically and Bolingbroke critically—played only once, his *Lear* (minus the Fool, with Cordelia married to Edgar and Lear given back most of his kingdom) was popular for the next half-century. His revision of *Coriolanus as The Ingratitude of a Commonwealth* (1681) heightened the play's violence. Tate was also commissioned by *Dryden to complete the second part of Dryden's anti-Whig satire *Absalom and Achitophel* (1682), with Dryden adding about 200 of his own lines. His most notable poem was *Panacea, A Poem on Tea* (1700); he also translated Ovid. He was named poet laureate after Thomas Shadwell in 1692. In 1696, he published *A New Version of the Psalms* with Nicholas Brady, containing the noted Christmas hymn "While Shepherds watched their flocks by night." He became historiographer-royal in 1702. Tate died in London on 30 July 1715, but he gained lasting notoriety as a butt of Pope's satire in *The Dunciad* (1728).

Bibliography: C. Spencer, *Nahum Tate*, 1972.

Christopher Baker

TAYLOR, EDWARD (?1642–1729). Born in rural Sketchley, Leicestershire, Edward Taylor emigrated to the Massachusetts Bay Colony after possibly attending Cambridge University for a short time in his twenties. He eventually earned a degree from Harvard in 1671, some thirty-five years after it had been established to educate young men such as he in "godliness." He accepted a Congregational pastorate in Westfield on the then frontier of western Massachusetts, 100 miles to the west of Boston and far from the more educated settlements of the East coast. However, he did not lose contact with such notables as Samuel Sewell and *Increase Mather (who had been president of Harvard at the time he matriculated). He married Elizabeth Fitch in 1674 and then Ruth Wyllys in 1693, three years after Fitch's death. He had a total of thirteen children, most of whom predeceased him. After an illness of several years, Taylor died on 24 June 1729.

Taylor had to wait two centuries for his writing to be discovered and appreciated; he is now regarded as the foremost poet of colonial America. For reasons still unclear, all but a few stanzas of his poetry remained unpublished until 1939, while the bulk of his sermons have been printed only within the last forty years. The two major groups of Taylor's poetry are *God's Determinations Touching His Elect*, a poetic sequence setting forth the Christian's tension between the poles of sin and redemption, and the approximately 200 *Sacramental Meditations*, written between 1682 and 1725, which Taylor termed his "Preparatory Meditations Before My Approach to the Lord's Supper." Taylor's poetry, clearly in the metaphysical tradition, shares with the work of *John Donne and *George Herbert a love of paradox, conversational spontaneity, unconventional diction ("tweedle," "Bemidnighted," "Bubs"), arrestingly unconventional metaphors ("Who in this bowling alley bowled the sun?"), and an allusive style drawing upon biblical, literary, scientific, and historical sources, which reveal his roots in a Caroline Christian humanism as well as in a dissenting sensibility.

Bibliography: T. Davis, *A Reading of Edward Taylor*, 1992; J. Hammond, *Edward Taylor: Fifty of Years of Scholarship and Criticism*, 1993.

Christopher Baker

TAYLOR, JEREMY (1613–1667). Born in Petty Cury, the son of Nathanial and Mary Taylor, Jeremy Taylor was baptized 15 August 1613 in Cambridge, where his father was a barber. Educated at King's College School, the Perse School, and Gonville and Caius College, Cambridge, he received the M.A. and ordination in 1633. Taylor's precosity attracted the attention of *Archbishop

Laud, who made him M.A. at University College, Oxford (October 1635), fellow at All Souls, Oxford (January 1636), and sometime later, chaplain to *Charles I. Taylor became rector of Uppingham, Rutlanshire (1638), where he married Phoebe Langsdale (1639). Pleased with Taylor's defense of episcopacy in *Episcopacy Asserted* (1642), the King bestowed upon him the Doctor of Divinity from Oxford and the sinecure living at Overstone (1643). Taylor supported the royalist cause during and after the civil war and was imprisoned three times during the interregnum. Although the exact date in relation to Charles's defeat remains disputed, Taylor went to Wales, helped establish a school at Newton Hall, served as domestic chaplain to Lord Carbery at Golden Grove, and produced an impressive body of prose works. Taylor's anonymous *A Discourse Concerning Prayer Extemporary*, republished under his name as *An Apology for Authorized and Set Forms of Liturgie* (1649), opposed the *Directory of Public Worship* with which Parliament had replaced *The Book of Common Prayer* in 1645. *A Discourse of the Liberty of Prophesying* (1647) supported the exercise of right reason in matters of interpretation and urged toleration for differing religious opinions. In addition to treatises on various doctrinal matters, Taylor's *Eniautos, A Course of Sermons for All the Sundays of the Year* (1653) and his devotional works, *The Great Exemplar* (1649), *The Rule and Exercises of Holy Living* (1650), and *The Rule and Exercises of Holy Dying* (1652), also found an appreciative audience among disenfranchised Anglicans.

Taylor's wife having died in 1651, he married Joanna Bridges (1655) and lived at Mandinan for the next two years. Ironically, a chapter on Original Sin in his *Unum Necessarium* (1655), intended to help Anglicans reconcile the practice of repentance with the belief in total depravity, infuriated orthodox Anglican leaders, some of whom branded him dangerous and prevented his receiving ecclesiastical preferments in England. When Taylor returned to London (1657) to minister secretly to royalist congregations, the controversy continued. Having no other options, he moved to Northern Ireland (1658), where he accepted a part-time lectureship in Lisnagarvey (Lisburn) parish and a chaplaincy to Lord Conway at nearby Portmore. Despite its theological strengths, *Ductor Dubitantium* (1660), which he dedicated to the newly restored *Charles II, gained Taylor only undesirable appointments in Northern Ireland as bishop of Down and Connor and vice-chancellor of the University of Dublin, and in 1661, as administrator but not bishop of the Dromore diocese and member of the Irish Privy Council. Although successfully reorganizing the University of Dublin, he met fierce diocesan opposition from Presbyterians and Catholics, who challenged his authority and disrupted all efforts to secure Anglican hegemony. Among his last prose works, *Rules and Advices to the Clergy of the Diocese of Down and Connor* (1661) and *Dissuasive from Popery* (1664) reflect the frustrated efforts of his beleaguered Anglican ministry in Northern Ireland. Denied permission to leave in 1664, he died in Ireland 13 August 1667 and was survived by his wife and three daughters.

Taylor's ability to voice a moderate, conciliatory approach to major theological concerns of the day stands as one of the hallmarks of his works even though some of them earned the hostility of certain Anglican clergy. He has been called the best prose writer of his time. Cited for its qualities of lyricism and its rich and multileveled imagery, his style, particularly in his sermons and devotional writing, often reflects the harmony and beauty he desired for the religious life. Certainly *Holy Living* and *Holy Dying*, widely popular in his own day, remained so in subsequent centuries, and some twentieth-century writers deemed *Liberty of Prophesying* on a par with Milton's *Aeropagitica* (1644) among masterworks of intellectual freedom.

Bibliography: R. Heber, ed., *Jeremy Taylor: The Whole Works*, 10 vols., rev. ed., 1969; F. Huntley, *Jeremy Taylor and the Great Rebellion*, 1970.

Sallye Sheppeard

TEMPLE, SIR WILLIAM (1628–1699). Eldest son of an estimable couple, Sir John Temple, master of the rolls in Ireland, and Mary, daughter of John Hammond, M.D., and sister of the renowned divine, Dr. Henry Hammond, Temple's life and influence came to exceed his family's. Reared mostly by his uncle, Dr. Hammond, Temple was educated at the Bishop Stortford School and then, without taking a degree, Emmanuel College, Cambridge. Expressing athletic interests more than academic, Temple left Cambridge for the Continent, on the way to which he met and fell in love with Dorothy Osborne on the Isle of Wight. Since the nineteenth century, the letters of Dorothy Osborne to Temple have drawn more attention than Temple's work, but before the nineteenth century and the defamation of Temple's character accomplished by Thomas Babington Macaulay, Temple's reputation as a diplomat and man of integrity was as admired as his work was influential.

Temple's reputation for integrity, courtesy, and incomparable civility was borne out by several major events. First, the seven-year romance he conducted with the royalist Dorothy Osborne suggested commitment and cultivation. If the literary results of the courtship are any indication, the two lovers were ardent and distracted by their love. Osborne's letters and Temple's romances testify to the attention each directed to the courtship, and such attention would not go unnoticed by others. During the courtship, 1648–1655, Temple traveled extensively on the Continent, learning French and Spanish. Between the romantic courtship and the second major event contributing to Temple's reputation—the negotiation of the Triple Alliance—Temple returned to Ireland. In 1668, he successfully and famously negotiated, mostly through the force of his personality and with unprecedented speed, the treaty among England, Sweden, and the Netherlands against *Louis XIV of France, known as the Triple Alliance, which ended the War of Devolution.

Though considered by many to be greatly serving the Protestant state, *Charles II was actually negotiating a counteragreement simultaneously with Louis XIV, who required Temple's recall in 1670. After attempting a retirement at Sheen, Temple agreed to a further diplomatic mission in 1674. This mission resulted in the treaty of Westminster but also, more importantly for Temple and for England, the marriage of Prince William of Orange (later *William III) and Princess Mary, Charles II's niece and Protestant daughter of the future *James II. Temple was intimate with William, while Lady Temple was intimate with Mary's noble guardian, the Lady Villiers, a combination that seems to have promoted the union and certainly made the Temples confidants of William and Mary after the revolution of 1688, the period of Temple's unretractable, final retirement. Before this period, though, from his not insignificant disgrace early in 1681 to his death on 27 January 1699, Temple effected one further major political achievement: the reinstitution of a reformulated Privy Council in April of 1679, meant to bridge the gap between the courts of the Parliament and the King. Though this feat was short-lived for Temple and the source of his final disappointment with and humiliation by Charles II, specifically, and politics, generally, it is important for its effect on the country, Temple, and his contemporaneous and subsequent reputation.

Such a man naturally drew the attention of readers in his day, so that even his manuscript "Essay on the Present State and Settlement of Ireland" (composed in 1667–1668) would be talked about, especially when his *Essay on the Original and Nature of Government* was published in 1671. The latter is remarkable for anticipating in print many of the key arguments of *Sir Robert Filmer's *Patriarcha* published in 1679. Other lesser though notable works, such as the gracefully executed "Letter to the Countess of Essex" (1672), *An Introduction to the History of England* (1695), and the *Poems of Sir W.T.* (privately printed, no date), remain as notable today as they were in their own time. Other interesting products of the author were *Observations upon the United Provinces of the Netherlands* (1672 and several subsequent printings and translations), *Miscellanea in Two Parts*, the second more remarkable than the first (first part, 1680; second part 1692; two parts together 1693; several further editions), the *Memoirs of What Past in Christendom, from the War Begun 1672 to the Peace Concluded 1679* (usually referred to as *Memoirs, Part II*, 1691) and the *Memoirs, the Third Part. From the Peace Concluded 1679 to the Time of the Author's Retirement from Public Business*, edited by Jonathan Swift (1709; the first part of the memoirs was burnt by Temple). The *Observations upon the United Provinces* and the *Memoirs* were especially keenly read in Temple's day as they gave firsthand testimony to the personalities and actions of many key political figures. Especially adept at writing "characters," the seventeenth-century penchant for detailed description of actual individuals and a genre combining features of modern history and biography, part II of the *Memoirs* was considered his best, though today part I's treatment of the exclusion crisis and its nature as political memoir (and perhaps the first such) is more interesting to those concerned more with England and its literature than with English-Continental relations.

Most influential, though, then as now, is the *Miscellanea* especially the second volume, which includes Temple's best writing, his essays on gardens, heroic virtue, poetry, and most famous, ancient and modern learning. These essays are marked by clarity of expression, wide-ranging and elegantly integrated associations, and promotion of expansive, skeptical views of knowledge or "science" and humanity. Like his political achievements, his essays promote civilized cultivation of individuals in the context of a society more spiritually or mentally comfortable than materially saturated, especially if the material gains are at the exploitative expense of other peoples. In the essay "Upon the Gardens of Epicurus; or, of Gardening, in the Year 1685," Temple surveys the world and the Bible to indicate how the work and enjoyment of the garden is the perfect interrelation between intelligent, feeling humanity and animate, divine nature: "For this reason," Temple explains, "Epicurus passed his life wholly in his garden; there he studied, there he exercised, there he taught his philosophy; and, indeed, no other sort of abode seems to contribute so much to both the tranquillity of mind and indolence of body which he made his chief ends" (10). In the ill-fated essay "An Essay upon the Ancient and Modern Learning," which is most remembered today for the vituperative glee it drew from its detractors as they communicated their detection of certain errors of attribution and dating, Temple advances his humanistic and cyclical historiographical and political assumptions: "Science and arts have run their circles and had their periods in the several parts of the world; they are generally agreed to have held their course from east to west, to have begun in Chaldaea and Egypt, to have been transplanted from thence to Greece, from Greece to Rome; to have sunk there, and, after many ages, to have revived from those ashes, and to have sprung up again both in Italy and other more western provinces of Europe" (52).

Progress is not Temple's organizational concept for history; rather, whatever condition creates the advanced possibilities for the individual tranquility of mind and indolence of body forms Temple's organizational concept for tracing the high and low points of history. This cosmopolitan attitude was and is despised by positivistic historiographers, in the arts as much as in the sciences, and may be Temple's most enduring influence. In contrast to his modernist detractors, Temple aligned himself solidly with Alphonsus, King of Aragon, whom he quotes at the close of the essay on "Ancient and Modern Learning"; "That among so many things as are by men possessed or pursued in the course of their lives, all the rest are baubles, besides old wood to burn, old wine to drink, old friends to converse with, and old books to read" (71). Swift's notation in his journal of Temple's death is a fitting close: "He died at one o'clock this morning (27th January 1698–1699), and with him all that was good and amiable among men."

Bibliography: S. Monk, ed., *Five Miscellaneous Essays by Sir William Temple*, 1963; H. Woodbridge, *Sir William Temple: The Man and His Work*, 1940.

Jesse G. Swan

THOU, JACQUES AUGUSTE DE (1553–1617). Born in Paris 8 October 1553, the nephew of Nicolas, bishop of Chartres and the son of Christophe, first president of the Parlement of Paris, Jacques de Thou grew up in a family attuned to the religious and political questions of the day, a background that equipped him to become an important historian and magistrate. In 1570, Jacques de Thou began his study of law at Orléans, moved to Bourges to continue, and thence to Valence. When Nicolas de Thou was raised to the episcopate, Jacques succeeded his uncle as canon of Notre Dame. Finding that his talent was not for clerical life, young de Thou began to move among influential secular circles. In 1572–1576, he accompanied Paul de Foix to the various courts of Italy, and in 1578, de Thou entered Parlement. In 1582, he began a trip through southern France during which he met Montaigne and Henry of Navarre, the eventual Henri IV of France. In 1586, de Thou became president de mortier of the Parlement of Paris and in 1588, councillor of state. However, de Thou had not left either the world of books or of religion. In the 1570s he had begun a collection of books, which led, in 1587, to the establishment of a library with Pierre and Jacques Dupuy as librarians.

Tolerant in his Catholicism and loyal to France, de Thou aided in the formulation of the 1598 Edict of Nantes, which granted equality to the Huguenots and held with the Gallican Church against the decrees of the Council of Trent. In 1593, using his own books and his continuing correspondence with scholars throughout Europe, de Thou began *Historia sui temporis* (*The History of his Own Times*), which he chose to write in Latin, a dead language, to avoid implications or connotations, positive or negative, embedded in the living vernacular. Although de Thou's desire was to produce an unbiased account, the first part of the work, dealing with the years 1545–1560 and published in 1604, drew some criticism, and the second part, which covered the years of the religious wars (1560–1572), made its way onto the Roman Index in 1609. The remaining parts of the work, published in 1607 and 1608, also provoked furious outcry; to address the criticism, de Thou wrote his *Memoirs*. After the death of Henri IV, de Thou continued to serve Marie de' Medici, the Queen Regent. In his later years, he became a member of the conseil de finances and assisted in negotiating the treaties of Ste. Menehould (1614) and Loudun (1616), dying on 7 May 1617 in Paris.

Bibliography: S. Kinser, *The Works of Jacques-Auguste de Thou*, 1966; A. Soman, *De Thou and the Index*, 1972.

Martha Oberle

TIRSO DE MOLINA (ca.1584–1648). "Tirso de Molina" was the pseudonym assumed in 1616 by Gabriel Téllez, a Mercedarian monk born in Madrid. A student of the University of Alcalá de Henares, Tirso entered the Mercedarian order in 1600 and professed in 1601. Between the years 1601 and 1614, he continued his studies in Guadalajara, Toledo, and Salamanca and lived briefly in Galicia and Portugal before settling in Toledo. In June 1616, Tirso accompanied members of his order to its province in Santo Domingo (Dominican

Republic), where he taught courses on theology and wrote several poems, all of which won prizes in poetry competitions. Upon his return to Spain in 1618, Tirso lectured in Segovia until 1620, the year in which he moved to his order's house in Madrid, which was the center of literary activity in Spain. While there he joined some of Spain's greatest literary figures: *Luis de Góngora (1561–1627), *Francisco de Quevedo (1580–1645), *Lope de Vega (1562–1635), and *Calderón de la Barca (1600–1681). In 1625, the Committee for Reform of the Council of Castile condemned Tirso for his writing of profane plays and demanded that he leave Madrid. The reason for this punishment, however, is more closely tied to Tirso's unflattering portrayal in his plays of the count-duke of Olivares, Philip IV's (1621–1665) royal favorite who ran the government. The Mercedarian Order assigned Tirso to its monastery in Trujillo, where he spent three years and then returned to Castile. In 1632, the order appointed Tirso its chronicler, a position that required him to return to Madrid. Tirso completed *The History of the Mercedarian Order* in 1639. While living in Madrid, four volumes of his collected plays were published. The vicar provincial of the order removed Tirso from his duties as chronicler and sent him to the monastery of Cuenca for writing satires against the government. The following year, Tirso moved from Cuenca to Toledo. Tirso died in February 1648 in Almazán, where he had retired after his last assignment in Soria.

Tirso's literary production consists of prose, poetry, and more than eighty works of drama, including four *autos sacramentales*. Best known for his plays, Tirso followed Lope de Vega's model of writing for popular appeal while eschewing the classical rules of drama. The greatest of Lope's disciples, Tirso surpassed Lope in his creation of characters. Tirso's most famous character is don Juan Tenorio, the legendary figure whom he brought to worldwide attention in *El burlador de Sevilla* (*The Trickster of Seville*, 1630), in which don Juan appears as a literary character for the first time. Another one of his innovations was the creation of intelligent, decisive, and spirited female characters. Tirso's male characters are timid and indecisive, often the victims of the female characters' cunning and daring behavior. His religious training and strict adherence to his vocation enabled him to produce the best religious plays of his day. Although Tirso is best known for his *comedias*, which are rich in psychology and symbolism, Tirso wrote several *autos sacramentales* (theological allegories of one act) and two prose miscellanies titled *Los cigarrales de Toledo* (*The Country Houses of Toledo*, 1624) and *Deleitar aprovechando* (*Pleasure with Profit*, 1635). Tirso's plays fall into different categories, including plays based on Biblical stories (*La venganza de Tamar*; *Tamar's Vengeance*, 1623); plays about saints' lives (*La santa Juana*; *Santa Juana*, 1614); plays of a philosophical and theological nature (*El burlador de Sevilla*; *The Trickster of Seville* and *El condenado por desconfinado*; *The Man of Little Faith*, both from 1630); and plays based on national themes (*La prudencia en la mujer*; *Prudence in Woman*, 1633). Tirso had a profound influence on writers of world renown, including *Molière (1622–1673), Lord Byron (1788–1824), and George Bernard Shaw

(1856–1950). The Spanish dramatist José Zorrilla (1817–1893) reintroduced don Juan to the Spanish stage in his play *Don Juan Tenorio*, 1844.

Bibliography: J. Albrecht, *Irony and Theatricality in Tirso de Molina*, 1994; M. Wilson, *Tirso de Molina*, 1977.

Michael J. McGrath

TOMPION, THOMAS (1639–1713). Tompion has been called "the father of English watchmaking." He is thought to have been born at Northhill in Bedfordshire; by 1664, he had become an apprentice to a London clockmaker, and he completed his formal apprenticeship on 4 September 1671. Within five years, the well-known quality of his work led to his being chosen as clockmaker for the Royal Observatory at Greenwich. He made the precision parts for the ten-foot quadrant located there, which *Robert Hooke had designed for the research of *John Flamsteed. About 1660, Tompion also worked with Hooke on the design of an improved balance spring, which was part of a watch presented to *Charles II. This watch was the subject of close scrutiny by others, as the search for a patentable balance spring to improve timekeeping had led to a competition between Hooke and Tompion's device and a competing design by *Christiaan Huygens; even Flamsteed expressed irritation at not being allowed to examine the workings of the Hooke-Tompion watch, whose cover Tompion had rendered unopenable! Tompion constructed four clocks for the observatory, two with thirteen-foot pendulums for greater precision. He enabled the construction of more compact pocket watches by inventing the cylinder escapement in 1695 and also constructed a barometer and sundial for monarchs *William and Mary at Hampton Court. It was reported in 1700 that Tompion planned to construct a clock for St. Paul's Cathedral, which would run for a century without rewinding, but this project was never completed. Several Tompion clocks are still in operation, such as his 1709 timepiece in the Pump Room at Bath. Tompion died on 20 November 1713 and was buried in Westminster Abbey. His career is an excellent example of the emerging role that expert "artificers" or precision technicians had begun to play in the growth of early modern science.

Bibliography: L. Jardine, *Ingenious Pursuits: Building the Scientific Revolution*, 1999; W. Milham, *Time & Timekeepers*, 1945.

Christopher Baker

TORRICELLI, EVANGELISTA (1608–1647). Italian mathematician and physicist Evangelista Torricelli was born and educated at the Jesuit college in Faenza, where his scientific aptitude led him to Rome in 1626. He studied the physics of motion under Benedetto Castelli (1577–1644), *Galileo's pupil and professor of mathematics at the Collegio di Sapienza. Castelli highly recommended him to Galileo, but Torricelli did not work with the great scientist until late in the year of his death (1641). Torricelli assumed Galileo's position as professor of mathematics in the Academy at Florence. He went on to solve the most difficult mathematical problems of the time, including the quadrature of the cycloid space (finding area and center of gravity of a cycloid) and the cubature of acute hyperbolic solids. He also ably defended Galileo's mechanics, developed concepts of momentum and impetus, and discovered the law of efflux of fluids.

Torricelli is best known for his study of air pressure and discovering the mercury barometer. Galileo believed that air had weight but could not explain why a siphon pump could not lift a column of water more than thirty-two feet, other than proving an atmospheric vacuum condition above the point. Torricelli reasoned that atmospheric weight provided a counterbalancing force to the rise of the water (1643). This was tested (1644) with the denser medium of "quicksilver" (mercury), requiring a much smaller tube. Sealing one end of a three-foot glass tube and filling it with mercury, its opened end was submerged in a small bath of mercury. The column of mercury stabilized at about twenty-eight inches by the external "heaviness of the air." The evacuated space at the top was a vacuum, the first sustained example, called the "Torricellian Vacuum" (a method of forming vacuums used today). The apparatus would become the mercury barometer, its applications providing one of the most practical group of scientific instruments. Torricelli developed other instruments and a technique for producing telescope lenses.

Bibliography: Faculty of Letters and Science of Nice, *The Works of Torricelli: Galilean Science and New Geometry*, 1987; P. Robinson, "Evangelista Torricelli," *Mathematical Gazette* 78 (1994): 37–47.

William J. McPeak

TOURNEUR, CYRIL (?1580–1626). Little is known of Cyril Tourneur's family. Born likely between 1575 and 1585, he may have been related to others in the service of Sir Francis Vere or Lord Marshall, Sir Edward Cecil, both of whom befriended and financially supported Tourneur. Tourneur's eventual receipt of an annuity from the United Provinces supports the claim that he served Vere and the Cecils during military campaigns and provided political service in the Netherlands. He left civil service around 1604 to begin a serious writing career.

Tourneur's first publication, a lengthy poem titled *The Transformed Metamorphosis* (1600), satirically depicts a world dominated by greedy and lustful men. Another work attributed to Tourneur, *Laugh and Lie Down; or, The World's Folly*, bearing only the initials "C.T.," appeared in 1600. Its satiric tone remains consistent with that of his other work, and thus most scholars agree on its attribution. Such agreement, however, no longer supports Tourneur's authorship of *The Revenger's Tragedy*, the drama that, ironically, secured his fame. Some now attribute the play, written and produced probably between 1606 and 1608, to Thomas Middleton, based on certain linguistic elements. In 1609, Tourneur wrote *A Funeral Poem upon the death of the Most Worthy and True Soldier, Sir Francis Vere*.

Tourneur retains undisputed authorship of *The Atheist's Tragedy*, performed in 1610 and entered on the Stationers' Register in 1611. Of autobiographical interest remains Tourneur's placing of his hero, Charlemont, at the siege of Ostend in the Netherlands, an action in reality led by Sir Francis Vere in 1604 during Tourneur's service. The drama's central character, D'Amville, labels himself an atheist, professing belief only in wealth and felicity. As D'Amville's foil, Charlemont represents valor, while his love interest, Castabella, symbolizes chaste affection of a spiritual sort. The movement of the play propels D'Amville through great personal loss resulting in suicide, while Charlemont's shaky spiritual faith strengthens, motivating him not to seek revenge upon D'Amville, an act resulting in the best type of revenge available. Critics find the play lacks the passion and vision of *The Revenger's Tragedy*. However, by offering an original treatment of the theme of revenge, it gained an important place in the revenge tragedy tradition.

Tourneur's *The Nobleman* remains lost, but likely enjoyed court perfomances on 23 February 1612 and again the following Christmas. Its recording by the Stationers' Company reads, "A play booke beinge a TrageComedye called The Noble man written by Cyrill Tourneur." That same year he wrote *The Character of Robert, Earl of Salisbury* and also *A Griefe on the Death of Prince Henrie*, included in 1613 with similar poems by *John Webster and *Thomas Heywood in *Three Elegies on the most lamented Death of Prince Henrie*. He may have later written an act in a lost play identified as *The Bellman of London*. In 1625, Tourneur accompanied Edward Cecil on his ill-fated Cadiz military expedition. Put ashore by Cecil along with many ill and wounded at Kinsale, Ireland, Tourneur died in February 1626.

Bibliography: S. Schuman, *Cyril Tourneur*, 1977.

Virginia Brackett

TRAHERNE, THOMAS (1637–1674). Born the son of a shoemaker in Hereford, England, in October 1637, Thomas Traherne wrote lyrical poems that were ideal precursors for Romantic poets such as Blake and Wordsworth. Traherne's devotional or mystical poetry is composed of standardized seventeenth-century metaphysical tropes such as conceit, dramatization, irony or paradox, catachrestic words and phrases, and spiritual catharsis. These Traherne used to express exegetical themes such as deism and the sensuality of nature, spiritual intuition, grace of innocence (childhood), and sincerity of devotion.

It is speculated that an elder Philip Traherne, twice mayor of Hereford, sponsored young Thomas and his brother Philip's education. Traherne entered Brasenose College, Oxford, in March 1652. He received his B.A. in 1656 and his M.A. at Brasenose by decree in 1661. In 1669, he received a Bachelor of Divinity degree, also from Brasenose. Supported by the Restoration, Traherne was ordained an Anglican priest on 20 October 1660. He was appointed rector of Credenhill in approximately 1661. Later that year, Traherne left Credenhill for a position in London as domestic chaplain to the Lord Keeper of the Great Seal, Sir Orlando Bridgeman. In 1672, Traherne accepted the appointment of minister at the Teddington Church. It is said he was buried under a reading desk at Teddington in October 1674.

The only book to be published during Traherne's lifetime was *Roman Forgeries* (1673), and this he printed anonymously. Published posthumously were *Christian Ethics* (1675) and a collection of rhythmical prose entitled *Thanksgiving* (1699). The latter work was first published anonymously by the Reverend George Hickes as *A Serious and Pathetical Contemplation of the Mercies of God, in Several Most Devout and Sublime Thanksgivings for the same.* Nathaniel Spinckes then published Traherne's *Meditations on the Six Days of the Creation* in 1717. In 1895, a Victorian bookseller named W.T. Brooke found a Traherne prose manuscript entitled *Centuries of Meditations* in a London bookstall; these were edited and published in 1908. The manuscript of *Poems of Felicity* was discovered in the British Museum and published in 1910. As late as 1964, James Osborn discovered a Traherne manuscript entitled *Select Meditations*. Although critics have suggested that Traherne's poetry is not as sophisticated as that of his contemporaries Vaughan and Herbert, it does possess a simple charm and lyrical quality as beautiful as that of his more renowned counterparts.

Bibliography: M. Day, *Thomas Traherne*, 1982; D. Dickson, *The Fountain of Living Waters: The Typology of the Waters of Life in Herbert, Vaughan, and Traherne*, 1987.

David M. Rosen

U–V

URBAN VIII, POPE (1568–1644). Maffeo Barberini was born into a noted Florentine family and was elected as Pope Urban VIII on 6 August 1623; he died on 29 July 1644. Urban received a Jesuit education as a youth, and he later earned a Doctor of Laws degree from the University of Pisa; he was also a recognized classical scholar and Latin poet. He rose in the church ranks through a variety of assignments, becoming nuncio to France 1604 (where he was a confidant of Henry IV) and was also named archbishop of Nazareth. In 1606, he became a cardinal and bishop of Spoleto two years after that. Seventeen years later, he was elected pope by a nearly unanimous vote. A strong advocate for missions, he permitted all religious orders—not just Jesuits—to evangelize in China and Japan. He forbade the enslavement of natives in Brazil, Paraguay, and the West Indies in his bull of 1639. Urban handed *Galileo his second papal censure, seventeen years after *Paul V had also condemned Galileo's advocacy of Copernicanism. Like other popes of his era, Urban felt it necessary to condemn as heresy *Cornelius Jansen's *Augustinus*; his bull to this effect ("In eminenti") was promulgated in 1642. His diplomatic initiatives included support for *Richelieu's anti-Hapsburg stance, but he also irritated the French minister by trying to block a French-Swedish alliance in 1631. He attempted to foster the Roman church in England through close contacts with Henrietta Maria, wife of *Charles I. Urban was a generous patron of *Bernini, whom he designated as architect of St. Peter's Basilica in 1629 and who also designed Urban's tomb. Nepotism, a not infrequent papal failing, tainted his pontificate also. After he removed the bronze beams from the Pantheon to melt them into cannons, it was said that "What the barbarians did not do, the Barberini did."

Bibliography: L. Pastor, *The History of the Popes from the Close of the Middle Ages*, vols. 28–29; 40 vols., 1891–1954.

Christopher Baker

URQUHART, SIR THOMAS (1611–1660). Urquhart's achievement of highest merit is his translation, published in 1653 and again in 1664, of the first two books of the work of Rabelais. The third book, translated by Urquhart but edited by Pierre Motteux, was published in 1693 by Motteux who then himself published, in 1693–1694, the final two books of Rabelais' works. Although the two books done by Motteux are entirely acceptable, those done by Urquhart are considered among the masterpieces of translation. A member of the ancient but financially strapped Scottish Urquhart family, Thomas was educated at Aberdeen, traveled through Europe, collected a substantial library, married a daughter of the fourth Lord Elphinstone, and found himself, in 1637, protected by *Charles I from creditors, who nevertheless did get Urquhart's books. On 14 May 1639, Urquhart took part in the "Trot of Turiff," a significant Royalist victory at Towie Barclay Castle near Turiff, Scotland; the result of his action was to be knighted by Charles I in 1641. However, in 1649, his support for Charles I caused Urquhart to be considered a traitor. Imprisoned in the Tower and at Windsor, Urquhart was released by *Cromwell's order. Urquhart's writings include a collection of *Epigrams*, a tract on logarithms published in 1645, and three unusual works published in 1652 and 1653. The first of the three is a genealogy of the house of Urquhart to the time of Adam; the second involves the search for a jewel, which turns out to be an Urquhart manuscript—perhaps lost after the rising at Inverness in 1649—together with a defense of a universal language and a eulogy of the Scottish character; the third deals once again with the idea of a universal language. A cavalier, Urquhart is thought to have gone into exile on the Continent before his death in 1660.

Bibliography: H. Brown, *Rabelais in English Literature*, 1933.

Martha Oberle

USSHER, JAMES (1581–1656). Scholar and Church of Ireland archbishop, James Ussher was born and educated in Dublin, at age thirteen becoming one of the first undergraduates at Trinity College, Dublin. He graduated B.A. (1597) and M.A (1600), and held the post of Professor of Theological Controversies (1607–1621). Strongly Calvinist, Ussher helped draft the Church of Ireland's articles of belief in 1615. He was named bishop of Meath in 1621, privy councilor in 1623, and archbishop of Armagh in 1625.

In sermons and print, Ussher repeatedly condemned Catholicism, particularly after a government order in 1603 granted Catholics free exercise of worship. Ussher also opposed Irish language religious instruction and the translation of the Bible into Irish. His *Discourse on the Religion Anciently Professed by the Irish and British* (1623) argued that the Church of Ireland was not a colonial imposition, but derived from a Celtic Christianity that was doctrinally Protestant. Never one to limit his theological targets, Ussher also criticized the "Arminians" (followers of *Jacob Arminius) within the Church of England for abandoning the doctrine of predestination.

Ussher's scholarly specialties were early church history and biblical chronology. He contributed to knowledge of St. Patrick and early practices of the Irish and British churches. His timelines, published as *Annalium Pars Prior* (1650) and *Annalium Pars Posterior* (1654), were collected posthumously in *The Annals of the World to the Beginning of the Emperor Vespasian's Reign* (1658). These works detail Ussher's famous calculation that Creation began at sundown on 22 October (thus, theologically, 23 October) 4004 B.C. (Old Style). Quaint though Ussher's assertion may now seem, it was based on meticulous scholarship in the best available sources, and not, as is sometimes misstated, merely on the "generations" in the book of Genesis.

Ussher had a wide circle of acquaintances and correspondents that included *Archbishop Laud, *Thomas Bodley, *William Camden, and *John Selden, and he frequently visited England. He did not return to Ireland after one such trip in 1640, becoming enmeshed in events leading to the civil war. He prepared a plan to reconcile the episcopal and presbyterian structures for the Long Parliament, witnessed the execution of Strafford, was offered a seat in the Westminster assembly of divines in 1643 and 1647, and as late as November 1648 was publicly advocating the divine right of kings. It is a mark of the respect accorded Ussher that when this loyal monarchist died in March 1656, *Cromwell ordered a public funeral in Westminster Abbey.

Bibliography: R. Knox, *James Ussher: Archbishop of Armagh*, 1967; H. Trevor-Roper, "James Ussher, Archbishop of Armagh," in *Catholics, Anglicans and Puritans*, 1989.

Michael R. Hutcheson

VANBRUGH, JOHN (1664–1726). Sir John Vanbrugh, a Restoration playwright, used his family connections to become the "Shakespeare of Architects." Vanbrugh was born in London in 1664 to Giles Vanbrugh of Flemish ancestry and his wife, Elizabeth, the daughter of Sir Dudley Carleton. The family moved to Chester in 1667, and little is known about Vanbrugh's early life. He returned to London in 1681 and subsequently came under the patronage of Robert Bertie. On the eve of the Glorious Revolution in 1688, he accompanied Bertie to Holland and subsequently to France where he was imprisoned. His travels had a direct impact on both his future literary and architectural works. While imprisoned in the Bastille, he outlined the comedy *The Provok'd Wife*, which with *The Relapse* established his reputation as a playwright. He also became familiar with the works of architect Louis Le Vau and other French Baroque architecture during this time. He returned to London in the 1690s and launched his dramatic career, producing ten plays in a decade.

His architectural career began around 1700 with no formal training. His first commission, the country house Castle Howard (1700–1712), came through his mother's family connections to the third earl of Carlisle. Carlisle also secured Vanbrugh's appointment as comptroller of the Royal Works, which introduced him to two major influences, *Sir Nicholas Hawksmoor and *Sir Christopher Wren. Vanbrugh was also named to the royal posts of Surveyor of Gardens and Water (1713–15) and Surveyor of Royal Hospital at Greenwich (1716). He was knighted by George II in 1714. Vanbrugh's work for Carlisle led to his most famous commission, Blenheim Castle (1705–1716), for the Duke of Marlborough. In his designs for Castle Howard and Blenheim, Vanbrugh assimilated French Baroque, Palladian, and Picturesque influences into his own distinct style. Vanbrugh gave both country houses dramatic skylines reminiscent of Tudor country houses while his other works often made references to English medieval architecture. Chief among these later works is his own house (1718–1721) in Greenwich, which reinterprets crenellation and turrets of a castle. Although Vanbrugh's architecture fell out of favor by the mid-eighteenth century, a new appreciation for his work began in the early nineteenth century. This revival is characterized by words of the architect John Soane who called him the "Shakespeare of architects."

Bibliography: B. Dobrée and G.F. Webb, eds., *Complete Works of John Vanbrugh*, 1928; K. Downes, *Vanbrugh*, 1977.

David D. McKinney

VAN DYCK, ANTHONY (1599–1641). By joining nobility to beauty in an attenuated, elegant portrait style, this painter from Antwerp found renown in Italy and at court in Brussels and London. The son of a prosperous cloth merchant and amateur painter, he trained with Hendrick van Balen from 1609 until 1618 to become a history painter. He then joined *Peter Paul Rubens's burgeoning shop as an independent master, assisting him in painting ceilings for the new Jesuit Church in Antwerp. Van Dyck so excelled at imitating his manner that Rubens described him as "his best pupil."

A successful trip to London in 1620 came before a more lengthy stay in Italy (1621–1627). There he developed a style of full-length portrait with refined, elongated proportions, a response to those Rubens portraits he saw in the houses of the Genoese nobility. He added staffage, such as the adoring servant holding a parasol in *Marchesa Elena Grimaldi Cattaneo* (1623), to lend an air of majesty and authority to his subjects. When he returned to Antwerp, he continued to produce portraits and such shimmering, refined history paintings as the 1629 *Rinaldo and Armida*, a gift ordered by Endymion Porter for *Charles I in London. His manner also lent itself to quite personal interpretations of religious subjects, such as the 1629 *Vision of St. Anthony*. The exiled Elector Palatinate Frederick V of Bohemia and Elizabeth Stuart sat for him in The Hague, as did stadholder Frederick Hendrick (1631–1632) and Amalia van Solms.

Although he established a London residence in Blackfriars and commenced service for the king and queen in 1632, he returned to Flanders for a year (1634–1635) where he portrayed the new Governor, the Cardinal-Infante Ferdinand (1634). His final period in London was marked by the broader, more colorful palette and growing informality in pose and composition exemplified by *Charles I at the Hunt* (ca.1635) and *Thomas Pettigrew and William, Lord Crofts* (1638). The exceptional beauty of his portraits has been related to the neo-Platonism of the Cavalier poets. It certainly lent visual support to the self-aggrandizement of the Stuart court. Although he knighted Van Dyck in 1635, the monarch was often slow to pay for his work and was occasionally difficult. Van Dyck's 1640 marriage to Lady Mary Ruthven failed to stem a growing unrest that prompted an unsuccessful journey to Paris in 1640. After seeming to recover from a long illness, he died at Blackfriars before his expected return to Flanders.

The most ambitious project he completed was the *Iconographia*, a volume of etchings of contemporaries, like an ancient gallery of worthies. Most of the important artists of the day were included, emphasizing the high status of their vocation. His painted self-portraits reveal a vain and slightly neurotic personality consistent with accounts of an indulgent lifestyle in London. His capacity for sweetness and beauty peaks in his portraits of children. He is credited with introducing the "Grand Manner" of portraiture to England, profoundly influencing Gainsborough and Reynolds.

Bibliography: C. Brown et al., *Van Dyck 1599–1641* (Exhibition Catalogue), 1999; M. Jaffé, *Van Dyck's Antwerp Sketchbook*, 1966; E. Larsen, *The Paintings of Anthony van Dyck*, 1988; A. Wheelock Jr. et al., *Anthony Van Dyck* (Exhibition Catalogue), 1990.

Lloyd DeWitt

VAUGHAN, HENRY (1621/22–1695). The Anglo-Welsh poet and country doctor Henry Vaughan is truly a poet whose work shows the influences of his contemporaries. Vaughan was artistically open to the inspirational influences of a wide variety of metaphysical and cavalier poets. As a result, he wrote some of the Commonwealth and Restoration's more original and eloquent mystic poetry by combining traditional devotional poetic conventions such as introspective conceits of individuality, epiphany, and a sense of the sublime in the presence of the divine with abstract influences of philosophy, the occult, and alchemy. Vaughan attributed his devotional inspiration to the works of George Herbert; he credited his twin brother, Thomas, as his esoteric influence.

Vaughan was born in the vicinity of Llansantffraed, Breconshire, Wales in about 1621–1622. He spent two years at Oxford; he left, however, without a degree to pursue law studies in London. He was called home from law school in 1642, prior to graduation, due to the outbreak of civil war. Vaughan most probably served with the Royalist forces. After the war ended, he returned home to his native Wales and, without any formal medical education, established a local medical practice. In 1646, he published a small volume of secular verse entitled *Poems, With the Tenth Satyre of Juvenal Englished*. This collection reflects the spiritual influence of *John Donne, "and followers of Johnson ("Sons of Ben") such as *Thomas Randolph." His most renowned work is Silex *Scintillans*, subtitled *Sacred Poems and Private Ejaculations*. The title is an homage to Herbert's *The Temple: Sacred Poems, and Private Ejaculations*. *Silex Scintillans* is a lyric sequence of devotional meditations loosely representing mankind's inherent quest for the divine within. The original text was published in 1650 and expanded with additional poems in 1655. The title *Silex Scintillans* ("burning stone") allegorically encapsulates the religious ideology that Vaughan respectfully appropriated from Herbert, that is, only God's love can rekindle the divine fire in a sinner's stone-cold heart. The emblem on the book's title page visually depicts this same religious ideology. In 1651, Vaughan published a book of occasional poetry, *Olor Iscanus*. The name *Olor Iscanus* ("The Swan of Usk") is an homage in itself to a little river that flows near Scethrog, close to Llansantffraed, Vaughan's hometown. Vaughan's brief poetic career culminated with the publication of an assorted collection of poetry and prose entitled *Thalia Rediviva*. Vaughan spent the rest of his life practicing medicine in Wales, where he died in April 1695.

Bibliography: L. Martz, ed., *Henry Vaughan*, 1995; J. Post, *Henry Vaughan: The Unfolding Vision*, 1982.

David M. Rosen

VEGA, LOPE FÉLIX DE (1562–1635).

Lope de Vega Carpio is the creator of the national Spanish drama of action and intrigue. Born in Madrid, Lope was a precocious child and professed an extraordinary knowledge of literature. He was a man of action and of letters serving as a soldier in the military expeditions and as a secretary to various noblemen. He had a tumultuous love life with several marriages and affairs, suffering exile twice for his amorous scandals. He professed a deep religious conviction and received Holy Orders in 1614, but he later again resorted to his amorous liaisons. Lope cultivated every genre of his time. His religious poetry is some of the best of his age. *La Dorotea* (1632) is an example of a lengthy, dialogued novel. Foremost, Lope was a prolific playwright, claiming to have written 1,500 plays of which only around 500 are preserved. Lope established the standard format of three acts and a specific type of verse for each situation. He abolished the classical unities of time and place, and blended tragedy and comedy. He defined the role of the comic servant type (*gracioso*) and elaborated on the honor theme. His plays deal with themes taken from Spanish and foreign chronicles, legends, Spanish customs, and pastoral, mythological, and religious sources. His most popular staged play is *Fuenteovejuna* (1612–1614), which presents the conflict between an abusive noble and honorable peasants. *Arte nuevo de hacer comedias* (1609) is his theoretical treatise that influenced the Spanish theater for a century and later influenced Spanish and European theater during the nineteenth-century Romanticism. Lope's work combines the vitality and sensuality of the Renaissance with the Spanish religious mentality of the Counter Reformation.

Bibliography: F. Carreter, *Lope de Vega: Introducción a su vida y obra*, 1966; M. Menéndez y Pelayo, *Estudios sobre el teatro de Lope de Vega*, 1964.

M. Ellen Blossman

VELÁZQUEZ, DIEGO RODRÍGUEZ DE SILVA (1599–1660).

Diego Rodríguez de Silva Velázquez, Spain's greatest portraitist, was born in Seville on 5 June 1599. The oldest of seven children born to Juan Rodríguez de Silva and Jerónima Velázquez, both of whom were of minor nobility, the painter served from 1611 until 1617 as an apprentice to Francisco Pacheco, a Mannerist painter who would later become Velázquez's father-in-law. His first paintings while living and studying in Seville were still-life pictures of everyday objects (fruit, vegetables, kitchen utensils, jars). Velázquez's most important paintings from this period include *El Aguador de Sevilla* (*The Water Carrier of Seville*, ca. 1619) and *Vieja Friendo Huevos* (*Old Lady Frying Eggs*, 1618). Velázquez left Seville for Madrid in 1622. Philip IV (1621–1665) named Velázquez in 1623 one of his official painters after Velázquez painted a portrait of the king. While living in Madrid from 1623 until 1629, Velázquez experimented with chiaroscuro, the distribution of light and shade. His most famous paintings during this time are portraits of the royal family and the mythological work *Los Borrachos* or *El Triunfo de Baco* (*The Drunkards* or *The Feast of Bacchus*). Velázquez's friendship with the Flemish painter *Peter Paul Rubens (1577–1640), who lived in Madrid from 1628 until 1629, influenced the painter's developing style. Conversations with Rubens about art collections in Italy inspired Velázquez to travel to Italy in 1629, where he visited Genoa, Milan, Venice, Florence, Rome, and Naples before returning to Spain in 1631. Two of Velázquez's paintings that clearly exhibit the influence of Italian Renaissance art are from 1630: *La Fragua de Volcano* (*Vulcan's Forge*) and *La Túnica de José* (*Tunic of Joseph*).

Velázquez returned to Madrid in 1631 and served under Philip IV as court portraitist, painting many portraits of the royal family during this time, including a poignant one of *Prince Balthasar Carlos with a Dwarf* in 1631 and a group of hunting portraits of the royal family for the Torre de la Parada, a hunting lodge near Madrid. In 1634, Velázquez supervised the decoration of the throne room of the Buen Retiro, a new royal palace. The most prestigious painters of the day, including *Francisco de Zurbarán (1598–1664), Vicente Carducho (1557–1638), Juan Bautista Maino (1558–1648), and Eugenio Cajés (1574–1634), contributed battle scenes of victorious Spanish troops. Velázquez's *La Rendición de Breda* (*Surrender of Breda*), a portrait of a Spanish general accepting the surrender of a Flemish leader in 1624, and five equestrian portraits of the royal family were among the paintings displayed. Portraits of mythological subjects, including *Marte* (*Mars*) and *Esopo* (*Aesop*), and court dwarfs, whom Velázquez depicted in a respectful manner and not as court jesters, date from the early 1640s. In 1649, Velázquez returned to Italy to purchase works of art for *Philip IV's collection. While living in Rome from 1649 until 1650, he painted protraits of Juan de Pareja, an Arab slave who accompanied Velázquez to Italy, and *Pope Innocent X. He also painted during this time one of his masterpieces, *Venus del espejo* (*Venus of the Mirror*). Velázquez returned to Spain in 1651. In 1656, Velázquez finished his masterpiece, *Las Meninas* (*The Maids of Honor*), a group portrait of the royal family and of Velázquez himself in the act of painting. Philip IV decided to reward Velázquez for his service and friendship in 1658 by inviting the painter to become a member of the military order of his choice, an honor that presupposed noble rank. Veláz-quez chose the Order of Santiago, the most prestigious of the three orders. Spain's most celebrated painter died on 6 August 1660 after a brief illness.

Seventy-five percent of Velázquez's 200 paintings are portraits. The royal family of Philip IV is the subject of many of his portraits, including more than twenty of the king himself. Velázquez takes his realism to a new level with the theme of instantaneous motion. He embraces time and re-creates it much like a photograph. Examples of this aspect of the painter's genius include *El Fragua de Vulcano* (*Vulcan's Forge*), when Apollo informs the smith-god of his wife's infidelity; Minerva's spinning wheel in *Las Hilanderas* (*The Spinners*); and the moment in which a chambermaid hands a drink to Philip IV's daughter Margarita in *Las Meninas* (*The Maids of Honor*). Velázquez was not the typical Baroque painter, for his painting exhibited one moment and one vision, and the manner in which he captured the fleeting moment makes his art eternal.

Bibliography: D. Brown, *The World of Velázquez, 1599–1660*, 1969; J. Brown, *Velázquez, painter and courtier*, 1986.

Michael J. McGrath

VERMEER, JOHANNES (1632–1675). Born in Delft to Digna Baltens and Reynier Janszoon Vermeer, Johannes followed the artistic inclinations of his father, a picture dealer and weaver of fine satin. In 1653, Vermeer married Catharina Bolnes, converting to Catholicism in the process; they eventually had fourteen children. Although his training remains conjectural, the artist was admitted to the Guild of St. Luke in Delft that same year and began to sell his work.

Vermeer's early paintings were of religious or mythological subjects, but within a few years, he shifted, like other Dutch artists, to scenes of everyday life. These works were far from ordinary, though, due to his ability to make the quiet activities of typically female protagonists transcendent through the evocative play of light. While Vermeer used the *camera obscura* as an aid in capturing these fleeting optical effects, he seems to have further abstracted rather than merely copied the reflected image. The settings of his interior scenes often employ a rectilinear framework that enhances their calm, introspective mood, as in *Woman Holding a Balance* (ca. 1664). This work is also indicative of the profound moral content believed to underlie many of Vermeer's paintings and more overtly demonstrated in the *Allegory of Faith* (ca. 1671). Townscapes also came under the artist's purview, most notably the incomparable *View of Delft* (ca. 1661). Somewhat unusually for a period when portraits were in demand, Vermeer may have painted only one; instead, the subjects of his portraitlike images, such as the mesmerizing *Girl with Pearl Earring* (ca. 1665), remain enigmatic, undoubtedly heightening their elusive charm. Vermeer's immediate influence was limited because of the rarity of his creations: It has been estimated that he produced only a few works per year and perhaps no more than sixty in his career (of which only thirty-five can be reliably identified today). Technical analysis has revealed an extremely meticulous method in which compositions were constantly revised until a harmonious perfection was achieved. To supplement his income, Vermeer worked as an art dealer, but remained deeply in debt upon his death at the age of forty-three. Although highly esteemed by his contemporaries in Delft, he apparently had no students. It was not until the nineteenth century that the timeless beauty of Vermeer's poetic realism was rediscovered.

Bibliography: A. Wheelock et al., *Johannes Vermeer*, 1995.

Julia K. Dabbs

VIAU, THÉOPHILE DE (1590–1626). In spite of its overall notoriety, parts of Théophile de Viau's life remain obscure. He was born in April 1590 in Clairac, the son of a Protestant lawyer. In his youth, he studied at various academies and reputedly worked as a playwright for the Hôtel de Bourgogne.

Théophile was exiled in June 1619 for having written un-Christian, libertine poetry. He returned the next year and joined the army of Louis XIII, fighting first against the Queen Mother, then against the Huguenots. He converted to Catholicism around September 1622; this display of faith did not prevent a second accusation of philosophical and moral libertinism. These allegations stemmed from the publication of the *Parnasse satyrique*, whose obscene opening sonnet was attributed to Théophile. A zealous denunciation by Father Garassus, author of the *Doctrine curieuse*, led to an order for Théophile's arrest and the latter's flight to the home of his protector, the duke of Montmorency, at Chantilly. In August 1623, Théophile was convicted of lese majesty and condemned to burn alive. He was instead burned in effigy, then arrested while trying to flee. In September 1625, after nearly two years of imprisonment in the Conciergerie, Théophile's sentence was changed to permanent exile; this was never enforced. Théophile died 25 September 1626 at Montmorency's Parisian residence.

Théophile's persecution sometimes unjustly eclipses his fame as a writer. He is noted for his independence, particularly in the face of classicism, and is often cited as an excellent "modern" poet in the seventeenth-century sense. He produced a popular tragedy, *Pyrame et Thisbé*, along with a rich, diverse, poetic corpus. He especially excelled at finding fresh combinations of the themes of love and nature, as can be seen in the odes *Le Matin* and *La Solitude*, and the collection *La Maison de Sylvie*.

Bibliography: A. Adam, *Théophile de Viau et la libre pensée française en 1620*, 1965; G. Saba, *Fortunes et infortunes de Théophile de Viau*, 1997.

Laura L. Dennis-Bay

VICTORIA, TOMÁS LUIS DE (1548–1611). Born in Avila, Victoria served the cathedral as a choirboy and was educated at the Jesuit school of San Gil. In his teens, he was sent to Rome for study at the Jesuit Collegium Germanicum, which trained both missionaries to Germany and seminarians from Italy, Spain, and England. At this time, Palestrina was maestro di cappella of the nearby Seminarium Romanum, and possibly Victoria studied with him. From 1569, Victoria served several Roman churches and chapels, and in 1571, the Germanicum enlisted him to teach music. In 1572, his first published music, a book of motets, was issued in Venice. In 1573, Pope Gregory XIII segregated the German students from the others and moved the Germanicum to the Palazzo di San Apollinare; Victoria was appointed its maestro di cappella. Following his ordination in 1575, Victoria joined the Congregation of the Oratory, a community of diocesan priests founded that year by St. Philip Neri. Victoria was a resident priest from 1578 until 1585 at San Girolamo della Carità, home of St. Philip's first oratory, built in 1554. These years saw many publications of Victoria's music—masses, motets, hymns, magnificats, and the Office for Holy Week. Victoria's mass settings were in demand all over Europe, even as far away as Mexico and South America.

After so much activity as composer, organist, and singer, Victoria longed to return home to "rest for a time in honest leisure and to be able to compose my soul in contemplation, as befits a priest," as he expressed it in the royal dedication of his Segundo Libro de Misas. In 1583, Philip II honored this request, appointing Victoria chaplain and maestro de capilla to his sister, the Dowager Empress Maria, widow of Maximilian II and daughter of Charles V. The empress, with her daughter, had entered in 1581 the monastery of the Descalzas Reales de Santa Clara in Madrid. Here, thirty-three cloistered nuns daily attended Mass in a chapel staffed by priests proficient in both chant and polyphony. After the Empress's death in 1603, Victoria served as choirmaster for another year and then became organist, a less demanding job. This he kept until his death, except for a leave of absence in Rome from 1592 until 1595, during which time he was part of the cortège at Palestrina's funeral (2 February 1594). Victoria died on 20 August 1611 in the chaplains' house of the monastery.

Bibliography: E. Cramer, "Some Elements of the Early Baroque in the Music of Victoria," *De musica hispana et aliis: Miscelanea en honor al Prof. Dr. Jose Lopez-Calo, S.J., en su 65° cumpleanos* (1990), 501–538; R. Stevenson, *Spanish Cathedral Music in the Golden Age*, 1961; R. Stevenson, "Tomás Luis de Victoria: Unique Spanish Genius," *Inter-American Music Review* 12 (1991): 1–100.

Susan Treacy

VINCENT DE PAUL (1581–1660). Christian spiritual writer and founder of a religious order, De Paul was canonized in 1828. Ordained in 1600 and made Bachelor of Theology (Toulouse, 1604) and Licenciate of Canon Law (Paris, 1623), Vincent de Paul was a contemporary and a companion of *Francis de Sales, and the pastoral successor of *Pierre de Bérulle, one of the leaders of the "French School" of mysticism. Seemingly destined for a vocation among the aristocracy and intelligentsia, in 1617 Vincent delivered his first missionary sermon and dedicated himself to charitable work among the poor. The urban poor had so increased in the wake of the Wars of Religion as to appear to be a new missionary field, and De Paul founded the "Congregation of the Mission" in 1625 to provide pastoral and educational care for them. The order quickly expanded throughout Europe. In addition to direct educational work within its mission field, the order also trained secular priests, conducting exercises and establishing seminaries. With Louise de Marillac (1591–1660 canonized 1934) he established in 1633 a sister order, known as Daughters of Charity of St. Vincent De Paul, for the care of the sick poor. These activities brought the famously modest "Monsieur Vincent," as he chose to be addressed, a reputation as the restorer of religious life among both clergy and laity in France. For the religious in these two orders, he wrote Rules and Constitutions, as well as numerous letters of spiritual guidance; many are exhortations to poverty and an active life of charitable works.

Bibliography: V. de Paul and L. de Marillac, *Rules, Conferences, and Writings*, ed. F. Ryan and J. Rybolt, 1995; A. Dodin, *Saint Vincent de Paul and Charity*, trans. J. Smith and D. Saunders, 1993.

Ralph Keen

VONDEL, JOOST VAN DEN (1587–1679). A contemporary of *Constantijn Huygens in Holland's "golden century" and sometimes called "the Dutch Shakespeare," Vondel was a poet and dramatist noted for his religious themes; he is generally regarded as the greatest Dutch author. He was born in Cologne on 17 November 1587 to an Anabaptist father exiled from Antwerp; his father relocated in Amsterdam in 1597 and established a hosiery business, which Joost, as eldest son, was expected to take over. But his wife instead directed the shop, while Joost immersed himself in French and German literature. His own church affiliations mirrored the changing religious climate of his day, as he moved from the doctrines of Calvin to *Arminius to Roman Catholicism. Vondel was prominent in the Chamber of the Eglantine, one of the "Rederijkers" or Dutch literary guilds of the day. He early on read the works of Du Bartas, whom he also translated (1607–1617). His practice of adopting biblical themes to illustrate moral qualities influenced the dramas of *Andreas Gryphius, and his first play, *Pascha of de Uyttocht der Kinderen Israels uit Egypte* (1612), a portrayal of the Exodus, helped to gain him the rank of national poet. However, in 1625, he incurred official criticism with the performance of *Palamedes*, which satirized the Calvinist theology of those who had ordered the judicial beheading of Johan van Oldenbarneveldt in 1619; Vondel was forced for a time to go into hiding. His anti-Calvinist and pro-Catholic sentiments grew, and he also formed a strong friendship with *Hugo Grotius. When Vondel formally became a Catholic in 1641, he was abandoned by many of his former Protestant friends, and when his son failed to keep the family business solvent, Vondel had to rescue it with his savings. He was compelled to take a minor government position, finally dying on 5 February 1679.

Vondel composed thirty-two plays, which tend to display Baroque features, his best-known being *Gijsbrecht van Aemstel* (1637); dedicated to Hugo Grotius (whose *Sophompaneas* Vondel had translated) and drawing upon Book II of Virgil's *Aeneid*, it uses a medieval Dutch theme to praise the city of Amsterdam upon the opening of a new theater. His dramas were increasingly colored by his Roman Catholic convictions, as in *De Gebroeders* (1640, on the fall of Saul's sons), *Joseph in Dothan* (1640), *Lucifer* (1654, which may have influenced *Milton's *Paradise Lost*), *Jephtha* (1659, which he felt to be his best work), and *Adam in Ballingschap* (*Adam in Exile*, 1664). Vondel wrote the first known refutation of *Spinoza's *Ethics*. He also composed songs; occasional poems; translations from the French, Latin, and Greek; a prose treatise on poetry, the *Aenleidinge ter Nederduitsche Dichtkunste*; and a brief epic on John the Baptist. His poetry, marked by a fluid euphony that may have been influenced by his study of French authors, is often inspired by his religious faith as well as his sorrow at the deaths of his wife and three of his five children.

Bibliography: A. Barnouw, *Vondel*, 1935.

Christopher Baker

VOSSIUS, GERHARDUS JOHANNES (VOSS, GERHARD JAN) (1577–1649).

A Dutch humanist and theologian, Vossius was the son of a Calvinist minister and was born in Heidelberg in March or April 1577. Lutheran opposition drove the family to Leiden, where Vossius studied the classics, theology, Hebrew, and church history; here he also befriended *Hugo Grotius. He published his *Institutiones oratoriae* (*Lessons in Rhetoric*) in 1606. For fourteen years (1600–1614), he was headmaster of the Latin school in Dordrecht, then became director of the Leiden theological college until 1619, when he was relieved of his post on suspicions of heresy first aroused in 1618 with the publication of his *Historia Pelagianismi* (*The History of Pelagianism*); the book was thought by some to advocate the ideas of *Arminius. He was named professor of rhetoric and history at Leiden in 1622, leaving ten years later to assume the rectorship of the new Amsterdam Atheneum, where he was also professor of history and politics. Rejecting an offer from Cambridge twice, he did accept a nonresidential prebend (benefice) from *Archbishop Laud at Canterbury Cathedral in 1629, receiving that year a Doctor of Law degree from Oxford as well.

Vossius was one of the earliest scholars to study the textual basis of religious history and to employ a historical approach to the examination of doctrine. For example, he contended that the apostles could not have been the actual authors of the Apostles' Creed, and he demonstrated that St. Athanasius was not the author of the creed attributed to him because it contained doctrinal expressions that postdated the saint; these issues were discussed in his *Dissertationes tres detribus symbolis* (*Three dissertations on the three early Christian creeds*, 1642). He wrote several influential books on language and grammar (e.g., *Aristarchus, sine de arte grammatica libri septem* [*Aristarchus or seven books on grammar*, 1635]; *Etymologicon linguae Latinae* [*An Etymological dictionary of the Latin language*, 1662]); on history (e.g. *Ars historia* [*The Art of History*, 1623]); and mythology (*Theological gentiles* [*The Theology of the Gentiles*, 1641]). He died in Amsterdam on 17 March 1649.

Bibliography: C. Rademacher, *Life and Work of Gerardus Joannus Vossius (1577–1649)*, 1981; N. Wickenden, *G.J. Vossius and the Humanist Concept of History*, 1993.

Christopher Baker

W–Z

WALLER, EDMUND (1606–1687). Edmund Waller was born in Hertfordshire, the eldest son of Robert Waller and Anne. After studying with private tutors, he attended Eton and King's College, Cambridge. He was admitted as a member of Lincoln's Inn in 1622. In 1631, he married Anne, the only daughter and heiress of John Banks, a wealthy mercer. She died three years later while giving birth to a daughter. About this time, Waller was introduced to Lucius Cary, Viscount Falkland, who presided over the Great Tew circle of intellectuals. In the 1630s, he wrote love lyrics (including a series addressed to Lady Dorothy Sidney as "Sacharissa"), poems of compliment to individuals, and public panegyrics that circulated in manuscript, securing his reputation as a poet well before his *Poems* appeared in print in 1645.

Waller said that he first sat in the House of Commons at the age of sixteen, but he did not rise to prominence until the Long Parliament when, after initially attempting to check the arbitrary power of the king, he concluded that the advocates of parliamentary privilege and radical church reform posed the greater threat to the liberties and property of the subject. When treating with *Charles I on behalf of Parliament, Waller became involved in an attempt to secure London for the king. Arrested on 31 May 1643, he revealed all that he knew of what came to be known as "Waller's plot." He was ejected from the House, fined £10,000, and banished from the realm. He likely married Mary Brace before leaving for France, where he associated with other royalist exiles. In 1651, thanks to the intercession of *Oliver Cromwell, Parliament pardoned him. Although he never celebrated the Commonwealth in verse, he praised Cromwell in three poems, and he served as a commissioner for trade under the Protectorate. In 1661, he was again elected to Parliament and served until his death, joining in the attack on *Edward Hyde, Earl of Clarendon, and serving on the Councils of Trade and Foreign Plantations. Waller adapted to political circumstances, but a common set of values does run through the verse and oratory that he produced under Charles I, Oliver Cromwell, and *Charles II. He supported the established church, but he wished to see its abuses reformed and dissenters tolerated. He could not imagine a government without a court at its center, but he wanted the court to unite grace and social tradition with an interest in economic and intellectual innovation. He championed naval and commercial expansion, but he sought to avert outright war.

By the twentieth century, Waller was remembered chiefly for his delicate lyrics, such as "Go, lovely rose" and "On a Girdle," and for his last, devotional poem, "Of the Last Verses in the Book." An aspect of his verse that remains underappreciated is his ability in *The Battle of the Summer Islands* to sustain a complex tone that leaves the reader uncertain whether to feel terror, pity, or amusement. But his chief importance is as a founder of literary Augustinism. *John Denham identified him as the "best of poets" because of his achievements in public verse. He was in fact so successful at establishing a graceful and politically effective poetic idiom—in *On his Majesties Repairing of Paul's* (1638?), *A Panegyrick to My Lord Protector* (1654), *On St. James's Park* (1661), and *Instructions to a Painter* (1666)—that we can forget that his style was an achievement born of political and discursive struggle. Praising the "sweetness" of Waller's poetry, *John Dryden said, "He first made Writing easily an Art." That was an achievement that required not only smooth versification but what Samuel Johnson termed "elegance of diction" and "propriety of thought."

Bibliography: W. Chernaik, *The Poetry of Limitation: A Study of Edmund Waller*, 1968; M. Donnelly, "Edmund Waller," in *Seventeenth Century British Nondramatic Poets*, 2nd ser., ed. M. Thomas Hester, 1993.

Blair Hoxby

WALLIS, JOHN (1616–1703). The discovery of the calculus by *Isaac Newton and *Gottfried von Leibniz was the single most important innovation in mathematics in the seventeenth century. The calculus provided powerful tools to those who practiced the new science and culminated centuries of mathematical work that had its roots in the mathematics of the ancient Greeks. John Wallis was an English mathematician who made major contributions to the progress toward the calculus in the decades prior to Newton's summary formulation in the last third of the century. Wallis was also an active participant in the weekly meetings that led to the formation of the Royal Society of London, writing on statics and mechanics, translating and editing texts of ancient Arabic and Greek mathematicians, and devising a system for teaching deaf-mutes.

Wallis was born on 23 November 1616, the eldest son of an English rector. Young Wallis excelled in the classical education that often led to appointment as a member of the English clergy, was ordained as a clergyman in 1640, and continued to serve the church throughout his life. His formal education included very little mathematics, but his exposure to mathematics and the natural sciences broadened when he joined a group of London scholars that met weekly to discuss and practice the new experimental science. Wallis was an enthusiastic and committed member of this colloquium, which was eventually chartered as the Royal Society of London.

Wallis's serious work in mathematics began in 1647 when he read *William Oughtred's *Clavis Mathematicae*, and by 1649 he had gained sufficient stature as a mathematician to be appointed the Savilian professor of geometry at Oxford. In 1655, he published *De Sectionibus Conicis*, in which he made important contributions to analytic geometry and, in 1656, *Arithmetica infinitorum*, in which he introduced the symbol ∞ for infinity and used arithmetic and algebraic reasoning to confirm and extend results leading to the calculus. Young Newton's reading of *Arithmetica infinitorum* a decade later inspired Newton's work in the calculus. Wallis's additional mathematical works included *Mathesis Universalis* (1657), which promoted and extended improvements in algebraic notation, and *Treatise of Algebra, Both Historical and Practical* (1685), which failed to be an accurate account of the historical development of algebra but succeeded as a textbook on algebra that was widely read over the next 100 years.

Much of Wallis's work on the calculus was inspired by his reading of the work of the Italian physicist *Evangelista Torricelli. Torricelli and his continental contemporaries made progress in understanding the areas and volumes of regions bounded by curves and curved surfaces by regarding those regions as made up of very small indivisible elements. Wallis replaced consideration of conglomerates of indivisible geometrical elements with sums of sequences of numbers, confirming and extending the observations of Torricelli. In so doing, he came very close to the evasive concept of limit that finally provided the careful analytic foundations of the calculus that were developed some 170 years later.

Bibliography: C. Boyer, *The History of the Calculus and Its Conceptual Development*, 1949; J. Scott, *The Mathematical Work of John Wallis (1616–1703)*, 1938.

Ed R. Wheeler

WALTON, IZAAK (1593–1683). Like *Bunyan a generation later, Walton was a largely self-educated author of working-class origins. His father had been a tavern keeper in his native Stafford, and he himself ran an ironmonger's shop in Fleet Street, London. This fact, whether by chance or providence, happened to place him in the parish of *John Donne, at that time vicar of St. Dunstan's church, and the two became good friends. This was but one of Walton's many personal links to the church of England; his first wife was a distant descendant of Archbishop Thomas Cranmer, architect of the 1549 *Book of Common Prayer*, and his second wife was the stepsister of *Thomas Ken, bishop of Bath and Wells. It is thus not surprising that Walton eventually became one of his era's foremost ecclesiastical biographers. He also became acquainted with Sir Henry Wotton, provost of Eton and minor poet, who, intending to publish a life of Donne, enlisted Walton to gather materials for the work. But Wotton died in 1639 and Walton himself finished it in 1642; it was published nine years later.

That same year, he also published a life of Wotton as well, and later went on to complete biographies of theologian Richard Hooker (1602), poet *George Herbert (1670), and Robert Sanderson, bishop of Lincoln, which appeared when Walton was eighty-five (1678). Though flawed by occasional factual errors and Walton's tendency toward hagiography rather than objectivity, these popular works nevertheless captured the devoted spirit of several of the era's Anglican notables and commanded several editions.

Walton, like Donne, Wotton, and Herbert, enjoyed fishing, the topic of the book for which he is best known, *The Compleat Angler, or the Contemplative Man's Recreation* (1653). This volume went through five editions in the seventeenth century, ten in the eighteenth, and 164 in the nineteenth. Adopting the format of a dialogue, Walton has Piscator and Viator (later a third character, Auceps, was added) discuss the pleasures and techniques of fishing in a relaxed and pastoral mood. The book's tone is a not insignificant Anglican response to the fact that it had only been four years since *Cromwell's revolution had executed *Charles I and further factionalized England. Walton's book is a kind of "anatomy" of fishing in *Robert Burton's sense: a repository of facts, hints, tips, lore, songs, myths, and stories about angling, the benefit of which is to teach one lesson: "Study to be Quiet." Fishing, one discovers, is only a metaphor for a frame of mind, an attitude named in the book's subtitle.

Bibliography: D. Novarr, *The Making of Walton's Lives*, 1958; P. Stanwood, *Izaak Walton*, 1998.

Christopher Baker

WATTEAU, ANTOINE (1684–1721).

Baptized Jean-Antoine on 10 October 1684, Watteau was the second son of Jean Philippe and Michele Lardenois. This is one of the few biographical certainties of an artist who remains an elusive figure in the history of art. He first apprenticed to an artist in his hometown of Valenciennes before arriving in Paris, possibly in 1702. There Watteau served as an assistant to Claude Gillot, whose renderings of theatrical scenes such as the *commedia dell'arte* profoundly influenced the young artist. After three years with Gillot, he assisted Claude Audran III in creating charming decorative scenes for the interiors of Parisian residences. Desiring a more serious artistic career, Watteau was accepted into the French Royal Academy of Painting and Sculpture in 1712. Yet it was not until 1717 that he completed the requisite reception piece: the magical *Pilgrimage to the Isle of Cythera*. Not a history painting nor a pure landscape, the *Pilgrimage* broke the academic boundaries and established a new genre of painting, the *fête galante*, or "fashionable party," in which aristocratic men and women enjoy intimate conversation, music, and love in a lushly verdant, parklike setting. At times criticized by his contemporaries for their lack of narrative content, Watteau's *fête galantes* were closer to painted poetry, designed to evoke a mood of reverie through a more sensual approach to color and brushwork. In this respect, the artist was clearly influenced by the works of

*Peter Paul Rubens, as well as Venetian Renaissance paintings, which Watteau could have readily studied in Paris (he never went to Italy). He further appealed to contemporary taste with his delicate depictions of actors or dancers in outdoor settings. These were generally small in scale, perfect for the more intimate salons of Parisian society. His only extant life-size figure painting is *Gilles* (1718), the tragic-comic actor whose isolation and weary expression suggest a melancholy wistfulness often associated with Watteau's works. Although his career was cut short by tuberculosis, Watteau left behind a legacy not only of paintings, but also drawings, which were particularly prized during the artist's lifetime (as they are today) for their spontaneity and delicacy of touch. His exploration of subjects reflecting the interests of contemporary aristocratic society had a lasting influence, most notably on Nicolas Lancret and William Hogarth.

Bibliography: M. Grasselli and P. Rosenberg, *Watteau*, 1984; D. Posner, *Antoine Watteau*, 1984.

Julia K. Dabbs

WATTS, ISAAC (1674–1748). Isaac Watts was born in 1674 at Southampton and was descended from a well-known and respected dissenting family. Watts's father was imprisoned twice for nonconformity, the first time soon after the younger Isaac's birth. Watts was the eldest of eight children, six of whom lived to maturity. As a child, Watts was given an excellent education at grammar school by John Pinhorne, rector of All Saints and master of the Free-School at Southampton. Watts was an excellent student with a recognizable talent for learning languages. Dr. John Speed, a physician at Southampton, recognized these gifts and offered Watts a scholarship to either Oxford or Cambridge. Preferring to continue in the dissenting schools, Watts declined the offer and entered Thomas Rowe's academy at Stoke Newington in 1690. Watts completed his studies in 1694 and returned home to work on the hymns that would later make him famous. In 1696, Watts became the tutor to the son of Sir John Hortopp and held the position for five years, dedicating all of his free time to the study of Hebrew and divinity. Watts accepted the pastorate at Mark Lane in 1702, but due to ill health was unable to fulfill his duties and instead became a copastor. In 1712, he was invited to live at the residence of Sir Thomas and Lady Abney, whose continued support allowed Watts to pursue his love of teaching and writing for the remaining thirty-six years of his life. In 1728, the University of Edinburgh granted Watts on honorary degree of D.D.

Watts is best remembered for his educational books and hymns, although he did write two successful books of poetry. Samuel Johnson would later cite Watts's *Horæ Lyricæ* (1706) as his favorite book of religious poetry. Among Watts's most popular educational manuals were *The Art of Reading and Writing English* (1721), *The Knowledge of the Heavens and the Earth* (1726), and *Catechisms* (1730). All of these books sought to provide young students with complex ideas presented in a more approachable language. His two books of hymns proved to be the best selling of his works, with the 1707 *Hymns* running through annual printings of around 50,000 copies. Aside from the popularity of his religious and poetic works, Watts was known in academic circles for his two books on informal logic: *Logick: Or, the Right Use of Reason in the Enquiry after Truth, with a Variety of Rules to Guard against Error, in the Affairs of Religion and Human Life* (1724) and *The Improvement of the Mind; Or, a Supplement to The Art of Logick: Containing a Variety of Remarks and Rules* (1741).

Watts's *Logick* is most immediately influenced by *Arnauld's and *Nicole's *The Art of Thinking* (1662), *Descartes' *Discourse on Method* (1637) and *Rules for the Direction of the Mind* (1628), as well as *Locke's *Essay Concerning Human Understanding* (1690). For Watts, logic is conceived as the relationship between perception (the objective contemplation of things), judgment (the joining of ideas as either affirmative or negative), reasoning (the inference of one proposition from two or more propositions premised), and disposition (the arranging of thoughts and memory to best suit the individual). This conception of logic is based on the inward operations of the mind to best support any outward presentation or communication to others. Watts's argument in *Logick* supports a personal understanding of the workings of the mind and offers strategies to help improve both the self-realization of these workings and ways to strengthen the mind's ability to work more efficiently. He has a tendency to treat formal logic as common sense, and while not discrediting the practice, he shows little enthusiasm for the subject. In 1741, Watts wrote *The Improvement of the Mind* as a sequel to *Logick*, and the new text offered a more practical and regimented program for the application of the ideas in the first book. The first part of *The Improvement of the Mind* offers suggestions on the most useful aspects of reading, hearing lectures, learning languages, conversation, debate, study habits, and improving memory to the development of the mind. The second part addresses such topics as teaching, rhetoric, the use and dangers of authority, instruction through preaching, and public speaking and writing. Basically, the first part presents a program for self-improvement, while the second part offers suggestions on how to best deliver the newfound knowledge to a public audience. Both of these books were successful with the public, but they found their most useful and influential audience as textbooks at both Oxford and Cambridge throughout the eighteenth century. Watts died in November 1748 as one of the most recognizable names of his time.

Bibliography: A. Brinton, "The *Logick* of Isaac Watts," *Historical Foundations of Informal Logic*, 1997; A. Davis, *Isaac Watts: His Life and Works*, 1943.

Michael W. Jackson

WEBSTER, JOHN (?1579–before 1637). Dramatist of horror, John Webster was the eldest son of John Webster, Senior, a successful and influential coachmaker, member of the Merchant Taylors' Company, and parishioner of St.

Sepulchre's, and Elizabeth Coates, daughter of a similarly situated blacksmith one parish over. Though his dates of birth and death remain speculative, Webster seems to have been born in 1578 or 1579, about two years after his parents' marriage. As circumstantially inferred as his dates is his education. The best suppositions have Webster educated at the humanist and Protestant academy, The Merchant Taylors' School, and then entering the Inns of Court, first the Chancery Court of New Inn and then the Middle Temple. The status of his father, the quality of his adult associates, the nature of his themes and apparent temperament, and the Inns of Court connections to the theater all point to such an education, in addition to the note in the Middle Temple rolls that on 1 August 1598 "Master John Webster, lately of New Inn, gentleman, son and heir apparent of John Webster of London, gentleman" was admitted. His marriage is a bit more certain. On 18 March 1605 he married Sara Peniall, who was seventeen, and less than two months later, on 8 May 1606, his son, John Webster III, was baptized in his wife's parish church, St. Dunstan-in-the-West, a fashionable area of town. About the time of his father's death, Webster claimed membership by birth in the Merchant Taylors' Company, on 19 June 1615, and became its playwright for the mayor's pageant upon a fellow company member's installation in 1624. The fellow member was Sir John Gore, and the pageant Webster wrote was *Monuments of Honor*, the most expensively produced such pageant of *James I's reign and Webster's most classically informed work. Webster seems to have died at about age sixty in the mid-1630s, perhaps before 7 November 1634, but perhaps just before 3 March 1637, the date "John Webster was buried," according to the parish records of St. James, Clerkenwell.

Webster began his writing career as a collaborator with several of the age's more remarkable playwrights, including *Thomas Dekker, *John Fletcher, *Philip Massinger, and *John Ford. He expanded *John Marston's *The Malcontent*, most notably by adding the Induction. The Induction emphasizes the satirical connection between the play's act of representation and what the play represents in one of the most admirable (because economically elaborate) ways in dramatic literature. Eventually moving out of collaboration into independent writing, Webster came to write two plays that, in many estimations, are superior to all but *Shakespeare's best in their theatrical energy and signal poetic passages. *The White Devil*, published in 1612 shortly after its first performance, sensationalizes the already scandalous story of the vengeance triggered when Brachiano, an Italian Duke, murders his wife and the husband of his lover in order to marry his lover, Vittoria, the honorably connected and outwardly honest "white devil." The final act is the horrific apogee of the period's drama, except for the final act of Webster's other singularly eventful play, *The Duchess of Malfi*, published in 1623, about a decade after its first production. Like *The White Devil*, *The Duchess of Malfi* concerns illicit love, intrigue, corporeal as well as psychological horror, and doubleness of character—here Ferdinand being the most grotesque in his lycanthropy. Distinguishing itself, though, *The Duchess of Malfi* introduces the specter of class-crossing into the love intrigue and tragedy with the Duchess's secret marriage to her servant, Antonio. To say that a grim outlook on human life informs the plays understates the fact. Antonio may express the typical attitude of Webster's plays: "Pleasure of life, what is't? Only the good hours / Of an ague." Psychologically troubled or dramatically decadent, Webster is certainly the primary playwright of horror of the early modern stage.

Bibliography: C. Forker, *Skull Beneath the Skin: The achievement of John Webster*, 1986; F. Lucas, ed., *Complete Works*, 1966.

Jesse G. Swan

WEISE, CHRISTIAN (1642–1608). When Weise died in his native Zittau, Saxony, he left behind an impressive oeuvre including poetry, handbooks for teachers, four novels, and sixty plays. Born at the end of the Thirty Years' War, he grew up during a postwar period marked by renewed optimism and an ethos based on common sense, which guided his life as a writer and teacher. Much criticized well into the twentieth century for being dry and dull, he is considered today the most interesting German writer at a time of transition at the end of the seventeenth century. A teacher's son, Weise began to study theology at the University of Leipzig in 1659, but he soon turned his attention with more enthusiasm to philosophy, history, politics, and law, while in his spare time he wrote poetry for cash. His dream of becoming a university professor did not materialize, a fact that facilitated his transition to a more practical life as secretary at the court of the duke of Saxe-Weimar. There he gained administrative experience that together with his years as senior teacher at the high school in Weissenfels (1670–1678) formed the basis for his teaching philosophy grounded in life's realities.

His most significant achievement is a cycle of novels that adopts the form of a journey to portray the follies of the world for didactic purposes. Though indebted to the satirical tradition of the sixteenth century, these works already announce the new spirit of the Enlightenment, and they position Weise as the most outstanding novelist since Grimmelshausen at the threshold of a new era. In 1678, he was offered the position of headmaster in his hometown, where he organized theatrical productions performed annually by his students. The plays were part of his didactic efforts to entertain (*delectare*) and instruct (*prodesse*), and they contributed to his fame as an outstanding educator with great influence on subsequent generations.

Bibliography: G. Burgess, *"Die Wahrheit mit lachendem Mund": Comedy and Humor in the Novels of Christian Weise*, 1990; P.N. Skrine, "Christian Weise," *Dictionary of Literary Biography* 168 (1996): 391–400.

Josef K. Glowa

WHICHCOTE, BENJAMIN (1609–1683). Whichcote was the founder of the Cambridge Platonists or Latitudinarians, who, drawing from Plato, the neo-

Platonists, and the early moderns, sought to establish an approach to religion that depended neither on absolute authority nor on rigid lines of argument. The members of the group included *Ralph Cudworth, Richard Cumberland, *Joseph Glanvill, *Henry More, and John Norris. Although tolerant in their thinking, all held for the absolute existence of right and wrong. Though without a direct influence upon the work of *John Milton, he and the Platonists shared a number of doctrines, especially regarding an Augustinian approach to Platonic concepts and a Protestant emphasis upon the role of reason in achieving the divine will. Whichcote matriculated at Cambridge's Emmanuel College in 1626, took his B.A. in 1629, and in 1633 earned his M.A. and became a fellow of his college. In 1636, he was ordained and appointed to the post of Sunday afternoon lecturer at Trinity Church in Cambridge. According to his letters, he sought to make his hearers consider the reasons for their beliefs and the foundational causes of phenomena. Despite success as a tutor and lecturer, Whichcote failed to win appointment to Gresham College's Divinity Chair and so retired, in 1643, to the rectory of North Cadbury in Somerset. He then married Rebecca Craddock, widow of Matthew Craddock, governor of Massachusetts. The year 1644 saw Whichcote back at Cambridge as Provost of King's College, a post he held until the Restoration when he was replaced. Complying with the Act of Uniformity brought Whichcote back into favor, and he spent the rest of his life in active ministry. During his years as provost, Whichcote had counseled *Cromwell on the question of toleration of the Jews (1655), and in 1659, Whichcote joined the group advancing Matthew Poole's idea that the university support and sustain gifted students, especially those studying for the ministry.

Bibliography: R. Colie, *Light and Enlightenment: A Study of the Cambridge Platonists and the Dutch Arminians*, 1957; C. Patrides, ed., *The Cambridge Platonists*, 1969.

Martha Oberle

WILLIAM III (1650–1702).

WILLIAM III (1650–1702). Son of Dutch William II of Orange and Mary, oldest daughter of *Charles I of England, William of Orange was raised by his mother after his father died only eight days after William's birth. When only twenty-two, he successfully defended Amsterdam against a combined attack by the English and French, though he was defeated in several other land engagements between 1674 and 1678. More adept diplomatically than militarily, he persuaded *Charles II of England to withdraw his troops from Dutch soil in 1674. Left to face the French troops of *Louis XIV alone, he sought to enlist English aid by marrying Mary, the king's niece and daughter of Charles's successor, *James II. However, James proved less inclined to aid William than to court favor with the Catholic French. Capitalizing on James's abysmal unpopularity in England, William agreed to become king when Parliament offered him the crown. On 11 April 1689, William and his wife Mary were crowned corulers of England and Scotland. He went on to defeat James's forces at the Battle of the Boyne in Ireland (1 July 1690), and he eventually wore down Louis XIV's troops on the Continent, culminating in his victory over the French at Namur in 1695. In the ensuing Treaty of Ryswick, Louis was forced to recognize him as king of England.

Domestically, the accession of William and Mary was termed a "Glorious Revolution" by which the remnants of Jacobite absolutism were eradicated and replaced by the firmer foundations of parliamentary authority. In 1689, the monarchs accepted the Bill of Rights, which assured members of Parliament freedom of speech and protection from prosecution for statements made in session. Parliamentary controls were placed on the king's authority to levy taxes and keep an army in peacetime, among other safeguards for individuals and limitations on kingly prerogatives. The Toleration Act of that year allowed broader freedom of religion to Dissenters and the Test Act, though still in effect, was largely side-stepped to the benefit of non-Anglican Protestants.

Mary died in 1694, and William governed alone until he died on 8 March 1702. Mary governed alone only during William's absences from England, and her role in the monarchy was largely one of loyal support for her husband's policies. Though popular, she suffered from the political contentions among her husband, father, and sister. The royal couple proved to be childless, but the harmony of their partnership made the Glorious Revolution a major step forward in the history of English parliamentary government.

Bibliography: T. Claydon, *William III and the Godly Revolution*, 1996.

Christopher Baker

WILLIAMS, ROGER (c.1603–1683). Religious reformer and founder of Rhode Island, Williams was born in London. Sponsored by Sir Edward Coke, Williams earned a bachelor's degree at Cambridge in 1627, but left before completing his master's degree. Williams married Mary Barnard in 1629 and fathered six children. A Puritan with increasingly Separatist leanings, Williams emigrated to Massachusetts in 1631. He was called to a church in Boston but refused because it was not Separatist, and took a church in Plymouth instead. In 1634, he became the pastor of a church in Salem where he became embroiled in controversy when he questioned the separation of church and state and the king's right to give patents to land not purchased from Indians. The Massachusetts General Court attempted to silence him and planned to banish him to England, but his friend Governor *John Winthrop warned Williams, and he left Salem during a blizzard, spending the winter in an Indian settlement. Joined by his family and other supporters, in 1636 Williams founded Providence, Rhode Island, on land obtained from the Narragansett Indians.

Learning that Massachusetts intended to claim the land around Narragansett Bay, Williams traveled to England and obtained a charter for the colony in 1644. While in England, he published his two most famous works, *A Key into the Language of America* (1643), a study of the Narragansett language, and *The Bloudy Tenent of Persecution* (1644), a declaration of freedom of conscience and a condemnation of Massachusetts' religious persecutions which aroused the opposition of *John Cotton. He returned to England in 1651 to confirm the charter, and during this trip became a friend of author *John Milton.

Williams grudgingly served in several capacities in the cantankerous world of Rhode Island politics, seeing the colony become a haven for Quakers, Baptists, and other persecuted religious groups. He briefly became Baptist long enough to found the first Baptist church in America before leaving organized religion permanently. As a political leader, he was able to persuade the Narragansetts not to join other Indian tribes during the Pequot war against Connecticut in 1637, but fought against them during King Philip's War in 1676. When he died in 1683, he was buried with honors at his home in Providence, but memory of him faded until the American Revolution when he gained a reputation as a father of religious freedom.

Bibliography: E. Gaustad, *Liberty of Conscience: Roger Williams in America*, 1991; O. Winslow, *Master Roger Williams: A Biography*, 1957.

Christopher E. Hendricks

WILMOT, JOHN, SECOND EARL OF ROCHESTER (1647–1680). The libertine court of *Charles II had no more rambunctious member than Rochester, who was as notorious for his wenching and profligacy as he was well known for his witty poetic satires. Born in Oxfordshire, he took an M.A. from Oxford at the age of fourteen, toured the Continent for three years, and returned to the Restoration court, becoming one of the king's gentlemen of the bed chamber. Among his numerous escapades was his kidnapping of the heiress Elizabeth Malet from the Tower of London; he married her in 1667. He was linked with a more violent incident twelve years later. After *John Dryden dedicated his play *Aurengzebe* (1675) to Lord Mulgrave, an enemy of Rochester's, relations between the two poets became very strained. When Dryden was severely beaten by thugs in a Covent Garden alley in 1679, Rochester expressed no regret over the event and was thought by many to have possibly organized it. By this time his health was broken as a result of his drinking and sexual adventures; he later admitted to *Gilbert Burnet that "for five years together he was continually drunk." His poetry, influenced by *Cowley and *Boileau, was largely unpublished until after his death but enjoyed wide circulation in manuscript, especially his pornographic verse; his social and political satires influenced Pope. His *Satire Against Reason and Mankind* depicts the conflict of the rational mind against sensual desires in a way that suggests the attitude of *Hobbes. He was also respected for his *Letter from Artemisia in the Town to Chloe in the Country*, a satire on social duplicity, and his *Allusion to Horace: The 10th Satire of the First Book*. The year before his death, Rochester requested a visit from Gilbert Burnet, later bishop of Salisbury, who helped him to a deathbed conversion. Recounted in Burnet's *Some Passages in the Life and Death of the Right Hon-*

orable John Earl of Rochester (1680), Rochester's final days became a cautionary tale of moral reprobation and divine mercy.

Bibliography: V. de sola Pinto, *Enthusiast in Wit*, 1962; D. Farley-Hills, *Rochester's Poetry*, 1978.

Christopher Baker

WINTHROP, JOHN (1588–1649). Winthrop was a leading figure in the creation of the Massachusetts Bay colony. He was born into the gentry class in Edwardstone, Suffolk, England, the son of Adam Winthrop and Anne Browne. In 1603, at the age of fifteen, he attended Trinity College, Cambridge, where he studied law. Two years later, he married Mary Forth, the first of four wives. In 1605, the couple had the first of Winthrop's sixteen children, only eight of whom reached adulthood. Winthrop lived the life of a country squire on Groton, his family's estate, serving as justice of the peace, and ultimately was appointed to government office as an attorney in the Court of Wards and Liveries in 1627.

Always a religious man, Winthrop became a Puritan. The increasing animosity against Puritans by the government of *King Charles I may have resulted in the loss of his government position in 1629. Faced with economic difficulties due to the lost appointment and his failing farm, Winthrop became a member of the Massachusetts Bay Company, a joint stock company that received a royal charter to establish a colony that same year. The charter differed from other colonial charters in that it did not specify where the company should be based. Winthrop and eleven other leaders signed the *Cambridge Agreement*, stating that the company would take the charter to New England, where they could establish a Puritan colony and ensure its independence. In 1630, Winthrop and some 700 colonists sailed for America. On board the ship *Arbella*, Winthrop composed the lay sermon *A Modell of Christian Charity*, calling for the Puritans to enter into a covenant with God to save the Christian church and build a community where the colonists would "be as a citty upon a hill." The Puritans stopped briefly at Salem, Massachusetts, but soon moved to Boston, which became the colony's capital and largest city.

The government of Massachusetts Bay consisted of a General Court, made up of shareholders or freemen, which chose assistants, who in turn elected a governor and deputy. This procedure violated the provisions of the charter, but when Winthrop was elected governor (the first of twelve terms), he kept the charter hidden and ruled somewhat independently, striving to guarantee the food supply, establish new towns, and see after the colony's defense. In 1632, the Watertown Protest, a controversy over tax issues, erupted into a two-year conflict that compelled Winthrop to produce the charter, whereupon the General Court reconstituted itself as a representative assembly. Winthrop was forced from the governor's office for three years.

As governor, Winthrop led the colony in the face of religious controversy. When the General Court ordered *Roger Williams banished to England, Winthrop, sympathetic to Williams, let the terms of the order be released, giving Williams time to flee to Rhode Island. Later Winthrop battled with *Anne Hutchinson during the Antinomian controversy. When he regained the governor's office in 1637, Winthrop oversaw Hutchinson's trial, which resulted in her expulsion from the colony. He later portrayed the stillborn birth of Hutchinson's child and her death four years later during an Indian attack on Long Island as proof of God's judgment against her.

Winthrop continued to lead the colony as governor, deputy, or assistant through the remainder of his life. He resisted Archbishop of Canterbury *William Laud's efforts as chairman of the Commission for Regulating Plantations to rein in the independent colony, preparing the colonists for armed combat if necessary and delaying requests to turn over the charter. Laud's efforts collapsed when the English Civil War broke out in 1642. Many New Englanders returned to England to aid the Puritan side, but Winthrop remained in Massachusetts writing political commentary criticizing the course the conflict was taking. Although he maintained a very conservative view of government, Winthrop was well loved and respected, and when he died 1649, his passing was mourned by the people on both sides of the Atlantic. Much of Winthrop's life and legacy are left in his prodigious output of correspondence, as well as a personal journal and a history of Massachusetts Bay.

Bibliography: E. Morgan, *The Puritan Dilemma: The Story of John Winthrop*, 1958; L. Schweninger, *John Winthrop*, 1992.

Christopher E. Hendricks

WITHER, GEORGE (1588–1667). Born in Hampshire and educated at Magdalen College, Oxford University, George Wither fought during England's civil war, eventually earning the rank of Major General in the Republican army. Throughout his life, he was a zealous Puritan and escaped execution at the Restoration due entirely to his reputation as a poet. Wither's poetry in earlier years had earned him this reputation after his student days at Oxford. His pastoral poetry—*The Shepherd's Hunting* (1615), *Fidelia* (1617), *Fair Virtue* (1622), and *Juvenilia* (1622)—remains his claim to fame if one views literature from an aesthetic standpoint. Equally as interesting from a biographical and historical perspective is his biting, Puritanical satire. Two particular works— *Abuses Stripped and Whipped, or Satirical Essays* (1613) and *Wither's Motto* (1621)—displeased authorities and earned him time in prison.

Later in life, Wither turned again to poetry, but of a more serious nature, as the title to the collection of 1624 indicates: *The Hymns and Songs of the Church*. Finally, returning to satire as the nation prepared for civil war, Wither wrote the work by which historians—particularly those studying bubonic plague— know him: *Britain's Remembrancer, Mr. Wither his prophesie of our present calamity and, except we repent, future misery* (1628), later followed (among other tracts and broadsides) by *Halelujah, or Britain's Second Remembrancer*

(1641). He exhorts his readers to heed his prophecy of impending doom and to amend their social and spiritual practices. Wither's exclamations earned him a lasting presence as a voice for England's future, with many works such as *Ecchoes from the sixth trumpet* . . . (1666) reissued posthumously as *Fragmenta prophetica, or, The remains of George Wither, Esq.: being a collection of the several predictions, dispers'd throughout his works: some verified in the au- thor's life-time, many yet unfulfill'd, the accomplishment whereof seem very probable: the last work of the author, and collected by his own hand a little before his death* (1669).

Bibliography: C. Hensley, *The Later Career of George Wither*, 1969.

Rebecca Totaro

WOOD, ANTHONY À (1632–1695).

Anthony Wood (he later added the à) was an affected eccentric who possessed a keen mind, excellent connections and, at the same time, a weak character and limited ambition. Born on 17 De- cember 1632 in Oxford, Wood spent almost his entire life in Oxford and in the two rooms that he had inherited from his father's estate. He was educated at the New College School, Thame Grammar School, and Merton College, Oxford. Wood developed an interest in English history and music and intended not to work. In 1660, he gained access to the archives of Oxford University and spent the next thirty-five years reading and writing on Oxford and Oxfordshire. His major work was *Historia et Antiquitates Universitatis Oxoniensis* (first two vol- umes, 1691–1692); additional volumes were published posthumously. Wood di- vided his work into three main sections: the city and the church in Oxford, the history of Oxford University, and the antiquities held by the various Oxford colleges. A gossip, Wood was involved in numerous personal disputes during his life; he had few friends and was criticized for his unorthodox political and religious views that were not deeply held but advanced to disrupt the common order. In addition to his publications on Oxford, Wood assisted other scholars on occasion but is remembered for gathering a wide range of documents and other source materials on Oxfordshire. Scholars in the nineteenth and twentieth centuries have used these materials in developing their understanding of Oxford life and history in the seventeenth century. Wood died on 29 November 1695 in Oxford; appropriately, he is buried in the Merton College Chapel.

Bibliography: L. Powis, ed., *The Life and Times of Anthony à Wood*, 1961.

William T. Walker

WREN, SIR CHRISTOPHER (1632–1723).

The English architect, astron- omer, and mathematician who designed St. Paul's Cathedral, London, and many other buildings after the Great Fire of 1666, was the most famous architect of the English Baroque. The son of the rector of East Knoyle, Wren was an An- glican, loyal to the monarchy during the civil war, the Restoration, and the

Glorious Revolution. Concomitantly, the monarchs *Charles II, *James II, and *William and Mary were among his patrons.

After attending the Westminster School, London, for five years and completing three years of private study, Wren studied at Wadham College, Oxford, 1649–1653. As a student, he read the Roman writer Vitruvius's *De architectura* and Renaissance treatises. Wren's early interests encompassed astronomy, anatomy, mechanics, and optics before he turned to architecture. Wren was professor of astronomy at Gresham College, London, in 1657 and Savilian professor of astronomy at Oxford (1661–1673). His achievements in mathematics and natural science allowed him to forge valuable friendships with members of the Royal Society.

As an architect, Wren applied principles of skeptical and experimental science popularized in England by *Francis Bacon, favoring architectural practice over theory and employing aesthetic judgment to modify geometrical form. In 1665, Wren studied for six months in Paris where he met with French and Italian architects. Upon returning to England, he proposed replacing the tower of Old St. Paul's Cathedral with a new dome, until the damage caused by the great fire convinced him the cathedral required total rebuilding. During the reconstruction of London, Wren helped establish standards for construction that were codified with the London Building Acts. Further, it was as Surveyor of the King's Works that Wren was appointed architect of St. Paul's in 1673. He received a knighthood in 1675.

Although Wren's city plan for rebuilding London was unexecuted, he shaped the skyline as the height of buildings was determined largely by the dimensions of St. Paul's. From 1670 to 1711, he was responsible for over forty London churches including St. Mary-le-Bow, one of his many spired buildings with spacious, unobstructed interiors for the Anglican liturgy. Among Wren's secular buildings are the Sheldonian Theater and Queen's College Library, Oxford; Temple Bar, London; the library for Trinity College, Cambridge; the Royal Hospital, Chelsea, and part of Greenwich Hospital; and the garden facade of Hampton Court Palace.

Bibliography: K. Downes, ed., *Sir Christopher Wren*, 1982; J. Summerson, *Sir Christopher Wren*, 1953.

Deborah H. Cibelli

WYCHERLY, WILLIAM (1641–1715). William Wycherly, dramatist and poet, was born in 1641 at Clive, near Shrewsbury, Shropshire. His father, Daniel Wycherly, the teller of the exchequer, sent him to the west of France for his education, where he converted to Catholicism under the influence of his mentor, Madame de Montausier. Upon returning to England, Wycherly reconverted to Anglicanism and enrolled at Queen's College, Oxford, although he never formally matriculated, and later at Lincoln's Inn. Indifferent to his studies, Wycherly preferred the life of a courtier and became a favorite of the duchess of Cleveland, mistress to *King Charles II.

Wycherly launched his literary career with *Hero and Leander*, a verse parody, but his reputation rests solely on his work as a dramatist. His first play, *Love in a Wood, or, St. James Park* debuted in 1671. Exploring the problems that arise from sexual jealousy, the play's central plot is the story of an heiress, Christina, and her obsessively jealous lover Valentine. The comedy features a fast-paced, complex, skillfully contrived, multilayered plot. It was an immediate success. Performed in 1672, Wycherely's next play, *The Gentleman Dancer*, gently attacks the notions of male intellectual supremacy and parental authority through the story of Hippolyta and her would-be husband Gerrard, who poses as a dance teacher to gain access to her. Lacking the complex plot structure of *Love in a Wood*, however, *The Gentleman Dancer* did poorly and closed after only a few performances. With *The Country Wife* (1675) Wycherly found success again. It recounts the exploits of Harry Horner, a rake who lets out the rumor that a cure for venereal disease has left him impotent, thus gaining him easy access to the wives of jealous husbands. With its skeptical but not cynical representation of human nature, the play is considered Wycherly's finest work. Wycherly's last play, *The Plain Dealer*, was performed in 1677. It is the story of Manly, a misanthropic but honest sea captain who returns home to find his beloved married to his best friend. Seeking revenge, Manly is aided by Fidelia, a woman who dresses as a man and follows him to sea and whom he eventually marries. *John Dennis credited the play for its excellent satire on hypocrisy and villainy. Comparing Wycherly's own forthright and plain-dealing manner with that of the play's protagonist, Dennis and *Dryden referred to the playwright as "Manly" Wycherly.

Thereafter, Wycherly wrote little of note and his fortunes declined. In 1678, he suffered a severe fever resulting in memory impairment. He journeyed to Montpelier to recuperate, with his expenses paid by the crown. Charles also offered Wycherly the position of tutor to Charles's son, the duke of Richmond, with a generous salary and pension. But Wycherly's secret marriage to Laetitia Isabella in 1680 cost him the favor of the king and the duchess, who expected their favorites to remain bachelors. He lost the tutorship and was banished from the court. Wycherly's marriage put him into debt, and when his wife died a year later, her estate was tied up in litigation. Wycherly was eventually imprisoned in Fleetwood prison for debt and remained there for three years. Finally, *James II pardoned him and granted him a pension of £200. With Wycherly's father, the king also paid the poet's remaining debt. In gratitude, Wycherly converted, once again, to Catholicism.

In his later years, Wycherly befriended Alexander Pope, who was fifty years his junior. The younger poet oversaw the 1704 publication of Wycherly's ill-received *Miscellany Poems*; relations between the two poets were strained, and Pope attacked Wycherly in *An Essay on Criticism*. Eleven days before his death in 1715, at age seventy-six, Wycherly married Elizabeth Jackson and left her his estate. He did so at the request of his cousin Thomas Shrimpton, who would become her second husband. Wycherly is ranked with *Congreve, *Etherege, and *Vanbrugh as one of the finest Restoration playwrights. Dark and satirical but not utterly cynical, his plays are noted for their craftsmanship and occasional brilliance.

Bibliography: B. McCarthy, *William Wycherly, A Reference Guide*, 1985.

Charles Pastoor

ZURBARÁN, FRANCISCO DE (1598–1664). Now recognized as one of the leading painters of the Spanish Baroque, Francisco de Zurbarán was born in Fuentes de Cantos, Badajoz. He was apprenticed from 1614 until 1616 to the *imaginero* Pedro Díaz de Villanueva of Seville, becoming familiar with polychrome sculpture, which influenced his sense of form. He also absorbed the naturalism that influenced contemporaries such as his friend *Diego Velázquez. Zurbarán left Seville in 1617, setting up a studio in Ilerena. In 1626, he contracted with the Dominicans of the monastery of S. Pablo El Real in Seville for a series of paintings. This led to work for the churches of San Pablo, the Trinitarians, and San Buenaventura, among others. His early experiments with dramatic contrasts of light and shadow, *tenebroso*, suggest the influence of *Caravaggio and the Italian Baroque, although Zurbarán's compositions are distinguished by drama within stillness.

He was invited by the municipality of Seville to take up permanent residence in 1629. This period saw an increasing realism in his work, most notably in the *Saint Bonaventure* series, and *Saint Hugh of Grenoble Visiting the Refectory*. Zurbarán spent a brief period in Madrid (1634–1635), where he produced his only secular works, *The Labours of Hercules* series, and *The Defense of Cádiz*. Returning to Seville, he executed a number of paintings for the Carthusian monastery at Jerez de la Frontera and the Hieronymite monastery at Guadalupe. He also exported a number of paintings to the New World. The works of his later career are marked by the influence of *Murillo, characterized by a lighter, more airy atmosphere. In 1658, he returned to Madrid in order to improve his fortunes, but without much success.

Zurbarán's subject matter was largely dictated by his monastic patrons, focusing on saints, monks, and the founders of religious orders. While his compositions were often awkward and his narrative frequently muddled, Zurbarán's draped figures were striking, fitting into a space that anticipates Cezanne and Picasso.

Bibliography: J. Gállego, *Zurbarán, 1598–1664*, 1987; P. Guinard, *Zurbarán et les peintres espagnols de la vie monastique*, 1960; M. Soria, *The Paintings of Zurbarán*, 1955.

Thomas L. Cooksey

Appendix A:
Entries Arranged by Subject

ART AND ARCHITECTURE

Bernini, Gian Lorenzo

Borromini, Francesco

Brueghel, Jan (the elder)

Brueghel, Pieter (the younger)

Carravaggio, Michelangelo Merisi da

Carriera, Rosalba

Churriguera, José de

Cuyp, Aelbert

Domenichino (Domenico Zampieri)

Finch, Anne

Gentileschi, Artemisia

Gibbons, Grinling

Giordano, Luca

Greco, El

Hals, Frans

Hawksmoor, Nicholas

Hilliard, Nicholas

Hollar, Wenceslaus

Honthorst, Gerrit van

Hooch, Pieter de

Jones, Inigo

Kneller, Sir Godfrey

La Tour, Georges de

Le Brun, Charles

Lely, Sir Peter

Leyster, Judith Jans

Lorrain (Lorraine), Claude

Maes, Nicolaes

Mansart (or Mansard), François

Murillo, Bartolome Esteban

Poussin, Nicolas

Rembrandt, Harmenszoon van Rijn

Reni, Guido

Ribera, Jusepe de

Rubens, Peter Paul

Ruisdael, Jacob Issackszoon van

Sirani, Elisabetta

Steen, Jan Havickszoon

Vanbrugh, John

Van Dyck, Anthony

Velázquez, Diego Rodríguez de Silva

Vermeer, Johannes

Watteau, Jean-Antoine

Wren, Sir Christopher

Zurbarán, Francisco de

EXPLORATION AND COLONIALISM

Bradford, William

Cadillac, Antoine de la Mothe

Champlain, Samuel de

Harvard, John

Hudson, Henry

Iberville et d'Ardillières, Pierre Le Moyne d'

Joliet, Louis

La Salle, Robert Cavalier, Sieur de

Marquette, Père Jacques

Minuit, Peter

Nicolet, Jean de Bellesborne

Penn, William

Pocahontas

Ralegh, Sir Walter

Smith, Captain John

Tasman, Abel Janszoon

Williams, Roger

Winthrop, John

LITERATURE

Addison, Joseph

Alarcón y Mendoza, Juan Ruiz de

Arbuthnot, John

Aubrey, John

Basile, Giovanni Battista

Boileau-Despréaux, Nicolas

Browne, Sir Thomas

Bunyan, John

Butler, Samuel

Campion, Thomas

Carew, Thomas

Cavendish, Margaret

Cervantes, Miguel de

Chapman, George

Coryate, Thomas

Cowley, Abraham

Crashaw, Richard

Cyrano de Bergerac, Savinien de

Daniel, Samuel

Denham, Sir John

Dennis, John

Donne, John

Drayton, Michael

Dryden, John

D'Urfé, Honoré

Earle, John

Evelyn, John

Fabricius, Johann Albert

Felltham, Owen

Fleming, Paul

Furetière, Antoine

Góngora y Argote, Luis de

Gracián y Morales, Baltasar

Grimmelshausen, Hans Jacob Christoffel von

Gryphius, Andreas

Herbert, George

Herrick, Robert

Hoffmanswaldau, Christian Hoffmann von

Howell, James

Huygens, Sir Constantijn

Johansson, Lars

Junius, Franciscus, the Younger

King, Henry

La Bruyère, Jean de

La Calprenède, Gautier de Costes

Lafayette, Marie-Madeleine (Pioche de la Vergne), Comtesse de

La Fontaine, Jean de

La Rochefoucauld, François VI, duc de

Le Bossu, René, Abbé

L'Estrange, Sir Roger

Lobo, Francisco Rodrigues

Lovelace, Richard

Marino, Giambattista

Marvell, Andrew

Milton, John

Opitz, Martin

Overbury, Sir Thomas

Parnell, Thomas

Pepys, Samuel

Perrault, Charles

Philips, Katherine

Prior, Matthew

Quarles, Francis

Quevedo y Villegas, Francisco Gomez de

Randolph, Thomas

Rowe, Nicholas

Rymer, Thomas

Savage, Richard

Savile, George, Marquis of Halifax

Scarron, Paul

Scheffler, Johann (Angelius Silesius)

Scudéry, Madeleine de

Sévigné, Marie (de Rabutin-Chantal), Marquise de

Steele, Sir Richard

Taylor, Edward

Traherne, Thomas

Urquhart, Sir Thomas

Vaughan, Henry

Viau, Théophile de

Vondel, Joost van den

Waller, Edmund

Walton, Izaak

Weise, Christian

Wilmot, John, Second Earl of Rochester

Wither, George

Wood, Anthony à

MATHEMATICS

Cavalieri, Bonaventura

Desargues, Girard

Fermat, Pierre de

Graunt, John

Gregory, James

Gunter, Edmund

Huet, Pierre-Daniel

Leybourn, William

Mersenne, Marin

Napier, John

Oughtred, William

Pell, John

Petty, William

Roberval, Gilles Personne de

Torricelli, Evangelista

Wallis, John

MONARCHS AND RULERS

Anne

Charles I

Charles II

Cromwell, Oliver

Frederick William

James I

James II

Louis XIV

Mary II

Peter I

Philip IV

Richelieu, Armand Jean du Plessis de

William III

MUSIC

Amati, Nicolo

Blow, John

Boyce, William

Bull, John

Buxtehude, Dietrich

Byrd, William

Carissimi, Giacomo

Charpentier, Marc-Antoine

Corelli, Arcangelo

Couperin, François

Dowland, John

Frescobaldi, Girolamo

Gabrieli, Giovanni

Gesualdo, Carlo

Gibbons, Orlando

Locke, Matthew

Lully, Jean-Baptiste

Monteverdi, Claudio

Morley, Thomas

Pachelbel, Johann

Pergolesi, Giovanni Battista

Peri, Jacopo

Praetorius, Michael

Purcell, Henry

Ravenscroft, Thomas

Scarlatti, (Pietro) Alessandro (Gaspare)

Schütz, Heinrich

Stradivari, Antonio

Sweelinck, Jan Pieterszoon

Victoria, Tomás Luis de

Watts, Isaac

PHILOSOPHY

Bacon, Francis

Bayle, Pierre

Campanella, Tommaso

Cudworth, Ralph

Descartes, René

Gassendi, Pierre

Geulincx, Arnold

Herbert, Edward, Baron of Chirbury

Hobbes, Thomas

La Mothe le Vayer, François de

Leibniz, Gottfried Wilhelm von

Locke, John

Malebranche, Nicolas

Sanches, Francisco

Shaftesbury, Anthony Ashley Cooper, Third Earl of

Spinoza, Baruch de

POPES

Alexander VII, Pope

Clement XI, Pope

Innocent X, Pope

Innocent XI, Pope

Paul V, Pope

Urban VIII, Pope

RELIGION

Amyraut, Moïse

Andreae, Johann Valentin

Andrewes, Lancelot

Arminius, Jacob

Arnauld, Antoine

Bellarmine, Cardinal Robert

Bérulle, Cardinal Pierre de

Boehme, Jakob

Bossuet, Jacques-Bénigne

Burnet, Thomas

Casaubon, Isaac

Chillingworth, William

Comenius, John Amos

Cotton, John

Du Vergier de Hauranne, Jean

Eliot, John

Episcopius, Simon

Escobar y Mendoza, Antonio

Fénelon, François de Salignac de la Mothe

Fox, George

Francis de Sales, St.

Gerhard, Johann

Gerhardt, Paul

Glanvill, Joseph

Hall, Joseph

Harris, John

Hooker, Thomas
Hutchinson, Anne
Jansen, Cornelius Otto
Ken, Thomas
La Salle, Jean-Baptiste de
Laud, William
Lilburne, John
Lobo, Jerónimo
Mabillon, Jean
Mather, Cotton
Mather, Increase
Molinos, Miguel de
More, Henry
Muggleton, Lodowicke
Nicole, Pierre
Nikon
Norton, John
Pascal, Blaise
Prynne, William
Shabbetai Zevi
Stillingfleet, Edward
Suárez, Francisco
Taylor, Jeremy
Ussher, James
Vincent de Paul
Vossius, Gerhardus Johannes (Voss, Gerhard Jan)
Whichcote, Benjamin

SCHOLARSHIP

Bodley, Thomas
Boulainvilliers, Henri de
Burnet, Gilbert
Byrd, William (the elder)
Camden, William
Carew, Richard

Colbert, Jean-Baptiste
Filmer, Robert
Garcilaso de la Vega
Grotius, Hugo
Harrington, James
Hearne, Thomas
Heinsius, Daniel
Hyde, Edward, Earl of Clarendon
Lipsius, Justus (Lips, Joest)
Maimbourg, Louis
Minsheu, John
Naudé, Gabriel
Pufendorf, Baron Samuel
Purchas, Samuel
Salmasius, Claudius
Selden, John
Speed, John
Temple, Sir William
Thou, Jacques Auguste de

SCIENCE

Baffin, William
Bernoulli, Jacques
Borelli, Giovanni Alfonso
Boyle, Robert
Burton, Robert
Cassini, Giovanni Domenico
Fahrenheit, Gabriel Daniel
Flamsteed, John
Galilei, Galileo
Gascoigne, William
Gilbert, William
Glauber, Johann Rudolf
Graaf, Regnier de
Grimaldi, Francesco Maria
Guericke, Otto van

Halley, Edmond

Harriot, Thomas

Hartsoeker, Nikolaas

Harvey, William

Haukesbee, Francis

Helmont, Johannes Baptista van

Hevelius, Johannes

Hooke, Robert

Horrocks, Jeremiah

Huygens, Christiaan

Kepler, Johannes

Kircher, Athanasius

Leeuwenhoek, Antony van

Lemery, Nicolas

Lilly, William

Lower, Richard

Malpighi, Marcello

Newcomen, Thomas

Newton, Isaac

Norden, John

Oughtred, William

Pacchioni, Antonio

Pecquet, Jean

Picard, Jean

Ray, John

Renaudot, Théophraste

Roemer, Olaus

Rudbeck, Olof

Sarpi, Paolo Pietro

Sauveur, Joseph

Scheiner, Christoph

Snell (or Snellius), Willebrord van Roijen

Stahl, Georg Ernst

Steno, Nicholas (Niels Stenson)

Swammerdam, Jan

Sydenham, Thomas

Tompion, Thomas

SOCIETY

Kidd, William

Lenclos, Anne de

Lloyd, Edward

MacGregor, Robert

Maintenon, Marquise de Françoise d'Aubigné

THEATER

Beaumont, Francis

Behn, Aphra

Calderón de la Barca, Pedro

Congreve, William

Corneille, Pierre

Davenant, Sir William

Dekker, Thomas

Etherege, George

Farquhar, George

Fletcher, John

Ford, John

Gay, John

Gwyn, Nell

Heywood, Thomas

Jonson, Ben

Killigrew, Thomas

Marston, John

Massinger, Philip

Middleton, Thomas

Molière, [Poquelin] Jean-Baptiste

Oldfield, Anne

Otway, Thomas

Quinault, Phillippe

Racine, Jean-Baptiste

Regnard, Jean-François

Rotrou, Jean de

Shakespeare, William
Shirley, James
Tate, Nahum
Tirso de Molina

Tourneur, Cyril
Vega, Lope Felix de
Webster, John
Wycherly, William

Appendix B:
Entries Arranged by Country

BELGIUM

Brueghel, Jan (the elder)

Brueghel, Pieter (the younger)

Geulincx, Arnold

Helmont, Johannes Baptista van

Rubens, Peter Paul

Van Dyck, Anthony

BOHEMIA

Hollar, Wenceslaus

CRETE

Greco, El

CZECHOSLOVAKIA

Comenius, John Amos

DANZIG

Hevelius, Johannes

DENMARK

Róemer, Olaus

Steno, Nicholas (Niels Stensen)

ENGLAND

Addison, Joseph

Andrewes, Lancelot

Anne

Aubrey, John

Bacon, Francis

Baffin, William

Beaumont, Francis

Behn, Aphra

Blow, John

Bodley, Thomas

Boyce, William

Boyle, Robert

Bradford, William

Browne, Sir Thomas

Bull, John

Bunyan, John

Burnet, Gilbert

Burnet, Thomas

Burton, Robert

Butler, Samuel

Byrd, William

Byrd, William (the elder)

Camden, William

Campion, Thomas

Carew, Richard

Carew, Thomas

Cavendish, Margaret

Chapman, George

Charles I

Charles II

Chillingworth, William

Congreve, William

Coryate, Thomas
Cotton, John
Cowley, Abraham
Crashaw, Richard
Cromwell, Oliver
Cudworth, Ralph
Daniel, Samuel
Davenant, Sir William
Dekker, Thomas
Denham, Sir John
Dennis, John
Donne, John
Dowland, John
Drayton, Michael
Dryden, John
Earle, John
Eliot, John
Etherege, George
Evelyn, John
Farquhar, George
Felltham, Owen
Filmer, Robert
Finch, Anne, Countess of Winchelsea
Flamsteed, John
Fletcher, John
Ford, John
Fox, George
Gascoigne, William
Gay, John
Gibbons, Grinling
Gibbons, Orlando
Gilbert, William
Glanvill, Joseph
Graunt, John
Gunter, Edmund
Gwyn, Nell
Hall, Joseph

Halley, Edmond
Harrington, James
Harriot, Thomas
Harris, John
Harvard, John
Harvey, William
Haukesbee, Francis
Hawksmoor, Nicholas
Hearne, Thomas
Herbert, Edward, Baron of Chirbury
Herbert, George
Herrick, Robert
Heywood, Thomas
Hilliard, Nicholas
Hobbes, Thomas
Hooke, Robert
Hooker, Thomas
Horrocks, Jeremiah
Howell, James
Hudson, Henry
Hutchinson, Anne
Hyde, Edward, Earl of Clarendon
James I
James II
Jones, Inigo
Jonson, Ben
Ken, Thomas
Killigrew, Thomas
King, Henry
Laud, William
L'Estrange, Sir Roger
Leybourn, William
Lilburne, John
Lilly, William
Lloyd, Edward
Locke, John
Locke, Matthew

Lovelace, Richard

Lower, Richard

Marston, John

Marvell, Andrew

Mary II

Massinger, Philip

Middleton, Thomas

Milton, John

Minsheu, John

More, Henry

Morley, Thomas

Muggleton, Lodowicke

Newcomen, Thomas

Newton, Isaac

Norden, John

Norton, John

Oldfield, Anne

Otway, Thomas

Oughtred, William

Overbury, Sir Thomas

Pell, John

Penn, William

Pepys, Samuel

Petty, William

Philips, Katherine

Prior, Matthew

Prynne, William

Purcell, Henry

Purchas, Samuel

Quarles, Francis

Ralegh, Sir Walter

Randolph, Thomas

Ravenscroft, Thomas

Ray, John

Rowe, Nicholas

Rymer, Thomas

Savage, Richard

Savile, George, Marquis of Halifax

Selden, John

Shaftesbury, Anthony Ashley Cooper, Third Earl of

Shakespeare, William

Shirley, James

Smith, Captain John

Speed, John

Stillingfleet, Edward

Sydenham, Thomas

Tate, Nahum

Taylor, Edward

Taylor, Jeremy

Temple, Sir William

Tompion, Thomas

Tourneur, Cyril

Traherne, Thomas

Vanbrugh, John

Vaughan, Henry

Waller, Edmund

Wallis, John

Walton, Izaak

Watts, Isaac

Webster, John

Whichcote, Benjamin

William III

Williams, Roger

Wilmot, John, Second Earl of Rochester

Winthrop, John

Wither, George

Wood, Anthony à

Wren, Sir Christopher

Wycherly, William

FLANDERS

Lipsius, Justus (Lips, Joest)

FRANCE

Amyraut, Moïse

Arnauld, Antoine

Bayle, Pierre

Bérulle, Cardinal Pierre de

Boileau-Despréaux, Nicolas

Bossuet, Jacques-Bénigne

Boulainvilliers, Henri de

Cadillac, Antoine de la Mothe

Champlain, Samuel de

Charpentier, Marc-Antoine

Colbert, Jean-Baptiste

Corneille, Pierre

Couperin, François

Cyrano de Bergerac, Savinien de

Desargues, Girard

Descartes, René

D'Urfé, Honoré

Du Vergier de Hauranne, Jean

Fénelon, François de Salignac de la Mothe

Fermat, Pierre de

Francis de Sales, St.

Furetière, Antoine

Gassendi, Pierre

Huet, Pierre-Daniel

Iberville et d'Ardillières, Pierre Le Moyne d'

Joliet, Louis

La Bruyère, Jean de

La Calprenède, Gautier de Costes

Lafayette, Marie-Madeleine (Pioche de la Vergne), Comtesse de

La Fontaine, Jean de

La Mothe le Vayer, François de

La Rochefoucauld, François VI, Duc de

La Salle, Jean-Baptiste de

La Salle, Robert Cavalier, Sieur de

La Tour, Georges de

Le Bossu, René, Abbé

Le Brun, Charles

Lemery, Nicolas

Lenclos, Anne de

Lorrain (Lorraine), Claude

Louis XIV

Lully, Jean-Baptiste

Mabillon, Jean

Maimbourg, Louis

Maintenon, François d'Aubigné, Marquise de

Malebranche, Nicolas

Mansart (or Mansard), François

Marquette, Père Jacques

Mersenne, Marin

Molière, [Poquelin] Jean-Baptiste

Naudé, Gabriel

Nicole, Pierre

Nicolet, Jean de Bellesborne

Pascal, Blaise

Pecquet, Jean

Perrault, Charles

Picard, Jean

Poussin, Nicolas

Quinault, Phillippe

Racine, Jean-Baptiste

Regnard, Jean-François

Renaudot, Théophraste

Richelieu, Armand Jean du Plessis de

Roberval, Gilles Personne de

Rotrou, Jean de

Salmasius, Claudius

Sanches, Francisco

Sauveur, Joseph

Scarron, Paul

Scudéry, Madeleine de

Sévigné, Marie (de Rabutin-Chantal), Marquise de

Thou, Jacques Auguste de

Viau, Théophile de

Vincent de Paul

Watteau, Jean-Antoine

GERMANY

Andreae, Johann Valentin

Boehme, Jakob

Fabricius, Johann Albert

Fahrenheit, Gabriel Daniel

Fleming, Paul

Gerhard, Johann

Gerhardt, Paul

Glauber, Johann Rudolf

Grimmelshausen, Hans Jacob Christoffel von

Gryphius, Andreas

Guericke, Otto von

Junius, Franciscus, the Younger

Kepler, Johannes

Kircher, Athanasius

Kneller, Sir Godfrey

Leibniz, Gottfried Wilhelm von

Opitz, Martin

Pachelbel, Johann

Praetorius, Michael

Pufendorf, Baron Samuel

Scheffler, Johann (Angelus Silesius)

Scheiner, Christoph

Schütz, Heinrich

Stahl, Georg Ernest

Weise, Christian

IRELAND

Parnell, Thomas

Steele, Sir Richard

Ussher, James

ITALY

Alexander VIII, Pope

Amati, Nicolo

Basile, Giovanni Battista

Bellarmine, Cardinal Robert

Bernini, Gian Lorenzo

Borelli, Giovanni Alfonso

Borromini, Francesco

Campanella, Tommaso

Caravaggio, Michelangelo Merisi da

Carissimi, Giacomo

Carriera, Rosalba

Cassini, Giovanni Domenico

Cavalieri, Bonaventura

Clement XI, Pope

Corelli, Arcangelo

Domenichino (Domenico Zampieri)

Frescobaldi, Girolamo

Gabrieli, Giovanni

Galilei, Galileo

Gentileschi, Artemisia

Gesualdo, Carlo

Giordano, Luca

Grimaldi, Francesco Maria

Innocent X, Pope

Innocent XI, Pope

Malpighi, Marcello

Marino, Giambattista

Monteverdi, Claudio

Pacchioni, Antonio

Paul V, Pope

Pergolesi, Giovanni Battista

Peri, Jacopo

Reni, Guido

Sarpi, Paolo Pietro

Scarlatti, (Pietro) Alessandro (Gaspare)

Sirani, Elisabetta

Stradivari, Antonio

Torricelli, Evangelista

Urban VIII, Pope

MASSACHUSETTS BAY COLONY

Mather, Cotton

Mather, Increase

NETHERLANDS

Arminius, Jacob

Cuyp, Aelbert

Episcopius, Simon

Graaf, Regnier de

Grotius, Hugo

Hals, Franz

Hartsoeker, Nikolaas

Heinsius, Daniel

Honthorst, Gerrit van

Hooch, Pieter de

Huygens, Christiaan

Huygens, Sir Constantijn

Jansen, Cornelius Otto

Leeuwenhoek, Antony van

Lely, Sir Peter

Leyster, Judith Jans

Maes, Nicolaes

Minuit, Peter

Rembrandt Harmenszoon van Rijn

Ruisdael, Jacob Isaackszoon van

Snell (or Snellius), Willibrord van Roijen

Spinoza, Baruch de

Steen, Jan Havickszoon

Swammerdam, Jan

Sweelinck, Jan Pieterszoon

Tasman, Abel Janszoon

Vermeer, Johannes

Vondel, Joost van den

Vossius, Gerhardus Johannes (Voss, Gerhard Jan)

NORTH AMERICAN ALGONQUIN

Pocahontas

POLAND

Hoffmannswaldau, Christian Hoffman von

PORTUGAL

Lobo, Francisco Rodrigues

Lobo, Jerónimo

PRUSSIA

Frederick William

RUSSIA

Nikon

Peter I

SCOTLAND

Arbuthnot, John

Gregory, James

Kidd, William

MacGregor, Robert ("Rob Roy")

Napier, John

Urquhart, Sir Thomas

SPAIN

Alarcón Mendoza, Juan Ruiz de
Calderón de la Barca, Pedro
Cervantes, Miguel de
Churriguera, José de
Escobar y Mendoza, Antonio
Garcilaso de la Vega
Góngora y Argote, Luis de
Gracián y Morales, Baltasar
Molinos, Miguel de
Murillo, Bartolome Esteban
Philip IV
Quevedo y Villegas, Francisco Gomez de
Ribera, Jusepe de
Suárez, Francisco
Tirso de Molina

Vega, Lope Félix de
Velázquez, Diego Rodríguez de Silva
Victoria, Tomás Luis de
Zurbarán, Francesco de

SWEDEN

Buxtehude, Dietrich
Johansson, Lars
Rudbeck, Olof

SWITZERLAND

Bernoulli, Jacques
Casaubon, Isaac

TURKEY

Shabbetai Zevi

Bibliography

GENERAL

Anderson, V. *New England's Generation: The Great Migration and the Formation of Society and Culture in the Seventeenth Century*. Cambridge: Cambridge University Press, 1993.

Beik, W. *Urban Protest in Seventeenth-Century France: The Culture of Retribution*. Cambridge: Cambridge University Press, 1997.

Brace, L. *The Idea of Property in Seventeenth-Century England*. New York: St. Martin's, 1998.

Collinson, P. *Puritan Character: Polemics and Polarities in Early Seventeenth Century English Culture*. Los Angeles: William Andrews Clark Memorial Library, 1989.

Darby, G. *Spain in the Seventeenth Century*. London: Longman, 1995.

Davis, N. *Society and Culture in Early Modern France*. Stanford: Stanford University Press, 1975.

DiVitiis, G. *English Merchants in Seventeenth-Century Italy*. Cambridge: Cambridge University Press, 1997.

Fox, E., and S. Shefter. *The Emergence of the Modern European World: From the Seventeenth to the Twentieth Century*. Oxford: Blackwell, 1991.

Greer, A. *The Jesuit Relations: Natives and Missionaries in Seventeenth-Century North America*. New York: St. Martin's, 2000.

Harth, E. *Ideology and Culture in Seventeenth-Century France*. Ithaca: Cornell University Press, 1983.

Hill, C. *Change and Continuity in Seventeenth-Century England*. Rev. ed. New Haven: Yale University Press, 1991.

Jones, J. *Britain and Europe in the Seventeenth Century*. New York: Norton, 1966.

Maland, D. *Culture and Society in Seventeenth-Century France*. New York: Charles Scribner's Sons, 1970.

Miller, J. *Absolutism in Seventeenth-Century Europe*. New York: St. Martin's, 1990.

Parker, G., and L. Smith. *The General Crisis of the Seventeenth Century*. London: Routledge, 1997.

Pennington, D. *Europe in the Seventeenth Century*. London: Longman, 1989.

Price, J. *The Dutch Republic in the Seventeenth Century*. New York: St. Martin's, 1999.

Safley, T. *Matheus Miller's Memoir: A Merchant's Life in the Seventeenth Century*. New York: St. Martin's Press, 1999.

Scarre, G. *Witchcraft and Magic in Sixteenth- and Seventeenth-Century Europe*. New York: St. Martin's, 1997.

Sella, D. *Italy in the Seventeenth Century*. New York: Addison-Wesley, 1997.

Sharpe, K. *Remapping Early Modern England: The Culture of Seventeenth-Century Politics*. Cambridge: Cambridge University Press, 2000.

Wedgewood, C. *The Thirty Years War*. New York: Anchor Books, 1961.

ART AND ARCHITECTURE

Alpers, S. *The Art of Describing: Dutch Art in the Seventeenth Century*. Chicago: University of Chicago Press, 1983.

Bold, J. *John Webb: Architectural Theory and Practice in the Seventeenth Century*. Oxford: Oxford University Press, 1990.

Brown, J. *The Golden Age of Painting in Spain*. New Haven: Yale University Press, 1991.

———. *Images and Ideas in Seventeenth-Century Spanish Painting*. Princeton: Princeton University Press, 1993.

Canaday, J. *Baroque Painters*. New York: Norton, 1969.

Chapman, H. *Rembrandt's Self-Portraits: A Study in Seventeenth-Century Identity*. Princeton: Princeton University Press, 1990.

Cliffe, J. *The World of the Country House in Seventeenth-Century England*. New Haven: Yale University Press, 1999.

Colantuono, A. *Guido Reni's Abduction of Helen: The Politics and Rhetoric of Painting in Seventeenth-Century Europe*. Cambridge: Cambridge University Press, 1997.

Duro, P. *The Academy and the Limits of Painting in Seventeenth-Century France*. Cambridge: Cambridge University Press, 1997.

Earls, I. *Baroque Art: A Topical Dictionary*. Westport: Greenwood, 1996.

Ford, B. *The Cambridge Guide to Arts in Britain: The Seventeenth Century*. Cambridge: Cambridge University Press, 1989.

Franits, W. *Paragons of Virtue: Women and Domesticity in Seventeenth-Century Dutch Art*. Cambridge: Cambridge University Press, 1993.

———, ed. *Looking at Seventeenth Century Dutch Art: Realism Reconsidered*. Cambridge: Cambridge University Press, 1997.

Freedberg, D., and J. de Vries, eds. *Art in History. History in Art: Studies in Seventeenth-Century Dutch Culture*. Los Angeles: Getty Trust, 1996.

Hersey, G. *Architecture and Geometry in the Age of the Baroque*. Chicago: University of Chicago Press, 2000.

Jordan, W. *Spanish Still Life in the Golden Age*. New York: Harry Abrams, 1988.

Kahr, M. *Dutch Painting in the Seventeenth Century*. Boulder: Westview Press, 1992.

Kaufmann, E. *Architecture in the Age of Reason: Baroque and Post-Baroque in England, Italy and France*. New York: Dover, 1990.

Merot, A. *French Painting in the Seventeenth Century*. New Haven: Yale University Press, 1995.

Millon, H., ed. *The Triumph of the Baroque: Architecture in Europe, 1600–1750*. New York: Rizzoli, 1999.

North, M., and C. Hill. *Art and Commerce in the Dutch Golden Age: A Social History of Seventeenth-Century Netherlandish Painting*. New Haven: Yale University Press, 1997.

Ridgeway, C., and R. Williams, eds. *Sir John Vanbrugh and Landscape Architecture in Baroque England 1690–1730*. Stroud: Sutton Publishing, 2000.

Southorn, J. *Power and Display in the Seventeenth Century: The Arts and Their Patrons in Modena and Ferrara*. Cambridge: Cambridge University Press, 1988.

Stoichita, V. *Visionary Experience in the Golden Age of Spanish Art*. London: Reaktion, 1996.

Summerson, J. *Architecture in Britain: 1530–1830*. New Haven: Yale University Press, 1990.

Varriano, J. *Italian Baroque and Rococco Architecture*. Oxford: Oxford University Press, 1990.

Wind, B. *Velasquez's Bodegones: A Study in Seventeenth-Century Spanish Genre Painting*. Lanham: George Mason University Press, 1987.

Wittkower, R., and J. Montague. *Art and Architecture in Italy, 1600–1750*. New Haven: Yale University Press, 1999.

Wölfflin, H. *Renaissance and Baroque*. Ithaca: Cornell University Press, 1994.

Wright, C. *French Painters of the Seventeenth Century*. Boston: Little Brown, 1985.

LITERATURE

Beasley, F. *Revising Memory: Women's Fiction and Memoirs in Seventeenth-Century France*. New Brunswick: Rutgers University Press, 1991.

Brownlee, M., and H. Gumbrecht, eds. *Cultural Authority in Golden Age Spain*. Princeton: Princeton University Press, 1998.

Contag, K. *Mockery in Spanish Golden Age Literature: Analysis of Burlesque Representations*. Lanham: University Press of America, 1996.

Gillespie, G. *Garden Labyrinth of Time: Studies in Renaissance and Baroque Literature*. Bern: Peter Lang, 1988.

Goldberg, J. *James I and the Politics of Literature: Jonson, Shakespeare, Donne and Their Contemporaries*. Stanford: Stanford University Press, 1989.

Guibbory, A. *Ceremony and Community from Herbert to Milton: Literature, Religion, and Culture in Conflict in Seventeenth-Century England*. Cambridge: Cambridge University Press, 1998.

Hill, C. *Milton and the English Revolution*. New York: Viking Press, 1978.

Hunter, W., ed. *A Milton Encyclopedia*. Lewisburg: Bucknell University Press, 1978–1983.

Lewalski, B. *Protestant Poetics and the Seventeenth-Century Religious Lyric*. Princeton: Princeton University Press, 1979.

Mariscal, G. *Contradictory Subjects: Quevedo, Cervantes, and Seventeenth-Century Spanish Culture*. Ithaca: Cornell University Press, 1991.

Martz, L. *The Poetry of Meditation*. New Haven: Yale University Press, 1962.

Moriarty, M. *Taste and Ideology in Seventeenth-Century France*. New York: Cambridge University Press, 1988.

Newman, J. *Pastoral Conventions: Poetry, Language, and Thought in Seventeenth-Century Nuremburg*. Baltimore: Johns Hopkins Press, 1990.

Pascal, R. *German Literature in the Sixteenth and Seventeeth Centuries: Renaissance, Reformation, Baroque*. Westport: Greenwood, 1979.

Rothstein, E. *Restoration and Eighteenth-Century Poetry, 1660–1780*. Boston: Routledge and Kegan Paul, 1981.

Rubin, D. *Knot of Artifice: A Poetic of the French Lyric in the Early Seventeenth Century*. Columbus: Ohio State University Press, 1981.

Skrine, P. *The Baroque: Literature and Culture in Seventeenth-Century Europe*. London: Methuen, 1978.

Smith, P. *Writing in the Margin: Spanish Literature of the Golden Age*. Oxford: Oxford University Press, 1988.

Soufas, T. *Melancholy and the Secular Mind in Spanish Golden Age Literature*. Columbia: University of Missouri Press, 1990.

Stone, H. *The Classical Model: Literature and Knowledge in Seventeenth-Century France*. Ithaca: Cornell University Press, 1996.

Wardropper, B., ed. *Spanish Poetry of the Golden Age*. New York: Irvington Publishers, 1977.

Wilding, M. *Dragon's Teeth: Literature in the English Revolution*. Oxford: Oxford University Press, 1987.

Young, R. *Doctrine and Devotion in Seventeenth-Century Poetry*. Cambridge: D.S. Brewer, 2000.

————. *Richard Crashaw and the Spanish Golden Age*. New Haven: Yale University Press, 1982.

MUSIC

Apel, W., and T. Binkley, eds. *Italian Violin Music of the Seventeenth Century*. Bloomington: Indiana University Press, 1990.

Bianconi, L. *Music in the Seventeenth Century*. Cambridge: Cambridge University Press, 1995.

Gouk, P. *Music, Science, and Natural Magic in Seventeenth-Century England*. New Haven: Yale University Press, 1999.

Herrisone, R. *Music Theory in Seventeenth-Century England*. Oxford: Oxford University Press, 2001.

Ledbetter, D. *Harpsichord and Lute Music in Seventeenth-Century France*. Bloomington: Indiana University Press, 1987.

Neumann, F., and J. Stevens. *Performance Practices of the Seventeenth and Eighteenth Centuries*. New York: Schirmer Books, 1993.

Rosand, E. *Opera in Seventeenth-Century Venice: The Creation of a Genre*. Berkeley: University of California Press, 1990.

Stein, L. *Songs of Mortals, Dialogues of the Gods: Music and Theatre in Seventeeth-Century Spain*. Oxford: Oxford University Press, 1993.

Stevenson, R. *Spanish Cathedral Music in the Golden Age*. Westport: Greenwood, 1976.

Thompson, T. *The Seventeenth-Century English Hymn: A Mode for Sacred and Secular Concerns*. Bern: Peter Lang, 1989.

Wainright, J. *Musical Patronage in Seventeenth-Century England: Christopher, First Baron Hatton (1605–1670)*. Aldershot: Ashgate, 1997.

Worsthorne, S. *Venetian Opera in the Seventeenth Century*. New York: Da Capo Press, 1984.

PHILOSOPHY

Barnes, T., and G. Feldman, eds. *Rationalism and Revolution, 1660–1815*. Lanham: University Press of America, 1979.

Deleuze, G. *Leibniz and the Baroque*. Minneapolis: University of Minnesota Press, 1992.

Gaukroger, S. *The Soft Underbelly of Reason: The Passions in the Seventeenth Century.* London: Routledge, 1998.

James, S. *Passion and Action: The Emotions in Seventeenth-Century Philosophy.* Oxford: Oxford University Press, 1999.

Meyer, R. *Leibniz and the Seventeenth Century Revolution.* Leiden: Brill, 1952.

Parkinson, G., ed. *Routledge History of Philosophy, Volume IV: The Renaissance and Seventeenth-Century Rationalism.* London: Routledge, 1993.

Pereboom, D., ed. *The Rationalists: Critical Essays on Descartes, Spinoza, and Leibniz.* Lanham: Rowman and Littlefield, 1999.

Reiss, T. *Knowledge, Discovery and Imagination in Early Modern Europe: The Rise of Aesthetic Rationalism.* Cambridge: Cambridge University Press, 1997.

Schiffhorst, G., and K. Keller, eds. *The Witness of Times: Manifestations of Ideology in Seventeenth-Century England.* Pittsburgh: Duquesne University Press, 1993.

Stewart, M., ed. *Studies in Seventeenth-Century European Philosophy.* Oxford: Oxford University Press, 1997.

Twersky, I., ed. *Jewish Thought in the Seventeenth Century.* Cambridge: Harvard University Press, 1987.

Wiley, M. *Subtle Knot: Creative Scepticism in Seventeenth-Century England.* Westport: Greenwood, 1969.

Willey, B. *The Seventeenth Century Background.* New York: Columbia University Press, 1942.

Yolton, J. *Philosophy, Religion and Science in the Seventeenth and Eighteenth Centuries.* Rochester: University of Rochester Press, 1990.

RELIGION

Almond, P. *Adam and Eve in Seventeenth-Century Thought.* Cambridge: Cambridge University Press, 1999.

Armstrong, B. *Calvinism and the Amyraut Heresy: Protestant Scholasticism and Humanism in Seventeenth-Century France.* Madison: University of Wisconsin Press, 1969.

Bond, E. *Damned Souls in a Tobacco Colony: Religion in Seventeenth-Century Virginia.* Macon: Mercer University Press, 2000.

Campbell, K. *The Intellectual Struggle of the English Papists in the Seventeenth Century: The Catholic Dilemma.* Lewiston: Edwin Mellen Press, 1986.

Campbell, T. *The Religion of the Heart: A Study of European Religious Life in the Seventeenth and Eighteenth Centuries.* Columbia: University of South Carolina Press, 1991.

Carden, A. *Puritan Christianity in America: Religion and Life in Seventeenth-Century Massachusetts.* Grand Rapids: Baker Book House, 1990.

Cunningham, B. *The World of Geoffrey Keating: History, Myth and Religion in Seventeenth-Century Ireland.* Portland: Four Courts Press, 2000.

Davies, H. *Worship and Theology in England.* Grand Rapids: Eerdmans, 1996.

de Certeau, M. *The Mystic Fable: The Sixteenth and Seventeenth Centuries.* Chicago: University of Chicago Press, 1992.

Deursen, A. van. *Plain Lives in a Golden Age: Popular Culture, Religion, and Society in Seventeenth-Century Holland.* Cambridge: Cambridge University Press, 1991.

Harline, C. *A Bishop's Tale: Mathias Hovius among his Flock in Seventeenth-Century Flanders*. New Haven: Yale University Press, 2000.

Hill, C. *The English Bible and the Seventeenth-Century Revolution*. London: Penguin, 1993.

Miskimin, P. *One King, One Law, Three Faiths: Religion and the Rise of Absolutism in Seventeenth-Century Metz*. Westport: Greenwood, 2001.

Modena, L. *The Autobiography of a Seventeenth-Century Venetian Rabbi: Leon Modena's Life of Judah*. Princeton: Princeton University Press, 1988.

Morgan, E. *The Puritan Family: Religion and Domestic Relations in Seventeenth-Century New England*. New York: Harper & Row, 1966.

Phillips, H. *Church and Culture in Seventeenth-Century France*. Cambridge: Cambridge University Press, 1997.

Rapley, E. *The Dévotes: Women and Church in Seventeenth-Century France*. Vol. 4. Montreal: McGill-Queen's University Press, 1990.

Thomas, K. *Religion and the Decline of Magic: Studies in Popular Beliefs in Sixteenth and Seventeenth Century England*. New York: Oxford University Press, 1997.

Weddle, M. *Walking in the Way of Peace: Quaker Pacifism in the Seventeenth Century*. New York: Oxford University Press, 2001.

SCIENCE AND MATHEMATICS

Bedini, S. *The Pulse of Time: Galileo Galilei, the Determination of Longitude, and the Pendulum Clock*. Florence: Leo S. Olschki, 1991.

Biagoli, M. *Galileo Courtier: The Practice of Science in the Culture of Absolutism*. Chicago: University of Chicago Press, 1993.

Brockliss, L., and C. Jones. *The Medical World of Early Modern France*. Oxford: Clarendon Press, 1997.

Burke, J., ed. *The Uses of Science in the Age of Newton*. Berkeley: University of California Press, 1983.

Cook, H. *The Trials of an Ordinary Doctor: Joannes Groenevelt in Seventeenth-Century London*. Baltimore: Johns Hopkins University Press, 1994.

Dear, P. *Mersenne and the Learning of the Schools*. Ithaca: Cornell University Press, 1988.

French, R., and A. Wear, eds. *Medical Revolution in the Seventeenth Century*. Cambridge: Cambridge University Press, 1989.

Goodman, D., and C. Russell. *The Rise of Scientific Europe 1500–1800*. London: Hodder and Stoughton, 1991.

Hall, A. *From Galileo to Newton 1630–1720*. London: Collins, 1963.

Hall, M. *Robert Boyle and Seventeenth-Century Chemistry*. Cambridge: Cambridge University Press, 1958.

Hunter, M. *Archives of the Scientific Revolution: The Formation and Exchange of Ideas in Seventeenth Century Europe*. Woodbridge: Boydell Press, 1998.

———. *Science and the Shape of Orthodoxy: Intellectual Change in Late Seventeenth-Century Britain*. Woodbridge: Boydell Press, 1995.

Jacob, M. *The Cultural Meaning of the Scientific Revolution*. New York: Knopf, 1988.

Mancosu, P. *The Philosophy of Mathematics and Mathematical Practice in the Seventeenth Century*. Oxford: Oxford University Press, 1999.

Rigaud, S., ed. *Correspondence of Scientific Men of the Seventeenth Century*. 2 vols. Oxford: Clarendon Press, 1891.

Shapin, S. *A Social History of Truth: Civility and Science in Seventeenth-Century England*. Chicago: University of Chicago Press, 1994.

———. *The Scientific Revolution*. Chicago: University of Chicago Press, 1996.

Stroup, A. *A Company of Scientists: Botany, Patronage, and Community at the Seventeenth-Century Parisian Royal Academy of Sciences*. Berkeley: University of California Press, 1990.

Thrower, J., ed. *Standing on the Shoulders of Giants: A Longer View of Newton and Halley*. Berkeley: University of California Press, 1990.

Wilson, C. *The Invisible World: Early Modern Philosophy and the Invention of the Microscope*. Princeton: Princeton University Press, 1995.

THEATER

Arnott, P. *An Introduction to the French Theatre*. Lanham: Rowman & Littlefield, 1997.

Bentley, G., ed. *The Seventeenth-Century Stage*. Chicago: University of Chicago Press, 1968.

Bevis, R. *English Drama: Restoration and Eighteenth Century, 1660–1789*. New York: Longman, 1998.

Canfield, J. *Heroes and States: On the Ideology of Restoration Tragedy*. Lexington: University Press of Kentucky, 1999.

Fisk, D., ed. *The Cambridge Companion to English Restoration Theatre*. Cambridge: Cambridge University Press, 2000.

Ganelin, C., and H. Mancing, eds. *The Golden Age Comedia: Text, Theory, and Performance*. West Lafayette: Purdue University Press, 1994.

Howarth, W. *French Theatre in the Neo-Classical Era, 1550–1789*. New York: Cambridge University Press, 1997.

Howe, E. *The First English Actresses: Women and Drama, 1660–1700*. Cambridge: Cambridge University Press, 1992.

Hughes, D. *English Drama, 1660–1700*. Oxford: Oxford University Press, 1996.

Lawrenson, T. *The French Stage in the Seventeenth Century: A Study in the Advent of the Italian Order*. Rev. ed. New York: AMS, 1986.

MacLean, G. *Culture and Society in the Stuart Restoration: Literature, Drama, History*. Cambridge: Cambridge University Press, 1995.

McBride, R. *Aspects of Seventeenth Century French Drama and Thought*. Lanham: Rowman & Littlefield, 1979.

McKendrick, M. *Theatre in Spain, 1490–1700*. Cambridge: Cambridge University Press, 1989.

———. *Women in Society in Spanish Drama of the Golden Age: A Study of the Mujer Varonil*. Cambridge: Cambridge University Press, 1974.

Moore, R. *Henry Purcell and the Restoration Theatre*. Westport: Greenwood, 1995.

Phillips, H. *The Theatre and Its Critics in Seventeenth-Century France*. Oxford: Oxford University Press, 1980.

Quinsey, K. *Broken Boundaries: Women and Feminism in Restoration Drama*. Lexington: University Press of Kentucky, 1996.

Shergold, N. *A History of the Spanish Stage from Medieval Times until the End of the Seventeenth Century*. Clarendon: Oxford University Press, 1967.

Stoll, A., and D. Smith. *The Perception of Women in Spanish Theater of the Golden Age*. Cranbury: Associated University Presses, 1991.

Turnell, M. *The Classical Moment: Studies of Corneille, Molière, and Racine*. New York: New Directions, 1963.

Weber, H. *The Restoration Rake-Hero: Transformations in Sexual Understanding in Seventeenth-Century England*. Madison: University of Wisconsin Press, 1986.

Wiley, W. *The Early Public Theatre in France*. Cambridge: Harvard University Press, 1960.

Ziomek, H. *A History of Spanish Golden Age Drama*. Lexington: University of Kentucky Press, 1984.

WOMEN AND GENDER

Amussen, Susan. *An Ordered Society: Gender and Class in Early Modern England*. London: Basil Blackwell, 1988.

Brailsford, M. *Quaker Women*. London: Duckworth, 1915.

Clark, A. *The Working Life of Women in the Seventeenth Century*. London: Routledge, 1982.

Davis, N. *Society and Culture in Early Modern France*. Stanford: Stanford University Press, 1975.

Davis, N., and A. Farge, eds. *Renaissance and Enlightenment Paradoxes*. Vol. 3 of *A History of Women in the West*. Cambridge: Harvard University Press, 1993.

Dugaw, D. *Warrior Women and Popular Balladry 1650–1850*. Cambridge: Cambridge University Press, 1989.

Erickson, A. *Women and Property in Early Modern England*. London: Routledge, 1993.

Fraser, A. *The Weaker Vessel: A Woman's Lot in the Seventeenth Century*. New York: Knopf, 1984.

Gibson, W. *Women in Society in Seventeenth Century France*. New York: St. Martin's, 1989.

Hufton, O. *The Prospect Before Her: A History of Women in Western Europe, 1500–1800*. New York: Knopf, 1996.

Irwin, J. *Womanhood in Radical Protestantism*. Lewiston: Edwin Mellen Press, 1979.

Levin, C., ed. *Extraordinary Women of the Medieval and Renaissance World: A Biographical Dictionary*. Westport: Greenwood Press, 2000.

Monson, C. *The Crannied Wall: Women, Religion and the Arts in Early Modern Europe*. Ann Arbor: University of Michigan Press, 1992.

Perry, M. *Gender and Disorder in Early Modern Seville*. Princeton: Princeton University Press, 1990.

Prior, M., ed. *Women in English Society 1500–1800*. London: Routledge, 1985.

Schiebinger, L. *The Mind Has No Sex? Women in the Origins of Modern Science*. Cambridge: Cambridge University Press, 1989.

Thompson, R. *Women in Stuart England and America: A Comparative Study*. London: Routledge & Kegan Paul, 1974.

Turner, J., ed. *Sexuality and Gender in Early Modern Europe*. Cambridge: Cambridge University Press, 1993.

Wiesner, M. *Women and Gender in Early Modern Europe*. Cambridge: Cambridge University Press, 1993.

———. *Working Women in Renaissance Germany*. New Brunswick: Rutgers University Press, 1986.

Index

Boldface page numbers indicate location of main entries.

About the Editor and Contributors

MELISSA D. AARON, Department of English Language and Foreign Languages, California State Polytechnic University at Pomona

MARIA ARCHETTO, Division of Humanities, Oxford College of Emory University

PATRICIA ARMSTRONG, McGraw Center for Teaching and Learning, Princeton University

CHRISTOPHER BAKER, Department of Languages, Literature and Philosophy, Armstrong Atlantic State University

RENZO BALDASSO, Department of Art History and Archaeology, Columbia University

JANET BERTSCH, Department of English, University of Lethbridge

JEAN-VINCENT BLANCHARD, Department of Modern Languages and Literature, Swarthmore College

M. ELLEN BLOSSMAN, Department of Languages, Literature and Philosophy, Armstrong Atlantic State University

VIRGINIA BRACKETT, Department of English, Triton College

PAMELA MERRILL BREKKA, independent scholar, New York, New York

DEBORAH H. CIBELLI, Department of Art, Nicholls State University

PATRICK COOK, Department of English, The George Washington University

THOMAS L. COOKSEY, Department of Languages, Literature and Philosophy, Armstrong Atlantic State University

JULIA K. DABBS, Department of History, University of Minnesota, Morris

LAURA L. DENNIS-BAY, Department of English and Modern Foreign Languages, Cumberland College

LLOYD DEWITT, Department of Art History and Archaeology, University of Maryland

CONNIE S. EVANS, Department of History, Baldwin Wallace College

YAEL EVEN, Department of Art and Art History, University of Missouri, St. Louis

JAMES P. FAIRLEIGH, Department of Music, Jacksonville State University

ENRIQUE FERNÁNDEZ, Department of French, Spanish, & Italian, University of Manitoba

SARA V. FINK, Department of English, Michigan State University

MARK FINLAY, Department of History, Armstrong Atlantic State University

DONALD R. FRANCESCHETTI, Department of Physics, University of Memphis

ROSI PRIETO GILDAY, Department of Art, California State University, San Bernardino

JOSEF K. GLOWA, Department of Foreign Languages, Moravian College

MARGARET GOSCILO, Department of English, University of Pittsburgh

RACHEL HAMMERSLEY, History Subject Group, University of Sussex

MARGARET HARP, Department of Foreign Languages, University of Nevada, Las Vegas

CHRISTOPHER E. HENDRICKS, Department of History, Armstrong Atlantic State University

PETER HESS, Department of Germanic Studies, University of Texas, Austin

NICOLE HOWARD, Department of History and Philosophy of Science, Indiana University

BLAIR HOXBY, Department of English, Yale University

MICHAEL R. HUTCHESON, Humanities Department, Landmark College

CHRISTOPHER IVIC, Department of English and Communication, State University of New York, Potsdam

MICHAEL W. JACKSON, Department of English, St. Bonaventure University

CAROL JAMISON, Department of Languages, Literature and Philosophy, Armstrong Atlantic State University

RALPH KEEN, School of Religion, University of Iowa

WALTER H. KEITHLEY, Department of English, Arizona State University

MATTHEW KOCH, Department of History, Queen's University

JOYCE LINDORFF, Esther Boyer College of Music, University of Calgary

MICHAEL R. LYNN, History Department, Agnes Scott College

KEN MACMILLAN, Department of History, University of Calgary

MICHAEL J. MCGRATH, Department of Foreign Languages, Georgia Southern University

DAVID D. MCKINNEY, Virginia Museum of Fine Arts

WILLIAM J. MCPEAK, independent scholar, Aliso Viejo, California

ANGELA MORALES, Department of Modern Languages, Central Connecticut State University

GERALD MORTON, Department of English and Philosophy, Auburn University, Montgomery

JAMES NORTON, Department of English, Marian College

HILLARY NUNN, Harlan Hatcher Graduate Library, University of Michigan

MARTHA OBERLE, independent scholar, Laytonsville, Maryland

CHARLES PASTOOR, Department of English, John Brown University

CASSANDRA L. PINNICK, Department of Philosophy and Religion, Western Kentucky University

JOAN KLOBE PRATT, Department of History, University of Northern Colorado

DALE G. PRIEST, Department of English and Foreign Languages, Lamar University

WILLIAM B. ROBISON, Department of History and Government, Southeastern Louisiana University

JOSÉ MARÍA RODRÍGUEZ-GARCÍA, Department of Romance Studies, Cornell University

ANNA MARIE ROOS, Department of History, University of Minnesota, Duluth

DAVID M. ROSEN, Department of English, Murray State College

MARTINE SAURET, Department of Foreign Languages and Literatures, Western Michigan University

JOHN SCHMIDT, Department of Art, Music and Theatre, Armstrong Atlantic State University

LESLIE A. SCONDUTO, Department of Foreign Languages, Bradley University

SALLYE SHEPPEARD, Department of English and Foreign Languages, Lamar University

KEIR B. STERLING, Command Historian, Fort Lee, Virginia

JESSE G. SWAN, Department of English Language and Literature, University of Northern Iowa

PAUL C. L. TANG, Department of Philosophy, California State University, Long Beach

OLGA TERMINI, Department of Music, California State University, Los Angeles

REBECCA TOTARO, College of Arts and Sciences, Florida Gulf Coast University

SUSAN TREACY, Department of English and Fine Arts, Franciscan University of Steubenville

WILLIAM T. WALKER, Vice President for Academic Affairs and Dean of Faculty, Chestnut Hill College

BRIAN WEISER, Department of History, University of New Hampshire

ED R. WHEELER, Dean of the College of Arts and Sciences, Armstrong Atlantic State University

MARK WILLIAMSON, Department of Art, Broome Community College

KRISTEN L. ZACHARIAS, Department of Philosophy, Albright College